About the Author

Iain E. Richardson is an internationally recognised expert on video compression and digital video communications. He is the author of four other books about video coding, which include two widely cited books on the H.264 Advanced Video Coding standard. For over thirty years, he has carried out research in the field of video compression and video communications, as a Professor at the Robert Gordon University in Aberdeen, Scotland, and as an independent consultant with his own company, Vcodex. He advises companies and delivers lectures on video compression technology and is sought after as an expert witness in litigation cases involving video coding.

Acknowledgements

This book would never have left my desk without the dedication, support and encouragement of my colleague Naomi Reid, who read many drafts, acted as a sounding board and kept the manuscript moving forwards until it was complete. Thank you, Naomi! I would also like to thank the ever-amazing Pat Ballantyne for reading and commenting on the final draft and for her constant support. Thanks to the editorial team at John Wiley & Sons, to Elecard, Parabola Research and Solveig Multimedia for providing access to their visualisation tools and to everyone who has asked me to explain video coding to them over the years.

About the Companion Website

This book is accompanied by a companion website

www.wiley.com/go/richardson/codingvideo1

1

Introduction

The scene: my video coding research lab at the Robert Gordon University, Aberdeen, Scotland, 1994. 'You're working on video compression?' asked my visitor. 'Isn't that going to be unnecessary in a few years? After all, network bandwidth is increasing every year. We've already got 10 Mbps Ethernet and we'll soon have 100 Mbps. Video compression will be redundant by the year 2000'.

My visitor predicted that research into video compression would fade away as ever-increasing bandwidth and storage capacity made it possible to send uncompressed, full-quality video with ease. He offered me a wager, that my research topic, video coding, would be redundant by the year 2000. If only I had taken that bet! In fact, at the present time, the need for efficient and effective video coding is perhaps greater than ever. Generation and consumption of video content have grown more quickly than available bandwidth has in recent years. This is because of the emergence of user-generated content that comes most notably from mobile devices. Consumption has also increased through changing viewing patterns – often with multiple simultaneous streams being viewed in a single household – and the move to higher-resolution content, such as High Definition and beyond.

This book is about video coding, also known as video compression. A video encoder converts digital video into a compressed form, that is, a form that takes up less storage or transmission capacity, known as coded video. A video decoder does the opposite, converting coded video into uncompressed digital video. The combination of a video encoder and a video decoder is known as a video codec, and the processes of compression and decompression are often described as video coding.

1.1 Why Write This Book?

I have written four books about video coding. Why write another one?

It is now 2024, and video coding is embedded in modern life as never before. Compressed video made a major impact in the 1990s with the advent of digital television and DVD video. Other applications followed throughout the early 2000s, and now compressed video

Coding Video: A Practical Guide to HEVC and Beyond, First Edition. Iain E. Richardson.
© 2024 John Wiley & Sons Ltd. Published 2024 by John Wiley & Sons Ltd.
Companion website: www.wiley.com/go/richardson/codingvideo1

is a fundamental part of an ever-expanding list of consumer and business applications. These include, but are not limited to:

- Recording, playing and sharing video on mobile devices.
- Video calling and videoconferencing, both of which received a massive boost following the COVID lockdowns of 2020.
- Video streaming, which is now overtaking digital TV broadcasting as the dominant way in which we watch video.
- Video in social media, with platforms such as Facebook, Instagram and TikTok increasingly relying on video as well as still images.
- Security and surveillance, from commercial security cameras to police bodycams and video-enabled doorbells.
- Automotive video, from dashcams to in-car entertainment.

Since I wrote my last book in 2010, we have seen a significant increase in the use of video compression and a shift in the way coded video is used, as well as the release of a series of new industry standards and formats for video compression. These formats include H.265/High Efficiency Video Coding (HEVC), H.266/Versatile Video Coding, VP9, AV1 and more.

These new standards and formats are publicly available documents. However, the published standards are not intended for the casual reader and can be challenging to understand. This is due in part to their focus on highly detailed technical descriptions of how standard-based video codecs work. Even a reader with a solid background in computer science or electronic engineering may need a more approachable way into the video coding standards.

I have spent much of my career explaining video coding to students, researchers, engineers and professionals. I have always tried to find approachable and intuitive explanations of how video coding works. I wrote this book primarily to gather this material in one place, and I hope that it can provide a more user-friendly guide to video coding that will enable readers to engage with the video coding standards and understand how to get the most out of video compression.

1.2 What Is in the Book?

In this book, I give an overview of digital video and video compression and a short history of some of the key concepts and standards. I then discuss each of the main core concepts of video coding. I explain each concept in general terms and discuss how each is put into practice in certain video coding standards. I have chosen to focus on the widely used H.26x series of standards, with a particular focus on H.265/HEVC. Each chapter is liberally illustrated with diagrams and examples.

In Chapter 2, I provide an overview of a video codec and introduce the concepts of performance and visual quality.

Chapter 3 is a short history of video coding. This chapter begins with a description of the basic concepts that developed between the 1950s and 1990s and goes on to describe the major video coding standards from H.261 through to H.266 and AV1.

Chapter 4 explains the structural elements used in video coding. A video codec processes a video by breaking it down into structures such as Groups of Pictures, pictures, slices, tiles, basic coding units, prediction blocks and transform blocks.

Chapters 5 and 6 deal with prediction. The information in a block of video pixels can be significantly reduced if we can predict the pixels from previously sent information. Intra prediction (see Chapter 5) involves predicting blocks using information that is available in the same frame of video. This is known as intra-frame or spatial prediction. Inter prediction (see Chapter 6) creates predictions using information that is available in other, previously coded frames. This is known as inter-frame or temporal prediction.

Chapter 7 examines transforms and quantisation. A transform converts a block of pixels, or difference samples, into another domain in which the important visual information is concentrated into a smaller number of values. Quantisation removes some of the resulting information, deliberately reducing the precision of the data in order to compress it, hopefully without sacrificing too much visual quality.

Chapter 8 looks at the final stage in a video encoder, entropy coding, in which data are converted into a compressed bitstream. I discuss the main types of entropy coding, with a particular emphasis on arithmetic coding, which has become the method of choice for entropy coding in recent video coding standards.

Chapter 9 deals with filtering, in particular the types of video filtering that are carried out as part of a video encoding or decoding process. In this context, video pixels are filtered to try to improve the performance of all the other compression processes.

In Chapter 10, I discuss how coded video can be stored and transmitted. Video is often stored together with audio in container files such as an MP4 file. Transport or transmission of video is an area that continues to develop, with the emergence of adaptive bitrate streaming as one of the most important ways of sending video across the internet.

Chapter 11 considers how video codecs are implemented in software or hardware and how the performance of video codecs can be measured and compared. I provide some suggestions for trying this yourself, for example, by experimenting with the various public-domain software video codecs and by analysing coded video to see how it is actually put together.

1.3 How Should You Use This Book?

I wrote this book with several audiences in mind. Each chapter starts by setting out basic concepts and continues into more technical detail. I have deliberately not assumed specialised background knowledge on the part of the reader. I hope that the book will be useful to:

- Students who are studying multimedia processing and communications.
- Academics and researchers, as the book presents a way to understand and approach the extensive research literature on video coding and provides a platform to develop ideas and research topics.
- Engineers and implementers, as a bridge between the basic concepts of video coding and the often challenging density of the video coding standards.
- Other professionals, as a more accessible way into this important technical topic.

I have written the book so that it can be read from start to finish, and I hope that at least some readers will do just that. Of course, you may want to dip in and out and concentrate on particular topics of interest. I would recommend reading Chapter 2, which will give an overview of how each topic fits into the overall video codec system or model. From Chapter 4 onwards, each chapter begins with an explanation of the basic concepts of the video compression process such as prediction and transform. The chapter then explains how the process is put into practice according to specific standards with a particular focus on H.265/HEVC.

At some point, you may want to go deeper into the topics presented here. For example, you might want to find out exactly how motion vectors are communicated in H.265 or H.266. There are many excellent articles in journals that can explain topics further, such as the *IEEE Transactions on Circuits and Systems for Video Technology*, which has published several special issues and special sections on H.265/HEVC [1] and a special section on H.266/Versatile Video Coding [2]. If there is any doubt or ambiguity about a particular process that has been specified in one of the standards, the standard document itself should be consulted.

Experimenting is an excellent way to deepen your understanding of a topic. Chapter 11 gives practical guidance on how to start experimenting with standard-based codecs. Examples of experiments can include changing encoding parameters to understand how they can affect the compressed video or using syntax analysers to visualise the way the video is encoded. Please visit the companion website, https://www.vcodex.com/coding-video-book, for further examples and resources.

I hope you are now ready to walk through the processes of coding video, starting with an overview of the video codec and the visual information it is designed to compress.

References

1 Gharavi, H. (2012). Combined issue on high efficiency video coding (HEVC) standard and research. *IEEE Transactions on Circuits and Systems for Video Technology* 22 (12): 1646–1646. doi: https://doi.org/10.1109/TCSVT.2012.2226073.

2 Boyce, J.M., Chen, J., Liu, S. et al. (2021). Guest editorial introduction to the special section on the VVC standard. *IEEE Transactions on Circuits and Systems for Video Technology* 31 (10): 3731–3735. doi: https://doi.org/10.1109/TCSVT.2021.3111712.

2

Video Coding and Video Quality

2.1 Introduction

Video coding or video compression bridges the gap between displayable or uncompressed video and practical methods of storing or transmitting data. Digital video is made up of a series of frames or pictures, each of which contains hundreds of thousands or millions of picture elements or pixels. Storing or transmitting video in its original, uncompressed form can require an impractically large amount of storage capacity or bandwidth. A video encoder compresses video into a smaller and more compact form known as coded video, suitable for storage or transmission. A video decoder reverses this process, extracting displayable video from coded video (see Figure 2.1). The complementary pair of an encoder and a decoder is a video codec.

In this chapter, I will introduce each of the following aspects of a video codec, before covering them in further detail throughout the book:

Inputs and outputs: The data inputs and outputs for a video codec. The input to an encoder and the output of a decoder are uncompressed or raw video data, i.e. video that can be displayed. The output of an encoder and the input to a decoder are an encoded, compressed bitstream that is suitable for storage or transmission.

Data structures: A video codec processes data in a hierarchy from a complete video sequence, e.g. a video scene or programme, through groups of coded frames and individual coded frames, down to basic coding units (Coding Tree Units [CTUs], Macroblocks [MBs] or Superblocks) and sub-blocks.

Prediction: Some or all of the data to be coded are predicted from previously coded data. We can predict a block of video data from neighbouring data in the same video frame. This is known as intra prediction. Inter prediction uses video data from previously coded frames to predict a block of video data. Once the prediction is created, it is subtracted from the original block to form a residual block, which typically contains less information than the original block.

Transform and quantisation: Blocks of video data or residual data are transformed into another form, such as a frequency domain representation, and then quantised to reduce the data precision. This has the effect of removing or discarding much of the information in a block, at the expense of a loss of visual quality.

Coding Video: A Practical Guide to HEVC and Beyond, First Edition. Iain E. Richardson.
© 2024 John Wiley & Sons Ltd. Published 2024 by John Wiley & Sons Ltd.
Companion website: www.wiley.com/go/richardson/codingvideo1

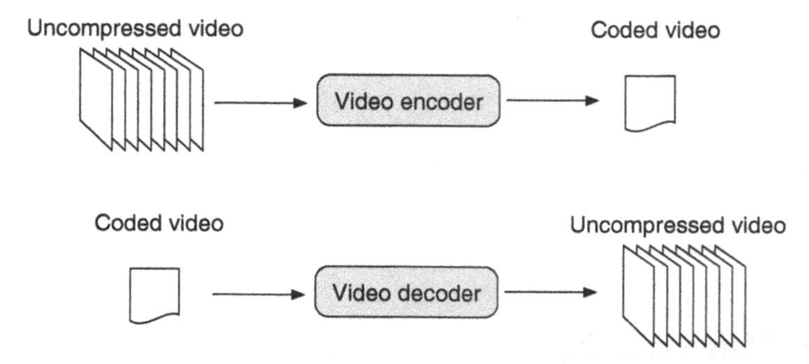

Figure 2.1 Video encoder and decoder

Bitstream coding: Quantised blocks and information needed for decoding are converted into a compressed digital bitstream, using a form of entropy coding such as variable-length coding or arithmetic coding.

Transport: A coded video bitstream is arranged in a suitable form for storage and/or transmission. This might include splitting the data into packets of a convenient size for network transmission or streaming, or placing the video data in a container file, along with other information such as an audio track.

Decoding: A video decoder reverses the encoding processes to create a video sequence that can be displayed. Depending on the encoding process, the decoded version may not be identical to the original video sequence.

Performance measurement: We can measure the performance of a video codec by considering how much it compresses video, i.e. how small the compressed bitrate is; by considering how good the decoded video looks, i.e. how much it is distorted by, compared to the original video; and by considering how much computation is required at the encoder and the decoder. These three dimensions of compression, quality and computation can be used to compare the effectiveness of different video codecs.

2.2 An Overview of Video Coding

2.2.1 Just How Much Data Are We Talking About?

Figure 2.2 shows the same frame of a video clip at four different resolutions:

1) **Standard Definition** (SD): Contains 720 × 576 pixels
2) **720p Definition**[1]: Contains 1280 × 720 pixels
3) **High Definition (1080p)**: Contains 1920 × 1080 pixels
4) **Ultra-High Definition** (UHD/**4K**): Contains 3840 × 2160 pixels.

One second of original, uncompressed SD video, captured at 25 frames per second, takes up around 15.5 Mbytes of storage space. This means that this video clip would require

1 720p is an intermediate resolution between Standard Definition and 1080p (Full HD).

Figure 2.2 Video resolutions

124 Mbits/second of bandwidth to transmit over a network or broadcast channel in real time, i.e. with one second of playable video sent every second. One second of uncompressed UHD video[2], captured at 50 frames per second, takes up around 620 Mbytes of storage space and would require a substantial 5 Gbits/second of transmission bandwidth to send in real time.

2.2.2 Bandwidth Is Increasing Every Year, But...

Whilst it is true that we have more storage capacity and network bandwidth available than ever before, the demand for storing and sending video continues to outstrip the available capacity. By 2023, around two-thirds of consumer TV sets were 4K resolution or higher [1]. The ability to build high-performance video codecs into consumer devices such as smartphones and TVs and the expectation of high-resolution video have made it the norm to compress or encode video before storing or transmitting it and to decode video before displaying it. Compressing video before storing or sending it means that:

- More videos can be sent or stored using the same capacity
- Videos can be sent or stored in higher resolutions (e.g. UHD/4K instead of SD)
- Videos can be uploaded or downloaded much more quickly.

2 Assuming 4:2:0 colour sampling.

Example:

Consider what happens when a user records a video clip using a smartphone camera. When the Record button is pressed, the smartphone captures frames of video, compresses them using a video encoder and saves the coded (compressed) file to the smartphone's built-in storage. When the clip is sent to a friend, the coded file is transmitted over the internet, received by the friend's smartphone, decoded and displayed (see Figure 2.3). Storing and sending the video file in compressed form saves storage capacity on each smartphone and reduces the time taken to send the file across the internet.

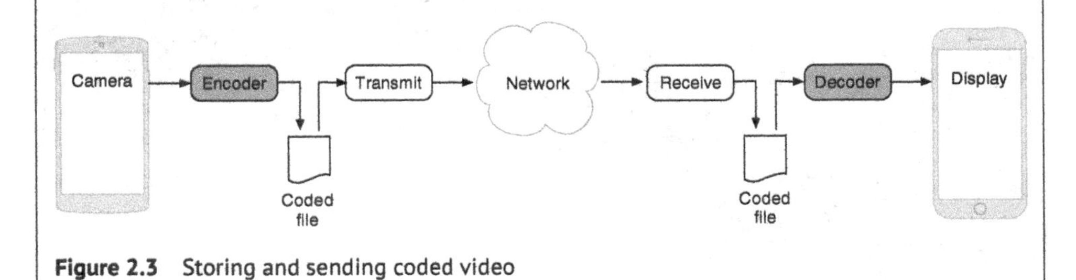

Figure 2.3 Storing and sending coded video

A video codec can be a software program, a hardware module on a chip (integrated circuit), or a combination of the two. The codec does a computationally intensive job, which is to convert many millions of pixels per second to and/or from a compressed or encoded form. When a video is recorded on a smartphone, a video encoder built into the smartphone compresses and stores the video from the camera in real time. When the video is played back, a video decoder built into the device reverses the process in real time to decode the clip into a playable, uncompressed form (see Figure 2.3). The process of encoding and decoding is highly repetitive and takes a lot of processing. In order to do this efficiently in real time, it is often necessary or desirable to do some or all of the processing steps in dedicated electronic hardware on a chip inside a smartphone or other device. Carrying out processor-intensive steps in hardware can make it possible to encode and decode higher-resolution video in real time, reduce battery usage and/or free up processing capacity for other applications and services.

There are many different codec formats and many different designs of video codecs, both in software and hardware. Most practical video codecs since the early 1990s have adopted the same basic approach to coding and decoding video. For each frame of video, this process generally works as follows. The encoder carries out the following basic steps (see Figure 2.4) to create a compressed file:

1) **Partition**: Decompose the video frame into basic regions, usually square or rectangular blocks of pixels.
2) **Predict**: For each region or block, create a prediction or an estimate of the block using information such as neighbouring pixels or previously coded frames. Subtract this prediction from the actual block to create a residual or difference block.

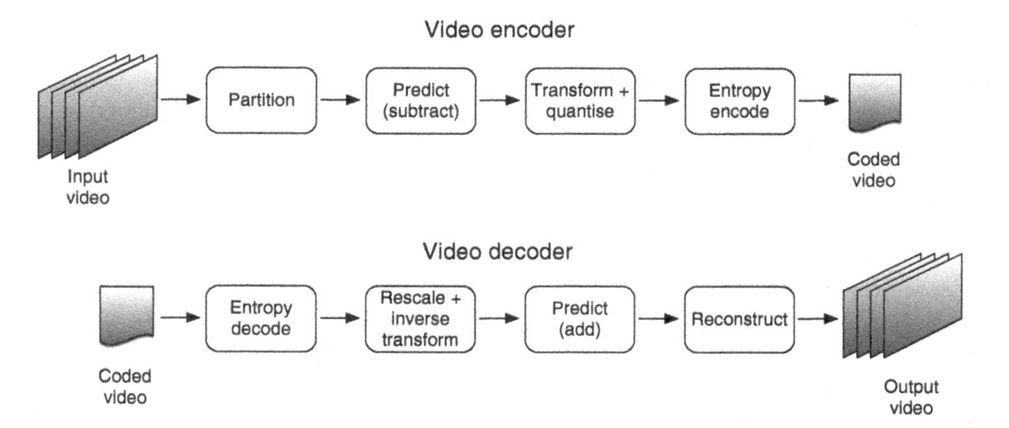

Figure 2.4 Processing steps in a video encoder and video decoder

3) **Transform and quantise**: Transform the residual block into a frequency domain. Quantise the transformed block, which has the effect of discarding less-important components of the data and retaining only the visually important features.
4) **Entropy encode**: Convert the result of the quantise step, together with any other information needed to decode the frame, into a compressed format.

The decoder reverses the steps to decode and extract video, i.e. entropy decode, rescale ('inverse quantise') and transform, add the prediction and reconstruct the video frame.

2.3 Inputs and Outputs

A video encoder takes in uncompressed video and outputs a compressed (coded) bitstream. As Figure 2.5 illustrates, other information may go into the encoder, such as parameters to control its operation and/or information to be embedded in the video bitstream.

The video decoder takes in a coded bitstream and produces uncompressed, displayable video as its output. Further information, such as status, timing and side information, may also be output (see Figure 2.5). In this section, we will look at each of these types of information.

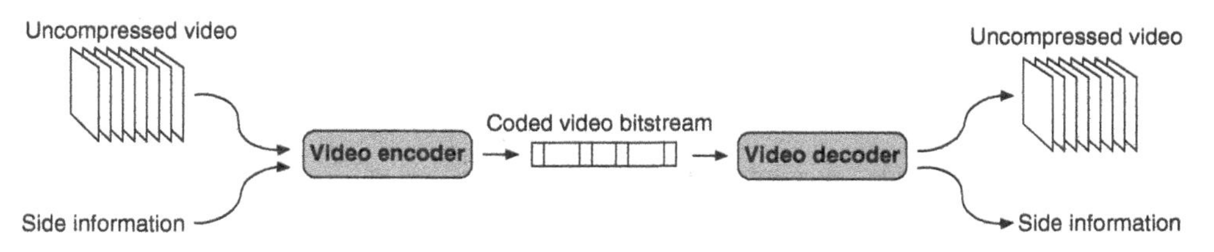

Figure 2.5 Video codec inputs and outputs

2.3.1 Digital Video

Uncompressed or raw digital video is stored without any compression. The function of a video encoder is to convert this uncompressed video into a coded form that is suitable for storage or transmission. The function of a video decoder is to convert coded video into an uncompressed form that is suitable for display or further processing.

Sources of uncompressed digital video include:

- **Camera**: Captures video directly from a scene. Stores or outputs video in uncompressed or compressed form.
- **RAW file**: Samples of video are stored without compression.
- **Rendering or editing process**: Intermediate or final output of processes such as video editing, animation and post-production.
- **Video decoder**: Compressed video is decoded into displayable frames.
- **Video encoder**: A video encoder typically produces decoded frames as part of the coding process (see Section 2.11). These decoded frames are stored so that the encoder can make predictions. The encoder may output these frames, for example, as a preview of the coded and decoded video clip.

Destinations for decompressed digital video include:

- **Display**: Decoded frames may be written into a frame buffer, an area of memory from which the display is generated.
- **RAW file**: Uncompressed video.
- **Editing process**: The input to editing or production tools.
- **Video encoder**: Decompressed video may be re-encoded, for example, to change the bit rate, frame size or streaming characteristics. This is known as transcoding (see Figure 2.6).

2.3.2 Uncompressed Video Formats

2.3.2.1 Frames, Fields and Pixels

A digital video camera captures a real-world image and converts it to a sampled form. The moving image is sampled at certain time instants to produce frames or fields of video. A frame is a snapshot of the scene at a particular time instant, and a field captures half the visual information in a frame, for example, the odd lines or even lines, at each time instant (see Figure 2.7). Interlaced video, which consists of a series of fields, is largely obsolete by the time of writing (2023). Modern-day capture devices and displays record and show video as a series of complete, progressive frames, at a particular number of frames per second, known as the video frame rate.

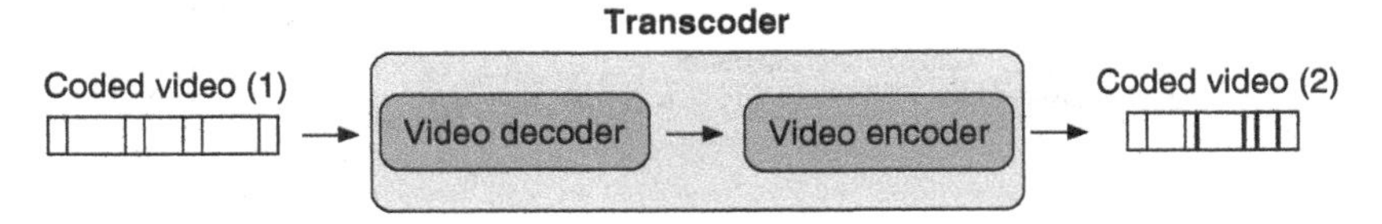

Figure 2.6 Transcoding

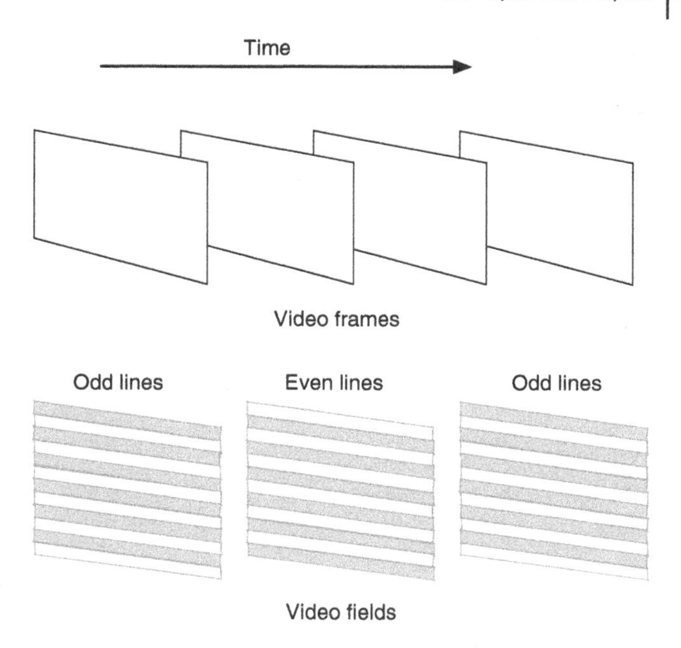

Figure 2.7 Video frames and fields

Each video frame is sampled, usually on a rectangular grid, forming an array of pixels. A displayable pixel may be monochrome or colour. A monochrome pixel can be represented with a single number, indicating the brightness of the pixel. Larger values represent brighter pixels that are closer to white, and smaller values represent darker pixels that are closer to black.

2.3.2.2 Colour and Colour Spaces

Video captured by a camera sensor or displayed on a screen may be in RGB format, which means that each captured or displayed pixel has a Red, Green and Blue component. Red, Green and Blue are additive primary colours, i.e. they can be mixed to produce a wide range of colours. Each of these can be represented as an N-bit number, where N is an integer. For example, Table 2.1 lists several common colours in 8-bit RGB representation. A value of 0 means the component is completely absent, and a value of 255 (binary 11111111) means that the component is at its maximum. In this format, each pixel is represented using $3 \times N$ bits, i.e. 24 bits in the 8-bit per component example.

A complete image or video frame can be represented as three colour planes, each containing all the pixel values of the red, green and blue components. Figure 2.8 shows an original image (top left) and its individual colour components. The red colour component, for example, represents the 'redness' of each pixel, so that 'redder' pixels are lighter and 'less red' pixels are darker. The word 'red', which is written in red ink, therefore appears very light. In the green colour component, the word 'green' (written in green ink) appears light and the word 'red' appears dark. A similar effect is observed in the blue component.

RGB is a three-component system, i.e. a colour pixel can be represented by a weighted combination of three components (R, G and B). Many other component systems may be used. A minimum of three components are required to represent full colour. YUV or YCrCb is a popular colour space for image and video data. In this system, the luminance or luma

Table 2.1 Common colours in 8-bit RGB format

Colour	Red	Green	Blue
White	255	255	255
Mid-grey	128	128	128
Black	0	0	0
Red	255	0	0
Green	0	255	0
Blue	0	0	255
Cyan	0	255	255
Magenta	255	0	255
Yellow	255	255	0
Orange	255	127	0
Pink	255	0	127
. . .etc.			

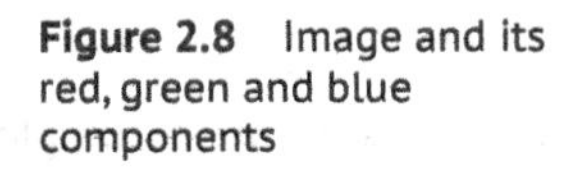

Figure 2.8 Image and its red, green and blue components

component Y represents the brightness of each pixel position, corresponding to a monochrome version of each image or video frame. The red component R can be expressed as $R = Y + Cr$, where Cr is the red chroma or colour difference, representing the positive or negative difference between each luma sample Y and each Red pixel R. Similarly, the blue component B = Y+Cb and G = Y+Cg. A complete, in fact, overcomplete representation of the image is Y, Cr, Cb and Cg (see Figure 2.9). In this figure, Cr represents the difference between the luma Y and the red component R. Hence, the word 'red' and the apple and orange appear lighter, indicating a strong red content. The words 'green' and 'blue' have very little red content and so appear dark in the Cr component. The background is near-monochrome and so appears mid-grey in the Cr component. Similar behaviours are observed in the Cg and Cb components.

Only three components are necessary to fully represent the image. The fourth component can be calculated by combining the other three. In the popular YCrCb format, the green chroma component Cg is omitted. An image can be converted from RGB to YCrCb or vice versa using a set of mathematical operations. Hence, the R, G and B representations of our test image can be converted to or from the Y, Cr and Cb representations without any loss of information (see Figure 2.10).

One advantage of a colour space such as YCrCb is that human observers are less sensitive to colour than to brightness or luma. The chroma components (Cr and Cb) can be subsampled with respect to the luma component (Y) without having an obviously detrimental effect on image quality. Figure 2.11 shows the so-called 4:2:0 subsampling in which the chroma components (Cr and Cb) have half the horizontal and vertical resolution of the luma component. Some information is lost by reducing the resolution of the chroma components, but the perceived quality is likely to be almost as good as the original, as a human observer tends not to notice the reduction in chroma resolution.

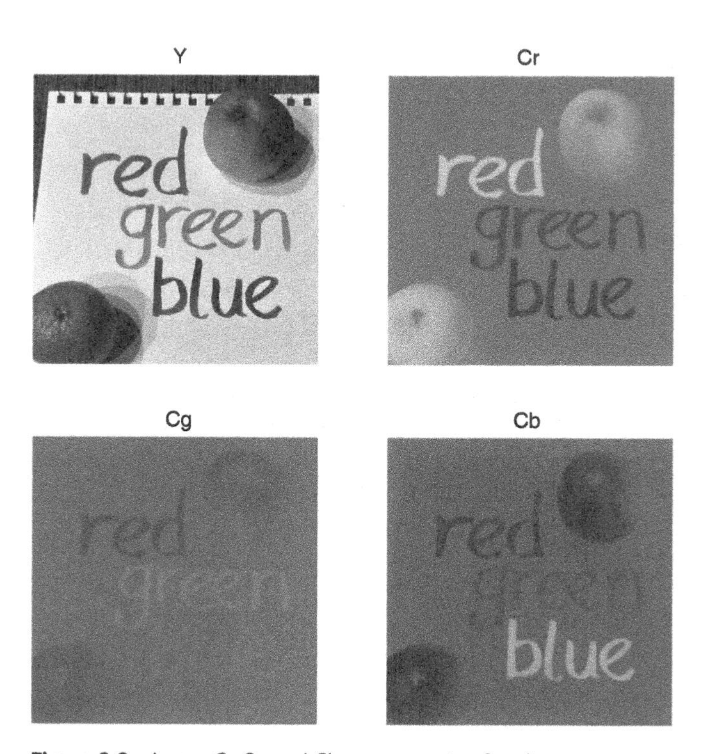

Figure 2.9 Luma, Cr, Cg and Cb components of an image

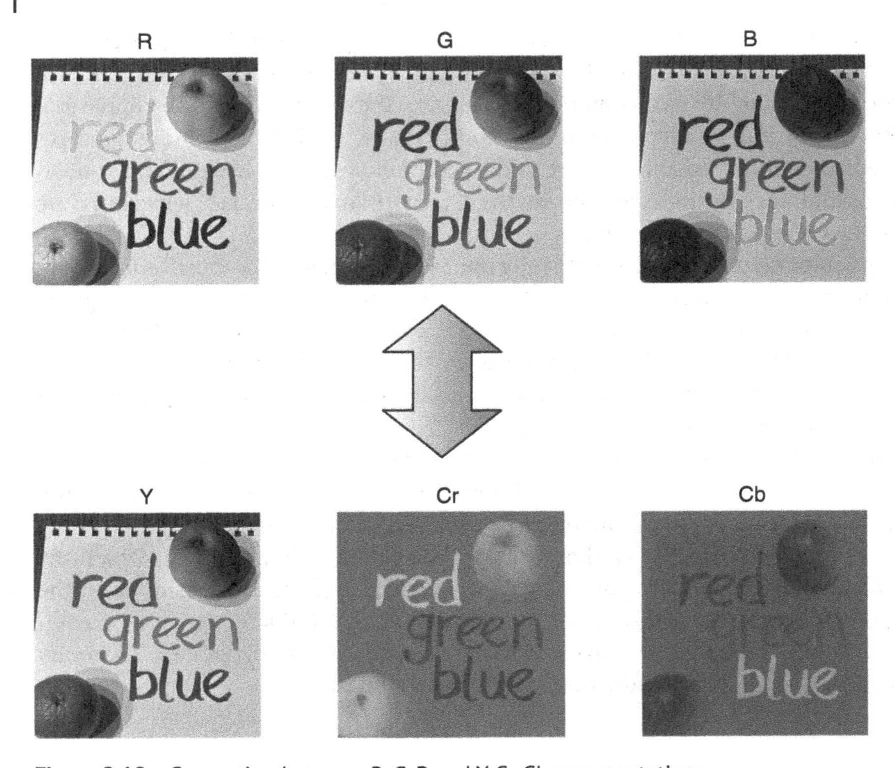

Figure 2.10 Converting between R, G, B and Y, Cr, Cb representations

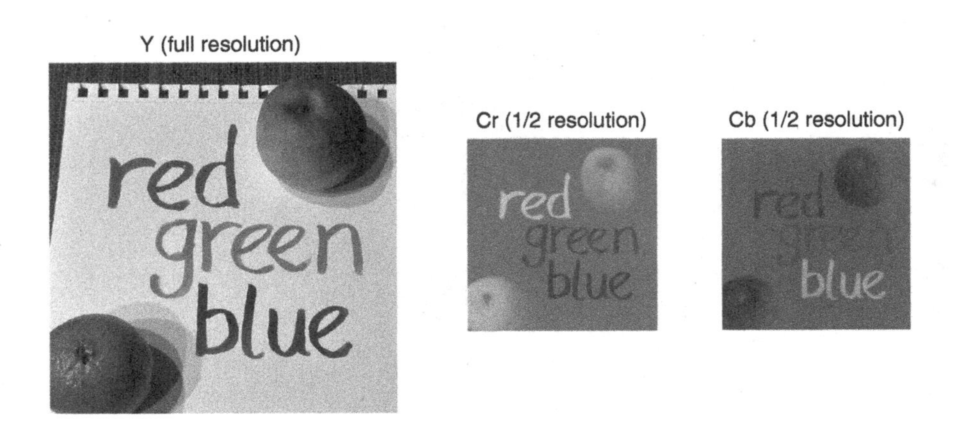

Figure 2.11 4:2:0 sampling of Y, Cr and Cb representation

2.3.3 Side Information

A video encoder makes many decisions while compressing a video sequence. Control parameters or values may be input to the encoder, such as target bit rates or file sizes and information about the format of the video input. These parameters may be:

i) selected by the user, for example, as a collection of parameters known as a preset,
ii) hard-wired into the encoder application or device, and/or
iii) calculated or derived automatically during encoding.

Along with encoding parameters, further side information may be passed to the encoder for embedding in the coded bitstream, such as text or metadata about what the video clip contains, information to help the playback device display the video and so on.

A video decoder may output statistics and other side information together with the decoded video.

2.3.4 Coded Video

Coded or compressed video is a sequence of bits. Encoded in this sequence is a complete compressed representation of the original video clip. The sequence of bits is put together in a specific way that is usually defined by a standard or a specification, such as High Efficiency Video Coding (HEVC)/H.265 or AV1. When the structure of the coded video sequence matches the requirements of the specification and the video can be decoded according to the specification, we would say that it conforms to the specification or standard.

The coded sequence may be stored in one or more video or media files or it may be streamed or transmitted directly. The way the coded video sequence is organised for storage or transmission may itself be defined as part of one or more standards.

2.4 Structural Elements

An encoder or decoder handles the video data in a structured and organised way. The general approach is to break the video clip into a hierarchy of elements, such as:

Sequence: A complete video clip, scene or program, consisting of a series of frames.

Group of Pictures: A set of coded video frames, with the decoder having the ability to start decoding within the Group of Pictures.

Picture: A coded video frame that may be inter-related to other pictures in a Group of Pictures.

Figure 2.12 shows a complete video sequence (top), with a Group of Pictures containing a number of coded frames (middle) and a single frame (bottom).

Within each coded picture, the structure continues:

Slice or tile: A subset or a region within a picture. The example in Figure 2.13 shows a slice, which in this case is a continuous series of Macroblocks (MBs) or Coding Tree Units (CTUs) (see below). Each picture is split up or partitioned into one or more slices. Alternatively, or perhaps simultaneously, the picture may be split up into Tiles, each of which is a square or rectangular region of basic units. The H.266/VVC standard introduces a further partitioning, the subpicture.

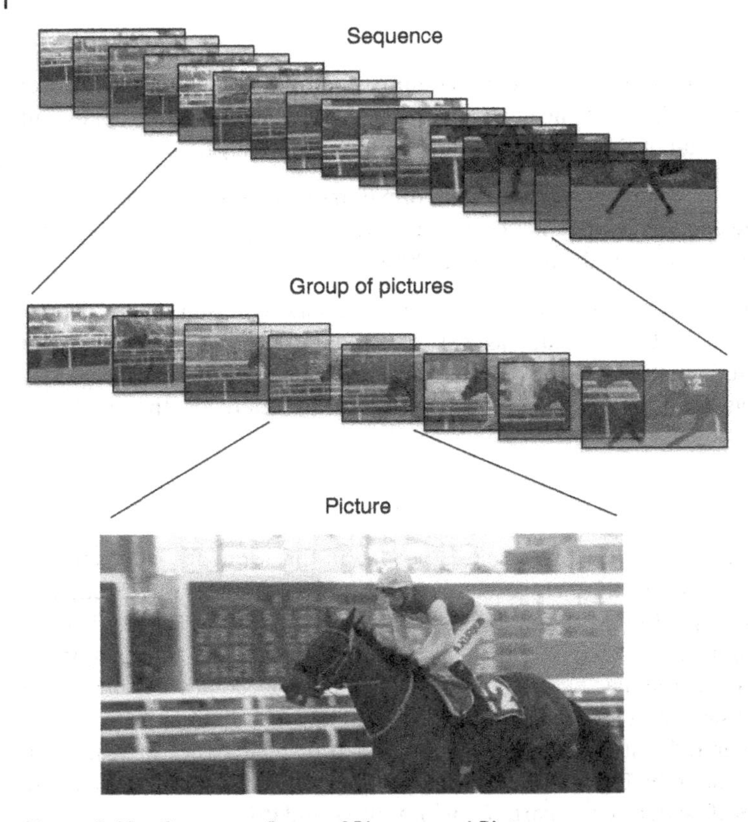

Figure 2.12 Sequence, Group of Pictures and Picture

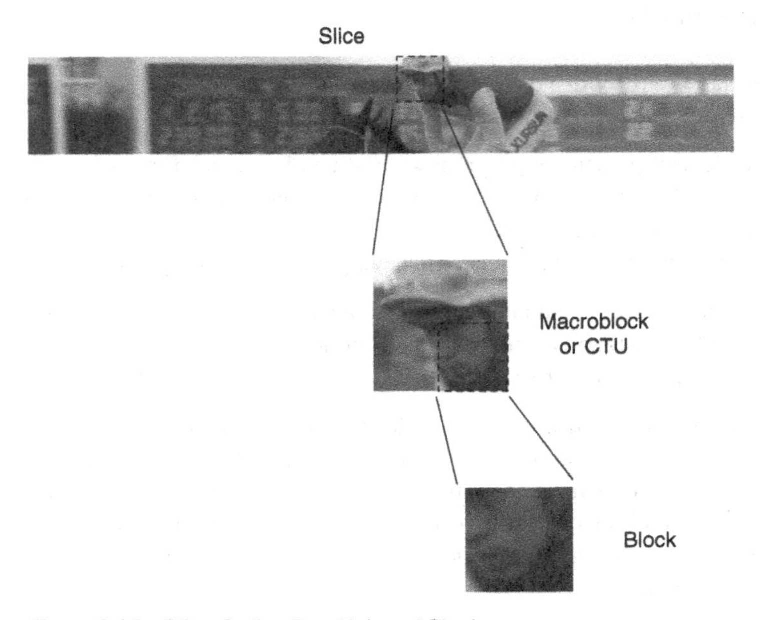

Figure 2.13 Slice, Coding Tree Unit and Block

Basic unit (MB, CTUs or Superblock): The basic unit of processing is typically a square region of video samples, as shown in Figure 2.13. In older standards such as H.264/AVC and MPEG-2, the basic unit or MB was set at 16×16 pixels. Formats such as H.265/HEVC and AV1 cater for larger basic units of up to 64×64 pixels, and the H.266/VVC standard supports a basic unit size of up to 128×128 pixels. Larger basic units tend to be more effective for compressing higher-definition content but can take more processing power and memory to encode or decode.

Block: The block is a square or rectangular region within a basic unit and is a grouping of samples for prediction, transformation or processing.

There is no fundamental restriction that the structures have to be rectangular, and, in fact, objects and regions in a video scene rarely match up with neat rectangular boundaries. However, rectangular structures are particularly well suited to efficient processing and are relatively simple to signal or denote in a bitstream. For example, a square region such as a basic unit can be identified simply by (1) noting the coordinates of the block in the frame and (2) noting the size of the block, e.g. 16×16 pixels. Identifying the location and shape of an irregular-shaped region would typically require many more parameters.

2.5 Prediction

The purpose of compression is to reduce the number of bits required to store or send a video clip. A video clip is made up of multiple frames, each of which contains thousands or millions of pixels. Predicting each pixel in the frame makes it possible to reduce the information in the frame, which in turn makes the frame easier to compress.

2.5.1 Prediction in a Video Encoder

Consider a single pixel within a video frame (see Figure 2.14). The encoder creates a prediction pixel and subtracts it from the original pixel. The result is a difference or residual pixel. If the prediction is accurate, the difference is small or zero, which means that it contains less information than the original pixel.

It is possible to create a new prediction for every pixel. However, it can be more efficient to process a group of pixels at a time. A video codec may split each frame up into regions or blocks and process each block as a unit. For each block of original pixels or samples, the video encoder creates a prediction block and subtracts this from the original block to generate a residual block (see Figure 2.15).

The encoder sends at least two things to the decoder (Figure 2.16):

i) A compressed version of the residual block, and
ii) The instructions for creating the prediction.

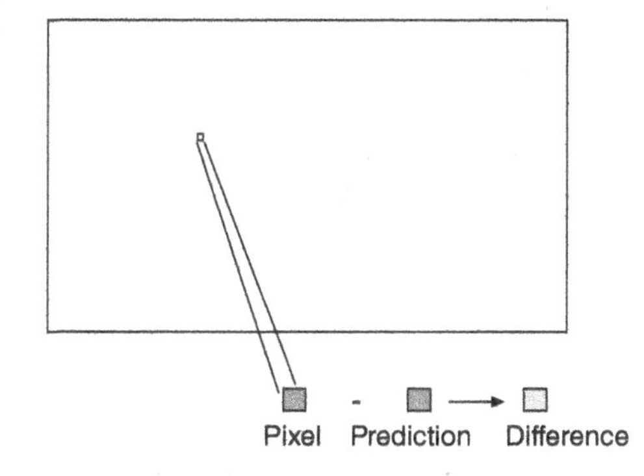

Figure 2.14 Predicting a single pixel (encoder)

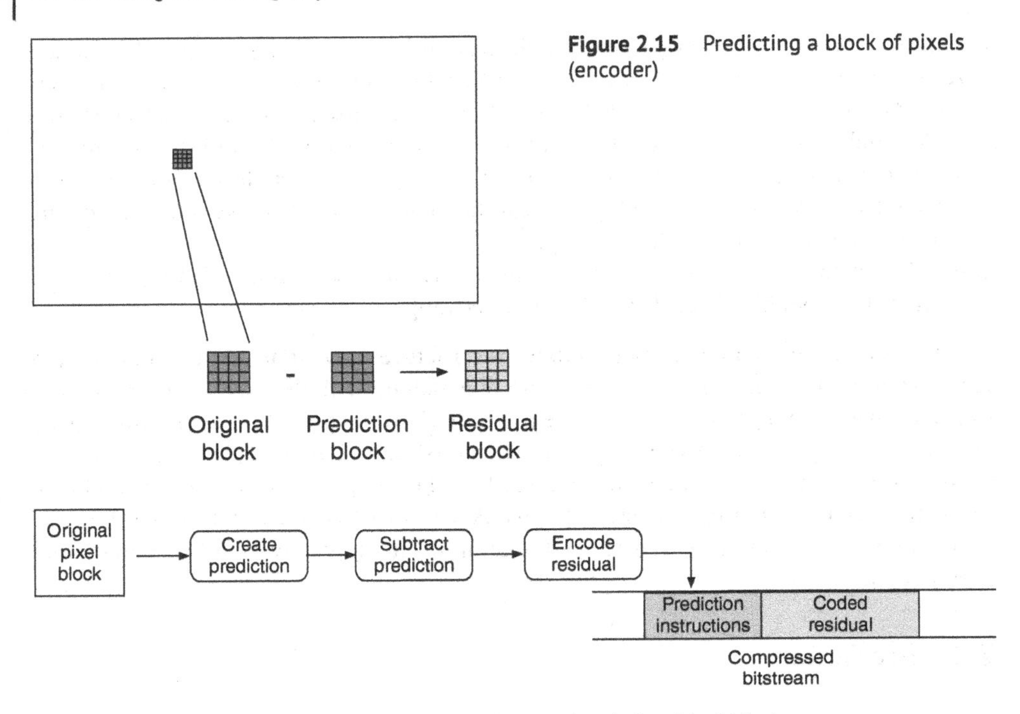

Figure 2.15 Predicting a block of pixels (encoder)

Figure 2.16 Encoder sends prediction instructions and coded residual block

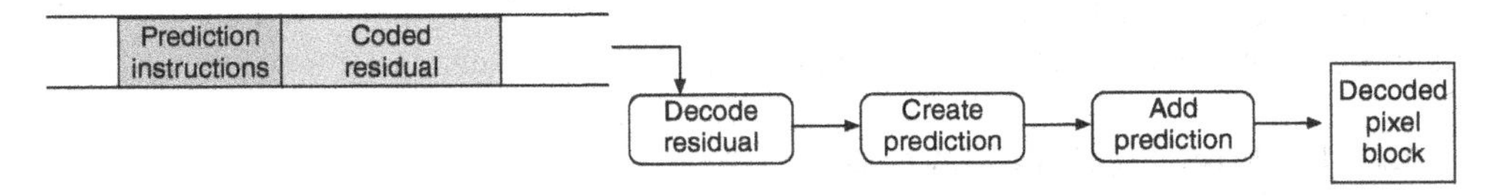

Figure 2.17 Decoder receives prediction instructions and coded residual block

2.5.2 Prediction in a Video Decoder

The decoder receives:

1) The compressed residual block and
2) The instructions for creating the prediction.

The decoder decompresses the residual block and creates the same prediction as the encoder by following the instructions (2) (Figure 2.17). The decoder adds the prediction block to the decompressed residual block to create a decoded block (see Figure 2.18). By doing this for every block in the frame, the decoder reconstructs a complete video frame.

2.5.3 Prediction Example

We will now look at a single 16×16 pixel block within a frame of the 'Foreman' sequence (Figure 2.19).

If we look at a close-up of this block, we can see the individual pixels (see Figure 2.20a). The greyscale or luma component of each pixel is a number in the range 0–255, where

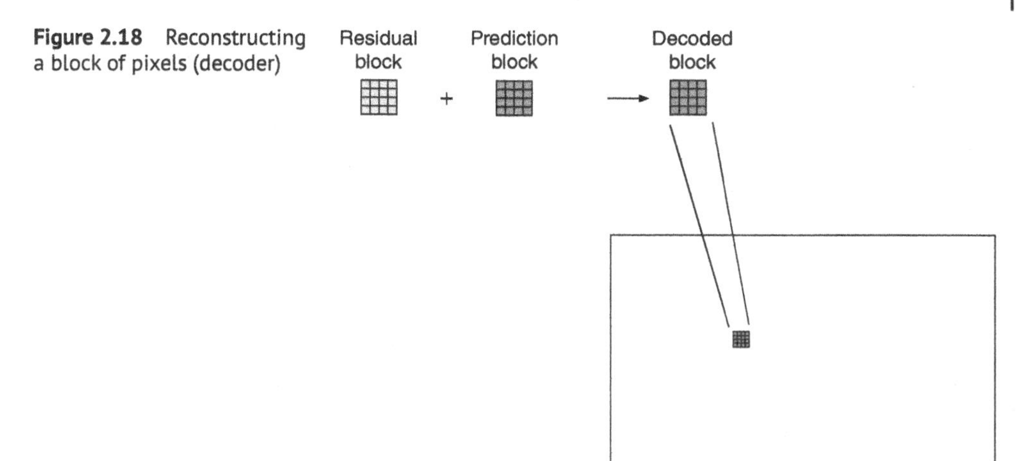

Figure 2.18 Reconstructing a block of pixels (decoder)

Figure 2.19 Single frame with a 16 × 16 block highlighted

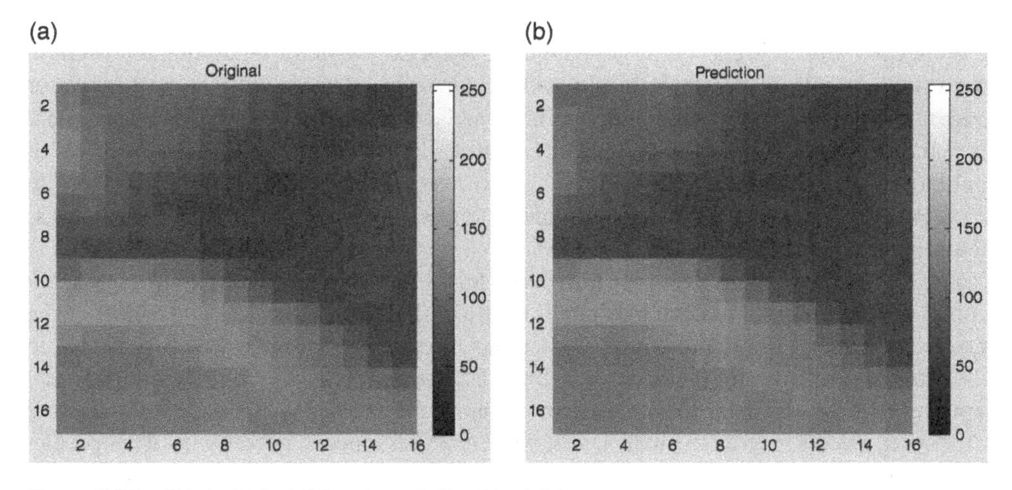

Figure 2.20 Original block (a) and prediction block (b)

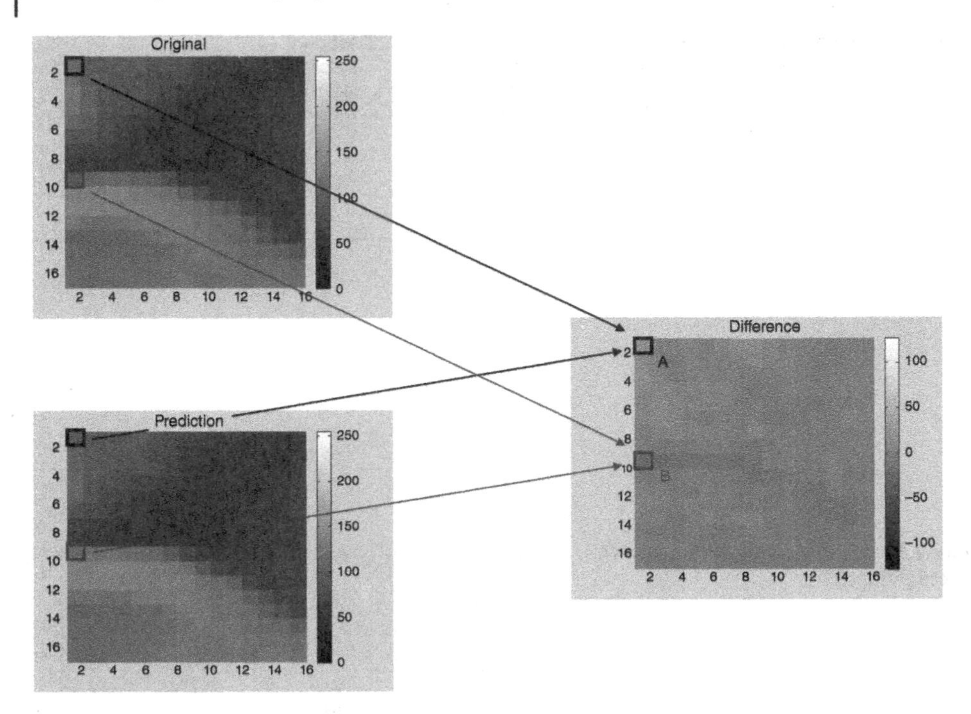

Figure 2.21 Encoder subtracts prediction block from original block, sample by sample

255 represents white, 0 represents black and the numbers in between represent shades of grey.

The encoder creates a prediction block, as shown in Figure 2.20b. We will consider later exactly how the encoder creates a prediction. Notice that the prediction block looks similar to the original block but is not identical. The encoder subtracts the prediction block from the original block, pixel by pixel, to create the residual block (see Figure 2.21). For example, residual sample A (top left of the residual) is created by subtracting the top-left pixel of the prediction from the top-left pixel of the original block.

Notice that the difference or residual block appears to have much less visual information than the original. Each sample in the residual block has a value that can be positive or negative. For example, residual sample A is a small positive number and residual sample B is a small negative number. Because the prediction block is reasonably similar to the original block, most of the residual samples are small positive or negative numbers, close to zero. As we will see later, small-magnitude or zero samples are much easier to compress than the original pixel values.

The encoder repeats this process for every block in the frame. Putting it all together, a complete frame of residual samples might look like this (see Figure 2.22). For most of the pixels, the encoder has found a good prediction, i.e. a prediction pixel that is very similar to the original pixel. Most of the residual frame is a flat grey colour, which represents residual samples that are very close to zero. In some areas of the frame, especially around the edges of moving objects such as the man's face, the prediction is not so good because the prediction pixels are not a good match for the original pixels. In these areas, the residual samples have larger-magnitude positive (light grey) or negative (dark grey) values.

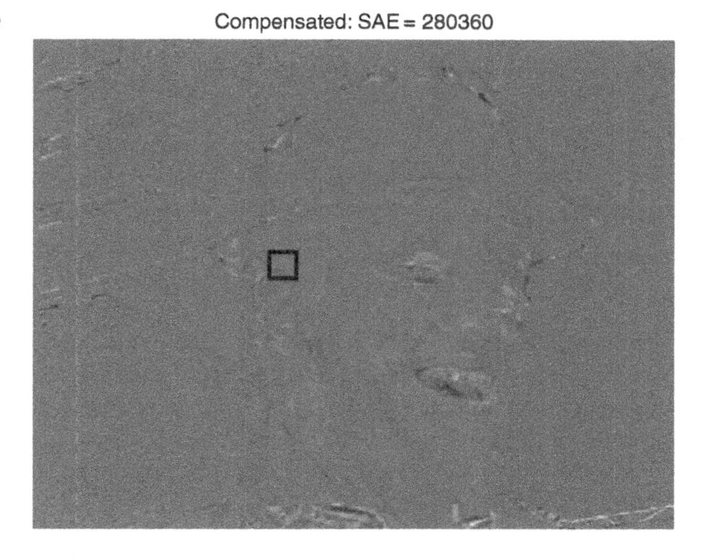

Figure 2.22 Complete frame of residual samples after prediction of every block

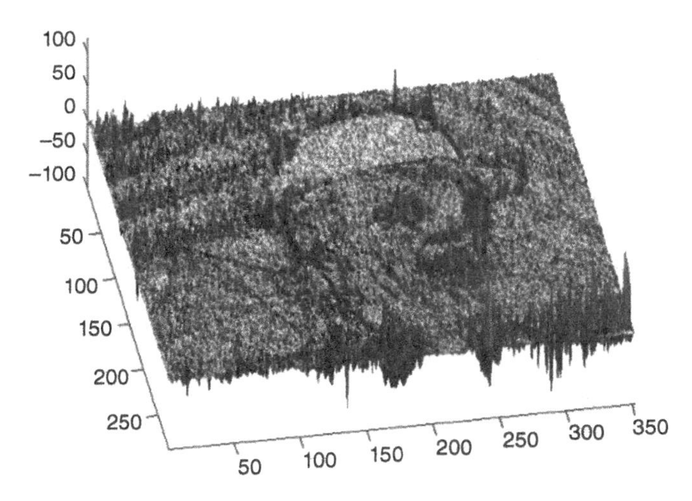

Figure 2.23 Surface plot of residual samples

If we view the residual frame as a surface plot (see Figure 2.23), we can see the positive and negative residual samples as peaks and troughs in the surface. The goal of the prediction process is to create a residual that is as close as possible to zero, subject to some other constraints that we will consider next.

2.5.4 Choosing a Prediction

The encoder has a number of choices for creating each prediction block, which we will examine in detail later. A video coding standard such as H.264/AVC or H.265/HEVC typically does not specify exactly how an encoder should choose a prediction. However, the Standard does set out the range of options available to the encoder. For example, an MPEG-2 video encoder has a relatively limited range of options for predicting each

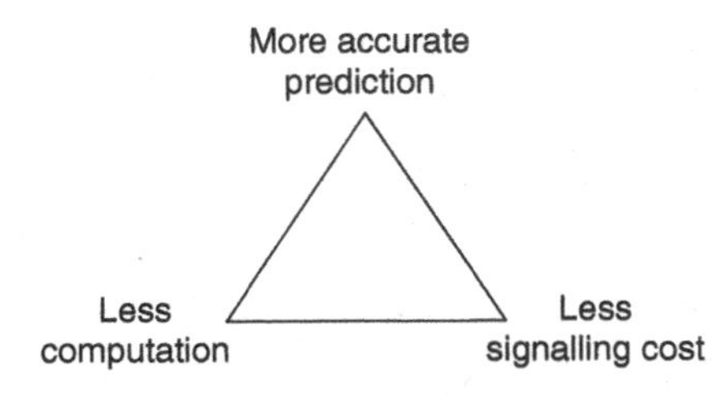

Figure 2.24 Factors involved in choosing a prediction (encoder)

block, whereas an HEVC/H.265 encoder has a much larger number of prediction choices. The Standard sets out the range of options for prediction, and a video encoder decides how to create the prediction, from within the range of options specified by the standard.

For every pixel in the frame, we generally want to create a prediction that is similar to the pixel value. For each block of pixels, the first criterion for choosing a prediction is that a 'good' prediction creates a prediction block that is similar to the original block and a 'bad' prediction creates a prediction block that is dissimilar to the original block. If the encoder can create a prediction that is similar to the original block, i.e. an accurate prediction, then the residual block will contain small values that are easy to compress.

However, there are at least two other factors to consider besides the desire to create an accurate prediction (see Figure 2.24):

a) How much information is required to tell the decoder how to create an identical prediction?
b) How much computation is required to generate the prediction?

The encoder has to communicate (signal) the prediction to the decoder so that the decoder can create an identical prediction. It takes a certain number of bits to signal the prediction, so it is desirable to reduce or minimise the cost of signalling the prediction.

As we will see later, finding a near-optimal prediction can take a lot of computation. For example, an encoder may have many different choices of prediction available for each block of pixels. Checking every choice is often a computationally intensive task. If an encoder is working in real time and is processing 30 or more frames of video every second, then it has a limited time in which to find a good prediction. Hence, it is often desirable to find a prediction that does not take too many computational resources to calculate.

Prediction is often a trade-off that involves:

a) Trying to find the most accurate prediction, that
b) Does not take too many bits to communicate, and
c) Does not take too much computation to calculate.

Once the encoder has chosen a prediction and encoded the residual, it sends the prediction instructions and the coded residual in the compressed video bitstream or file.

In a standards-based video codec, the choice of prediction for every pixel block is communicated to the decoder as part of the compressed bitstream or compressed file. The decoder does not need to choose a prediction; instead, it creates a prediction according to the instructions sent by the encoder[3]. The Standard specifies exactly how the choice of prediction is communicated, and the decoder interprets the prediction instructions to create the same prediction as the encoder. We could say that the encoder chooses the prediction for the decoder.

3 Unless the decoder is carrying out prediction or refinement itself.

2.5.5 Inter and Intra Prediction

For every block in a video frame, the encoder can create a prediction block. The prediction block should be:

- A good match for the current block, i.e. similar to the current block.
- Created using the information that is already available to the decoder, so that the decoder can create the same prediction. This usually means that the prediction should be based on previously coded information.
- Capable of being communicated to the decoder.

How can the encoder satisfy these requirements? If we want to find pixels that are similar to our current block, a good place to start is to look at nearby pixels in the same frame or previous/future frames.

Example

Consider a frame of the 'Foreman' clip. We want to find a prediction for the 16 × 16 pixel block marked in Figure 2.25.

If we look inside the same frame, we can see that the pixels *around* the block are quite similar to the pixels *inside* the block (see Figure 2.26). This is often the case. Unless there is a big discontinuity, such as the boundary between the man's face and the background, we can usually find similar pixels immediately next to the block. The further away from the block we go, the less likely we are to find similar pixels.

If we look at the previous and next frames, i.e. frames N − 1 and N + 1, we will

Figure 2.25 Frame N (Foreman)

Figure 2.26 Similar pixels in frame N

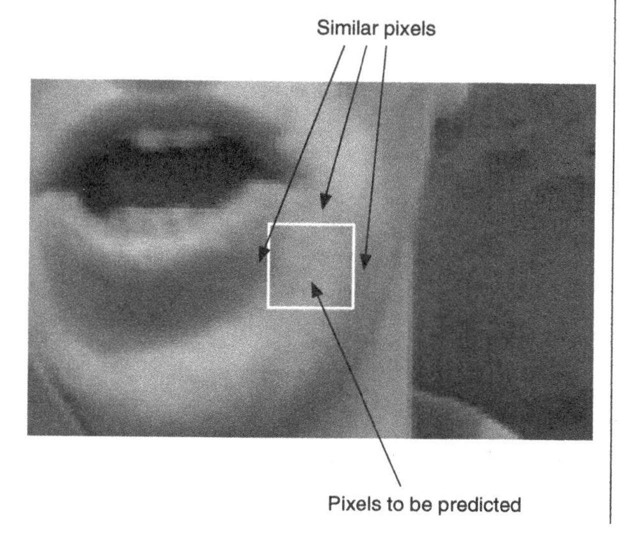

see that the pixels *nearby* the same position in the past and future frames are quite similar to the current block in frame N (see Figures 2.27 and 2.28).

Predicting the block from pixels in the same frame (see Figure 2.26) is known as **intra prediction**. Predicting the block from pixels in a different frame (see Figure 2.28) is known as **inter prediction**. In Chapters 5 and 6, we will look at each of these prediction methods in detail and consider how and why the encoder chooses a particular prediction for a block.

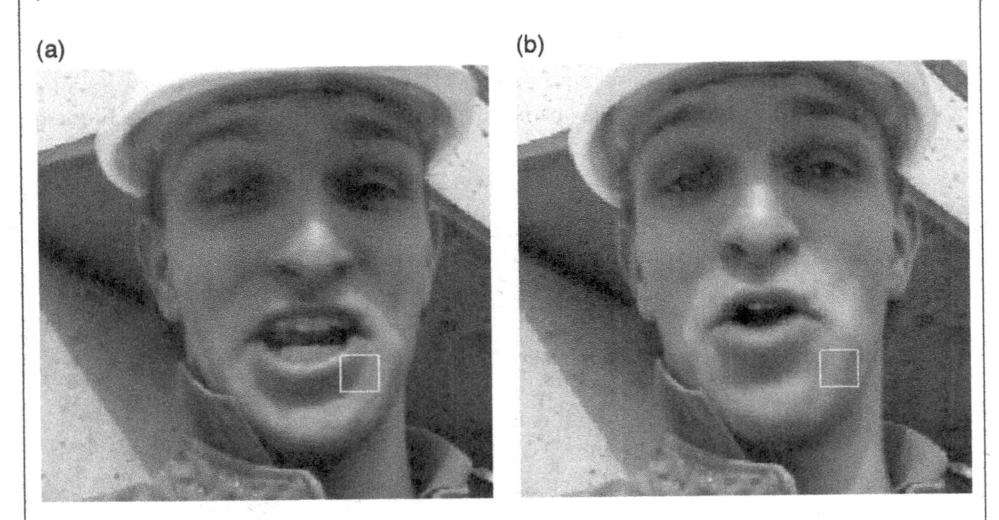

Figure 2.27 Frame N − 1 (a) and frame N + 1 (b)

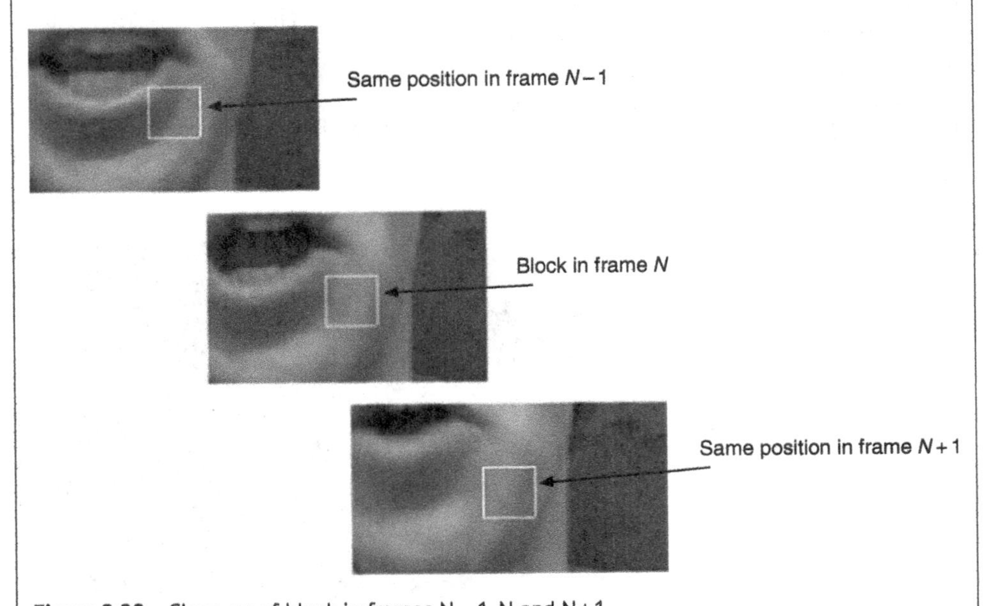

Figure 2.28 Close-up of block in frames N − 1, N and N + 1

2.6 Transform and Quantisation

The output of the prediction process is a residual block, which is also the input to the transform and quantisation process. The transform process converts this block of residual samples into a block of transform coefficients. In itself, the transform does not remove or reduce information. Instead, it represents the block in a different form or domain. The quantisation process reduces the precision and the dynamic range of each of the transform coefficients. Quantisation is a lossy process, i.e. it removes information and is not reversible.

2.6.1 Transform

A two-dimensional transform converts a block of image samples into a block of transform coefficients (see Figure 2.29). The transform coefficients represent the spatial frequency content in the original block of samples.

Each type of transform has its own set of **basis patterns**. Any image or residual block can be formed by weighting and combining or superimposing these patterns. The forward transform can be considered as a process that finds the weighting of each basis pattern required to form the image block.

Figure 2.30 shows the basis patterns for a particular 4×4 transform, a version of the Discrete Cosine Transform (DCT). There are 16 samples in a 4×4 image block, and there are 16 coefficients of the transform, one corresponding to each of the 16 basis patterns. The top-left pattern is a flat block, corresponding to zero frequency or 'DC'. Moving to the right, each pattern increases in frequency in the horizontal direction. Moving down, each pattern increases in frequency in the vertical direction. The lowest-right pattern is a chequerboard of horizontal and vertical frequencies. Figure 2.31 shows the corresponding one-dimensional signals. For example, a horizontal slice through the first row of basis patterns would have intensities corresponding to Figure 2.31: DC, half a cycle, one cycle, 1.5 cycles.

Any 4×4 block of pixels or samples can be created by weighting and combining the 4×4 basis patterns of Figure 2.30. Different weights (i.e. coefficients) produce different blocks of samples when combining the patterns. Figure 2.32 shows an example. An input block (left) is transformed (right). The transformed block consists of coefficients, which are weighting values for each of the 4×4 basis patterns (see Figure 2.30). For example, a coefficient of −41 means that we scale the corresponding basis pattern by 41, and because of the − sign, invert it. The shaded top-left value is the DC coefficient, and this has the largest magnitude (+961). Moving to the right and down, the transform coefficient magnitudes tend to reduce. This is typical for a block from a natural image. Most of the larger valued spatial frequencies are clustered

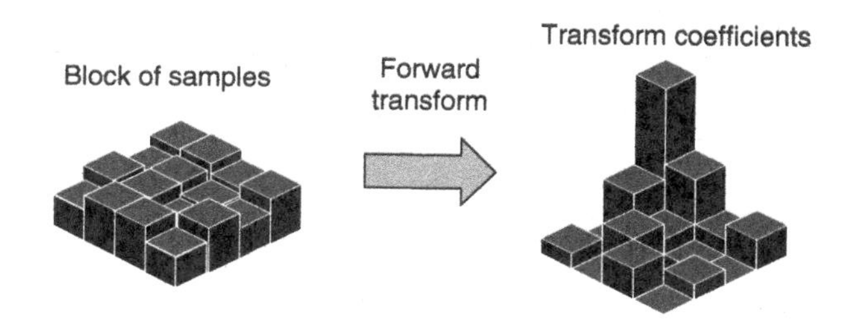

Figure 2.29 Forward transform process

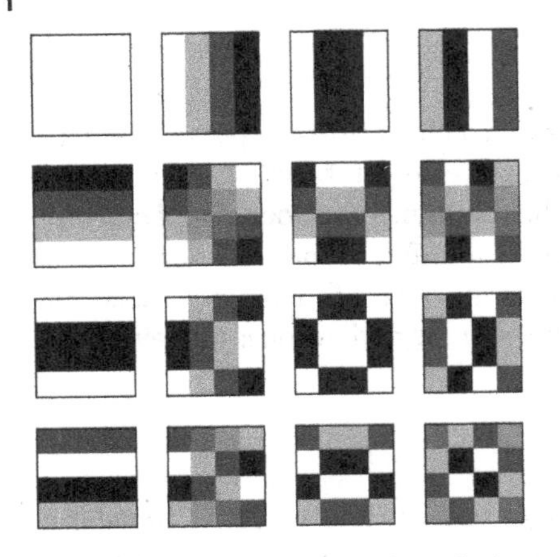

Figure 2.30 Basis patterns for a 4×4 transform

around the top left or DC coefficient. Note that some of the coefficients are negative, which means that the corresponding basis pattern is inverted.

The transform does not compress data. Instead, it represents the information from the original block in a form that may be more easily compressible. In Figure 2.31, every sample in the original block has a similar magnitude, around 60, and makes a similar contribution to the appearance of the image. In contrast, a few of the samples in the transformed block have large magnitudes and the rest are relatively insignificant. If we discard the small-magnitude coefficients and retain only the significant transform coefficients, using quantisation, we will still retain most of the important visual information in the block. This will be discussed further in Chapter 7.

2.6.2 Quantisation

A transformed block is converted into a quantised block. Quantisation, or 'forward quantisation', reduces the precision of every coefficient in the block, which has the effect of setting small-valued or insignificant coefficients to zero. Figure 2.33 illustrates an example of a scalar quantisation process. Unquantised values (left) are converted to quantised values (right). In this example, unquantised values in the range −1.0 to +1.0 are quantised to 0. Unquantised values in the range +1.0 to +3.0 are quantised to +1, and so on.

In Figure 2.34, a 4×4 block of transform coefficients is quantised. Larger coefficients are reduced in magnitude, losing some precision, whereas smaller coefficients are set to zero. In this example, 12 of the 16 coefficients become zero after quantisation.

Figure 2.31 One-dimensional basis signals

58	64	51	58
52	64	56	66
62	63	61	64
59	51	63	69

Block of samples

961	−41	15	−48
−34	72	−30	−104
−15	3	15	24
13	81	−5	8

Transformed block

Figure 2.32 Block of samples and forward transform output

The behaviour of this process can be controlled by a quantisation parameter (QP), which affects the strength of quantisation. A larger QP results in more quantisation, i.e. a greater loss of precision, but also results in more compression. QP is an important control parameter in a video codec because it has a direct effect on compression and quality (see Figure 2.35). Two decoded versions of the same frame are shown in the left and right halves of Figure 2.36. Both have the same spatial resolution (number of pixels) but have been coded with different QPs:

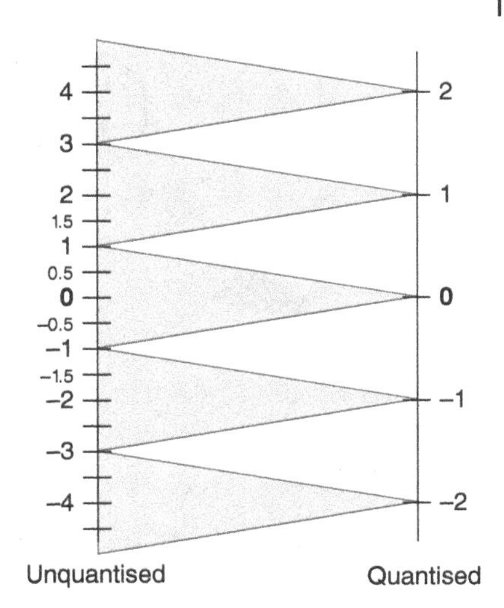

Figure 2.33 Scalar quantisation

- Low QP (left half)
  - Better-decoded video quality
  - Less compression = higher coded bitrate
- High QP (right half)
  - Worse decoded video quality
  - More compression = lower coded bitrate.

Quantisation is an irreversible or 'lossy' process. The information removed during quantisation cannot be restored later.

2.7 Bitstream Coding

All the information necessary to decode the video sequence must be encoded, i.e. packaged into a bitstream that is (1) compressed, (2) suitable for transmission and (3) decodable.

The information that needs to be sent may include:

- Quantised transform coefficients
- Prediction parameters, such as motion vectors, prediction mode choices, reference frame choices and prediction block sizes
- QP choice
- MB information
- Picture information
- Sequence information, i.e. high-level parameters that affect the entire video clip.

Figure 2.34 Transform coefficients before and after quantisation

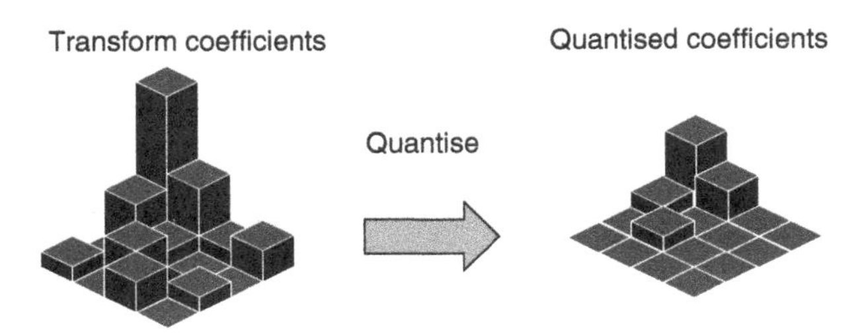

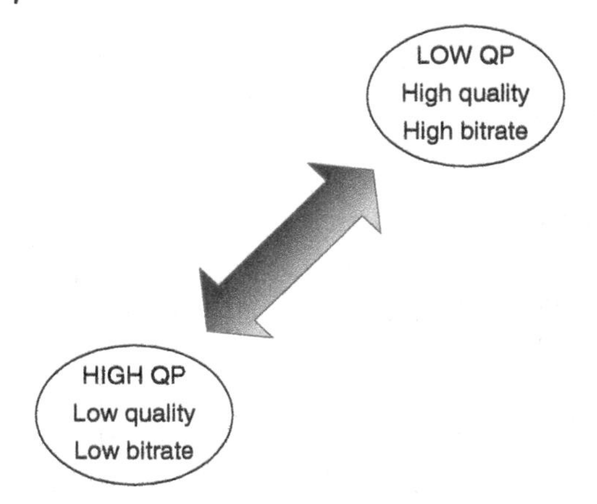

Figure 2.35 Quantisation, quality and bitrate

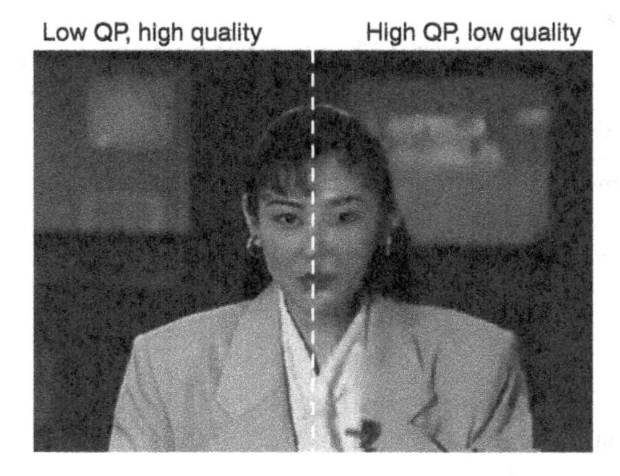

Figure 2.36 Low versus high quantisation

First, certain values are **pre-processed**, i.e. converted into a suitable form for compression. Then, the processed values are entropy encoded, i.e. coded into a compressed bitstream. Two commonly used forms of entropy coding are **variable-length coding** and **arithmetic coding**.

2.7.1 Pre-processing

Pre-processing takes advantage of the statistical behaviour of certain values produced by the video encoder.

Coefficient re-ordering: Quantised transform coefficients tend to follow particular distributions. Typically, larger non-zero values are clustered around the top left or DC coefficient (see Figure 2.34), with smaller or zero values towards the lower right or highest frequencies. The two-dimensional block of transform coefficients is converted to a one-dimensional array in a scanning pattern such as a zig-zag scan (see Figure 2.37). The effect is generally to concentrate the significant non-zero coefficients together.

Predictive coding: Prediction parameters and QP tend to be highly correlated amongst neighbouring blocks in a coded frame. These parameters can be efficiently coded using differential prediction. For example, the encoder forms a predicted QP based on previously transmitted QP values. Instead of sending the QP itself, the encoder calculates and sends the difference between the predicted and actual QP. The decoder receives the differential value, forms the same prediction and adds the differential to the prediction to construct the QP. A similar approach can be taken for motion vectors, prediction modes, prediction block sizes, etc.

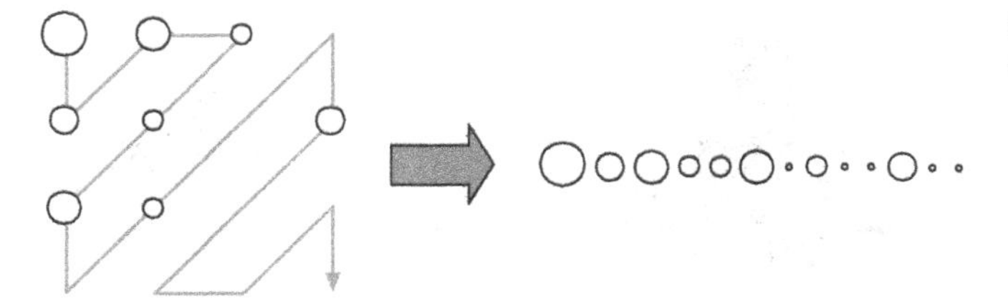

Figure 2.37 Zig-zag scan of a block of coefficients

2.7.2 Entropy Encoding

By this stage, the original video sequence has been converted into a set of parameters and values, sometimes described as symbols. The entropy encoding process converts the symbols comprising a coded video sequence into an efficient binary form, a compressed bitstream. Entropy encoding is a mapping between symbols and bits. The following list describes popular mappings in video encoders.

Fixed-length encoding: Each of the possible values of a symbol is represented with a binary code with the same number of bits, regardless of the value. This mapping may be suitable for symbols that occur infrequently, e.g. a symbol that occurs once or only a few times per video sequence, such as a start code or a sequence-level parameter. A fixed-length encoding is perhaps the simplest mapping but is not particularly efficient in terms of compression.

Variable-length encoding: A symbol is mapped to a binary code that may have a variable number of bits, depending on the value of the symbol. Typically, shorter binary codes are assigned to values that occur frequently (with a high probability), and longer binary codes are assigned to values that occur less frequently (with a low probability). Variable-length codes may be structured, i.e. capable of algorithmic construction, such as the Exponential-Golomb (Exp-Golomb) codes shown in Table 2.2. Exp-Golomb codewords can be constructed according to a pattern: n leading zeros, followed by a 1, followed by an n-bit binary word. An Exp-Golomb codeword can be constructed for any index value. Alternatively, variable-length codewords may be designed specifically for a particular value or set of values. For example, Context-Adaptive Variable-Length Coding (CAVLC) for transform coefficients in H.264/AVC uses a mapping that is specific to the structure and values found in a block of transform coefficients.

Variable-length codewords for video coding should be uniquely decodable so that there is no ambiguity in the value that is to be decoded. Each of the Exp-Golomb codewords shown in Table 2.2 is uniquely decodable from left to right. As long as the decoder starts decoding at the beginning of a codeword, there is no possibility of confusing one codeword with another.

Table 2.2 Exp-Golomb variable-length codes

Index	Codeword
0	1
1	010
2	011
3	00100
4	00101
5	00110
6	00111
7	0001000
8	0001001
...	...

A limitation of variable-length encoding is that each symbol must be represented by an integral (whole) number of bits. In fact, the most efficient mapping for a symbol with probability of occurrence P is $\log_2(1/P)$ bits, which may be a fractional number rather than an integer.

Binary arithmetic encoding: In an arithmetic coder, each symbol is mapped to a range or interval, so that a series of symbols can be represented by a fractional number. This fractional number is in turn mapped to a binary code. Arithmetic coding does not require that each symbol be individually mapped to an integer number of bits. Instead, symbols are represented by a fractional number of bits, and as the number of symbols represented by a single number increases, arithmetic coding can approach the ideal mapping of $\log_2(1/P)$ bits per symbol.

In a practical video codec, several compromises are made when implementing arithmetic coding. General arithmetic coding, in which a symbol can take any one of multiple values, can be replaced with binary arithmetic coding, in which symbols are binarised, i.e. converted into a binary representation, prior to encoding, such that the arithmetic coder only works on binary values. Computationally intensive multiplications or divisions can be replaced with approximations, and the range of potential values for each interval may be restricted in order to simplify processing.

Context adaptation: Variable-length encoding and arithmetic encoding rely on knowledge of symbol probabilities to provide maximum compression. Early video coding standards such as MPEG-2 and H.263 used pre-determined probability tables derived from analysis of generic video sequences. More recently, H.264 and HEVC employ context adaptation, in which symbol probabilities are adapted based on the actual statistics of the current video sequence. The video encoder and decoder measure the frequency of occurrence of symbol values and maintain a running total of the probability of each value occurring. Along with statistics from neighbouring, previously coded values, these probability totals can be used to derive the codewords (VLC) or ranges (arithmetic coding).

2.8 The Coded Bitstream

A coded video sequence consists of a hierarchy of parameters or **syntax elements** (see Figure 2.38). At the **sequence** level, a header contains coded parameters that are necessary for decoding subsequent coded frames or **pictures**. Each picture may or may not have a header and may be made up of a number of **slices**, each of which corresponds to an area of the coded picture. Each slice is made up of a number of coded basic units, such as **MBs**, Superblocks or CTUs. Finally, each coded unit comprises header information, prediction parameters such as motion vectors or intra prediction modes, and coded transform coefficients.

2.9 Storing and Transmitting the Coded Bitstream

The coded bitstream (see Figure 2.38) is suitable for efficient storage or transmission.

File storage: A complete coded sequence can be stored in a file. A media file containing compressed video is known as a **container** and may contain extra information such as a header, timing or synchronisation parameters, one or more audio tracks and subtitles. These other data may be interleaved with the video information, so that the video file may be played by extracting and decoding video frames, audio chunks and associated information.

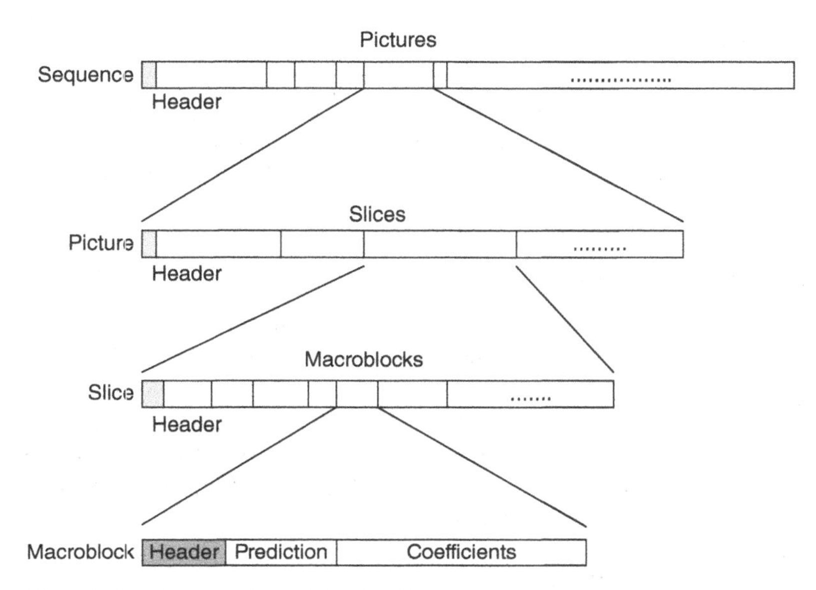

Figure 2.38 Syntax of a coded video bitstream

Network transmission: The coded sequence may be transmitted over a packet network such as the Internet. The sequence must be packetised, i.e. split up and placed in network packets for transport. The syntax of a coded video sequence is typically designed to accommodate packetised transmission. For example, important header information such as sequence parameters may be sent in an initial packet and at regular intervals. Each slice or picture may be sent in a separate packet, or multiple slices may be sent in a packet (see Figure 2.39). Re-synchronisation points, e.g. intra coded pictures that can be decoded in isolation from other data, are inserted at frequent intervals so that the clip can be played from random points and so that any transmission losses can be recovered.

Broadcast: Transmission over a broadcast medium such as cable, terrestrial or satellite TV involves mapping coded video and audio into a sequence of transmission packets, which is associated with a particular programme or channel. Transmission packets from multiple programmes are multiplexed together and carried via the broadcast medium. Issues such as random access, recovery from transmission errors and switching between channels or programmes are handled in a similar way to network transmission, i.e. by inserting regular synchronisation points in the coded sequence.

Figure 2.39 Packetising a coded video sequence

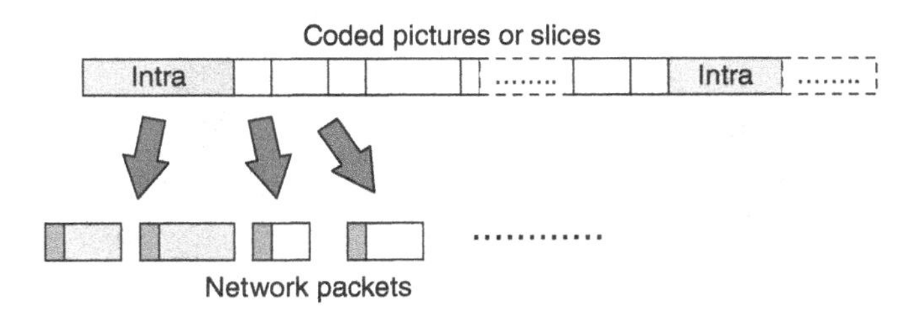

2.10 The Decoder

Decoding the compressed video sequence involves reversing the steps of the encoder (see Figure 2.4). Typically, each step of the decoding process is defined in a video coding standard or format specification, so that any compatible decoder is capable of correctly decoding a standard-compliant sequence.

2.10.1 Entropy Decoding

An entropy decoder reads the compressed bitstream, parses the arithmetic, variable-length or fixed-length coded data and recovers the sequence of symbols that comprise the coded video sequence.

Pre-processing steps such as predictive coding and coefficient scanning are reversed. The decoder constructs the same predictors as the encoder, e.g. motion vector predictors, and uses these to recover the coding parameters. Decoded coefficients are inverse scanned to reconstruct blocks of quantised transform coefficients.

2.10.2 Rescaling and Inverse Transform

The reverse of the forward quantisation process, i.e. the quantisation of transform coefficients carried out during encoding, is **rescaling** (see Figure 2.40). This is often referred to as inverse quantisation, but it is important to note that this is not a fully reversible process. Because forward quantisation removes information, rescaling does not recover the original transform coefficients but instead produces a decoded set of coefficients with some loss of precision. Transform coefficients that were originally low-valued and were set to zero by the quantisation process will remain zero after rescaling. Other coefficients will be restored to their approximate magnitude but may have lost precision.

The inverse transform (see Figure 2.41) converts the block of decoded transform coefficients into a block of decoded images or residual samples. Because of the loss of information in the transform coefficients, the decoded image block will not be identical to the original image block.

Figure 2.42 shows the result of decoding the block of quantised coefficients shown earlier (see Figure 2.32). The received coefficients are inverse-transformed to produce a decoded image block. Comparing this with the original image block, we can see that most of the image sample values are similar but not identical. The difference or loss between the original and decoded blocks is due to quantisation in the encoder.

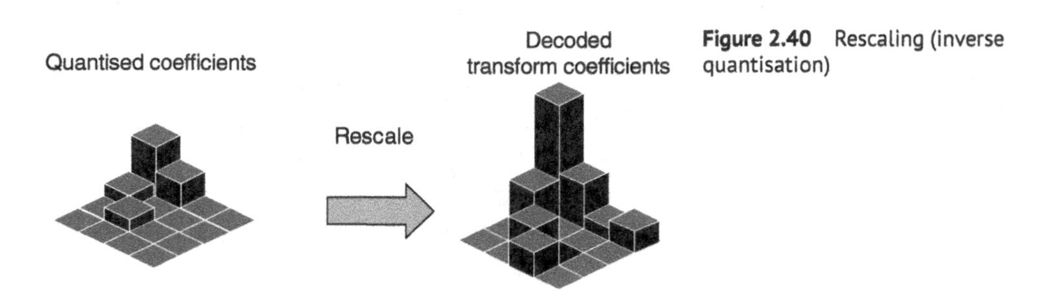

Figure 2.40 Rescaling (inverse quantisation)

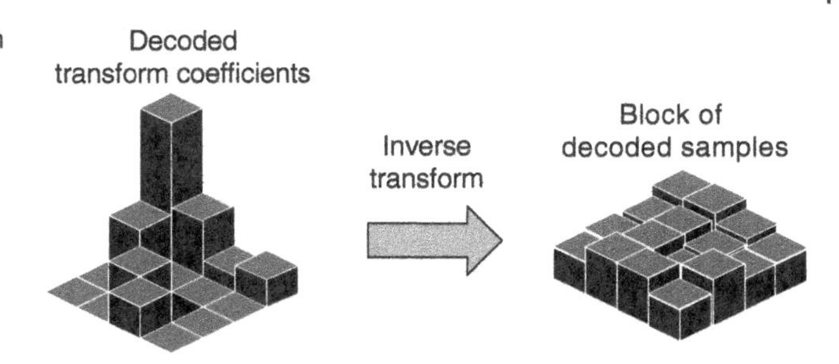

Figure 2.41 Inverse transform

2.10.3 Reconstruction

The decoder creates the same prediction block or MB that the encoder created, by generating a spatial or motion-compensated temporal prediction using the same prediction parameters – modes, motion vectors, etc. – as the encoder.

This prediction block is added to the decoded residual samples from the output of the inverse transform. The result is a decoded block or MB that can be displayed and/or used for further predictions (see Figure 2.17).

58	64	51	58
52	64	56	66
62	63	61	64
59	51	63	69

Block of samples

961	–41	15	–48
–34	72	–30	–104
–15	3	15	24
13	81	–5	8

Transformed block

48	–1	0	–1
–1	1	–1	–2
0	0	0	0
0	1	0	0

Quantised block

55	66	54	58
54	62	58	63
61	59	61	62
60	55	65	67

Rescaled and inverse transformed

Figure 2.42 Example: original, transformed, quantised, decoded

2.10.4 Decoder Output

As each MB is reconstructed and decoded, it can be placed in its correct position in a decoded frame (see Figure 2.43). Once a complete decoded frame has been built up, it is available for output and/or for storing in a decoded frame buffer to form further predicted frames.

2.11 The Video Codec Model

Since the 1990s, popular or mainstream video codecs have all adopted the same basic structure. The encoder (see Figure 2.44) applies the following steps to convert basic units, e.g. a MB, 16 × 16 pixels, into a compressed bitstream:

1) Motion compensated prediction from one or more previously encoded frames and/or intra prediction from previously coded samples in the same frame[4].
2) Subtraction of a motion-compensated prediction from the current basic unit to form a residual unit, such as a residual MB.

4 For simplicity, intra prediction is shown based on the current frame. In practice, intra prediction is formed from decoded, reconstructed samples from the current frame.

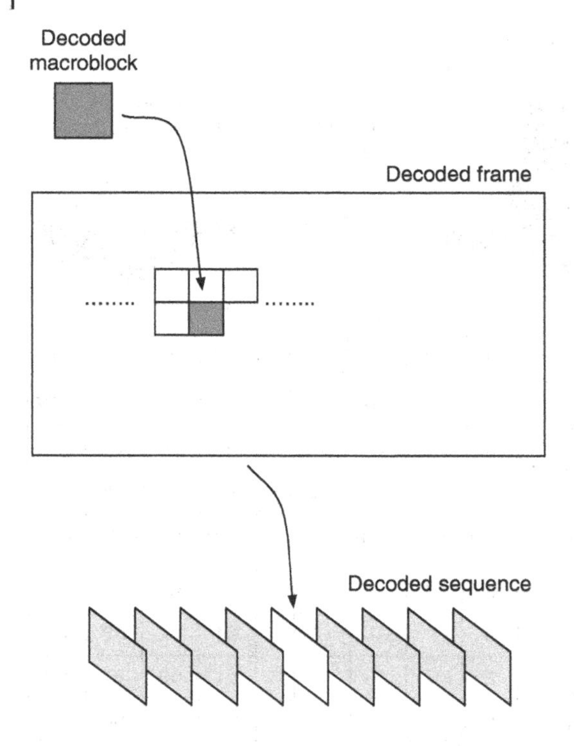

Figure 2.43 Decoding a frame and a sequence

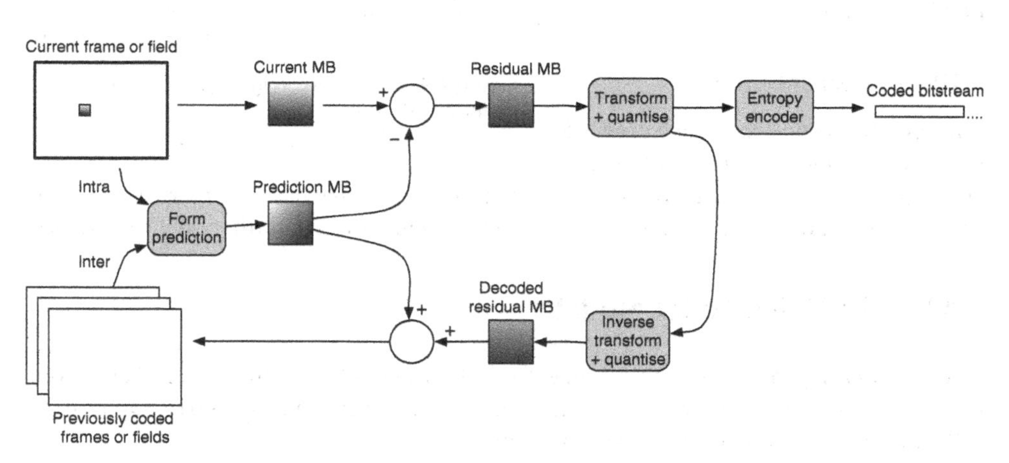

Figure 2.44 Video encoder model

3) Block transform and quantisation to form blocks of quantised coefficients.
4) Entropy encoding of quantised coefficients and side information such as motion vectors and headers.
5) Rescaling (inverse quantisation), inverse transform and addition of a motion-compensated prediction to reconstruct a local copy of the decoded frame or field.

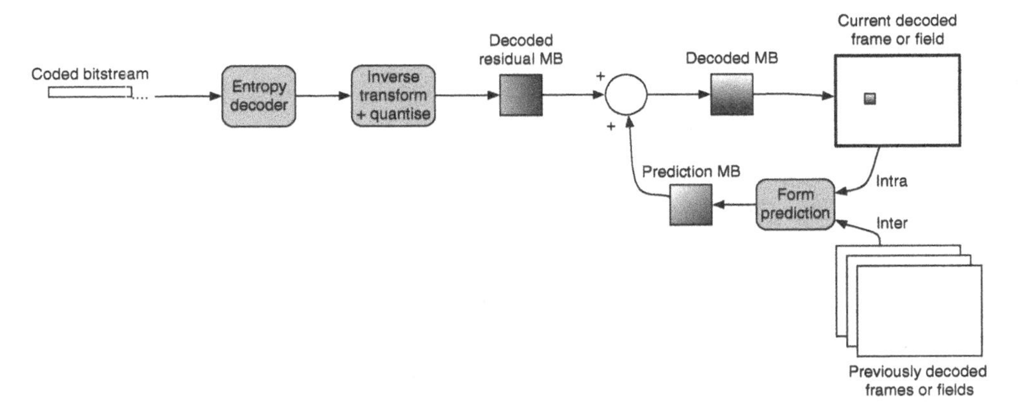

Figure 2.45 Video decoder model

The decoder (see Figure 2.45) carries out the following steps to convert a compressed bitstream into a decoded sequence of frames or fields:

1) Entropy decoding to extract quantised coefficients, motion vectors and headers.
2) Rescaling and inverse transform to form a residual unit.
3) Motion-compensated prediction and addition of the prediction to the residual.

2.12 Video Codec Performance

We have seen how a video encoder can compress a sequence of video frames to create an encoded bitstream and how a video decoder can decompress a bitstream to create a decoded video sequence. Most present-day codecs for consumer applications use lossy video compression, which means that the decoded video is not identical to the original video. Typically, the encoder introduces this loss during the quantisation process (see Section 2.6.2). Once the visual data has been quantised, the lost data cannot be restored.

The performance of a video codec, i.e. how good it is at compressing video, can be evaluated along three main dimensions:

1) **Compression or rate**: How much is the video compressed by? How much space do I need to store the compressed video file? What bitrate do I need to transmit the compressed video?
2) **Computation**: How complex is the process of encoding or decoding? Can I encode or decode in real time on a particular processor? How much memory, e.g. RAM, do I need during encoding or decoding? How much power does encoding or decoding consume?
3) **Quality or distortion**: How does the decoded video look? Does it look indistinguishable from the original? (visually lossless). Are there obvious distortions such as blockiness or blurring? Does the decoded video look smooth or jerky? Is the quality good enough for the application?

When comparing video codecs or determining whether to include a proposed modification during the development of a video coding standard, it is common practice to measure

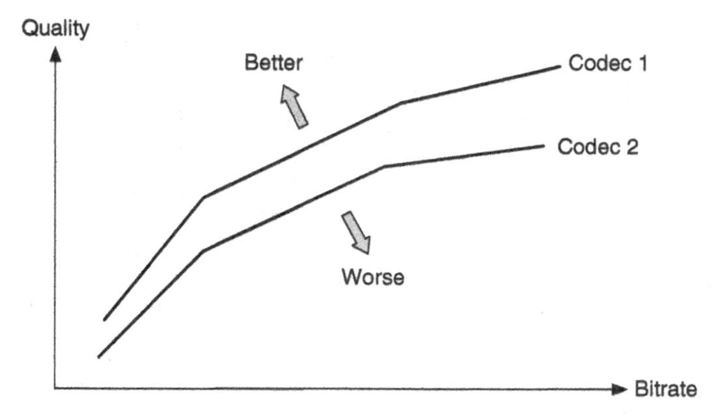

Figure 2.46 Rate–distortion performance

the rate–distortion performance (see Chapter 11). This involves comparing the decoded video quality across a range of compressed bitrates. The relationship between quality and coded bitrate is known as a rate–distortion curve (see Figure 2.46). A 'better' codec or coding method will tend to produce higher-quality decoded video across a range of bitrates. In the example of Figure 2.46, Codec 1 performs better than Codec 2 across all the tested bitrates. What this figure does not show is the computational performance. It is often the case that better rate–distortion performance comes at a cost of higher computation, at the encoder or decoder or both. For example, with suitable choices of encoding parameters, an H.265/HEVC encoder can have significantly better rate–distortion performance than an H.264/AVC encoder, but this might come at the cost of increased computation.

Bitrate or compressed file size is relatively easy to measure – simply count the number of encoded bits per second of video, i.e. encoded bitrate, or the number of bytes in the encoded video file. Computation can be measured or estimated, for example, by counting the number of seconds it takes to encode a complete video file. However, video quality is a different matter, because the question 'how good does this video look?' is fundamentally a subjective question.

2.12.1 Video Quality

Measuring visual quality is a challenge because quality is subjective. Ask any two people the question 'how good does this video look?' and you are likely to get two different answers. In fact, it has been shown that the same person may give a different answer to this question depending on many factors, such as the environment in which they are watching the video, the content of the video, their own experience and even their state of mind. How, therefore, can we measure video quality? The short answer is that we cannot fully measure or predict how a human observer will assess the quality of a video.

We can approximate or estimate visual quality in several ways, which can be categorised as subjective or objective methods. Subjective quality measurement involves asking human observers to rate visual quality, with the aim of compensating for the natural variations in opinions between observers. Objective quality measurement involves calculating a measure or metric that attempts to approximate the opinion or rating of a human observer.

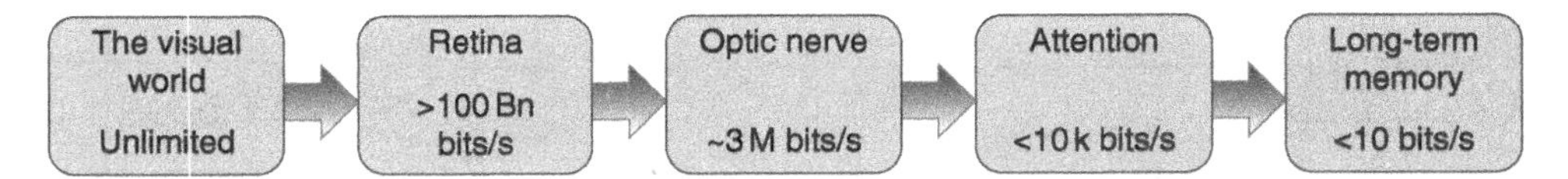

Figure 2.47 Visual attention and information content, adapted from Anderson, Van Essen and Olshausen

2.12.2 Subjective Quality Measurement

2.12.2.1 Factors Influencing Subjective Quality

Our perception of a visual scene is formed by a complex interaction between the components of the human visual system (HVS), including the eye and the brain. The perception of visual quality is influenced by spatial fidelity, i.e. how clearly parts of the scene can be seen, whether there is any obvious distortion, and temporal fidelity, i.e. whether motion appears natural and smooth. However, a viewer's opinion of quality is also affected by other factors such as the viewing environment, the observer's state of mind and experience and the extent to which the observer interacts with the visual scene. For example, it has been shown that a viewer's opinion of visual quality is measurably higher if the viewing environment is comfortable and non-distracting, regardless of the quality of the visual image itself.

Other important influences on perceived quality include visual attention, i.e. the way an observer perceives a scene by fixating on a sequence of points in the image rather than by taking in everything simultaneously [2], and the so-called recency effect, which means that our opinion of a visual sequence is more heavily influenced by recently viewed material than older video material [3, 4]. Whilst a very large amount of visual information reaches the retina (see Figure 2.47), the amount of information we can attend to at any one time is limited to less than 10 kbits per second, according to ref. [5]. Because of this, we fill in much of the visual scene based on our experience and expectation of how the visual world works. Hence, the human observer can only really attend to a small subset of a displayed video image. All of these factors make it difficult to quantitatively measure visual quality.

2.12.2.2 Subjective Quality Assessment Methods

Many different subjective quality assessment methods are described in the literature [6] and in several industry standards such as ITU-T BT.500 [7]. For example, the double stimulus continuous quality scale (DSCQS) method presents the viewer with a pair of images or short video sequences A and B, one after the other, and the viewer is asked to give A and B a quality score by marking on a continuous line with five intervals ranging from Excellent to Bad. In a typical test session, the assessor is shown a series of pairs of video sequences and is asked to grade each pair. Within each pair of sequences, one is an unimpaired or reference sequence and the other is the same sequence, modified by a system or process under test. Figure 2.48 shows an experimental set-up appropriate for the testing of a video codec in which the original sequence is compared with the same sequence after encoding and decoding. The selection of which sequence is A and which is B is randomised during the test session so that the assessor does not know which is the original and which is the impaired sequence. This helps prevent the assessor from

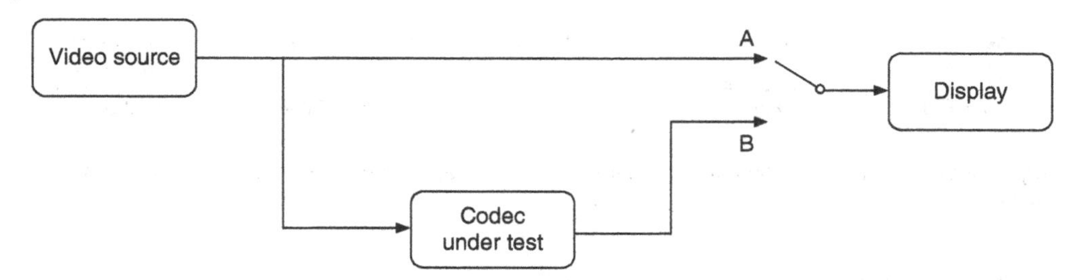

Figure 2.48 DSCQS testing system

pre-judging the impaired sequence compared with the reference sequence. At the end of the session, the scores are converted to a normalised range, and the end result is a score, sometimes described as a mean opinion score (MOS), that indicates the relative quality of the impaired and reference sequences.

Tests such as DSCQS are accepted as practical measures of subjective visual quality. However, the results can vary significantly depending on the assessor and also on the video sequence under test. This variation is compensated for by repeating the test with several sequences and several assessors. An expert assessor who is familiar with the nature of video compression distortions or artefacts may give a biased score since they are likely to pay close attention to distortions. It is recommended to use non-expert assessors. This means that a large pool of assessors is required because a non-expert assessor will quickly learn to recognise characteristic artefacts in the video sequences and so will become an expert. These factors can make it expensive and time-consuming to carry out subjective video quality testing.

2.12.3 Objective Quality Measurement

The complexity and cost of subjective quality measurement make it attractive to be able to measure quality automatically using an algorithm. Developers of video compression and video processing systems rely heavily on so-called objective (algorithmic) quality measures such as the following.

2.12.3.1 PSNR

Peak signal-to-noise ratio (PSNR) (Eq. (2.1)) is measured on a logarithmic scale and depends on the mean squared error (MSE) between an original and an impaired image or video frame, relative to $(2^n-1)^2$, the square of the highest-possible signal value in the image, where n is the number of bits per image sample. The result of Eq. (2.1) is a number expressed in decibels, where a higher number implies better visual quality.

$$\mathrm{PSNR_{dB}} = 10\log_{10}\frac{\left(2^n-1\right)^2}{\mathrm{MSE}} \qquad \text{Equation 2.1}$$

PSNR is a simple calculation and has been widely used to estimate the quality of compressed versus original video images. Figure 2.49 shows a still frame from the 'Akiyo' test sequence, under increasing levels of compression. The top-left frame is from a decoded

Figure 2.49 PSNR examples

sequence with an average PSNR of 45.8 dB. The frame is virtually indistinguishable from the uncompressed original, so the quality of the top-left frame could be described as 'visually lossless'. The frame at the top right is from a version with an average PSNR of 40.5 dB. You may or may not notice a very slight loss of sharpness. The lower-left clip has an average PSNR of 34.6 dB and is visibly distorted compared to the top-left version. The lower-right clip, with an average PSNR of 32.0 dB, is very obviously distorted.

The PSNR metric suffers from several limitations. PSNR requires an unimpaired original image for comparison, but this may not be available in every situation. PSNR does not correlate particularly well with subjective video quality measures such as ITU-R 500. For a given image or image sequence, high PSNR usually indicates high quality and low PSNR usually indicates low quality. However, a specific value of PSNR does not necessarily equate to an absolute subjective quality. PSNR and similar metrics are adequate for comparisons of different encodings of the same video sequence but not necessarily adequate for comparisons across different video materials.

2.12.3.2 Other Objective Quality Metrics

Because of the limitations of metrics such as PSNR, much work has been done to develop more sophisticated objective tests that correlate better with subjective test results. Many different approaches have been proposed. It has proved difficult to establish reliable objective video quality metrics that accurately predict the results of subjective tests. Some examples include just noticeable difference (JND) [8], digital video quality (DVQ) [9], Structural SIMilarity index (SSIM) [10], PSNRplus [11] and predicted mean opinion score (MOSp) [12]. These metrics have varying degrees of success in predicting subjective test scores, with reported correlations of between 70% and 90% between each objective metric and measured subjective quality scores.

ITU-T Recommendation J.247 was completed in 2008 and focused on full reference video quality measurement, i.e. quality metrics that require access to an original, uncompressed version of a video signal [13]. The Recommendation listed four objective quality metrics that could be 'recommended by ITU-T at this time':

A: NTT Full Reference Method.
B: OPTICOM Perceptual Video Quality Method.
C: Psytechnics Full Reference Method.
D: Yonsei Full Reference Method.

The general approach in these methods is as follows. First, the original (reference) and test (degraded) video sequences are aligned spatially and temporally. Next, a series of degradation parameters are calculated. Each of the four metrics calculates a different set of parameters such as blurring, edges introduced by compression and blockiness. Finally, these parameters are combined to produce a single number that is an estimate of subjective quality. The correlation between subjective and estimated quality is reported as ranging from 77% to 84% in the Recommendation.

The video quality model for variable frame delay (VQM-VFD) [14] and video multimethod assessment fusion (VMAF) methods have reported promising results, reporting correlations of above 90% with subjective test results [15]. VMAF is integrated into the FFMPEG software tool, and some examples of its usage are presented in Chapter 11.

An even more challenging task is to measure or estimate quality when a full reference, an unimpaired copy of the original video, is not available. This is the case in many practical applications. For example, the original source may not be available in the case of user-generated video content, or it may be desirable to monitor quality at the receiver, for example, in a customer's digital television receiver, where there is no access to the original video. In these situations, no reference (NR) or reduced reference (RR) quality estimation is required. NR metrics attempt to estimate subjective quality based only on characteristics of the decoded video clip. This is a difficult task, but some success has been reported using methods such as modelling typical image/video compression artefacts [16, 17]. RR metrics calculate a quality signature: typically a low-bitrate side signal, which is communicated to the decoder. A quality estimate is formed by processing the decoded video clip together with the side information [18].

2.13 Conclusion

In this chapter, we have examined the key building blocks of a video codec. An encoder exploits spatial and temporal redundancy in a video clip, predicting each block in a frame from information that has already been coded and is therefore available to the decoder.

Subtracting this prediction reduces the information content of the video scene. The difference or residual is transformed into a spatial frequency domain and quantised to reduce the precision of the information. The resulting data, together with headers and prediction information, are encoded into a binary stream that is a compressed version of the original video. A video decoder reverses these steps to recreate a version of the original video clip.

Present-day compression standards make it possible to reduce video file sizes by a factor of 100 or more, which helps to explain why video codecs are ubiquitous in devices with cameras or displays. The techniques used in today's standards have evolved over 50 or more years, from early experiments in the 1960s and 1970s to the first standards in the 1990s and onwards. Chapter 3 examines the historical development of video coding concepts and standards.

References

1. Cisco (2020). Annual Internet Report. https://www.cisco.com/c/en/us/solutions/collateral/executive-perspectives/annual-internet-report/white-paper-c11-741490.pdf.
2. Findlay, J.M. and Gilchrist, I. (2003). *Active Vision: The Psychology of Looking and Seeing*. Oxford: OUP.
3. Wade, N. and Swanston, M. (2001). *Visual Perception: An Introduction*. London: Psychology Press.
4. Aldridge, R., Davidoff, J., Hands, D., et al. (1995). Recency effect in the subjective assessment of digitally coded television pictures. *Proceedings of the Fifth International Conference on Image Processing and its applications*, Heriot-Watt University, Edinburgh, (July 1995). doi.org/10.1049/cp:19950676.
5. Anderson, C., Van Essen, D., and Olshausen, B. (2005). Directed visual attention and the dynamic control of information flow. In: *Neurobiology of Attention* (ed. L. Itti, G. Rees, and J.K. Tsotsos). Cambridge, MA: Academic Press/Elsevier https://redwood.berkeley.edu/bruno/papers/attention-chapter.pdf.
6. Pinson, M. and Wolf, S. (2003). Comparing subjective video quality testing methodologies. *Proceedings of SPIE, Visual Communication and Image Processing*. http://dx.doi.org/10.1117/12.509908.
7. Methodology for the subjective assessment of the quality of television pictures (2023). ITU-T Recommendation BT.500-15. https://www.itu.int/rec/R-REC-BT.500 (accessed 12 January 2024).
8. Lubin, J. and Fibush, D. (1997). Sarnoff JND vision model, Working group Document 97-612, ANSI T1 Standards Committee.
9. Watson, A.B., Hu, J., and McGowan, J.F. III (2001). Digital video quality metric based on human vision. *Journal of Electronic Imaging* 10 (1): 20–29.
10. Wang, Z. and Bovik, A.C. (2002). A universal image quality index. *IEEE Signal Processing, Letters* 9 (3): 81–84.
11. Oelbaum, T., Diepold, K. and Zia, W. (2007). A generic method to increase the prediction accuracy of visual quality metrics. *Picture Coding Symposium*, Lisboa, Portugal, (7–9 November 2007).

12 Bhat, A., Richardson, I.E. and Kannangara, C.S. (2009). A novel perceptual quality metric for video compression. *Proceedings of the International Picture Coding Symposium*, May 2009, pp. 1–4.

13 Objective perceptual multimedia video quality measurement in the presence of a full reference (2008). ITU-T Recommendation J.247. https://www.itu.int/ITU-T/recommendations/rec.aspx?rec=9497&lang=en (accessed 12 January 2024).

14 Wolf, S. and Pinson, M. (2011). Video quality model for variable frame delay (VQM_VFD). *NTIA Technical Memorandum TM-11-482*. https://its.ntia.gov/publications/details.aspx?pub=2556. (accessed 12 January 2024).

15 Rassool, R. (2017). VMAF reproducibility: validating a perceptual practical video quality metric. *IEEE International Symposium on Broadband Multimedia Systems and Broadcasting (BMSB)*, pp. 1–2.

16 Wang, Z., Sheikh, H. and Bovik, A. (2002). No-reference perceptual quality assessment of JPEG compressed images. *Proceedings of the International Conference on Image Processing*, vol. 1, pp. 477–480. https://doi.org/10.1109/ICIP.2002.1038064.

17 Dosselmann, R. and Yang, X. (2007). A prototype no-reference video quality system. *Proceedings of the Canadian Conference on Computer and Robot Vision*, Montreal, Canada (May 2007), pp. 411–417.

18 Wang, Z. and Simoncelli, E. (2005). Reduced-reference image quality assessment using a wavelet-domain natural image statistic model. *Proceedings, SPIE Human Vision and Electronic Imaging*, p. 5666. https://doi.org/10.1117/12.597306.

3

A History of Video Coding and Video Coding Standards

3.1 Introduction

The concept of sending video images to a remote location has captured the human imagination for centuries. The first demonstration of television in 1926 by the Scotsman John Logie Baird[1] led to the realisation of this concept over analogue communication channels. Analogue TV systems make use of a simple form of compression or information reduction to fit higher-resolution visual images into limited transmission bandwidths.

The emergence of digital video in the 1990s was made possible by compression techniques developed in the preceding decades. Even though the earliest videophones and consumer digital video formats were limited to very low-resolution images of 352 × 288 pixels or smaller, the amount of information required to store and transmit moving video was too great for the available transmission channels and storage media. Video coding or compression was an integral part of these early digital applications, and it has remained central to each further development in video technology since 1990.

By the early 1990s, many of the key concepts required for efficient video compression had been developed. From the 1970s onwards, industry experts recognised that video compression and digital video had the potential to revolutionise the television industry. Efficient compression would make it possible to transmit many more digital channels in the bandwidth occupied by the older analogue TV channels.

The development of industry standards for compression or coding and decompression or decoding set the scene for mass-market digital video in the 1990s, alongside improvements in integrated circuit technology that made it possible to encode and decode video in real time on consumer devices. Video DVDs and digital TV broadcasting rapidly replaced their analogue predecessors in the late 1990s and early 2000s, bringing digital video and video coding into many of the world's households. The years 2000–2010 saw the increasing use of video over the internet and on mobile devices and a gradual move from Standard Definition, around 720 × 480 pixels, to High Definition, 1920 × 1080 pixels. As of 2024, the quantity and resolution of transmitted video continues to increase, with 4K or Ultra HD increasingly the norm for many video applications.

1 TV was invented in Scotland, more than a year before Philo T. Farnsworth's first demonstration in the USA.

Coding Video: A Practical Guide to HEVC and Beyond, First Edition. Iain E. Richardson.
© 2024 John Wiley & Sons Ltd. Published 2024 by John Wiley & Sons Ltd.
Companion website: www.wiley.com/go/richardson/codingvideo1

This chapter will trace the development of video coding from the early theoretical concepts, many of which were first proposed in the 1970s and 1980s, through the first industry standards, the emergence of mass-market applications, the rise of internet and mobile video, to the present day.

In particular, I will look at:

- When and how were the early concepts of video coding developed?
- How did these concepts evolve into the well-known block-based video codec model?
- How did the video coding standards contribute to or influence the changing patterns of digital video usage, from television and DVDs to video streaming?
- Why are new video coding standards continuing to be developed?
- What is the best way to compare video coding standards?

3.2 The Foundations of Video Coding, 1950–1990

Many of the key components of the video codec model introduced in Chapter 2 were first developed before 1990. Table 3.1 and Figure 3.1 show a selection of key concepts with the dates of early papers on each topic. These are not necessarily the very first publications describing each concept but were all influential on later developments.

3.2.1 Entropy Coding: 1952–1987

David Huffman's paper on Minimum Redundancy Codes, published in 1952 [1], described a method of constructing binary codewords with multiple digits to efficiently encode a message with a finite alphabet of symbols. This is the basis of the well-known Huffman coding method, which represents symbols by variable length, uniquely decodable binary codewords. Versions of Huffman Coding were used in early video coding standards including H.261, H.263, MPEG-1, MPEG-2 and MPEG-4 Part 2.

Table 3.2 shows an example of Huffman-type codewords used in MPEG-4 Part 2. The parameter motion vector difference (MVD) is encoded as a variable-length codeword. The most common value, 0, is mapped to the shortest codeword, a single binary 1. Less probable

Table 3.1 Selected key concepts

Year	Concept	Authors of key publication(s)
1952	Variable-length coding	Huffman
1972	Frame differencing	Schroeder
1974	Discrete cosine transform	Ahmed, Natarajan and Rao
1976	Arithmetic Coding	Rissanen
1981	Motion-compensated prediction	Jain and Jain
1990	Variable block size motion compensation	Chan
1991	Bidirectional prediction	MPEG committee

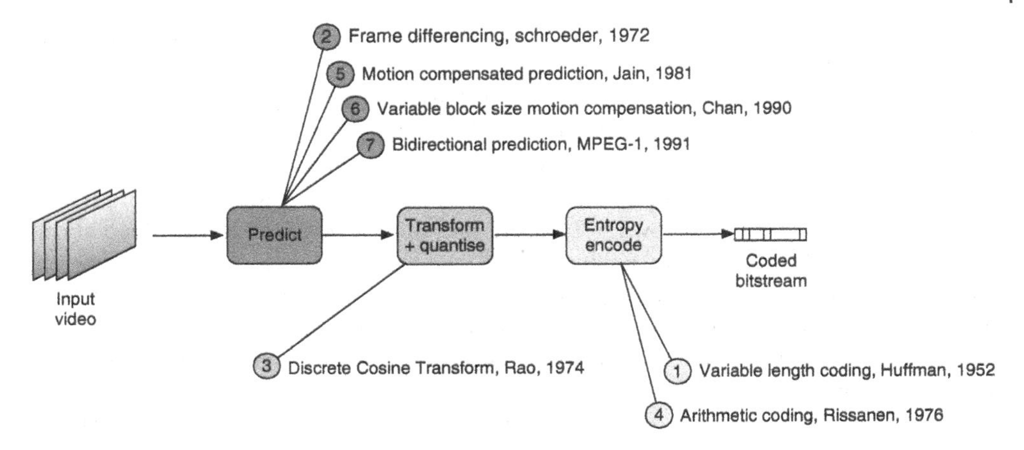

Figure 3.1 Key publications in the history of video coding

Table 3.2 Variable-length codes: MPEG-4 Part 2 motion vector difference

MVD	Code
0	1
+0.5	010
−0.5	011
+1	0010
−1	0011
+1.5	00010
−1.5	00011
+2	0000110
−2	0000111
+2.5	00001010
−2.5	00001011
+3	00001000
−3	00001001
+3.5	00000110
−3.5	00000111
...	...

values, i.e. larger MVDs, are mapped to longer codewords. Each codeword is a uniquely decodable binary string, which means that reading from left to right from the start of a codeword, no two codewords share the same sequence of bits.

Huffman coding can approach the theoretical optimum number of bits per coded symbol and is relatively simple to implement. However, a significant limitation is that each symbol must be represented with an integral number of bits. Symbols with a probability above 0.5, for example, must be coded as a single bit, rather than the theoretically optimum fractional number of bits $b < 1.0$.

Rissanen's 1976 paper on arithmetic coding [2] was extended by Witten in 1987 [3] to provide a practical alternative to Huffman variable-length coding. A strength of arithmetic coding is that symbols do not have to be represented with an integral number of bits, making it possible for arithmetic coding to outperform variable-length coding. For example, the single-bit code for MVD = 0, shown in Table 3.2, is optimal if MVD = 0 has a probability of exactly 0.5. If MVD = 0 has a probability P of 0.7, then ideally it should be represented with a codeword that is 0.5146... bits long, as its information content $\log_2(1/P) = 0.5146...$. A binary arithmetic coder can represent MVD = 0 using a fractional number of bits, less than one bit, and therefore can potentially compress more efficiently than a Huffman or other variable-length coder.

We will examine variable-length coding and arithmetic coding in detail in Chapter 8.

3.2.2 Frame Differencing: 1956–1972

The concept of prediction is to generate a prediction of the information to be coded, based on previously processed data, and to subtract this prediction from the information to be coded. The more accurate the prediction, the less information remains after subtraction.

An early contribution to this concept was a patent by Oliver, published in 1956 [4]. Oliver's patent was specifically directed towards reducing the channel capacity required to transmit signals such as television signals: 'It is the primary object of the present invention to increase the efficiency of transmission systems now commonly in use for the communication of intelligence by reducing the redundancy in the signals which are transmitted' (1:35–39) – essentially describing video compression.

Figure 3.2 is based on a diagram in Oliver's 1956 patent, see Figure 3.2, and bears a resemblance to the modern video codec model with a predictor, a subtractor, a quantiser and a reconstructor or adder. Oliver describes the general concepts of both spatial prediction and temporal prediction: 'Often, the previous samples used to determine a particular predicted value will comprise primarily *immediately preceding* samples' [describing a form of spatial

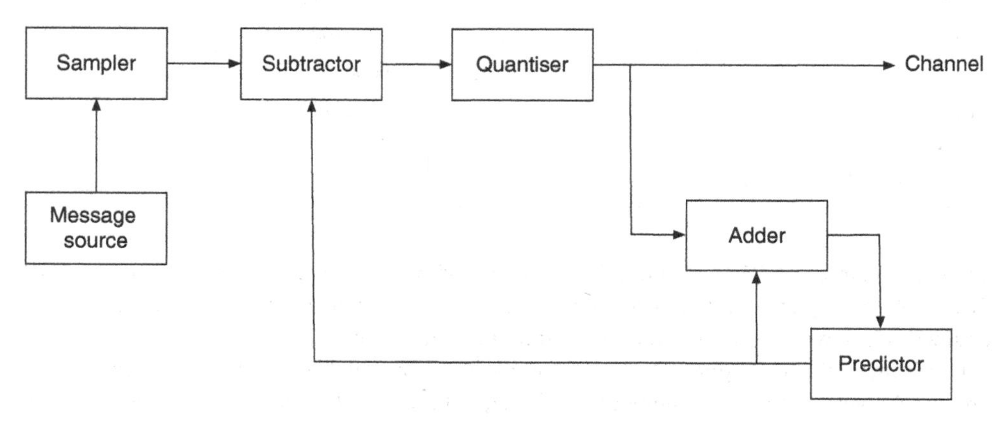

Figure 3.2 Reducing channel capacity for television signals. *Source:* Adapted from Oliver [4]

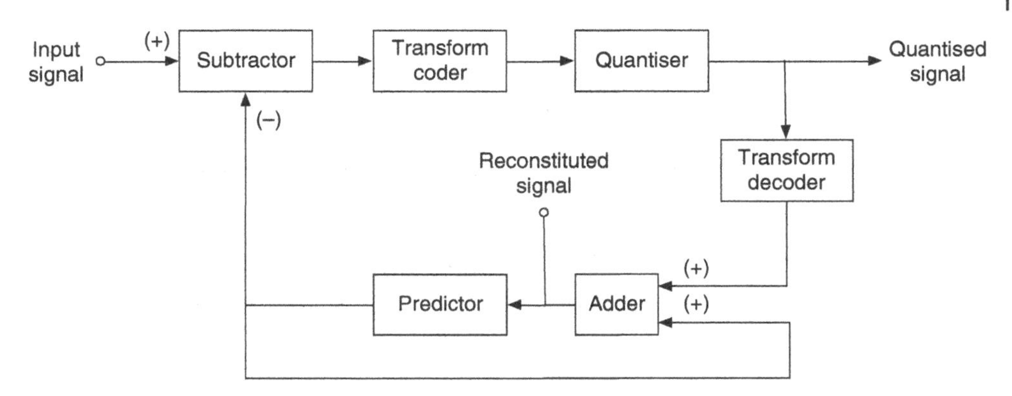

Figure 3.3 Frame differencing and transform coding. *Source:* Adapted from Schroeder [6]

prediction], 'but in certain cases, such as in a television signal [...] there may ... more advantageously be utilized samples a considerable number of sampling intervals *previous in time*' [describing or suggesting temporal prediction] (Oliver, Column 5, emphasis added).

Oliver's work was primarily directed towards data compression for the transmission of analogue signals, such as analogue television transmissions. In his 1970 patent [5], Robert Lucky of Bell Labs extended this concept towards the compression of digital signals via adaptive prediction. Manfred Schroeder, also of Bell Labs, applied the prediction concept specifically to video signals 'preferably... in digital form' in his 1972 patent [6]. Schroeder's model (redrawn as Figure 3.3) introduces a transform coder to 'distribute' the difference or error signal 'over the entire frame', so as to provide a relatively constant information content to the quantiser, Schroeder at abstract. Schroeder's 1972 patent provides early disclosure of frame differencing, transform coding and quantisation. There is no mention of motion-compensated prediction or entropy coding. A number of publications in the early 1970s investigated different aspects of frame differencing or conditional replenishment [7, 8].

See Chapter 6 for a detailed discussion of interframe coding.

3.2.3 The Discrete Cosine Transform: 1974–1984

The concept of transform coding for image compression was introduced by Andrews and Pratt [9]. The well-known Fourier and Hadamard transforms were shown to be sub-optimal for image compression. The theoretically ideal Karhunen–Loeve transform (KLT), whilst offering optimum compression performance, was too complex for implementation on practical devices or processors.

Schroeder [6] suggested applying two-dimensional transforms such as the Discrete Fourier Transform (DFT) or Hadamard transform to an entire residual image. However, neither approach was considered particularly efficient or practical for image or video compression.

In their classic 1974 paper, Ahmed et al. introduced a new transform for image processing, the Discrete Cosine Transform (DCT)[10]. The authors explained its potential benefits in applications such as pattern recognition and Wiener filtering. The DCT approaches the performance

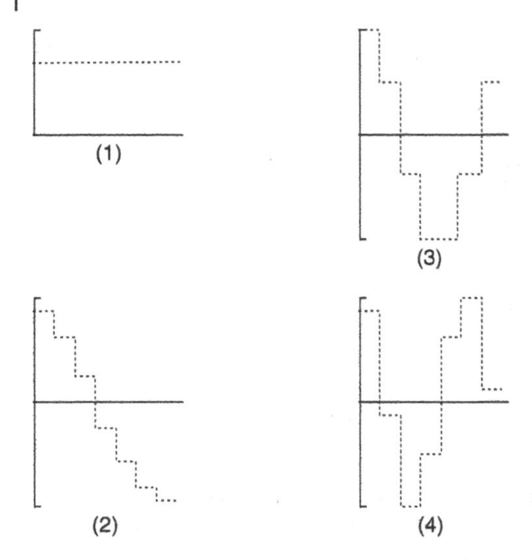

Figure 3.4 First four basis vectors of the DCT. *Source:* Adapted from Ahmed et al. [10]

of the KLT and has the significant advantage of being easier to compute. The basis vectors or patterns of the DCT (see Figure 3.4) are similar to the eigenvectors of the KLT except for phase shifts in certain vectors. The one-dimensional signals shown in Figure 3.4, the basis vectors of the DCT, correspond to a horizontal or vertical slice through the well-known basis patterns of a 4×4 DCT, which we introduced in Chapter 2 (see Section 2.5, Transform and Quantisation).

The authors mention that a two-dimensional transform can be computed using a one-dimensional orthogonal transform such as the DCT. After the publication of Ahmed, Natarajan and Rao's paper, the DCT was widely adopted for image compression research experiments. Chen, Smith and Fralick described a fast, low-complexity implementation of the DCT in 1977 [11], making it an attractive option for practical image and video codecs in terms of compression performance and computational efficiency. The DCT was to become the transform of choice for video codecs from the 1980s onwards.

Chen and Pratt's 1984 paper [12] brings together many of the elements of the modern-day image or video codec, with the exception of motion-compensated prediction. Their 'scene adaptive coder' processes blocks of pixels using a discrete cosine transform, threshold and quantisation, zig-zag scanning, run-level and Huffman coding of transform coefficients. An output rate buffer feeds back to the coefficient threshold and normalisation stages to control bitrate. As the output buffer fills up, the rate control feedback adjusts the thresholding to reduce the data rate and maintain a fixed transmitted bitrate. See Chapter 7 for an in-depth discussion of transforms and quantisation in video coding (Figure 3.5).

3.2.4 The Development of the Video Codec Model: 1977–1985

Three components that have been present in every popular video coding standard since the early 1990s are block-based motion-compensated prediction, block transform and entropy coding. The combined use of these components for video coding first occurred in the late 1970s/early 1980s. This was a period of intensive research to develop the first practical codecs

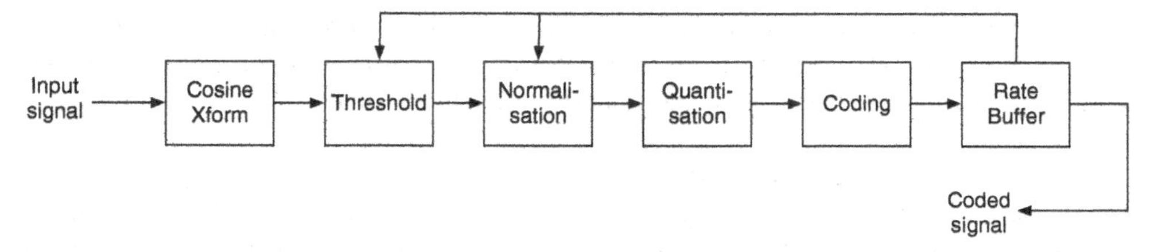

Figure 3.5 Scene adaptive coder. *Source:* Adapted from Ahmed et al. [10]

for digital television transmission. The motivation for this work was the potential for saving bandwidth compared with conventional analogue systems. Video compression offered the possibility to deliver multiple digital TV channels for every one analogue channel. Interestingly, this implies a motivation of cost and efficiency saving, rather than visual quality improvement.

Roese, Pratt and Robinson extended the concept of two-dimensional block transform coding to the third dimension – time – in 1977 [13]. They describe two methods of coding interframe video, i.e. using the relationship between successive frames to improve coding efficiency. First, they describe applying a three-dimensional DCT across an $N \times N \times M$ volume that contains $N \times N$ spatial pixels and M frames of video, where N and M are integers. Second, they apply temporal or interframe prediction in the transform domain, i.e. applying a predictor after a two-dimensional spatial transform. Motion compensation is absent. Another approach to interframe coding was described in the same issue of IEEE: 'Transactions on Communications' by Haskell, Gordon, Schmidt and Scattaglia [14].

Netravali and Stuller [15] presented the concept of motion-compensated prediction in 1979. In their paper, the first encoding step is a DCT, followed by prediction in the transform domain. Two motion-compensated prediction techniques are reported. In the first, estimates of displacement, which is a form of motion vector, are transmitted to the decoder. In the second, displacement estimates are calculated from previously transmitted transform coefficients, so that the estimates or motion vectors need not be transmitted.

In their 1981 paper, Jain and Jain [16] described a codec that is close to the modern-day model, at least at a high level (see Figure 3.6). Motion-compensated prediction of a

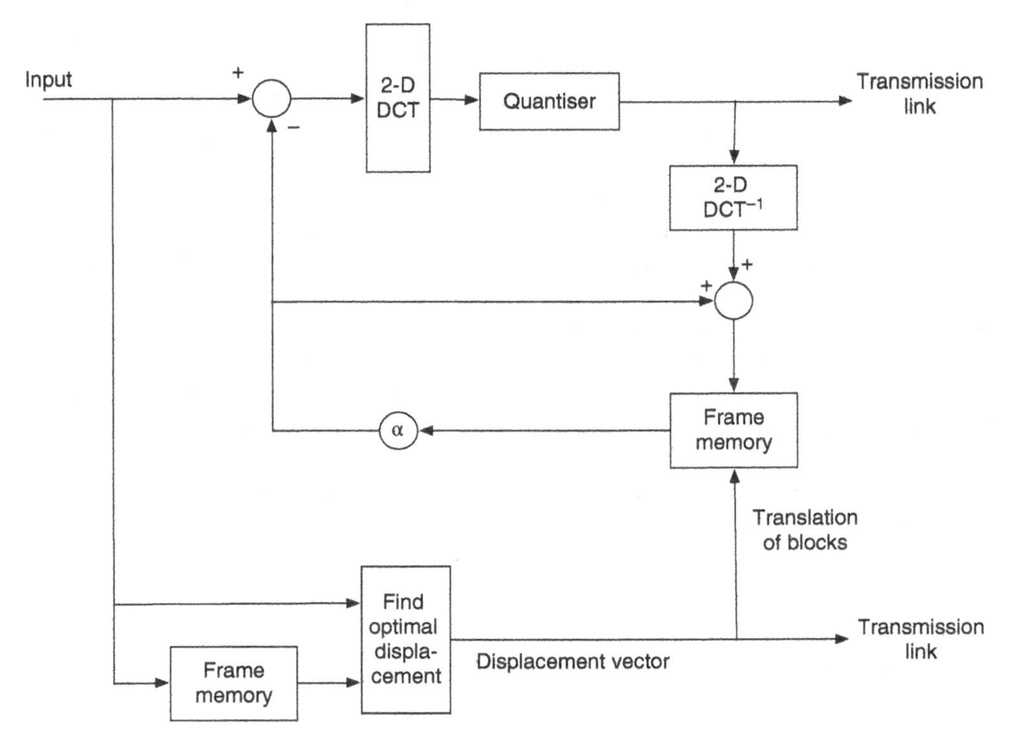

Figure 3.6 Motion-compensated interframe hybrid coder. *Source:* Adapted from Jain and Jain [16]

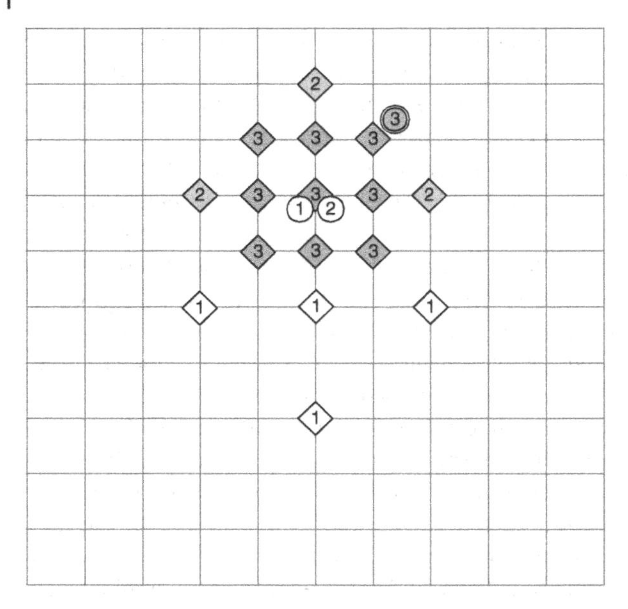

Figure 3.7 Logarithmic search. *Source:* Adapted from Jain and Jain [16]

block of pixels is carried out and the residual is transformed with a two-dimensional DCT, quantised and transmitted. A search unit finds the optimal displacement vector – a motion vector – which is transmitted along with the prediction residual. The quantised residual is inverse transformed and forms a reconstructed frame for further predictions. The authors also introduce the concept of computationally efficient motion estimation, describing a two-dimensional Logarithmic Search algorithm for locating a motion vector in a relatively small number of computational steps (see Figure 3.7). To find the matching block in a stored previous frame, the encoder first searches five locations in a cross shape, marked as diamonds (1) on the diagram, and compares each block with the current block. The best result, i.e. the search that returns the smallest difference between the search block and the current block, is marked as a circle (1) and is used as the centre for a smaller search, marked as diamonds (2) on the diagram. The best result from the second search, marked as a circle (2), is used as the centre for the third and final search, marked as diamonds (3) on the diagram. The final result, marked as a circle (3), is obtained after a maximum of 21 search comparisons, instead of the 121 comparisons required to search the full 11×11 grid of possible locations.

By 1985, all of the main elements had been brought together. Ericsson [17] describes a coder (see Figure 3.8) that incorporates:

- Motion-compensated prediction of rectangular blocks, using sub-pixel interpolation
- 16×16 DCT
- Quantisation of DCT coefficients
- Zig–zag diagonal scanning of quantised coefficients
- Run-level and Huffman coding
- Bit rate control by modifying the quantiser step size.

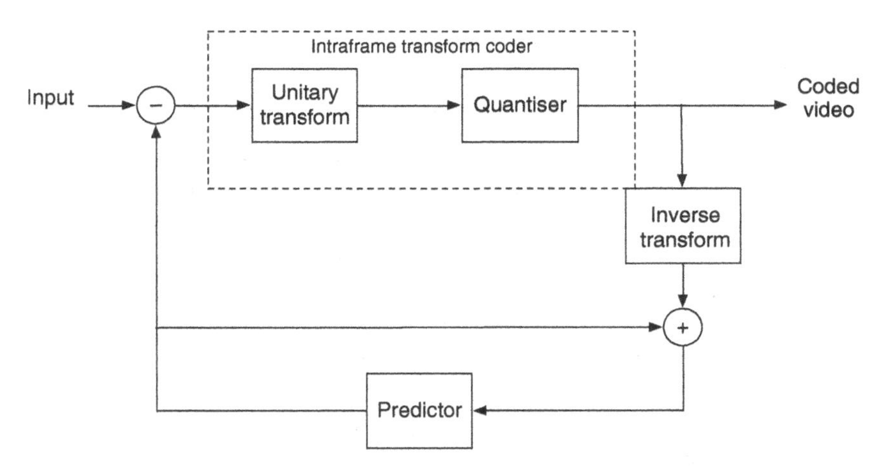

Figure 3.8 Hybrid video encoder. *Source:* Adapted from Ericsson [17]

The basic codec model was now complete and was to be used, with successive refinements of each component, in the video coding standards of the 1990s and 2000s.

3.2.5 Early Advances in Motion-Compensated Prediction: 1989–1995

3.2.5.1 Bidirectional Prediction

Thoma and Bierling introduced an early version of bidirectional prediction in their 1989 paper on motion-compensated interpolation [18]. They showed that it is possible to reduce the transmitted bitrate of a video sequence by skipping rather than transmitting frames and by replacing the dropped frames using motion-compensated interpolation at the decoder.

The MPEG-1 video coding standard introduced the concept of bidirectionally predicted B-pictures and a repeating Group of Pictures (GoP) structure in 1991 [19]. Each B-picture is predicted from two available reference pictures, an I or P-picture in the past and an I or P-picture in the future (Figure 3.9). For example, B-picture 2 is predicted from I-picture 1

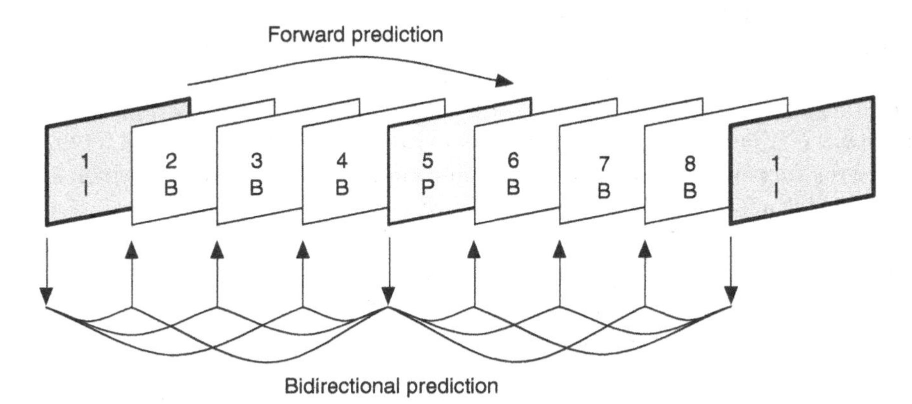

Figure 3.9 Forward + bidirectional prediction, MPEG-1 overview. *Source:* Adapted from Le Gall [19]

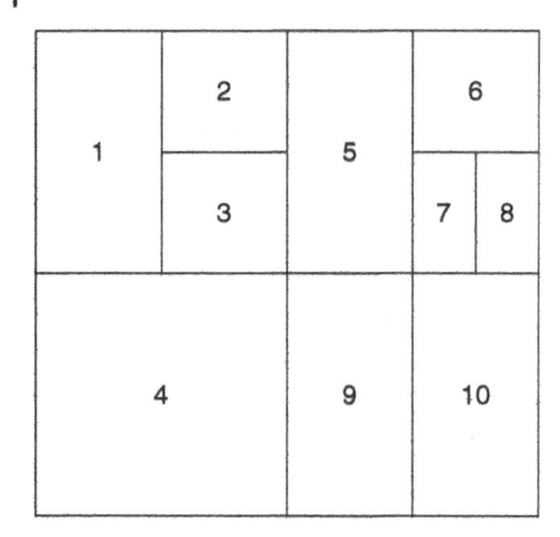

Figure 3.10 Variable block size motion compensation. *Source:* Chan et al. [20]/ The Institution of Engineering and Technology

and P-picture 5; B-picture 7 is predicted from P-picture 5 and I-picture 1 from the next GoP. Each macroblock in a B-picture can be predicted using:

- Forward prediction from the previous reference picture, with one forward motion vector
- Backwards prediction from the future reference picture, with one backward motion vector
- Interpolated prediction from the previous and the future reference picture, with one forward and one backward motion vector.

3.2.5.2 Variable Block Size Motion Compensation

Jain and Jain [16] describe motion-compensated prediction using a fixed block size of 16×16 pixels. Moving regions in a video sequence rarely match up with 16×16 block boundaries. In their 1990 paper, Chan, Yu and Constantinides describe a method of motion-compensated prediction with varying block sizes [20]. The encoder can choose from a range of square and rectangular block sizes for motion compensation (see Figure 3.10). This makes it possible to match irregular object motion more accurately than a fixed block size scheme can. The potential disadvantages of variable block size motion compensation include the increased complexity of the coding process, especially motion estimation and selection of block sizes, and the extra overhead required to encode and communicate the choice of block sizes to the decoder. As processor speeds and capabilities increased during the 1990s, variable block size motion compensation became an important feature of video coding standards such as MPEG-4 Visual, H.263 and H.264.

3.2.5.3 Multiple Reference Frames

In Figure 3.9, frames 3 and 4 are predicted from a reference picture that exists several frames in the past. A variation on this concept was described by Fukuhara et al. in their 1997 paper [21], in which current frame N may be predicted from a previous frame N-1 stored in a Short-Term Frame Memory and/or an older frame N-K in a Long-Term Frame Memory (see Figure 3.11). Flierl investigated the concept of multi-hypothesis long-term motion-compensated prediction in his 1997 diploma thesis [22]. Wiegand et al. (1997) paper describes a framework for long-term motion-compensated prediction, in which an encoder may predict a frame from multiple, previously coded video frames [23].

3.2.5.4 Motion Compensation with Fractional Pixel Accuracy

Just as video object motion rarely matches up to regular block boundaries, objects do not necessarily move in integer units between frames. The best match in the previous frame for the ball in the current frame (see Figure 3.12) is at position (−2.5, −1.5), i.e. a fractional number of pixels to the left and down. These fractional pixel positions do not exist in the previous frame but may be created by interpolating between the actual pixels of the previous frame.

Ericsson described the concept of motion compensation with a ‘fractional displacement estimate’, i.e. motion compensation with an accuracy of 1/8 pixel, in his 1985 paper [17].

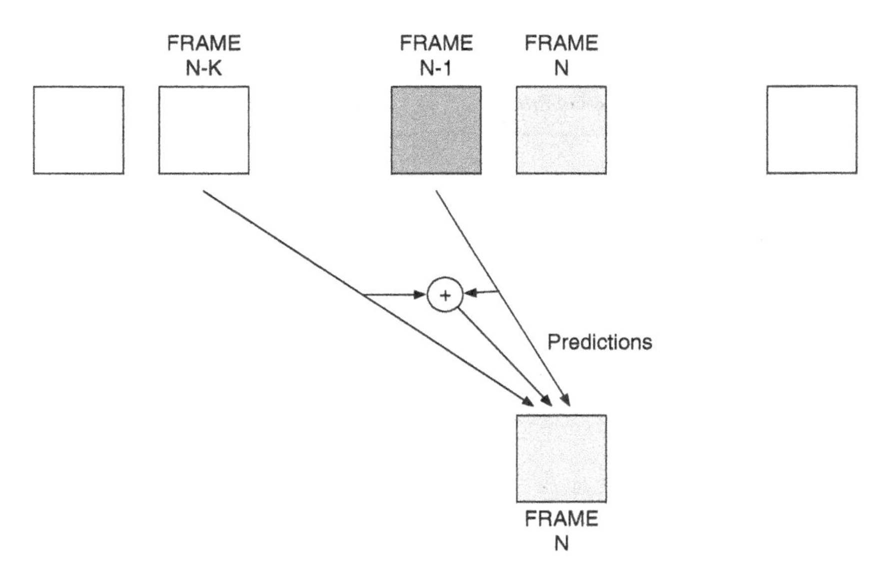

Figure 3.11 Prediction from multiple reference frames. *Source:* Adapted from Fukuhara et al. [21]

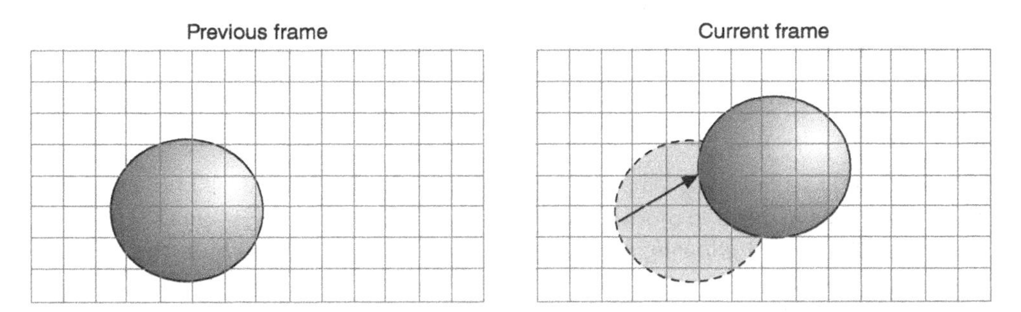

Figure 3.12 Prediction with fractional pixel accuracy

A 1993 paper by Girod analyses motion-compensated prediction with fractional pixel accuracy and describes a practical algorithm for motion estimation. Reference frame pixels are interpolated to sub-pixel positions, and integer motion estimation search results are refined to half-pixel and then to quarter-pixel accuracy [24]. Girod notes that significant performance gains may be achievable using sub-pixel motion compensation and that better results may be achieved using more sophisticated interpolation filters for the sub-pixel positions in the reference frame.

I will discuss inter prediction and coding in detail in Chapter 6.

3.3 Video Coding Standards and Formats: 1990–2021

Since 1990, a series of industry standards and formats for video coding have been published (see Table 3.3). The general intention of such standards and formats is to enable interoperability between devices and terminals and to support emerging technologies such as video communications, television broadcasting and video playback.

Table 3.3 Video coding standards and formats, 1990–2022

Year	ITU-T	ISO/IEC MPEG	Co-published	Non-ITU or ISO
1984	H.120			
1990	H.261			
1993		MPEG-1 Video		
1995			MPEG-2 Video/H.262	
1996	H.263			
1999		MPEG-4 Part 2		
2003			H.264/AVC	
2005				VC-1
2010				VP8
2013			H.265/HEVC	VP9
2018				AV1
2020		EVC		
2021			H.266/VVC	
2021		LCEVC		

MPEG, Moving Picture Experts Group; AVC, Advanced Video Coding; HEVC, High Efficiency Video Coding; VVC, Versatile Video Coding; EVC, Essential Video Coding; LCEVC, Low-Complexity Enhancement Video Coding.

A video coding standard specifies a bitstream format for compressed video and a method of decoding the bitstream to produce a decoded video sequence. This is intended to enable interoperability whilst allowing manufacturers quite a lot of flexibility in the way they design their encoders.

A video decoder that conforms to a standard must produce the same output as the decoding process specified in the standard. A video encoder must produce a bitstream that conforms to the bitstream format in the standard and is capable of correct decoding by the decoding process specified in the standard.

3.3.1 How Are the Standards Developed?

3.3.1.1 MPEG and VCEG

The International Organisation for Standardisation/International Electrotechnical Commission (ISO/IEC) and the International Telecommunication Union Telecommunication Standardisation Sector (ITU-T) are international organisations that develop and publish standards for information and communication technology. Each organisation has a working group that is responsible for image and video compression standards. The Moving Picture Experts Group (MPEG) is part of ISO/IEC's JTC1/SC29 subcommittee and the Video Coding Experts Group (VCEG) is the informal name for ITU-T's Study Group 16, Question 6/16. MPEG and VCEG are working groups comprising specialists in the field of video coding, including experts from industry and academia. Since the 1990s, MPEG and

VCEG have worked together on specific standards, setting up joint working groups including the Joint Video Team (JVT), responsible for developing H.264; the Joint Collaborative Team on Video Coding (JCT-VC), responsible for developing H.265; and the Joint Video Experts Team (JVET), which developed H.266.

3.3.1.2 The Development Process

The process of developing a new standard typically follows a path such as this:

- The working group identifies an industry need for a new standard. This, for example, might be due to emerging or changing patterns in delivering video to consumers, and/or improvements in hardware capabilities. The group sets objectives for the new standard, such as delivering twice the compression performance of the most recent standard. The working group issues a Call for Proposals.
- Organisations, such as consortia of companies, submit technology proposals in response to the Call for Proposals. For example, a proposal might be a complete video codec in the form of a software prototype, documentation and performance test results.
- The working group defines a Test Model, which is a prototype video codec comprising software and documentation based on elements of the proposals. Typically, the first Test Model needs to be developed further to meet all the aims of the standardisation effort.
- The working group meets every few months. At a typical meeting, group members submit technical contributions such as proposals to change an aspect of the Test Model to improve its performance. The working group selects certain proposals based on technical merit, which are then adopted into the Test Model. This continues over a period that may last for several years, until the Test Model document has developed into a complete draft specification for a new standard.
- Once the Test Model has reached a sufficiently mature stage, the working group proposes the draft specification for standardisation. The standards bodies (ISO/IEC and ITU-T) decide whether to approve publication as an industry standard.

3.3.1.3 Standards and Intellectual Property Rights

The process described above is likely to lead to a video coding standard that includes technical proposals from many different contributors. These may include companies active in the digital video industry, research centres, academics and so on. Some of the technologies in these proposals may be protected by patents. The ITU defines standard essential patents (SEPs) as 'patents that must be licensed in order to implement a given technical standard' [25]. The video coding standards discussed below each specify the format of a compressed bitstream and a method of decoding the bitstream. In this context, a patent that reads onto aspects of the decoding process and/or the bitstream defined in the standard might be an SEP.

The ITU and ISO/IEC address SEPs through their patent or Intellectual Property Rights (IPR) policies. These policies only allow the inclusion of patented technology in standards if patent holders agree to licence the relevant IPR to implementers of the standards on reasonable and non-discriminatory (RAND) terms [25].

What this means in practice is that implementers of video coding standards may be asked or expected to pay a royalty fee or licence fee to owners of SEPs. A patent pool can provide a single licence to a large number of patents that are claimed to be essential to a standard.

For example, the MPEG-LA/Via Licensing patent pool for the H.264/AVC standard lists many hundreds of patents, owned by around 30 companies and organisations, which are claimed to be essential to the H.264/AVC standard. The patent pool terms state that certain uses of the standard, such as larger-volume product sales or subscriptions, require a licence fee payment. In return for such a payment, the organisation using the standard has a licence to use all of the patents in the pool [26].

3.3.2 H.120, H.261 and MPEG-1: Early Video Coding Standards

H.120 (1984) [27] was an early attempt by the ITU-T, then known as the International Telegraph and Telephone Consultative Committee (CCITT), to standardise video coding for videoconferencing. The video compression part of the standard (see Figure 3.13) included the following features:

- Spatial prediction using differential pulse-code modulation (DPCM) from neighbouring pixels
- Optional motion-compensated prediction using a block size of 16×8
- Quantisation of prediction error samples – no transform
- Variable-length coding of prediction error, motion vectors and coding parameters.

H.120 was not widely implemented and did not have a significant industry impact.

H.261 (1990) [28] was the first video coding standard to adopt the now-familiar model of motion-compensated prediction, transform coding, quantisation and entropy coding. Designed to support two-way videoconferencing at bit rates of 64 kbps and upwards, the H.261 codec included the following components:

- Motion-compensated prediction and interframe coding of 16×16 blocks, with integer-pixel motion vectors.
- 8×8 DCT.
- Quantisation of DCT coefficients.
- Zig–zag scan and run-level coding of coefficients.
- Huffman-like entropy coding of coefficients, motion vectors and coding parameters.
- An in-loop filter that smoothed out blocking effects prior to prediction.

H.261 was the first of the industry standards for video coding to have an appreciable commercial impact. The standard achieved a balance between compression performance and implementation complexity and met an emerging need for videoconferencing terminal devices that could work effectively over the Integrated Services Digital Network (ISDN).

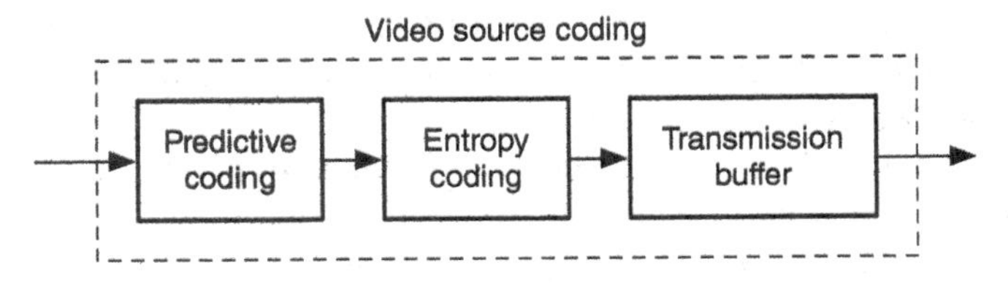

Figure 3.13 Video source coding in H.120

MPEG-1 (1993) [29] is a multi-part standard covering:

Part 1: Systems (multiplexing and storage)
Part 2: Video (video coding and syntax)
Part 3: Audio (audio coding and syntax)
Part 4: Compliance testing (procedures for checking compliance with the standard).

MPEG-1 Video (Part 2) was developed to provide a format for storing compressed video on digital storage media, particularly Compact Discs (CDs). The standard was specifically intended for use with the nowobsolete Video CD format. However, the compression techniques described in the standard are of more general application. MPEG-1 Video gained some popularity as a format for early computer-based and internet video applications, and its structure formed the basis for the widely used MPEG-2 Video standard. MPEG-1 Video includes the following features:

- Coding of progressive – non-interlaced – video, up to 768 × 576 pixels per frame, i.e. up to Standard Definition, 4:2:0 sampling format.
- Motion-compensated prediction – interframe coding – of 16 × 16 macroblocks, with half-pixel motion vectors
- 8 × 8 DCT
- Quantisation of DCT coefficients
- Zig–zag scan and run-level coding of coefficients
- Huffman-like entropy coding of coefficients, motion vectors and coding parameters
- I-, P-, B- and D-pictures
- Intra-coding in I-, P- or B-pictures
- Forward (in P- or B-pictures), backward and bidirectional motion prediction (in B-pictures)
- DC coefficient only (D) intra-coding (in D-pictures)
- Sequence and GoP structures, where a GoP is a repeating sequence consisting of an I-picture and one or more P- and/or B-pictures.

MPEG-1 Video was designed with relatively simple and low-cost decoders in mind. An MPEG-1 Video decoder (see Figure 3.14) could be implemented on a low-cost chipset or, by the mid-1990s, in software on a consumer PC. The main functional blocks are as follows:

Buffer: Store the incoming coded video bitstream
Mux^{-1}: Demultiplex coded coefficients, motion vectors and coding parameters
VLD: Variable-Length Decode
Q^{-1}: Rescale or inverse quantise

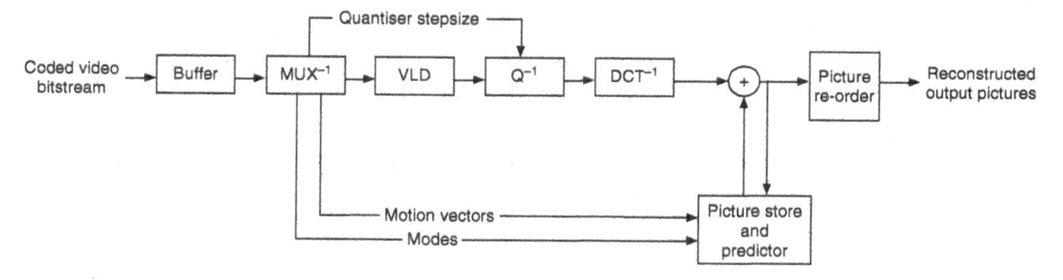

Figure 3.14 MPEG-1 Video decoder block diagram

DCT^{-1}: Inverse DCT
Picture store and predictor: Store previously decoded pictures, apply motion-compensated prediction based on decoded motion vectors
Reconstruction (+): Add the prediction to the decoded transform output
Picture re-order: Store and output decoded pictures in display order.

3.3.3 MPEG-2: Digital TV and DVD-Video

MPEG-2 [30] is a multi-part standard covering video coding, audio coding, multiplexing and transport or transmission. MPEG-2 Part 2 describes the coding of video and was co-published by the ISO/IEC as International Standard 13818-2 and by the ITU-T as Recommendation H.262. It is intended to provide a 'generic video coding scheme which serves a wide range of applications, bit rates, picture resolutions and qualities' (MPEG-2 Video: Summary, page iii).

MPEG-2 has been used in a wide range of applications but made a particular impact as the standard adopted for two important consumer applications: digital television broadcasting (DTV) and DVD-Video.

MPEG-2 Video is based on MPEG-1 Video and uses a superset of the MPEG-1 Video syntax. On top of the features of MPEG-1 Video, MPEG-2 Video adds:

- Coding of video up to resolutions of 1920×1152 samples – full High Definition
- Support for interlaced video formats, 4:2:2 and 4:4:4 video
- Motion compensation of interlaced and progressive macroblocks
- Profiles that define subsets of coding tools, and Levels that define maximum resolutions and processing rates
- Optional scalable coding tools.

The use of Profiles and Levels makes it possible to handle a range of decoding requirements within a single Standard. A Profile is a subset of the coding tools and features described in the standard. For example, a Simple Profile bitstream only requires a decoder to handle I and P pictures – not B-pictures, and 4:2:0 format video. A Main Profile bitstream requires a decoder to handle B-pictures, which necessitates more storage and processing power in the decoder. A High-Profile bitstream may include 4:2:2 format video and/or scalable coding.

A Level is a set of parameters that defines certain maximum values, such as the maximum resolution or bitrate of a coded video bitstream. A Low-Level bitstream has a maximum decoded resolution of 352×288 samples and a maximum frame rate of 30 frames per second. This is known as the common intermediate format (CIF). A Main Level bitstream has a maximum resolution of 720×576 and a maximum frame rate of 30 frames per second, known as Standard Definition. A High-Level bitstream has a maximum resolution of 1920×1152 and a maximum frame rate of 60 frames per second; this is High Definition.

Scalable coding enables a video sequence to be encoded into a number of layers, each of which is a coded bitstream. The base or lowest layer can be decoded to provide a low-fidelity version of the sequence, for example, at a lower spatial resolution or visual quality. Each higher layer can be decoded together with the base layer to provide progressively higher-fidelity video output. The idea is that transmission networks with lower capacity, or

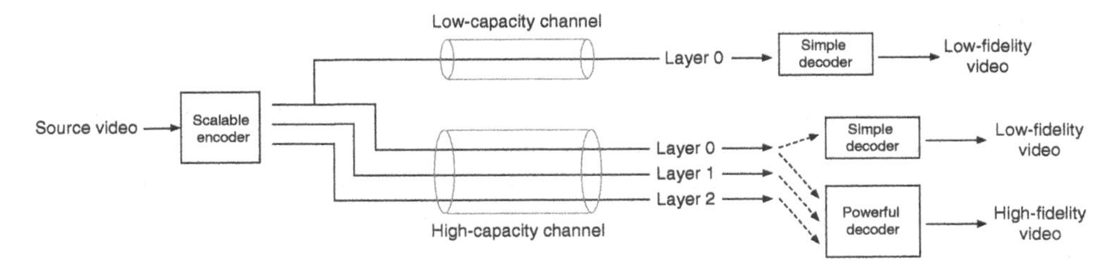

Figure 3.15 Scalable coding example

decoders with lower computational capabilities, may selectively discard layers to decode a good enough version, whilst a high-performance decoder with a good network connection may decode the full high-fidelity output (see Figure 3.15).

The scalable coding tools of MPEG-2 Video were not widely used in practical applications. This was perhaps due to the complexities of encoding, transporting and decoding multiple coded layers of video. However, MPEG-2's core tools for coding and transmitting television-quality video became a central component of the switch-over to digital television and DVD-Video in the 1990s and 2000s.

3.3.4 H.263 and MPEG-4 Part 2: Internet and Mobile Video

Two video coding standards released in the late 1990s, H.263 [31] and MPEG-4 Part 2 [32], share some common elements. Each of these standards provides a set of core coding tools that evolved from the earlier H.261 and MPEG-2 standards. H.263 supports additional functionality intended for video conferencing and video calling. MPEG-4 Part 2 extends the core coding tools to offer a wide range of functions that enable sophisticated interactions with coded video.

H.263 and MPEG-4 Part 2 share a common basic method of video coding, the so-called Short Header mode of MPEG-4 Part 2 and the Baseline Profile of H.263. This basic codec has similar features to earlier standards such as MPEG-1 and MPEG-2, including:

- Motion-compensated prediction of 16×16 macroblocks
- 8×8 DCT of motion-compensated residual and intra-coded macroblocks
- Quantisation and zig–zag coding of DCT coefficients
- Modified Huffman entropy coding.

H.263 adds a number of optional coding tools to the basic codec, each of which is described in an Annex to the standard. These tools can provide improved compression efficiency and include motion-compensated prediction using 8×8 blocks, arithmetic coding and/or improved support for applications such as two-way and multi-point videoconferencing.

MPEG-4 Part 2 defines a wide range of functionalities relating to the coding and representation of visual information. The standard provides coding tools that support many different types of visual data, including:

- Classic rectangular video scene
- Arbitrary-shaped regions of moving video, known as video objects
- 2D and 3D mesh objects

- Animated human faces and bodies
- Still images with static texture.

Despite the rather bewildering range of coding tools in the MPEG-4 Part 2 standard, only the tools and algorithms for coding rectangular video scenes were put into practice in commercial applications. The more esoteric tools for coding video objects, meshes and animation parameters did not find their way into products and were perhaps ahead of their time.

H.263 and MPEG-4 Part 2 found some success in internet and mobile video applications of the late 1990s and early 2000s, such as internet videoconferencing, internet video streaming, and mobile video capture, storage and streaming.

In these and other application areas, the improved compression performance of the next standard, H.264/AVC, led to H.263 and MPEG-4 Part 2 becoming largely superseded from the mid-2000s onwards.

3.3.5 H.264: Multi-Purpose Video Coding

The first edition of the H.264 Advanced Video Coding standard was co-published in 2003 by ISO/IEC as MPEG-4 Part 10 and ITU-T as Recommendation H.264 [33]. H.264 supports efficient coding and transport of conventional, rectangular video scenes. It provides broadly similar functionality to earlier standards such as MPEG-2 Video, MPEG-4 Part 2 and H.263, but with improved compression performance and support for transmission or storage in a wide range of application scenarios.

H.264 is specified in more technical detail than earlier standards in an effort to minimise the possibility of misinterpretation and to ensure reliable compatibility between conforming encoders, decoders and bitstreams. Key features of H.264/AVC include:

- A basic coding unit of a Macroblock, 16×16 pixels.
- Multiple colour spaces and Profiles that are subsets of coding tools, including support for different bit depths.
- Support for progressive and interlaced video.
- Partitioning of frames or fields into slices and macroblocks.
- Motion-compensated inter-prediction with square or rectangular block sizes from 16×16 down to 4×4 samples.
- Flexible biprediction, comparable to MPEG-2's B-pictures but with the ability for the reference pictures to be both before or both after the current picture.
- Sub-pixel prediction for motion compensation, with quarter-pixel resolution for luma samples.
- Intra-prediction from neighbouring samples of a 4×4 or 16×16 block, with a total of 13 prediction modes.
- 4×4 and 8×8 block-based transforms using specified integer approximations to the DCT.
- Combined quantisation and transform scaling.
- An adaptive loop filter applied to decoded/reference frame data.
- Variable-length coding with context adaptive coding of transform coefficients (Context Adaptive VLC).
- Optional Context-based Adaptive Binary Arithmetic Coding (CABAC).

- A bitstream syntax that is organised for efficient packetisation.
- Parameter Sets that communicate high-level syntax elements separately from coded pictures.

Certain Profiles of H.264/AVC, especially the Main and High Profiles, have become widely adopted in consumer applications such that at the time of writing, 20 years after its first publication, H.264/AVC remains the most widely used video coding standard in many application areas including streaming, storage and broadcasting.

The ITU-T has published several versions of the standard since its first edition in 2003, with later versions adding support for higher fidelity video, scalable video coding and multi-view video coding.

3.3.6 H.265/HEVC

H.265/High Efficiency Video Coding (HEVC) was developed between 2010 and 2013 by the JCT-VC, a joint committee of ISO/IEC MPEG and ITU-T VCEG, and was published jointly by ISO/IEC and ITU-T in 2013 [34]. H.265 was intended to be a successor to H.264, with the aim of providing better compression efficiency, exploiting the improvement in computational capability of consumer electronic devices and supporting usage cases that had emerged since the release of H.264.

Whilst not backwards compatible with H.264, H.265 shares some of the technical approaches of the earlier standard and extends its capabilities in a number of areas. We will cover H.265 in detail later. Its main features include:

- A basic unit of coding, the Coding Tree Unit (CTU), that is fixed throughout a sequence and can be 16 × 16, 32 × 32 or 64 × 64 pixels
- Partitioning of frames/pictures into slices, tiles and CTUs
- Quadtree partitioning of a CTU into smaller-sized Coding Units (CUs). The CUs can be as small as 8 × 8 pixels in area. Each CU can be coded with independent decisions
- Partitioning of each CU into Prediction Units (PUs) and Transform Units (TUs). PU sizes can be square for intra- or inter-prediction, or rectangular for inter-prediction only. TU sizes can range from 32 × 32 down to 4 × 4
- Intra prediction – square blocks, including 33 directional prediction modes
- Inter prediction with motion-compensated prediction, biprediction, sub-pixel interpolation and prediction copying or merging
- Variable block size transforms that use specified integer approximations, with combined quantisation and transform scaling
- Adaptive loop filters and sample adaptive offset filters that may be applied to decoded/reference frame data
- CABAC
- Parallel decoding support, for example, using tiles and/or wavefront parallel processing
- Flexible signalling of repeating picture structures such as GoP-type structures
- Parameter Sets and syntax with increased flexibility and functionality compared to H.264/AVC.

H.265/HEVC can provide more compression than H.264/AVC, particularly when coding video at higher resolutions such as HD and 4K. H.265 hardware support is now built into many consumer electronic devices. Despite H.265's improved compression efficiency over H.264/AVC, the older AVC standard continues to be widely used.

Consumer electronic chipsets, such as the integrated circuits in mobile devices and televisions, often support multiple video coding standards such as older standards including H.264/AVC and newer standards including H.265/HEVC. Adaptive streaming frameworks such as DASH and HLS can make it possible to use H.264/AVC as the codec for lower-bitrate streams, likely to be supported in most consumer devices, with more efficient formats such as H.265/HEVC being used if bandwidth is at a premium and/or for higher-resolution content (see Chapter 10).

3.3.7 H.266/VVC

H.266 Versatile Video Coding (VVC) was published by ISO/IEC and ITU-T in 2020 [35]. This new standard aimed to provide a better compression performance than its predecessors (including H.265/HEVC) at the expense of an increased computational cost and to add further support for emerging applications and usage scenarios. Key developments in H.266 include:

- Support for a wide variety of content including high dynamic range, very high-resolution video, computer-generated content and 360-degree video.
- The CTU can range from 32×32 up to 128×128 pixels.
- Picture partitioning into slices, tiles, sub-pictures and CTUs.
- CTU partitioning using quadtrees, which are square sub-blocks, and multi-type trees, which split a block into rectangles.
- Intra-prediction with more modes than HEVC, which include prediction of rectangular blocks.
- Inter-prediction with considerably more flexibility of prediction and signalling in comparison with HEVC. This includes affine motion vector derivation, adaptive motion vector resolution and decoder motion vector refinement.
- Variable block size transforms including non-square or rectangular transforms.
- Sophisticated in-loop filters.
- Context-based CABAC.
- Coding tools for screen content that can support screen sharing over network connections.

At the time of writing (2024), chip manufacturers are beginning to announce hardware support for H.266/VVC, i.e. built-in hardware decoding that can be integrated into consumer devices. If VVC becomes popular in the market, it will gradually be added to future generations of devices along with the earlier codecs such as H.264 and H.265 and other formats such as AV1, as discussed below.

3.3.8 Other Standards and Formats

3.3.8.1 VC-1

Based on Microsoft's Windows Media Video 9 (WMV9) video codec, the VC-1 specification was published by the Society of Motion Picture and Television Engineers (SMPTE) in 2006 [36]. VC-1 makes use of the same basic features common to earlier video coding standards, including 16×16 pixel macroblocks, block-based motion-compensated prediction, an integer approximation to the discrete cosine transform, quantisation and entropy coding. VC-1 also supports variable-sized square and rectangular transforms, a mix of intra- and inter-predicted blocks within a macroblock, intensity compensation and overlapped transform filtering. Entropy coding in VC-1 uses pre-defined variable-length coding tables.

VC-1 reportedly provides comparable compression performance to H.264, at least for Standard Definition video content [37]. However, the VC-1 codec is not widely used.

3.3.8.2 VP8, VP9 and AV1

VP8, VP9 and AV1 are publicly available specifications for video coding. Originally developed by On2 Technologies, Google released VP8 as an open-source video coding specification in 2010 [38]. Google continued to develop the technology further, releasing the VP9 specification in 2013 [39]. Google and other technology organisations founded the Alliance for Open Media (AOMedia) consortium in 2015, leading to the release of the AV1 video codec specification in 2018 [40].

In a similar way to the ISO/IEC/ITU standards, the VP8, VP9 and AV1 specifications each define a bitstream format and a method of decoding the bitstream but do not define an encoder. All three specifications follow the well-known video codec model that includes block-based processing, inter and intra-prediction, block transforms, quantisation and entropy coding, but with some differences to the H.26x standards. Table 3.4 summarises some of the key features of VP9 and AV1.

VP8, VP9 and AV1 are reportedly used for an appreciable proportion of streaming content encoding [41], with hardware support in an increasing number of consumer electronic devices.

Table 3.4 Selected features of VP9 and AV1

Feature	VP9	AV1
Basic coding block	Superblock, up to 64×64 pixels	Superblock up to 128×128 pixels
Coding block partitioning	Down to 4×4 blocks	Down to 4×4 blocks, more partitioning options than VP9
Intra-prediction	8 directional modes, DC mode, True Motion mode, prediction blocks up to 32×32	56 directional modes, DC mode, smooth modes, prediction blocks up to 128×128
Inter-prediction	Motion compensation with candidate motion vectors, multiple reference frames	More candidate motion vectors, motion field estimation, overlapped block motion compensation, affine motion compensation
Transform	DCT up to 32×32 size, hybrid DCT/Discrete Sine Transform	Multiple transform types, combined transforms, up to 64×64 size
Entropy coding	Binary arithmetic coding with per-frame probability updates	Multi-symbol arithmetic coder
Filtering	Deblocking filter	Deblocking filter, directional deringing filter, noise-reducing filter

3.3.8.3 Essential Video Coding and Low-Complexity Enhancement Video Coding

MPEG EVC was published in 2020 [42]. EVC specifies a royalty-free baseline codec and a set of optional coding tools for video compression [43]. The Baseline Profile defines a subset of EVC coding tools that are considered to be royalty-free. In this case, royalty-free means that either the coding method was first publicly disclosed more than 20 years ago and is therefore unlikely to be protected by an active patent, or the proposed method was accompanied by a declaration that the technology would be made available without any royalty claims for existing patent rights. The Main Profile comprises the Baseline Profile plus coding tools that can offer increased compression performance, subject to a potential licensing cost for commercial use.

EVC's Baseline Profile specifies a basic block size of up to 64 × 64 pixels, which is similar to HEVC's CTU, a basic form of intra prediction, inter prediction with motion-compensated prediction, variable-size DCT transform, quantisation, loop filtering and binary arithmetic coding based on the scheme proposed in the JPEG standard. The Main Profile adds functionality such as flexible partitioning of each basic block, variable-resolution motion vectors for motion compensation, affine motion compensation, decoder-side motion vector refinement, adaptive filtering, adaptive transform selection and bit-plane coding of transform coefficients.

Published in 2021, MPEG's LCEVC specifies an enhancement layer that can provide extra compression performance when combined with an existing video coding technique as a base layer [44, 45]. LCEVC, which should not be confused with EVC, operates in a way that is analogous to scalable or layered coding (see Figure 3.15). When using LCEVC, a source video signal is downsampled to a lower resolution and encoded using an existing coding format such as AVC or HEVC, to form the Base Layer. Any existing standard-compliant encoder includes a decoding loop that reconstructs each coded and decoded video frame. The reconstructed frames are upsampled, and the difference between the uncompressed input and these reconstructed frames is encoded into one or two enhancement layers. Figure 3.16 illustrates a simplified version of an LCEVC encoder, showing only a single enhancement layer. A decoder can decode the base layer

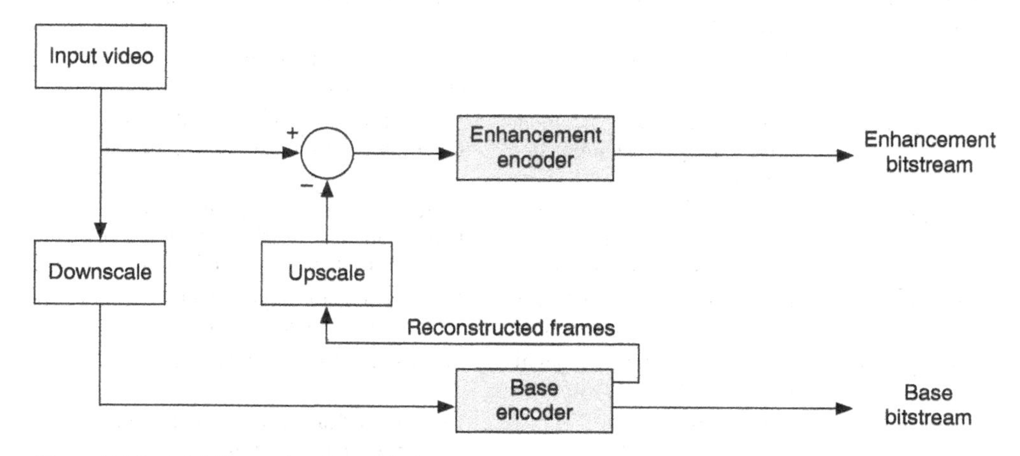

Figure 3.16 LCEVC encoder (simplified)

bitstream in the usual way, without using the enhancement bitstream, to obtain a basic-quality version of the original video. Alternatively, an LCEVC decoder can decode the base layer and the enhancement bitstream in order to reconstruct a higher-quality version of the video sequence.

3.4 Comparing Video Coding Standards

In Chapter 2, we saw how the performance of video codecs can be compared, by comparing the bitrate and quality performance and/or comparing the computational requirements of each codec. With an increasing number of video coding standards available, how should we compare them to each other? The standards themselves are merely documents, but it is possible to measure the bitrate, visual quality and computational performance of a codec that conforms to each standard. As the standards do not define an encoding method, it is important to consider the performance of an optimised encoder that conforms to each standard, i.e. an encoder that attempts to produce the best performance within the constraints of the standard.

We will look at measuring performance in more detail in Chapter 11. Comparing the performance of video coding standards can be challenging, not least because of the difficulties of measuring or estimating visual quality discussed in Chapter 2. We can reach a reasonable estimate of performance by measuring the reduction in or increase in bitrate at the same visual quality. If we assume that a video clip is encoded using Codec A, the quality is estimated using the PSNR metric (see Chapter 2) at 38 dB and the coded bitrate is 10 Mbps. The same video clip is then encoded using Codec B. The PSNR is the same (38 dB), and the coded bitrate is 5 Mbps. We can conclude that for this sequence and this estimated quality, Codec B is twice as efficient as Codec A, i.e. it reduces the coded bitrate by 50%.

Repeating this sort of experiment for multiple video sequences and multiple quality levels can give a reasonable estimate of the performance of video codecs conforming to different standards. Here are some examples from the research literature:

The authors of Ref. [46] concluded that HEVC Main Profile encoding resulted in average bitrate savings of 49.3% compared with H.264/AVC High Profile, at the same subjective video qualities and across a range of video sequences, with subjective video quality measured using Mean Opinion Score, MOS.

When comparing the VVC, EVC and AV1 standards with the older HEVC standard, one paper [47] found that on average, when considering the bitrate reduction for the same PSNR, VVC outperformed HEVC by around 36%, EVC outperformed HEVC by 22% and AV1 outperformed HEVC by around 17.5%.

Another paper [48] determined that the tested VVC encoder, VVenC-0.1, resulted in bitrate savings of over 50% compared with HEVC, when measuring subjective quality using MOS and when coding UHD video.

These and other studies demonstrate two points. The first is that newer standards tend to outperform older standards, with HEVC estimated to use around half the bitrate of AVC and VVC estimated to use around half the bitrate of HEVC. The second is that different research teams can end up reaching different conclusions on video codec performance, depending on the test material, codec implementations and test methodology.

3.5 Conclusions

In this chapter, we have seen how video compression coding developed from a set of core concepts, some of which were first introduced more than 70 years ago, through decades of experimentation, into a series of video compression standards that have continued to evolve from the early 1990s until the present day. These video coding standards have become increasingly important as digital video continues to pervade our working and leisure time. Every one of the popular video coding standards uses aspects of the familiar video codec model but of course, the details differ. In the following chapters, we will look at the technology behind each of the core video coding concepts and will examine the H.265/HEVC standard in more detail.

References

1 Huffman, D.A. (1952). A method for the construction of minimum redundancy codes. *Proceedings of the Institute of Radio Engineers* 40 (9): 1098–1101.

2 Rissanen, J.J. (1976). Generalized Kraft inequality and arithmetic coding, *IBM Journal of Research and Development*, 20.3: 198–203. doi: https://doi.org/10.1147/rd.203.0198.

3 Witten, I., Neal, R., Cleary, J. (1987). Arithmetic coding for data compression, *Communications of the Association for Computing Machinery*, 30.6:520–540. https://doi.org/10.1145/214762.214771.

4 Oliver, B.M. (1956). Linear predictor circuits. US Patent 2,732,424, issued 1956.

5 Lucky, R.W. (1970). Adaptive prediction for redundancy removal in data transmission systems. US Patent 3,502,986, issued 24 March 1970.

6 Schroeder, M.R. (1972). Transform coding of image difference signals. US Patent 3,679,821, filed 30 April 1970, issued 25 July 1972.

7 Haskell, B.G., Mounts, F.W., Candy, J.C., (1972). Interframe coding of videotelephone pictures, *Proceedings of the IEEE*, 23.7: 792–800. doi: https://doi.org/10.1109/PROC.1972.8778.

8 Mounts, F.W. (1969). A video encoding system using conditional picture element replenishment. *Bell System Technical Journal* 48 (7): 2545–2554.

9 Andrews, H.C. and Pratt, W.K. (1968). Fourier transform coding of images. Hawaii International Conference on System Science, 677–679.

10 Ahmed, N., Natarajan, T., Rao, K.R., (1974). Discrete cosine transform, *IEEE Transactions on Computers*, C-23.1: 90–93. doi: https://doi.org/10.1109/T-C.1974.223784.

11 Chen, W.-H., Smith, C., Fralick, S. (1977). A fast computational algorithm for the Discrete Cosine Transform, *IEEE Transactions on Communications*, 25.9: 1004–1009. doi: https://doi.org/10.1109/TCOM.1977.1093941.

12 Chen, W.-H. and Pratt, W.K. (1984). Scene adaptive coder. *IEEE Transactions on Communications* 32 (3): 225–232.

13 Roese, J., Pratt, W., Robinson, G., (1977). Interframe cosine transform image coding, *IEEE Transactions on Communications*, 25.11: 1329–1339. doi: https://doi.org/10.1109/TCOM.1977.1093761.

14 Haskell, B., Gordon, P., Schmidt R., Scattaglia, J. (1977). Interframe coding of 525-line, monochrome television at 1.5 Mbit/s', *IEEE Transactions on Communications*, 25.11: 1339–1348. doi: https://doi.org/10.1109/TCOM.1977.1093764.

15 Netravali, A.N., Stuller, J.A., (1979). Motion compensated transform coding, *Bell System Technical Journal*, 58.7: 1703–1718. https://doi.org/10.1002/j.1538-7305.1979.tb02277.x.

16 Jain, J.R., Jain, A.K., (1981). Displacement measurement and its application in interframe image coding, *IEEE Transactions on Communications*, 29.12: 1799–1808. doi: https://doi.org/10.1109/TCOM.1981.109495.

17 Ericsson, S., (1985). Fixed and adaptive predictors for hybrid predictive/transform coding, *IEEE Transactions on Communications*, 33.12: 1291–1302. doi: https://doi.org/10.1109/TCOM.1985.1096251.

18 Thoma, R. and Bierling, M. (1989). Motion compensating interpolation considering covered and uncovered background. *Signal Processing: Image Communication* 1 (2): 1911–1212. https://doi.org/10.1016/0923-5965(89)90009-X.

19 Le Gall, D. (1991). MPEG: a video compression standard for multimedia applications. *Communications of the Association of Computing Machines (ACM)* 34 (4): 46–58.

20 Chan, M.H., Yu, Y.B., Constantinides, A.G., (1990). Variable size block matching motion compensation with applications to video coding, *IEEE Proceedings*, 137.4: 205–212. https://doi.org/10.1049/ip-i-2.1990.0029.

21 Fukuhara, T., Asai, K., and Murakami, T. (1997). Low bit rate video coding with block partitioning and adaptive selection of two time-differential frame memories. *IEEE Transactions on Circuits and Systems for Video Technology* 7 (1): 212–220.

22 Flierl, M. (1997). Bewegungskompensierte Multihypothesen-Langzeitprädiktion. Diploma thesis. Telecommunications Laboratory, University of Erlangen-Nuremberg, Germany.

23 Wiegand, T., Zhang, X., Girod, B., (1997). Motion compensating long-term memory prediction, *Proceedings of the IEEE International Conference on Image Processing*, 2: 53–56. doi: https://doi.org/10.1109/ICIP.1997.638671.

24 Girod, B. (1993). Motion compensating prediction with fractional pel accuracy. *IEEE Transactions on Communications* 41 (4): 604–612.

25 Bekkers, R.R., Dalais, M., Doré, A., et al. (2014). Understanding patents, competition and standardization in an interconnected world. https://www.itu.int/en/ITU-T/Documents/Manual_Patents_Final_E.pdf.

26 Via, L.A. (2023). Patent portfolio license for AVC/H.264. https://www.via-la.com/licensing-2/avc-h-264/ (accessed October 2023).

27 Codecs for videoconferencing using primary digital group transmission (1984). ITU-T Recommendation H.120.

28 Video codec for audiovisual services at p x 64 kbits (1990). ITU-T Recommendation H.261. Approved 14 December 1990.

29 Information technology – coding of moving pictures and associated audio for digital storage media at up to about 1.5 Mbit/s – Part 2: Video (1993). MPEG-1 Video, ISO/IEC 11172-2.

30 Information technology – generic coding of moving pictures and associated audio information: Video (1995). MPEG-2 Video, ISO/IEC 13818-2 and ITU-T Recommendation H.262.

31 Video coding for low bit rate communication (1996). ITU-T Recommendation H.263, amended in 1997 and thereafter.

32 Information technology – coding of audio-visual objects – Part 2: Visual (1999). MPEG-4 Part 2: Visual. ISO/IEC 14496-2, amended in 2000 and thereafter.

33 Advanced video coding for generic audiovisual services (2003). H.264/AVC or MPEG-4 Part 10. ISO/IEC 14496-10 and ITU-T Recommendation H.264.

34 High Efficiency Video Coding (HEVC) Family, H.265 (2019). ITU-T.

35 Versatile video coding (2020). ISO/IEC 23090-3 and ITU-T Recommendation H.266.

36 SMPTE 421M-2006 (2006). *Standard for Television: VC-1 Compressed Video Bitstream Format and Decoding Process.* 10.5594/SMPTE.ST421.2006.

37 Srinivasan, S. and Regunathan, S. (2005). An overview of VC-1. *Proceedings of SPIE* 5960: 720–728. https://doi.org/10.1117/12.631574.

38 Bankoski, J., Wilkins, P. and Xu, Y. (2011). Technical overview of VP8, an open source video codec for the web. *Proceedings of the 2011 IEEE International Conference on Multimedia and Expo*, (11–15 July 2011). Barcelona, Catalonia, Spain, pp. 1–6.

39 Mukherjee, D., Bankoski, J., Grange, A., et al. (2013). The latest open-source video codec VP9-an overview and preliminary results. *2013 Picture Coding Symposium (PCS)*, pp. 390–393.

40 Chen, Y., Murherjee, D., Han, J., et al. (2018). An overview of core coding tools in the AV1 video codec. In *2018 Picture Coding Symposium (PCS)* pp. 41–45.

41 6th Annual Bitmovin Video Developer Report (2023). http://bitmovin.com.

42 Information technology – General video coding – Part 1: Essential Video Coding (2020). ISO/IEC 23094-1. https://www.iso.org/standard/57797.html (accessed 12 January 2024).

43 Choi, K., Chen, J., Rusanovskyy, D. et al. (2020). An overview of the MPEG-5 essential video coding standard [Standards in a Nutshell]. *IEEE Signal Processing Magazine* 37 (3): 160–167.

44 Information technology – General video coding – Part 2: Low complexity enhancement video coding (2021). ISO/IEC 23094-2:2021.

45 Battist, S. et al. (2022). Overview of the low complexity enhancement video coding (LCEVC) standard. *IEEE Transactions on Circuits and Systems for Video Technology* 32 (11): 7983–7995. https://doi.org/10.1109/TCSVT.2022.3182793.

46 Ohm, J.R., Sullivan, G.J., Schwarz, H. et al. (2012). Comparison of the coding efficiency of video coding standards, including High Efficiency Video Coding (HEVC). *IEEE Transactions on Circuits and Systems for Video Technology* 22 (12): 1669–1684. https://doi.org/10.1109/TCSVT.2012.2221192.

47 Grois, D., Giladi, A., Choi, K., and Park, M.W. (2021). Performance comparison of emerging EVC and VVC video coding standards with HEVC and AV1. *SMPTE Motion Imaging Journal* 130 (4): 1–12. https://doi.org/10.5594/JMI.2021.3065442.

48 Bross, B., Chen, J., Ohm, J.R., Sullivan, G.J., Wang, Y.K., (2021). Developments in international video coding standardization after AVC, with an overview of versatile video coding (VVC), *Proceedings of the IEEE*, 109.9: 1463–1493. 10.1109/JPROC.2020.3043399.

4

Structures

4.1 Introduction

4.1.1 How Does a Video Codec Use Structures?

A complete video clip or sequence is processed by a video encoder to create a compressed bitstream. In order to handle the large amount of image data in a sequence of video frames, the encoder splits it up into manageable structures. In this chapter, we will look at how a video encoder does this and how it bridges the gap between a source video, which contains multiple frames each made up of thousands or millions of pixels, and encoded data units. We will also look at the processing and storage capabilities of a practical codec that handles a relatively small region or block of data at a time. Figure 4.1 illustrates how a source video sequence is split up into manageable-sized structures, such as blocks of pixels. These structures are processed and encoded to produce the compressed bitstream.

A coded video sequence is a series of coded pictures that, when decoded, will play back as a complete video programme or clip. Coded frames or pictures may be organised into multi-picture structures during encoding, such as Group of Pictures (GoPs). Each picture may be coded as a single unit or in multiple sections known as slices or tiles. Each slice or tile contains one or more Basic Units, such as Macroblocks or Coding Tree Units.

The Basic Unit is a unit of data handled by the encoder and decoder. In present-day codecs, it is a square. In the older MPEG-2 and H.264/AVC standards, it is 16 × 16 pixels, up to 64 × 64 pixels in the H.265/High Efficiency Video Coding (HEVC) standard and up to 128 × 128 pixels in the H.266/VVC standard. The Basic Unit may be split up or partitioned into smaller blocks for processing steps such as prediction and transformation.

In this chapter, we will answer the following questions:

- How are coded pictures or frames organised into a coded video sequence?
- What is a GoP, and how does it help operations such as random access?
- What aspects of a coded video sequence are designed to support network transmission? What aspects can support parallel processing?
- What are the advantages and disadvantages of larger or smaller coding unit (CU), prediction unit (PU) and transform unit (TU) sizes?
- How are structures handled in the H.265 standard and other standards?

Coding Video: A Practical Guide to HEVC and Beyond, First Edition. Iain E. Richardson.
© 2024 John Wiley & Sons Ltd. Published 2024 by John Wiley & Sons Ltd.
Companion website: www.wiley.com/go/richardson/codingvideo1

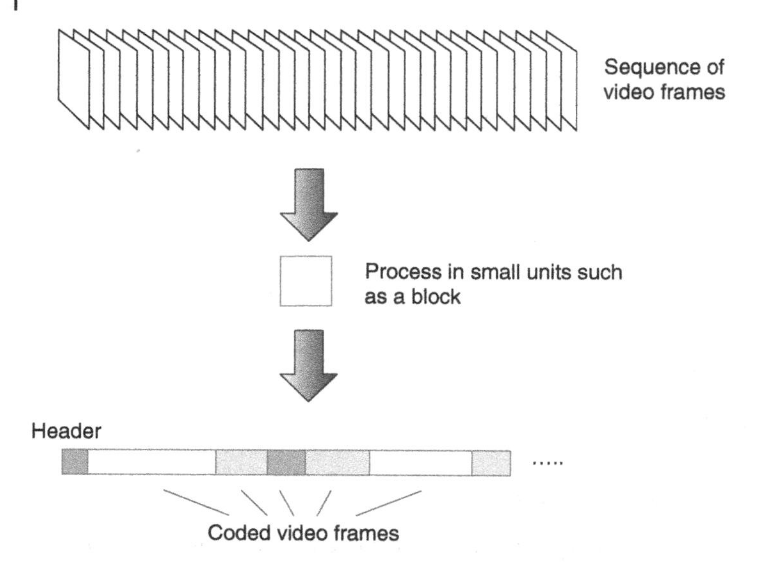

Figure 4.1 Mapping a video sequence to a coded bitstream

4.1.2 Separating a Sequence into Processing Units

Early video codecs had limited processing and storage capacity, and so processing was applied to small regions such as 8×8 blocks. Memory and processing capacity have improved significantly since the early 1990s, but nevertheless, a codec breaks a sequence down into progressively smaller regions for processing. A complete sequence is divided or partitioned into groups of coded frames or pictures, then into individual coded frames, then regions within a frame and then square or rectangular blocks, as illustrated in Figure 4.2.

4.1.3 Regular or Irregular Structures

A typical video scene is made up of irregular-shaped regions with varying textures, colours and movement behaviour. This does not correspond neatly to a regular block-based structure. A codec attempts to model and predict the behaviour of the video sequence, which implies that irregular structures might produce better compression efficiency. However, irregular or changing structures need to be communicated to the decoder, whereas a fixed or repetitive structure can often be inferred and may not require to be communicated. For example, if the Macroblock size is fixed at 16×16, there is no need to communicate the size from encoder to decoder.

There is a trade-off between using structures such as fixed-size blocks versus structures that can change size and perhaps shape. Fixed-sized blocks are simple to process and do not require many bits to communicate the structure to the decoder, but they may be a poor match to the actual video content. Structures that can change size and shape may be capable of more accurately reflecting actual video content, but they may be more complex to process and require a significant number of bits to communicate.

Most practical codecs make a compromise, which allows a certain amount of variation in the structures whilst basing the structures on rectangular or square blocks. Figure 4.3 shows a part of a coded video frame with an overlay of prediction block (PB) structures.

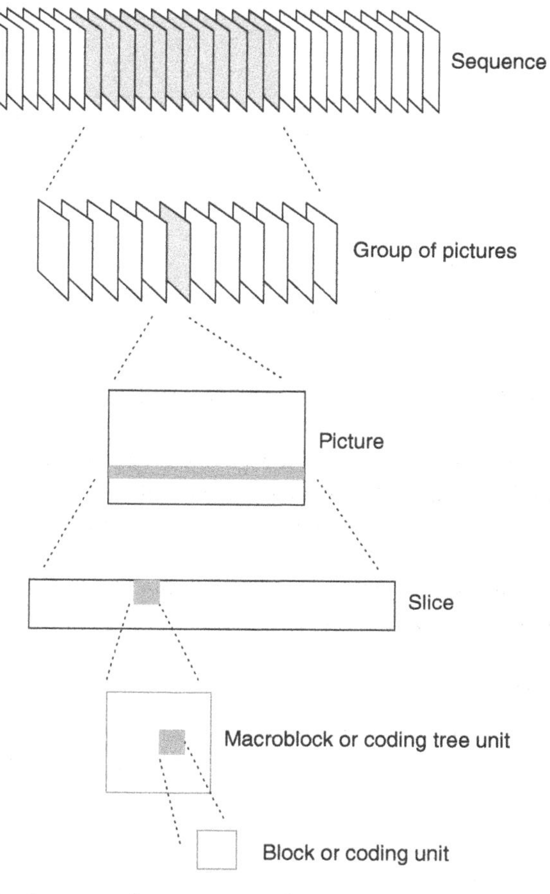

Figure 4.2 Structures from sequence down to coding unit

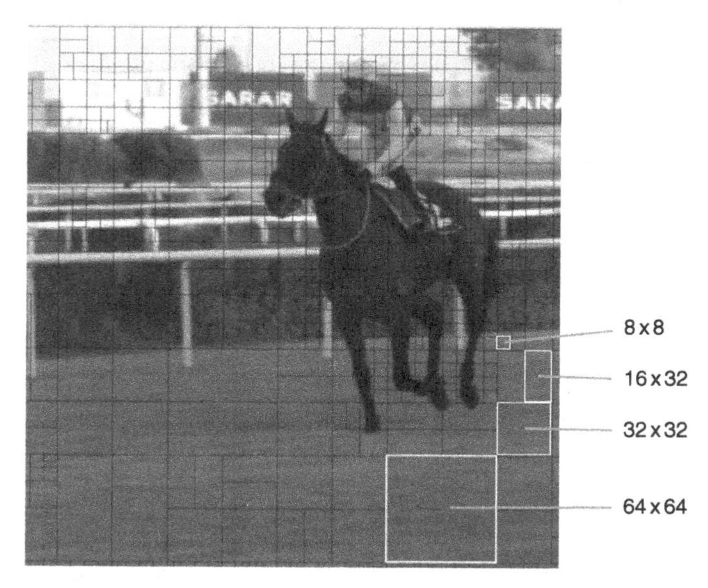

Figure 4.3 Portion of video frame showing HEVC prediction block structures. *Source:* Elecard StreamEye

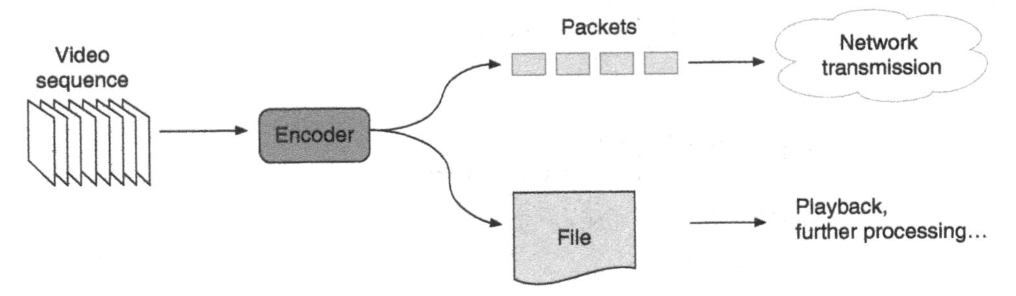

Figure 4.4 Coding a video sequence for network transmission or file-based output

In this example, the H.265/HEVC encoder chooses between a range of square and rectangular block sizes in order to optimise compression of regions with different amounts of detail. In a relatively homogeneous area such as the highlighted grass region at the lower right, the encoder chooses a large block size, 64 × 64 pixels in this example. In complex moving areas such as around the jockey and the horse, the encoder chooses smaller block sizes that more closely follow the movement and object boundaries.

4.1.4 Transmission and Access

A coded video sequence is stored or transmitted before decoding and playback. Transmission over a network requires an efficient mapping between the coded bitstream and network packets, as we will see in Chapter 10. Other challenges include:

- Supporting random access, so that a user can join part-way through a clip or navigate to a suitable playback point rather than always having to start from the beginning of the sequence,
- Streaming playback, achieving a compromise between good playback quality and avoiding stalling or rebuffering, and
- Handling errors, so that a lost packet or corrupted section does not ruin the playback of the entire clip.

A video encoder maps a video sequence to a series of coded data chunks that are suitable for storage or transmission, either in a single file or a sequence of files or network packets, as illustrated in Figure 4.4.

4.2 Coded Video: Sequence to Picture

A coded video sequence may correspond to a video clip, a programme or a section of a programme. A sequence is made up of a series of coded frames or coded pictures (see Figure 4.2).

Some of the parameters necessary to decode a sequence are contained within each coded picture. In addition, a coded sequence may include parameter sets, which are coded structures that contain the common parameters that are needed for decoding and that apply to multiple coded pictures. For example, an H.264 or H.265 Sequence Parameter Set (SPS) contains parameters that are common to an entire coded sequence. A Picture Parameter Set (PPS) contains further parameters that are common to one or more pictures in a coded sequence.

Example:

Figure 4.5 illustrates a coded video sequence. One SPS contains decoding parameters for the entire coded sequence. The first PPS (PPS0) inherits some common parameters from the SPS and includes further decoding parameters. In order to decode slices of coded video, a decoder needs parameters from both the SPS and the currently active PPS. When decoding pictures 1 and 2, the decoder uses parameters defined in PPS0 and the SPS. A second PPS (PPS1) also refers back to the SPS and specifies further parameters. The coded picture, Pic N, is decoded using the parameters from PPS1 and SPS.

A single encoded video frame is a coded picture or a coded frame. Decoding a coded picture should produce a complete, displayable video frame. Pictures or frames may be coded without prediction from other pictures or frames, using intra-coding, and/or with prediction from other pictures or frames, using inter-coding.

Pictures or frames may be grouped into multi-picture structures. For example, random access to a video sequence requires an independently decodable picture such as an intra-coded picture, and it is common to insert intra-pictures at intervals in a sequence. In many applications of video coding, coded pictures are organised as GoPs, where each GoP contains an intra-coded picture followed by a number of inter-coded pictures. This can make it easier for a user to join a video stream part-way through or to switch to an earlier or later part of a video clip – a process known as random access – and/or to recover from transmission errors by seeking to the next intra-coded picture.

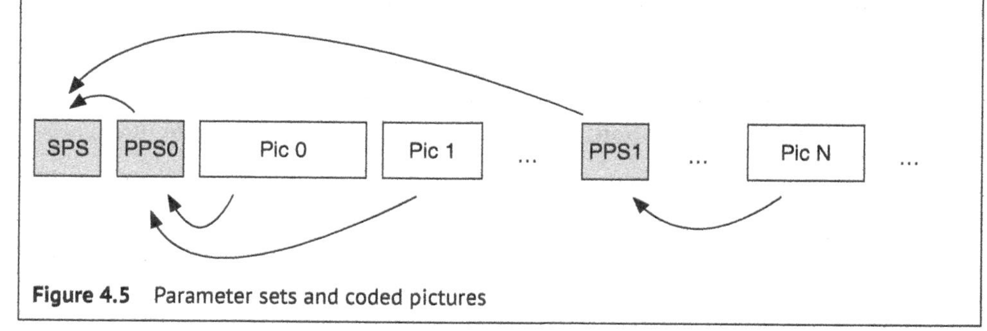

Figure 4.5 Parameter sets and coded pictures

4.2.1 Coded Sequence

A sequence is a series of coded frames or pictures representing a complete video. This could be, for example, a programme, a clip or a scene. Pictures or frames in the sequence will typically have common characteristics, such as:

- Spatial resolution, e.g. Standard Definition, High Definition or other resolutions.
- Frame rate, e.g. 50 or 60 frames per second.
- Colour depth and colour space, e.g. 8 bits per pixel, 4:2:0 sampling.
- Coding parameters such as Profile, Level and CU size.

Pictures in the sequence may be arranged with certain structural relationships. For example, a GoP is a repeating, multi-picture structure consisting of an intra-coded picture or I-picture, with a number of inter-coded pictures. The inter-coded pictures could be P-pictures, in which each block is predicted from one frame, or B-pictures, in which each block can be predicted from two frames.

While the coded frames are typically intended for playback and display, a coded sequence may include non-displayed coded data such as:

- Parameter sets. These are coded data structures containing information that may be common to multiple coded pictures in the sequence, such as the common characteristics described above.
- Redundant frames. These may contain redundant picture data that are not intended to be displayed unless the main or primary pictures are lost or damaged.
- Non-displayable reference frames. These are pictures that may be used for prediction but are not intended for display, such as automatically generated reference frames or background scene elements.

4.2.2 Coded Frame or Picture

A coded picture is a complete compressed frame, or a field or pair of fields in the case of an interlaced video sequence. It may be made up of multiple sections such as slices and/or tiles. Parameters required to decode the picture may be coded in a header, such as a picture header, in parameter sets and/or in one or more slice headers. A coded picture may have one of a number of types or characteristics, such as:

- **Intra-coded or I-picture**: The picture is coded without any prediction from other coded pictures, using intra-only prediction.
- **Inter-coded, P- or B-picture**: Each block in the picture may be coded using prediction from one (P) or more than one (B) previously coded pictures.
- **Used or not used for prediction**: The picture may be used or not used for inter prediction of further coded pictures.
- **Resynchronisation**: The picture triggers a reset of coding parameters at the decoder and can be used as a resynchronisation point.
- **Access/spicing point**: The picture is intended to provide an access point for joining or splicing video clips together.

4.2.3 Multi-picture Structures

Coded pictures in a video sequence may be arranged in multi-picture structures. The early MPEG-1 and MPEG-2 standards introduced the concept of GoPs, which historically consisted of an intra or I-picture followed by a series of inter-coded P- and/or B-pictures such as the example shown in Figure 4.6. Recent standards such as H.264, HEVC and VVC have introduced considerable flexibility in multi-picture structures.

Such structures can serve a number of purposes, including:

- Improved compression efficiency, by enabling an encoder to make efficient use of multiple reference pictures.

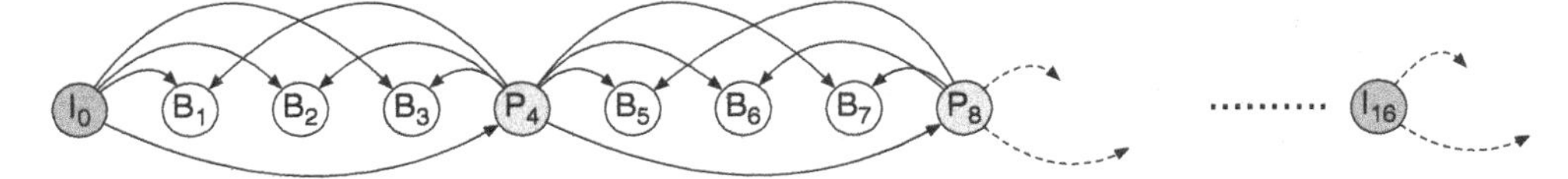

Figure 4.6 Group of Pictures example

- Limiting decoding delay, by restricting the number of pictures that need to be decoded before any given picture can be completely decoded and displayed.
- Providing resynchronisation points, by including intra-coded I-pictures at regular intervals.

Example: Group of Pictures

Figure 4.6 shows a series of coded pictures. The first, I_0, is coded using only intra prediction, without any references to other coded pictures. This means the decoder can start decoding once it receives I_0. The later pictures are coded using either inter prediction from one other picture (P_4, P_8, etc.) or using inter-prediction from two other pictures (B_1, B_2, etc.). As we will see later, single-reference prediction or P-picture prediction is often more efficient than intra prediction, and two-reference prediction using B-pictures can be more efficient than single-reference prediction. This GoP structure repeats after 16 pictures, which means that I_{16} begins the second GoP. Because I_{16} is coded using intra prediction, a decoder can choose to start decoding from I_{16}, i.e. I_{16} provides a random access point.

4.3 Coded Video: Picture to Basic Unit

A coded picture is a compressed representation of a single video frame, or a coded field or pair of fields for interlaced video. In standard-based video codecs, a picture is processed in Basic Units such as Macroblocks or Coding Tree Units.

The Basic Unit usually has a fixed size within a coded picture. The choice of Basic Unit size is a compromise: a larger size may be more efficient for representing large areas of low detail in frame but requires more memory and processing power, whereas a smaller size may be computationally simpler but less efficient for compression. Each picture is arranged as a regular grid of Basic Units.

The Basic Units within a picture may themselves be arranged into structures such as Slices and/or tiles. A Slice is a series of Basic Units in a raster scan order, and slices may contain an irregular number of Basic Units. A tile is a square or rectangular structure containing multiple Basic Units. Slices and/or tiles may be useful for:

- Mapping coded video data to transmission packets, for example, by sending each slice in a separate network packet.
- Handling errors, for example, resynchronising at the start of the next error-free slice or tile.
- Parallel processing, which could involve processing multiple slices or tiles simultaneously for faster encoding or decoding.

4.3.1 Basic Units

Each coded picture is made up of a series of Basic Units. A Basic Unit is a regular-sized square region. The terminology varies from standard to standard:

- **Macroblock**: A Basic Unit consisting of 16×16 luma samples and associated chroma samples, used in VP8, H.264 and older standards.
- **Supermacroblock**: A Basic Unit consisting of 64×64 or 128×128 luma samples and associated chroma, used in VP9 and AV1.
- **Coding Tree Unit**: A Basic Unit consisting of an N×N array of luma samples and associated chroma. In H.265/HEVC, N can be 16, 32 or 64 and is set at the sequence level via the SPS. In H.266/VVC, N can range from 32 to 128.

An encoder or decoder has to process and store all the data and parameters of a Basic Unit. A larger Basic Unit requires more computational power and more local storage and so a smaller Basic Unit may be more suitable for decoders with lower processing capabilities.

A certain number of parameters must be coded for each Basic Unit, unless the Basic Unit is skipped, i.e. not coded. It may be more efficient to minimise the number of Basic Units in a frame by using a larger Basic Unit size. Furthermore, the Basic Unit is typically the largest possible partition size for prediction and coding, so a larger Basic Unit gives more flexibility in assigning block sizes to regions of the frame.

Larger Basic Units can give better compression efficiency than smaller Basic Units can for higher-resolution video, e.g. HD and UHD/4K resolutions. This is because image features such as foreground and background objects are likely to take up a relatively larger number of pixels in a high-resolution video frame.

4.3.2 Slices

A picture may be coded as one or more slices, each of which is a contiguous series of Basic Units. Figure 4.7 shows the concept, in which a picture is divided into a set of slices in a raster scan order from top left to lower right. Each slice contains an integral number of Basic Units, and the end of one slice is immediately followed by the beginning of the next slice.

An encoder can use slices as a way of mapping coded data into packets or units for storage or transmission, for example, with each slice transmitted in a separate network packet. If a packet is lost or corrupted, decoding can restart at the beginning of the next slice, albeit with potential problems if other slices or other pictures are predicted from the lost slice. Predictions and inherited parameters may be restricted across slice boundaries. An encoder may choose a fixed size for each slice, for example, a single slice per picture, or a fixed number of Basic Units per slice, as shown in Figure 4.8. A potential disadvantage of fixed-size slices is that the amount of coded data per slice may vary considerably and can depend on whether the slice contains stationary background regions that are easy to predict or fast-moving foreground regions that are hard to predict. This means that a picture coded with a uniform number of Basic Units per slice may result in coded slices with a significant variation in coded size, as can be seen in Figure 4.9. It may be preferable to choose slice sizes to maintain a roughly constant coded slice size, as shown in Figure 4.10.

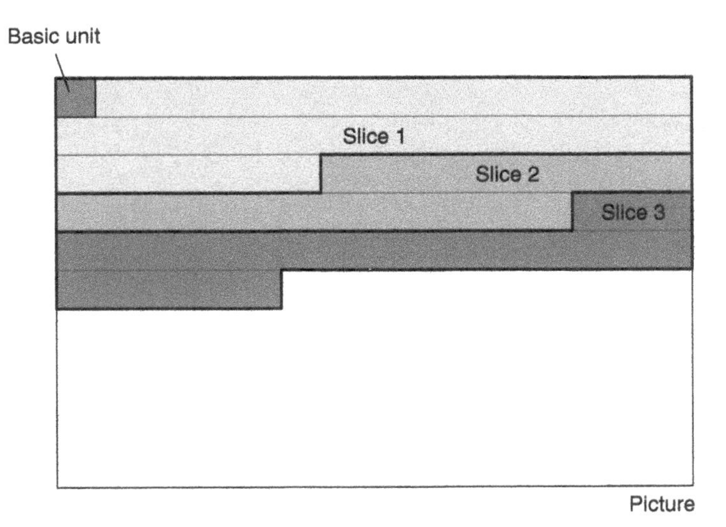

Figure 4.7 Slices in a picture

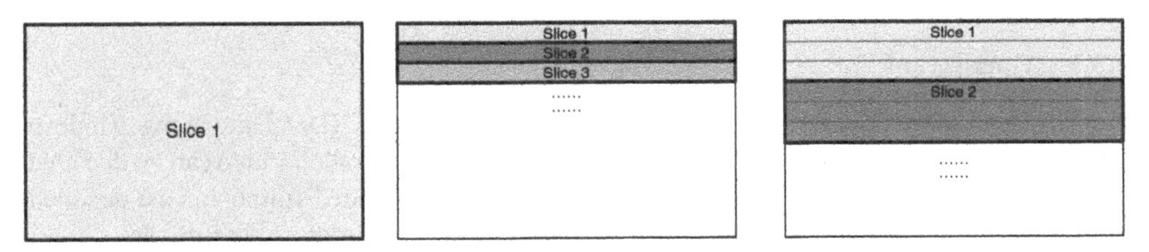

Figure 4.8 Fixed-size slices: one slice per picture, one slice per row, one slice per 3 rows of Basic Units

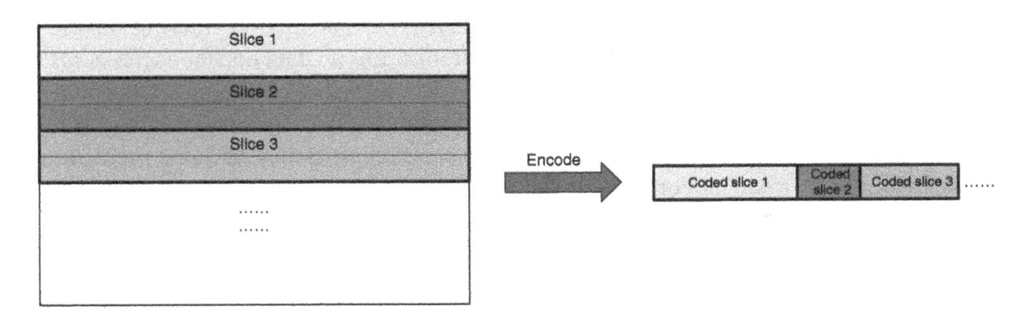

Figure 4.9 Fixed slice size in Basic Units, variable coded slice size

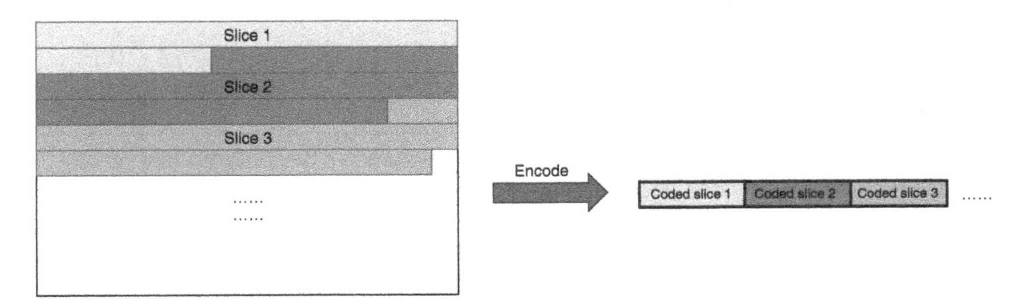

Figure 4.10 Variable slice size in Basic Units, approximately equal coded slice size

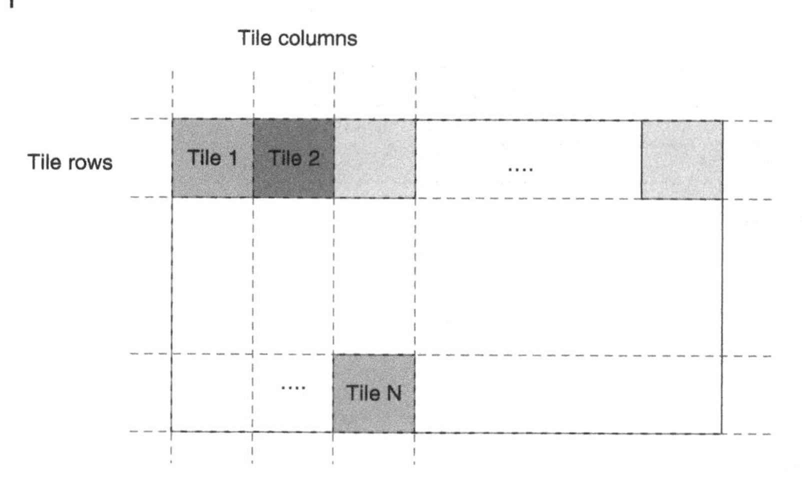

Figure 4.11 Picture divided into tiles

4.3.3 Tiles

Increasingly, video codecs are handling larger frame sizes such as HD, 4K and above. Modern processors can contain multiple cores capable of operating in parallel. Video can be decoded more quickly and more efficiently if multiple sections of a coded frame can be decoded simultaneously by multiple processing cores and/or in multiple processing threads.

A picture may be coded as one or more tiles. Unlike slices, which may or may not be rectangular, tiles are always rectangular. Figure 4.11 shows the general idea. A coded picture is partitioned into regular, equally sized tiles, which can be square as shown here, or rectangular. During encoding, certain operations such as intra prediction, filtering and prediction mode signalling may be constrained so that there are no data dependencies across tile boundaries. This means that a decoder can process multiple tiles in parallel, decoding each tile without waiting for data in another tile to be decoded. Constraining encoding dependencies in this way may cause a loss of compression efficiency, as some prediction choices will not be available for Basic Units next to a tile boundary. However, splitting a frame into regular-sized tiles that can be independently decoded makes it possible to decode large frames efficiently in parallel, which has the potential to increase decoding speed and efficiency.

As Figure 4.11 shows, rectangular tiles are arranged in rows and columns and the tiles are each bounded by row and column boundaries. This type of structure can be signalled relatively easily by identifying each of the internal row and column boundaries.

Example 1:

Figure 4.12 shows two possible ways that a coded picture can be split into four rectangular tiles. Consider a video decoder chip with four processing cores. Each tile may be allocated to a separate core for decoding. Assuming that the four cores operate

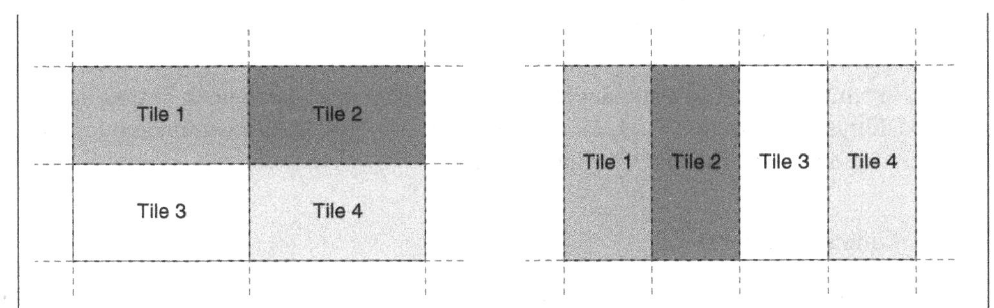

Figure 4.12 Picture divided into four tiles, two options

independently, each core processes one tile at an independent rate. Once all four cores have finished decoding, the complete frame has been decoded and is available for output or display.

This example illustrates why tiles should be independently decodable for efficient parallel operation. If the decoding of Tile 4 depended on data in Tile 3, then the core that is decoding Tile 4 could end up waiting for the processing of Tile 3 to catch up and make the data available.

Example 2:

In Figure 4.13, a coded picture is divided into 12 tiles. A four-core processor can allocate three tiles to each core, as illustrated in the Figure 4.13. Core 1 processes tiles 1, 2 and 3, while Core 2 processes tiles 4, 5 and 6, and so on. If each tile is independently decodable, then the four-core processor can decode the picture up to four times faster than an equivalent single-core processor.

Tiles are handled in a specific way in the HEVC/H.265 standard, as will be discussed later in this chapter in Sections 4.5.8 and 4.5.9.

Figure 4.13 Twelve-tile picture decoded using four cores

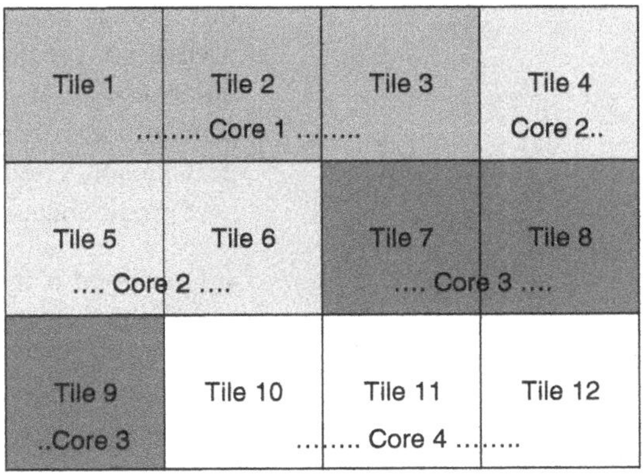

4.4 Coded Video: Basic Unit to Block

The Basic Unit of coding in a standard-based video codec is a Macroblock or Coding Tree Unit, as discussed in Section 4.3.1. Each Basic Unit may be split or partitioned into structures for coding, prediction and transformation.

4.4.1 Coding Structures

A video codec handles image data in units of square or rectangular blocks. How big should these blocks be? The answer may depend on the size and content of each video frame. As Figure 4.3 illustrates, the best size of coding units (CUs) might depend on the content of the frame. In this example, it may be more efficient to code the moving jockey and horse using smaller blocks and to code the more static background using larger blocks.

Early codecs such as H.261 and MPEG-1 adopted a one-size-fits-all approach, fixing the basic unit size – the Macroblock size – at 16×16 pixels and the basic transform block (TB) size at 8×8 pixels. Later, H.264 introduced some flexibility with a fixed basic unit size – a 16×16 Macroblock – but with flexible choices of block size for prediction and transform. H.265/HEVC and VVC have extended this flexibility and allowed the CU size to vary depending on the local scene content. In HEVC, as we will discuss later in Section 4.5.11, a fixed-size Coding Tree Unit (CTU) may be partitioned into multiple CUs with varying sizes, such as in the example of Figure 4.14. The fixed-size CTU allows a decoder to know exactly how much processing power and local storage are required to handle each CTU, whereas the varying-size CU makes it possible to adjust the CU size depending on image content.

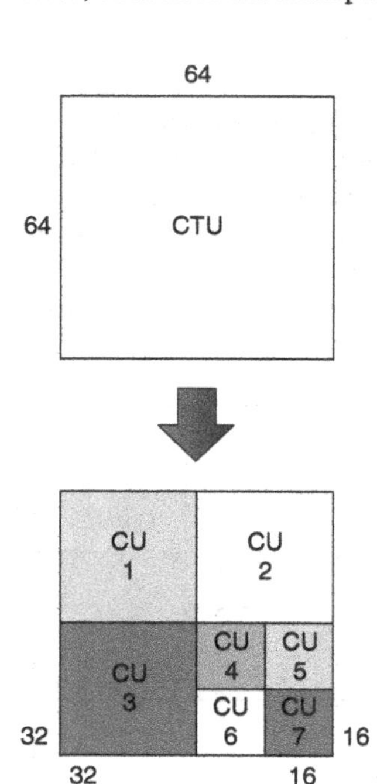

Figure 4.14 HEVC Coding Tree Unit partitioned into coding units

4.4.2 Prediction Structures

Pixels or samples are grouped into prediction areas such as Prediction Units (PUs), each of which is predicted by using previously coded data, such as neighbouring areas using intra-prediction, or previously coded frames using inter prediction. Prediction areas are typically square or rectangular blocks. All the samples in a PU are predicted in the same process. Codecs such as H.264, HEVC and VVC support flexible choices of PU sizes and types. Choosing the best prediction structures involves questions such as:

- How good is the prediction? In other words, how accurately does the prediction match the pixels or samples in the block?
- How many bits does it take to encode the choice of PU?
- How many bits does it take to encode the type of prediction and parameters, such as prediction mode and motion vectors?

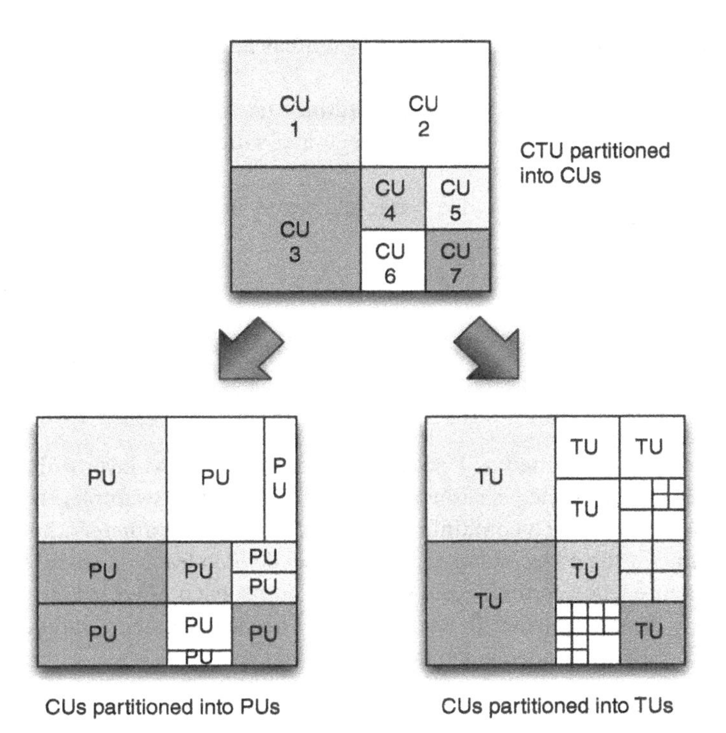

Figure 4.15 HEVC example: coding units partitioned into prediction units and transform units

PBs are usually square or rectangular blocks with dimensions that are powers of 2, making it relatively easy to signal the size and shape of the block with a minimal number of bits. In HEVC, an inter-predicted PU can be square or rectangular, with certain specified dimensions. This will be discussed in Section 4.5.12.1.

Each HEVC CU can be split up into one or more PUs and also into one or more TUs, as illustrated in Figure 4.15. As this example shows, some CUs may be partitioned differently for prediction and transformation, such as the top-right CU in Figure 4.15.

Signalling the prediction parameters for each PU may require a significant number of bits. Neighbouring PUs often have similar parameters, such as motion vectors, reference frame choice(s) and prediction mode(s), which can be used to reduce the number of bits required to signal prediction choices to the decoder.

4.4.3 Transform Structures

After prediction, blocks of residual samples are transformed, quantised and encoded. The size and shape of the region that is transformed, the Transform Unit (TU), affect compression performance. As with PUs, larger or smaller TUs may each be suitable for different areas of a frame. A larger TU size may be more efficient for homogeneous areas where all the residual samples in a large area share similar characteristics. This tends to be the case for areas of low detail and/or uniform texture, such as smooth background regions.

A smaller transform size may be more suitable for detailed areas and/or where there is complex motion.

HEVC makes it possible to split each CU into another quadtree, this time for the purpose of specifying transform units (TUs). Each TU is a square region of samples that are transformed and quantised (Figure 4.15).

We will consider the detailed operation of video codec structures in the context of the HEVC/H.265 standard.

4.5 HEVC Coding Structures

4.5.1 Coded HEVC Sequence

An HEVC sequence is stored or transmitted as a sequence of network access layer units (NAL units or NALUs). A typical sequence is shown in Figure 4.16. In this example, the sequence begins with parameter sets, each containing common decoding parameters that a decoder uses in the decoding of the coded video data. Each coded video frame or picture is sent as a series of one or more video coding layer (VCL) NAL units, each of which contains a coded slice segment, discussed further in Section 4.5.7. The slice segments can be decoded to generate the sequence of decoded video frames.

4.5.2 Parameter Sets

HEVC Parameter Sets are used to communicate coding parameters that may be common to many or all of the pictures in a video sequence. Each parameter set may be encoded in its own NAL unit, i.e. it is coded as a separate structure from the pictures in the sequence. The behaviour of parameter sets for a sequence coded as a single layer is described below. Things change somewhat for multiple layers if scalable or multiview coding is used. Parameter sets

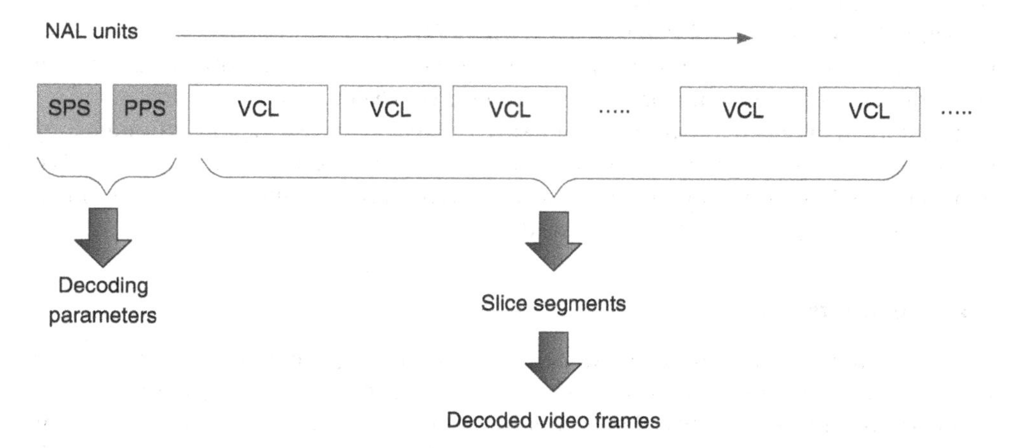

Figure 4.16 HEVC sequence structure

can be sent as part of the coded bitstream, or they may be communicated or defined separately and loaded as part of an agreed initialisation sequence, without actually transmitting the parameter set from encoder to decoder. The HEVC Parameter Sets are as follows:

Video parameter set (VPS): This may be sent once at the start of a coded sequence or communicated 'out of band', for example, a decoder could use a VPS that is predefined in the decoder software. Most of the parameters in the VPS relate to scalable or multiview coding and can be ignored for non-scalable coding with a single video layer. The Profile and Level discussed further below can be specified in the VPS and/or in the SPS.

SPS: This is sent once at the start of a coded sequence for non-scalable coding or communicated out of band. The parameters in the SPS are common to all the coded pictures in the sequence and include:

- VPS identifier, referring to the VPS in use for this sequence.
- Profile and Level identifiers, as part of the profile_tier_level syntax structure.
- SPS identifier, a number in the range 0 to 15 that identifies this SPS. PPSs use this identifier to indicate the inherited SPS.
- Spatial and temporal resolution, i.e. frame size and frame rate.
- Colour space format, 4:2:0, 4:2:2 or 4:4:4.
- Bit depths of luma and chroma samples, 8 or 10 bits per sample.
- Coding Tree Block (CTB) size, either 16×16, 32×32 or 64×64 luma samples.
- Minimum CB size.
- Minimum and maximum TB sizes.
- Enable or disable certain coding tools such as sample adaptive offset (SAO) filter and pulse code modulation (PCM) mode.

SPS parameters place certain requirements on the decoder such as the ability to handle a particular frame size, colour depth and CTU/CTB size. Such requirements may determine whether or not a particular decoder is capable of decoding this sequence. For example, a decoder with the ability to handle only 8-bit samples can determine whether the sequence is within its capability by parsing the SPS.

Sending these parameters just once in the SPS means that they do not need to be re-sent during the video sequence, which minimises the overhead in the coded bitstream.

PPS: The PPS contains parameters that may be common to some or all of the pictures in a sequence. Once a PPS has been received by a decoder, it is available for activation, i.e. for use in decoding. Each coded slice refers to one PPS identifier, and all slices in a picture must use the same PPS.

Parameters in the PPS include:

- SPS identifier, indicating the SPS in use for this sequence.
- PPS identifier, a number in the range 0 to 63 identifying this PPS. Coded slices can refer to any PPS that has already been received or is already available at the decoder.
- Enable or disable certain decoding features, such as dependent slices, sign data hiding, Context Adaptive Binary Arithmetic Coding (CABAC) initialisation per slice, transform skip and weighted prediction.

- Initialise the quantisation parameter (QP) for each slice using this PPS.
- Number of active reference pictures for inter-prediction.
- Wavefront parallel processing (WPP) enabled or disabled, using the flag entropy_coding_sync.
- Multiple tiles enabled or disabled. If enabled, the tile widths and heights are specified in the PPS.
- Parameters controlling the behaviour of the in-loop deblocking filter.
- Merge mode parameters, used in inter-coding.

In general, the PPS contains parameters that may be common to multiple pictures thus reducing overhead, but that may have to change during a coded sequence. For example, an encoder may choose to switch to a different default QP at some point during coding.

Parameter Set Activation: Each slice header contains a PPS identifier. When a given PPS is referred to in a slice header for the first time in a sequence, the PPS is activated and is then used for that slice and any further slices in the same coded picture. When an SPS is referred to in a PPS for the first time in a sequence, the SPS is activated. When the VPS is referred to in an SPS, the VPS is activated. When a slice is decoded referring to a different PPS identifier, the old PPS is deactivated and the new PPS is activated, as the following example illustrates.

Example:

Figure 4.17 shows a non-scalable bitstream starting with a VPS, VPS0, an SPS, SPS0 and a PPS, PPS0, each sent in a NAL unit. The single VPS0 is activated once a coded slice refers to it, via the PPS and SPS, and remains active for the entire video sequence. SPS0 is sent next and is not activated until a slice refers to it, via the PPS. PPS0 is sent and refers back to SPS0, i.e. any picture using the parameters of PPS0 also inherits the parameters of SPS0.

Each coded picture, such as Pic 0, etc., is sent as one or more slice segments, each contained in a Video Coding Layer NAL unit. The first slice of Pic 0 refers to PPS0. When this slice is processed by the decoder, PPS0, SPS0 and VPS0 are all activated.

Each slice within Pictures 0, 1 and 2 refers to PPS0. This means that these slices are decoded using the parameters of PPS0 and SPS0.

PPS1 is sent and refers back to SPS0. PPS1 is not active until it is referred to by a coded slice. Picture 3 refers to PPS1 and activates it, deactivating PPS0. Pictures 3 and 4 refer to PPS1 and hence use the parameters of PPS1 and SPS0. Picture 5 refers to PPS0, activating PPS0 again.

PPS2 is sent. Pictures 6 and 7 are decoded using the parameters of PPS2 and SPS0, and so on.

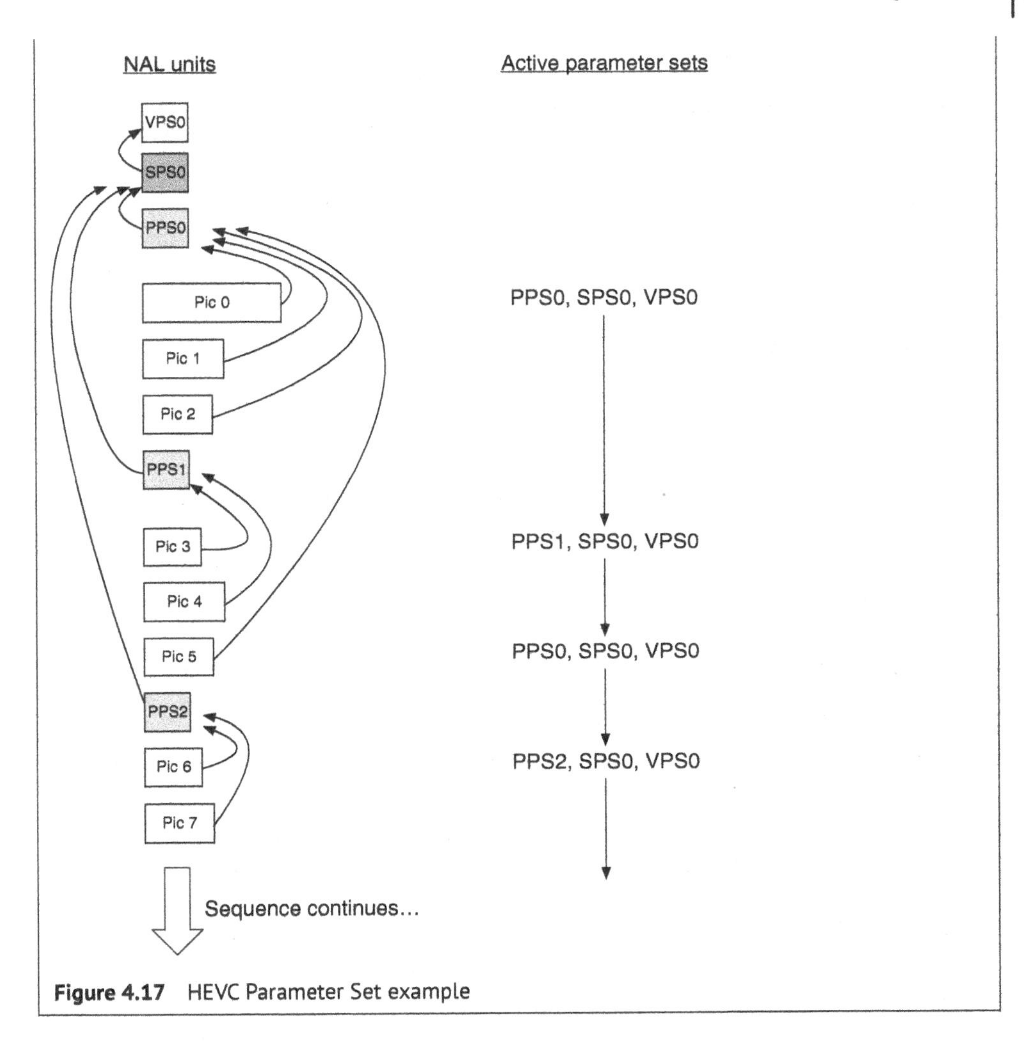

Figure 4.17 HEVC Parameter Set example

4.5.3 Profiles and Levels

Profiles and Levels specify certain restrictions on the coded bitstream in terms of the features and algorithms that may be in use and also the range of values that a decoder might encounter [1]. An HEVC bitstream contains an identifier of a Profile and Level in the SPS and an HEVC decoder can use this to determine whether it has the capability to actually decode the bitstream.

Table 4.1 lists some of the HEVC Profiles defined in Annex A of the standard. Main Profile is intended for general-purpose consumer video applications, using 8 bits per sample and 4:2:0 colour subsampling. Various permutations of bit depths, 8, 10 or 12 bits per sample and colour subsampling formats such as 4:2:0, 4:2:2 or 4:4:4 are supported by further Profiles. The Intra-Profiles specify only intra-coded pictures, and the Still Picture

Table 4.1 Selected HEVC profiles

Profile name	Bits per sample	Colour space	Applications
Main	8	4:2:0	General consumer video coding
Main 10	10	4:2:0	Video coding with 10 bits per sample
Main 12	12	4:2:0	Video coding with 12 bits per sample
Main 4:2:2	8	4:2:2	Video coding with higher chroma resolution
Main Intra	8	4:2:0	Intra-pictures only
Main still picture	8	4:2:0	One picture only – still image coding

Profiles specify that a coded sequence can contain only one intra-coded picture, i.e. it is a single coded still image.

A Profile indicates which subset of the HEVC features and algorithms may be present in the coded bitstream. A decoder that is capable of handling higher bit depths such as 12 bits per sample may be capable of decoding Main, Main 10 or Main 12 bitstreams. A decoder that is only capable of handling 8 bits per sample cannot decode higher bit-depth Profiles.

A Level, also defined in Annex A of HEVC, indicates maximum ranges of certain values that a decoder may encounter in a coded bitstream. Each Level defines a maximum number of luma samples per second, a maximum Coded Picture Buffer size, a maximum number of slice segments, a maximum number of tiles per picture and a maximum coded bitrate. Each Level, from 1 to 6.2, effectively limits the data rate and video resolution that a decoder can encounter. For example, Annex A specifies that:

- A decoder capable of decoding up to Level 2 should be able to handle up to 3,686,400 luma samples per second. This is slightly more than the sample rate of CIF (352 × 288) video at 30 frames per second (3,041,280 samples per second).
- A decoder that can handle Level 4.1 should be able to process up to 133,693,440 luma samples per second, which is slightly more than the sample rate of HD (1920 × 1080) video at 60 frames per second.
- A Level 5.1 decoder should be able to process up to 534,773,760 luma samples per second, which is enough to handle 4K (3840 × 2160) video at 60 frames per second.

This means that an HEVC decoder that has the computational capability to handle up to a maximum of Level 4.1 will be able to successfully decode any bitstream marked Level 4.1 or below, i.e. HD or lower resolution, but will not be capable of decoding a Level 5.1 (4K) bitstream.

4.5.4 Reference Picture Sets

HEVC's Reference Picture Sets (RPSs) can be used to signal repeating picture structures. The classic GoP, introduced with the early MPEG standards such as MPEG-1 and MPEG-2 Video, consists of an I-picture together with a number of P- and/or B-pictures. The RPS enables this and other structures to be efficiently communicated to the decoder.

Figure 4.18 shows an example of a simple multi-picture structure and corresponding RPS entries. The picture structure comprises an I-picture (I_0), three B-pictures (B_1, B_2 and B_3)

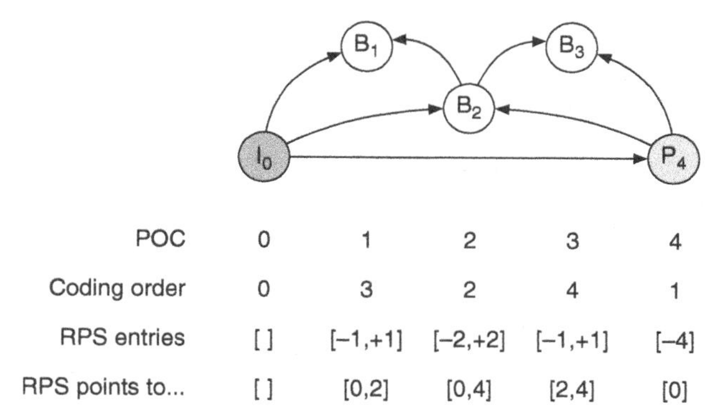

Figure 4.18 Part of a GOP example with Reference Picture Set

and a P-picture (P_4). The subscripts here refer to the picture order count (POC), which is the displayed order of the decoded pictures. The pictures are sent out of order, with I_0 sent first, followed by P_4, then B_2 and then the remaining B-pictures. This is indicated by the coding order shown in Figure 4.18. Sending coded pictures out of order can reduce the delay between the decoder receiving and displaying decoded pictures.

This simple hierarchical GoP structure has potential coding benefits, for example:

- Three out of the five pictures, the B-pictures, are coded using two reference pictures, i.e. bipredicted, which is an efficient form of inter prediction coding.
- Each B-picture is predicted from a relatively close temporal neighbour, which tends to lead to effective prediction.
- The pictures are arranged in a temporal hierarchy, making it easy for a decoder to skip decoding of certain pictures to save computation or to change the displayed frame rate. A decoder can decode every fourth frame by decoding I_0 then P_0, then skipping decoding of the B-pictures, or it can decode every second frame by skipping B_1 and B_3. This is straightforward, as none of I_0, B_2 or P_4 are predicted from B_1 and B_3 in this structure.

In an HEVC bitstream, the inter-dependencies between the coded pictures are captured by RPS entries for each picture, which indicate the choice of the reference picture(s) as an offset relative to the current picture. In this example, the RPS entry for I_0 is empty, as it is intra-coded and uses no other pictures for prediction. P_4 uses a single-reference picture, I_0, and so its RPS entry refers to POC [0], as only the picture with POC $= 0$ is used for prediction. This is captured as an RPS offset of −4, i.e. pointing to the picture whose POC is 4 in the past.

The RPS entry for B_2 is [−2,+2]. Its reference pictures are the previously decoded I_0 and P_4. Finally, the RPS entries for B_1 and B_3 are each [−1,+1], indicating that each is predicted from its immediate neighbours, i.e. I_0, B_2 and B_2, P_4, respectively.

A more complex RPS example is shown in Figure 4.19. Once again, this structure is arranged as a temporal hierarchy. At the lowest level, an I-picture (I_0) is followed in coding order by a P-picture (P_8), predicted from I_0. B-picture B_4 has two reference pictures, the preceding I-picture and the following P-picture. B_2 and B_6 each have two references, the preceding and following pictures in the lower level. Finally, the odd-numbered B-pictures

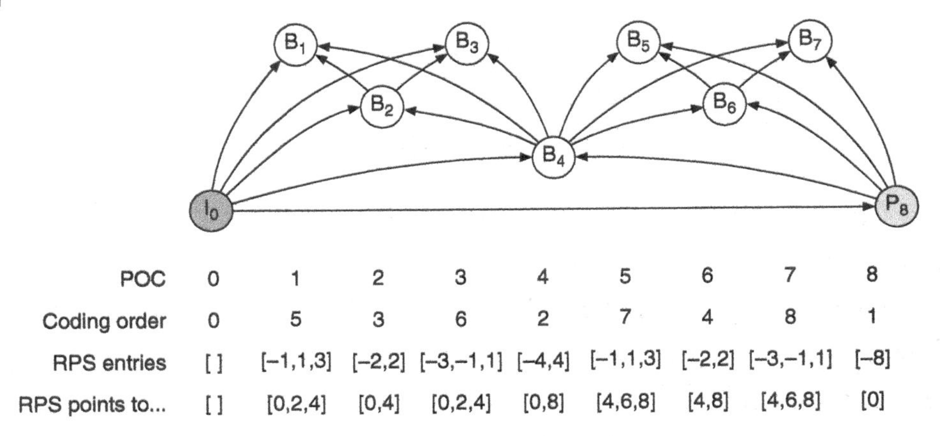

Figure 4.19 Part of a GOP example with Reference Picture Set

each have three references, all at lower levels in the hierarchy. Once again, a decoder can decode all the pictures to produce a full frame rate output sequence, or it can discard one or more layers, starting at the uppermost layer of odd-numbered B-pictures, to produce a lower frame rate output with less decoder processing.

Each of the relationships is captured in the RPS entries. P_8 has an offset of [−8], indicating that picture P_8 is predicted from the picture 8 pictures in the past, which is I_0. B_4 has an RPS containing offsets [−4,+4], which means that it is predicted from I_0 and P_8.

The decoder needs to know the complete list of reference pictures for each current picture. Communicating this for every picture may be expensive in terms of coding overhead. In practical scenarios, RPSs often repeat, which means that the RPS for picture N can be used again for picture N + M at some point in the future. Using HEVC, an encoder can specify a limited number of RPSs in the SPS. Individual coded pictures or slices refer to one of these RPSs. This means that it is not necessary to send a complete specification of reference pictures for every picture. Instead, the encoder only needs to communicate an index identifying a pre-existing RPS.

4.5.5 HEVC Picture Types and Random Access

In the older H.264/AVC standard, random access is supported using instantaneous decoder refresh (IDR) pictures. HEVC extends this concept, introducing random access points and associated leading pictures.

4.5.5.1 Random Access Picture

As the name suggests, decoding of an HEVC stream can start at a random access picture (RAP). An RAP contains only intra-coded slices, so it is sometimes described as an intra-RAP (IRAP). There are three types of RAP: IDR, clean random access (CRA) and broken link access (BLA). The difference between these RAP types relates to the use of leading pictures, as explained below. A picture that is coded only using prediction from previous pictures in display order is a trailing picture.

A picture that is predicted from a RAP, which is sent after the RAP in coding order but precedes the RAP in display order, is a leading picture.

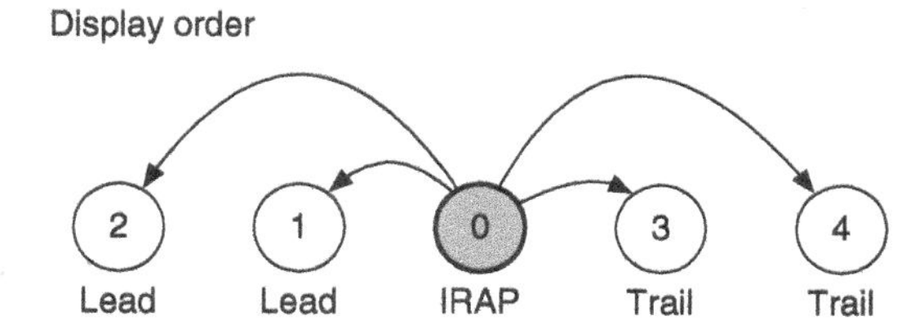

Figure 4.20 RAP, leading and trailing pictures

Figure 4.20 shows the general idea. The first picture coded in the bitstream is the IRAP, picture 0. In this example, two leading pictures, 1 and 2, are coded after IRAP 0 but will be displayed before it. Two trailing pictures, 3 and 4, are coded and displayed after IRAP 0. The pictures can be decoded in the order 0, 1, 2, 3, 4 and will be displayed in the order 2, 1, 0, 3, 4.

There are two types of leading pictures, decodable and skipped, denoted random access decodable leading (RADL) picture and random access skipped leading (RASL) picture, respectively. RADLs are coded without any prediction from reference pictures prior to the IRAP in coding order, whereas RASLs can be coded using prediction from reference pictures that are older than the IRAP.

Table 4.2 summarises the RAP types in HEVC. An IDR picture clears the decoded picture buffer. With an IDR, RADL-leading pictures may be allowed, depending on the NAL unit type, but RASL-leading pictures are not allowed. This effectively means that the IDR starts a new video sequence. A CRA picture does not clear the decoded picture buffer, and RADL and/or RASL-leading pictures are allowed. A CRA is changed to a BLA during a splicing operation (see Section 4.5.5.2), which causes RASLs to be ignored and skipped during decoding.

Table 4.3 summarises the non-RAP types. Here, the associated RAP is the most recent RAP in coding order. A trailing picture is coded and displayed after the associated RAP. A leading picture, RADL or RASL, is coded after the RAP but displayed before it. RADL or RASL pictures can use the RAP, and any pictures that do not refer to pictures coded before the RAP, as a prediction source. RASL pictures can use the RAP, and pictures coded before the RAP, as a prediction source. The use of RASLs can be beneficial to compression

Table 4.2 RAP picture types in HEVC

Picture type	NAL unit type	Clear DPB?	Start decoding here?	RADL allowed?	RASL allowed?
IDR	IDR_W_RADL or IDR_N_LP	Yes	Yes	Optional	No
CRA	CRA_NUT	No	Yes	Yes	Yes
BLA	BLA_W_LP, BLA_W_RADL or BLA_N_LP	No	Yes	Yes	Yes, will be skipped

Table 4.3 Non-RAP picture types in HEVC

Picture type	Code before RAP?	Display before RAP?	Predict based on RAP?	Predict based on references older than RAP?
Trail	No	No	Yes	No
RADL	No	Yes	Yes	No
RASL	No	Yes	Yes	Yes

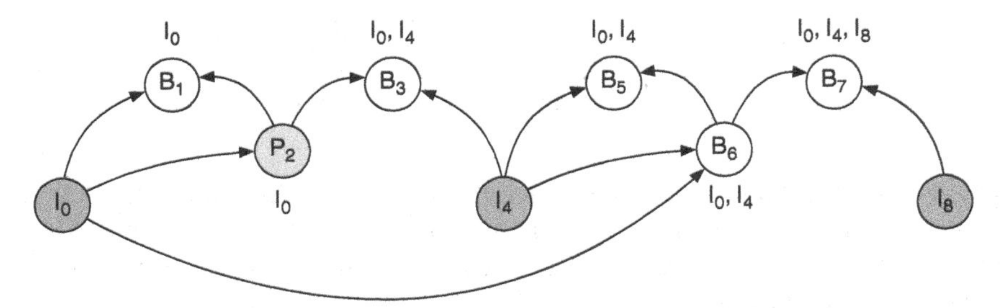

Figure 4.21 Predicted pictures and prediction sources, showing IRAP dependencies

efficiency, since some pictures can have extra prediction references and therefore more opportunities to create an optimal prediction, but can introduce some complexity.

What does it mean to use a picture as a prediction source or to predict based on a picture? Consider the picture structure shown in Figure 4.21. Nine coded pictures are shown, three of which are IRAP pictures (I_0, I_4 and I_8) and six of which are inter-coded pictures. Each inter-coded picture is predicted from one or two other pictures. Which IRAP pictures does each inter-picture depend on? Picture P_2 is predicted from I_0, so it only depends on one IRAP picture (I_0). B_1 is predicted from I_0 and P_2, so it also only depends on just one IRAP picture (I_0). B_6 is predicted from I_0 and I_4. B_3 and B_5 each depend on two IRAP pictures, I_0 and I_4, i.e. I_0 and I_4 are the only IRAP pictures that influence the prediction of B_3 and B_5. Finally, B_7 is predicted from B_6 and I_8. It is therefore dependent on a total of three IRAP pictures, I_0, I_4 and I_8. These dependencies are important when considering random access or error handling. For example, to successfully decode B_7, a decoder needs to receive and decode I_0, I_4, B_6 and B_8.

4.5.5.2 Examples: RAPs, Trailing and Leading Pictures

The behaviour of RAP, trailing and leading pictures is probably easiest to understand with the aid of a few examples.

a) IDR with no leading pictures (see Figure 4.22). An IDR picture is followed by three trailing pictures. The IDR is intra-coded, i.e. without temporal prediction. The trailing pictures, coded after the IDR and also displayed after the IDR, do not use any prediction from pictures before the IDR.
b) IDR with leading (RADL) pictures (see Figure 4.23). Depending on the NAL unit type, the IDR may have associated RADL pictures. Here, two RADL pictures are coded after the IDR but displayed before it. Because they are RADL, they cannot depend on any RAPs prior to the IDR. The IDR is followed by trailing pictures that are coded and

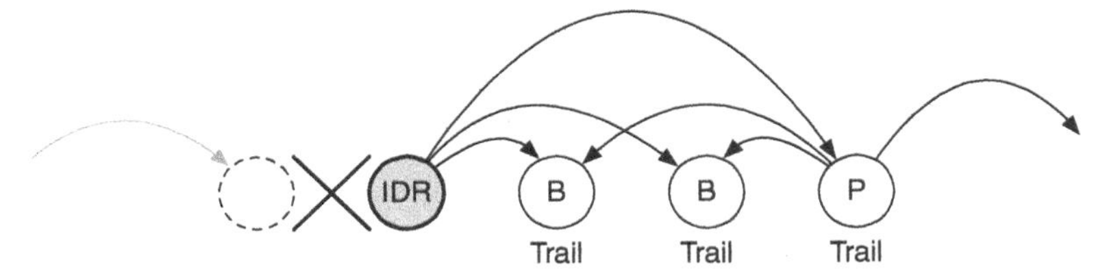

Figure 4.22 IDR with no leading pictures

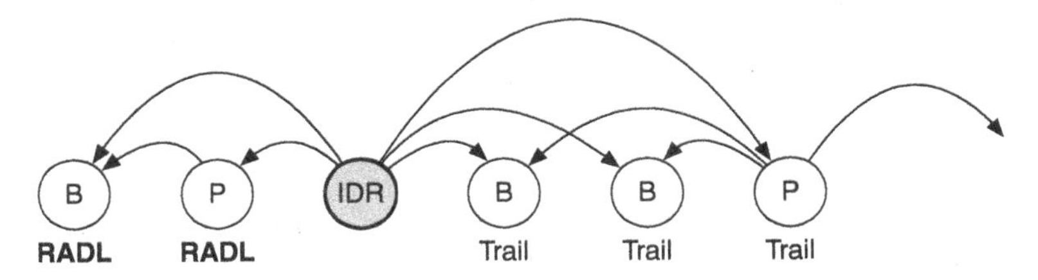

Figure 4.23 IDR with RADL pictures

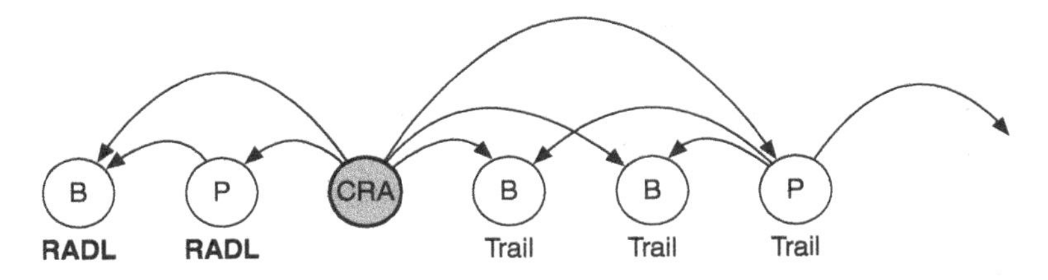

Figure 4.24 Clean random access picture with RADL pictures

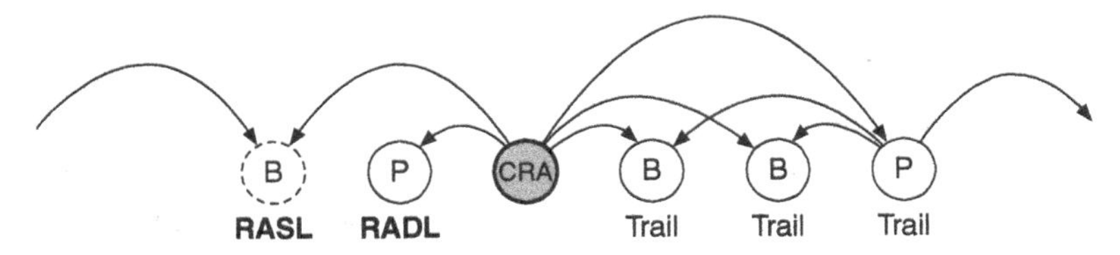

Figure 4.25 Clean random access with RADL and RASL pictures

displayed after the IDR. The IDR effectively forces the start of a new coded video sequence.

c) CRA with RADL pictures (see Figure 4.24). Two RADL pictures are coded after the CRA but displayed before it. This structure appears similar to example (b), but the CRA does not force the start of a new coded video sequence.
d) CRA with one RADL and one RASL picture (see Figure 4.25). This time, one of the leading pictures is a RASL picture, predicted with a dependency on an older reference picture, i.e. a picture preceding the CRA in coding order and display order. When continuously decoding this sequence, the RASL picture can be successfully decoded. When starting decoding at the position of the CRA picture, for example, during random access or error recovery, the RASL picture might not be successfully decoded.

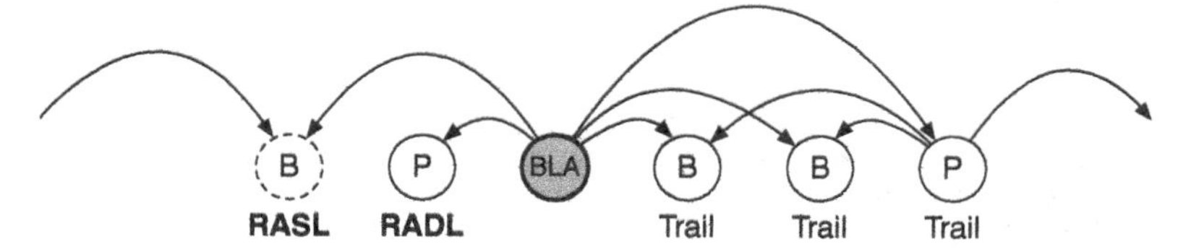

Figure 4.26 Broken link access with RADL and RASL

e) BLA with one RADL and one RASL picture (see Figure 4.26). This looks identical to example (d), except that the CRA is now a BLA. In fact, this is how a BLA is typically used. The CRA is changed into a BLA to enable bitstream splicing as we will see below.

4.5.5.3 Random Access and Error Recovery

RAPs can facilitate random access and error recovery. As with many features of HEVC, it is up to codec designers how they use RAPs. Random access occurs when a user joins a video stream part-way through, for example, by switching to a live TV channel or when a user performs a seek operation to jump to a new playback point. Decoding and playback can start at any RAP. If it is an IDR picture, decoding starts with the IDR and proceeds as normal. If it is a CRA picture, decoding starts with the CRA itself. Any leading pictures are coded after the CRA picture but are intended to be displayed before the CRA picture. RADL pictures may be successfully decoded because they do not depend on any earlier pictures. However, RASL pictures depend on earlier reference pictures that may not exist in the decoded picture buffer, so the decoder will discard RASL pictures, such as the example shown in Figure 4.27.

Similarly, if an error occurs such as one or more lost packets, a decoder may restart decoding at the next RAP. Once again, any RASLs will typically be discarded since they may depend on missing or errored older reference pictures.

4.5.5.4 Bitstream Splicing Operations

A RAP will normally be encoded as an IDR or CRA picture. When two separate coded sequences are joined together or spliced, say, in a video editing session, the video editing software may use an IDR or CRA as a suitable cutting point to start the second sequence. If the splice point starts with a CRA picture, the editing process may change its designation from CRA to BLA. Leading pictures may be left as they are to avoid the need to adjust the coded bitstream. Note that changing from a CRA to a BLA picture merely involves replacing the nal_unit_type of each NAL unit associated with the picture. There is no need to partially or fully decode any pictures.

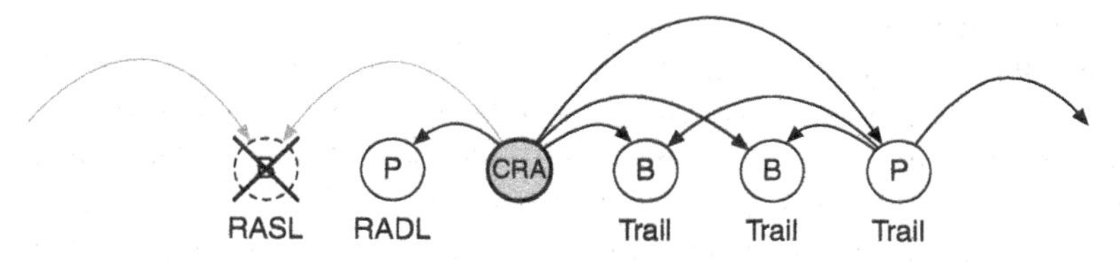

Figure 4.27 Random access starting with a CRA picture

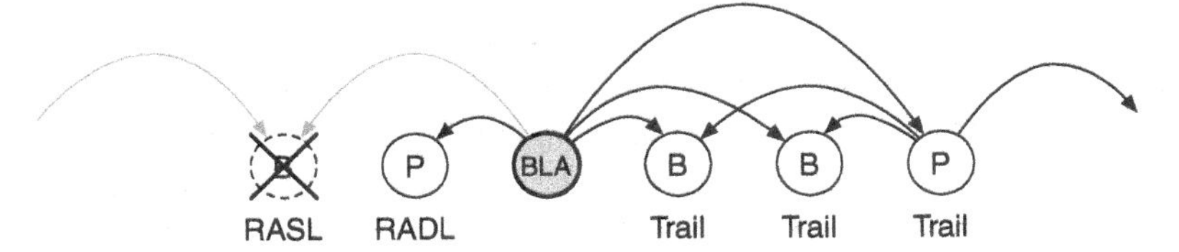

Figure 4.28 CRA changed to BLA after a splicing operation

Any RASLs associated with the BLA may not be successfully decodable, since they depend on reference pictures prior to the splice point that no longer exists. A decoder encountering a BLA in a coded video sequence should therefore ignore any RASLs associated with the BLA, i.e. those coded immediately after the BLA, which means that a certain number of original video frames might not be displayable. RADLs can be successfully decoded.

In the example of Figure 4.28, a decoder encounters a BLA picture within a coded sequence. Recognising that this was a splice point, the decoder discards any associated RASL pictures, one in this case, during decoding.

Example: Splicing two HEVC Sequences

Let us look at how an editing tool might splice two HEVC sequences together to form a new clip. Figure 4.29 shows the final pictures of the first clip, Clip 1, and the chosen splice point for the second clip, Clip 2, in display order. The editing tool chooses a RAP point to start Clip 2, in this case, CRA_8. The same clips are shown in coding order in Figure 4.30. Note that in coding order, Clip 2 starts with the chosen RAP point, CRA_8,

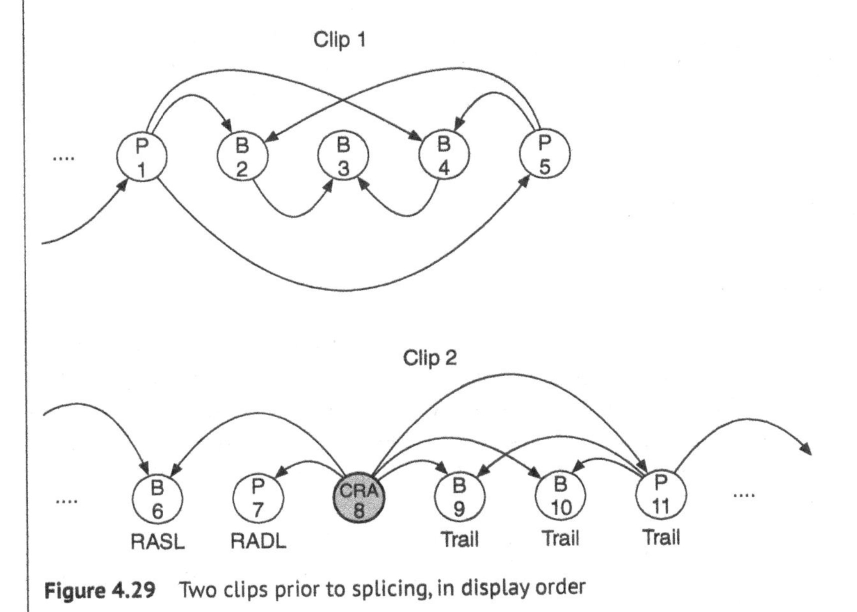

Figure 4.29 Two clips prior to splicing, in display order

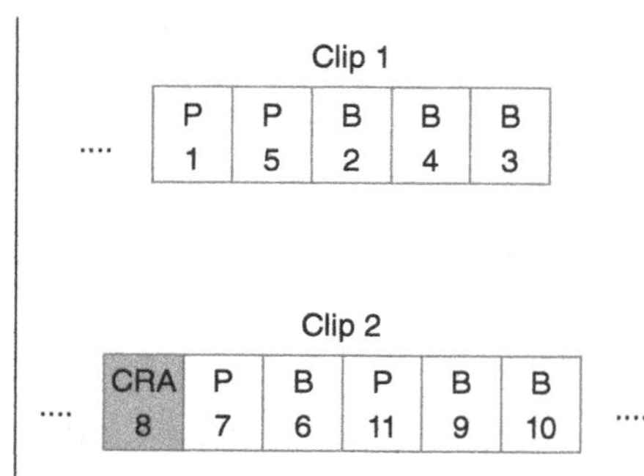

Figure 4.30 Two clips prior to splicing, in coding order

followed by leading pictures P_7 and B_6 then trailing pictures P_{11}, B_9 and B_{10}.

To splice the two coded clips together, the editing tool simply changes CRA_8 to a BLA picture and concatenates the two coded bitstreams (see Figure 4.31). This newly constructed coded video sequence can be transmitted or stored as a single video sequence. Figure 4.32 shows the prediction dependencies and display order of the spliced sequence. A decoder decodes P_1, P_5, B_2, B_4 and B_3. The next coded picture is BLA_8. The decoder recognises this as a splice point. P_7 is a RADL and can be successfully decoded because it depends on BLA_8, which has already been decoded. B_6 is a RASL and was originally predicted using older reference picture(s), which do not exist in the spliced bitstream. The decoder therefore skips B_6 and carries on to decode the trailing pictures.

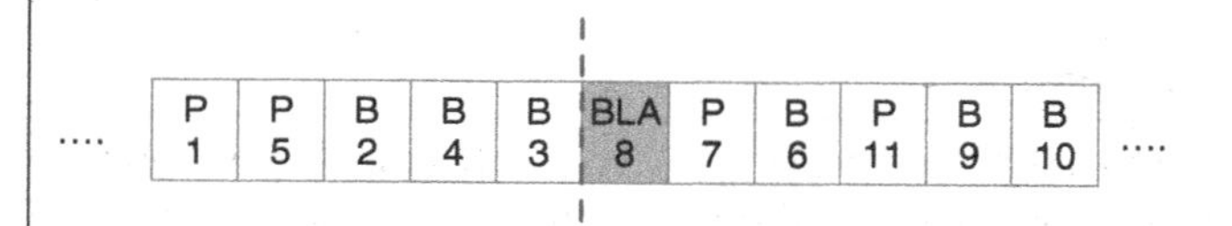

Figure 4.31 After splicing, coding order

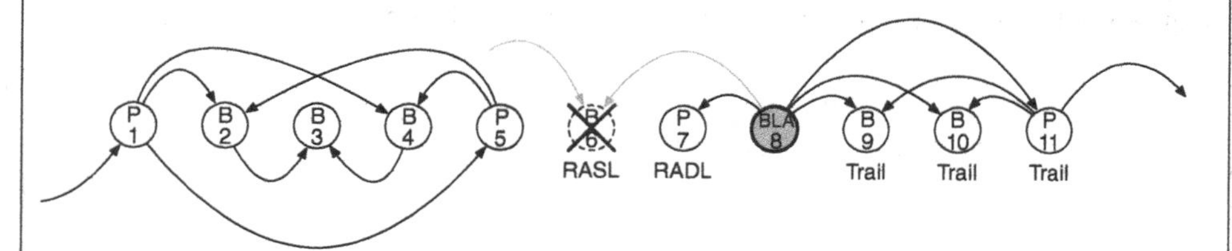

Figure 4.32 After splicing, display order

4.5.6 HEVC-Coded Pictures

Figure 4.33 is a simplified view of the syntax of an HEVC-coded picture. There is no separate picture header in the bitstream. Instead, what is sent is a series of one or more slice segments, each comprising a slice segment header and slice segment data. Each slice segment in a picture shares the same POC Value PicOrderCntVal and refers to the same PPS. When a decoder identifies a new PicOrderCntVal in a slice, this indicates that the slice is part of a new picture.

The slice segment data consists of a series of CTUs. Each CTU is decoded and reconstructed in the appropriate position within the slice segment and, if tiles are in use, within the appropriate tile, as discussed in Sections 4.5.7 and 4.5.8.

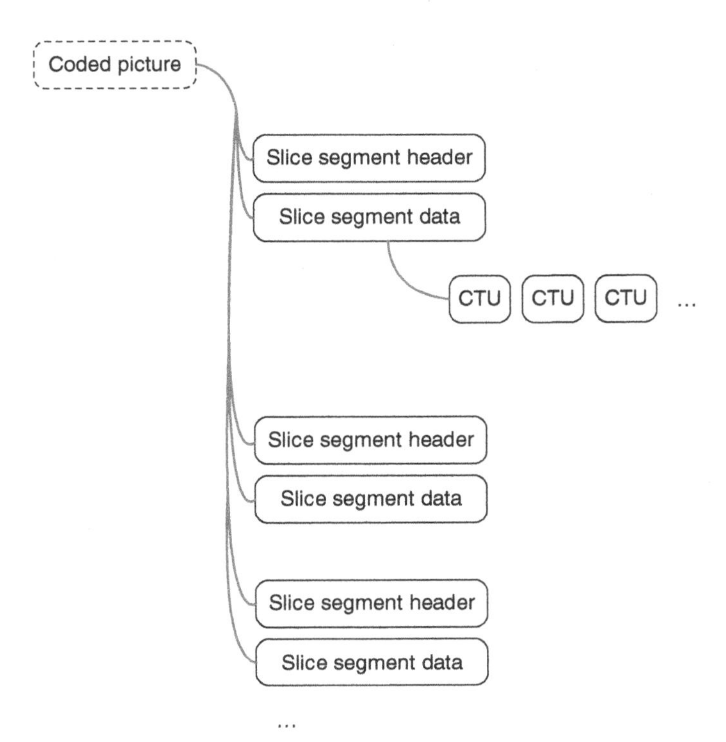

Figure 4.33 HEVC codec picture syntax

4.5.7 Slices and Slice Segments

H.265/HEVC introduces the concept of slice segments. These can be considered as a mechanism for subdividing slices into smaller segments and/or a way to reduce the size of the headers for each of a collection of slice segments. In HEVC, a slice consists of one independent slice segment, with a complete slice header, followed optionally by one or more dependent slice segments, which each inherit some header syntax values from the independent slice segment. Predictions cannot cross slice segment boundaries, as a CB/CU in one slice segment cannot be predicted from a CB/CU in a different slice segment. For intra-prediction, this means that samples outside the current slice segment cannot be used to predict the current CB/CU and the prediction mode cannot be predicted based on blocks outside the current slice segment (see Chapter 5). For inter-prediction, this means that prediction parameters cannot be predicted from outside the current slice segment. For example, motion vectors predicted using Merge mode or Advanced Motion Vector Prediction (AMVP) mode (see Chapter 6) cannot be predicted from outside the slice segment. Loop filtering may or may not cross slice segment boundaries, depending on parameter set choices.

Each picture is coded as a series of slice segments, as illustrated in Figure 4.34. Slice segments can be useful for packetisation, e.g. sending each slice segment in a separate network packet without the overhead of sending a full independent slice segment header in every packet. However, if the decoder encounters an error, it will need to wait for the next

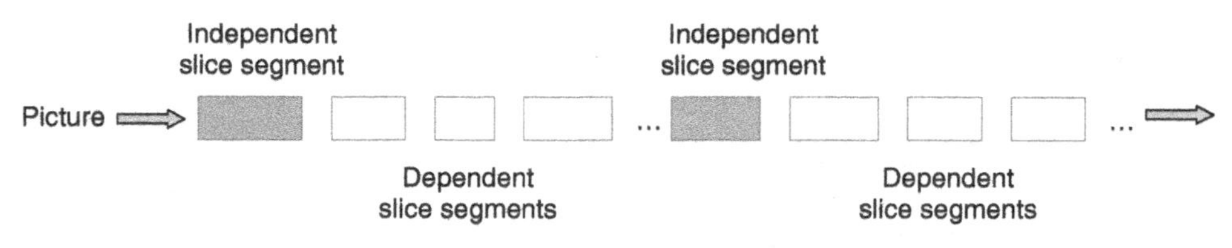

Figure 4.34 HEVC picture and slice segments

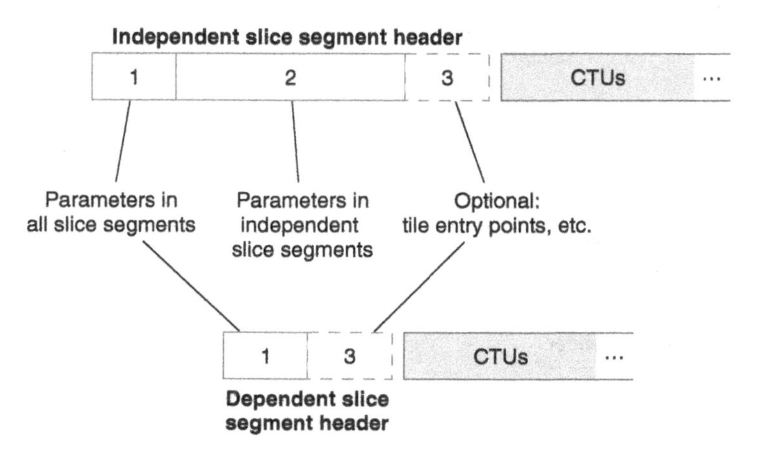

Figure 4.35 Slice segment headers

complete slice header, i.e. the next independent slice segment, to obtain all the parameters necessary to continue decoding.

An independent slice segment starts with a complete slice header, containing all the parameters necessary to decode the slice. Each dependent slice segment starts with a slice segment header that inherits some of its parameters from the slice header. All slice segment headers start with a section (labelled 1 in Figure 4.35) that includes a few basic parameters such as the PPS identifier and the slice segment address, indicating the position of the first CTU in the slice within the picture, in raster scan order. An independent slice segment header also includes a section (labelled 2 in Figure 4.35), which contains all the syntax elements that are necessary for decoding a slice. Tile entry points may optionally be included (labelled 3 in Figure 4.35). These are pointers that enable the decoder to process tiles in parallel. This will be discussed further in Section 4.5.8. After the slice segment header, the slice data consist of a series of CTUs in raster order.

A dependent slice segment header starts with the parameters in header Section 1 but does not include header Section 2. Instead, the parameters in Section 2 are inherited from the most recent independent slice segment header. If both independent and dependent slice segments are used within a coded picture, a series of dependent slices inherit parameters from the preceding independent slice segment header, until a new independent slice is received (see Figure 4.36).

Figure 4.37 shows an example of slice segments in a coded picture. The first independent slice segment is followed by three dependent slice segments. Together, these could be considered equivalent to Slice 1 in Figure 4.7, since they all share common parameters. A second independent slice segment is followed by one dependent slice segment, Slice 2, and so

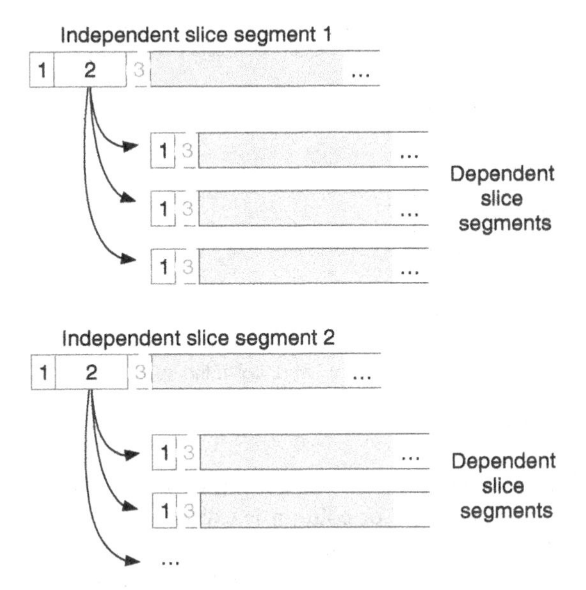

Figure 4.36 Independent and dependent slice segments

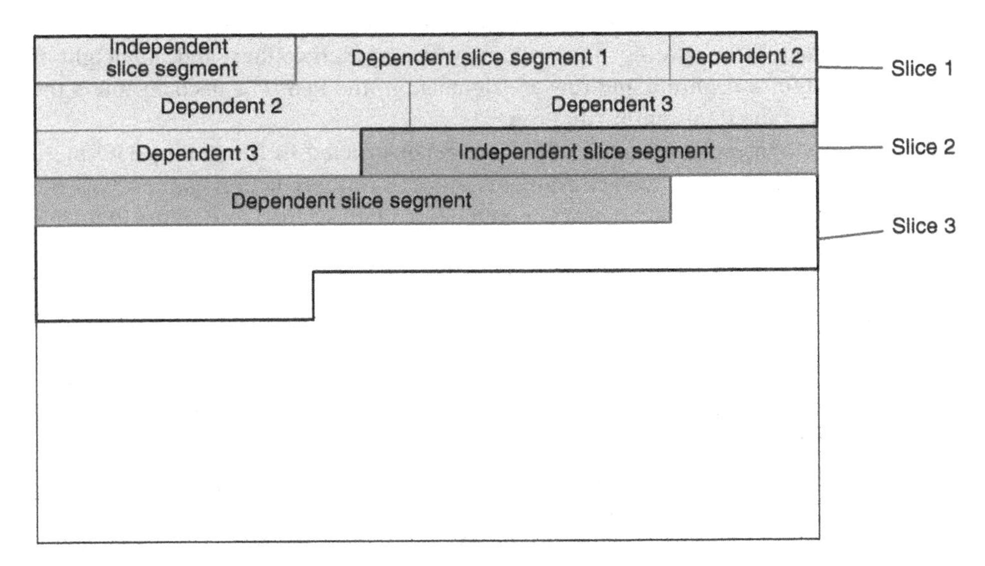

Figure 4.37 Independent and dependent slice segments in a picture

on. The first CTU at the top left of each slice segment is indicated by the slice segment address in the slice header.

4.5.8 Tiles

Section 4.3.3 introduced the concept of tiles. In HEVC, the use of tiles is signalled with the tiles_enabled_flag in the PPS. If the flag is 1, then tiles are present in each picture, which means that there may be more than one tile per picture, and the PPS further defines the

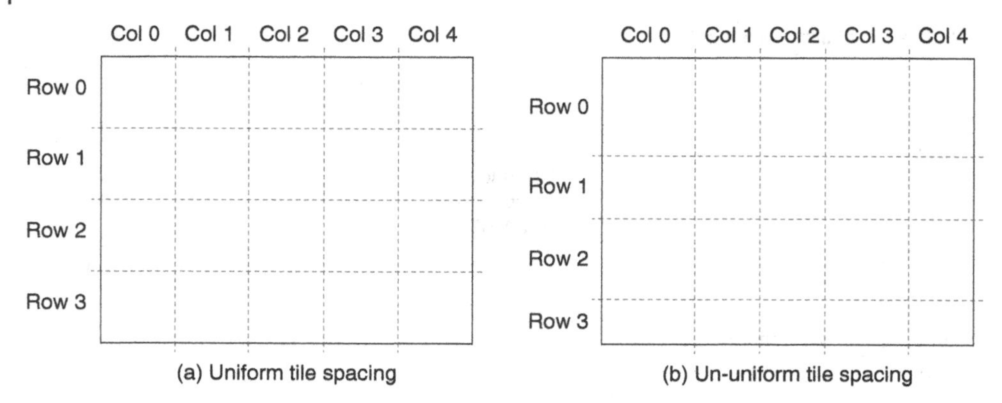

Figure 4.38 HEVC tiles: uniform or non-uniform spacing

number of tile rows and columns and indicates whether rows and columns are uniformly spaced. If not uniformly spaced, the width of each row or column is explicitly signalled. Figure 4.38 shows two examples. Example (a) is a picture divided into 20 tiles in five columns and four rows. The tiles are uniformly spaced as each row and column contains an equal number of CTUs[1]. In this example, the tiles are square in shape but they can also be equal-sized rectangles. Example (b) also contains twenty tiles, but the spacing is not uniform. For example (b), the specific, non-uniform width of each column and the height of each row except the last column and row are signalled in the PPS. The decoder infers the width and height of the final column and row.

When each CTU in a picture is decoded, it is reconstructed in the appropriate place within a Tile. In a similar way to slice segments, inter or intraprediction cannot cross tile boundaries. Loop filtering may or may not cross tile boundaries depending on the PPS. This makes it possible for tiles to be decoded in parallel.

Example: Intraprediction within a Tile

Consider the tile structure shown in Figure 4.39. Each tile consists of twelve CTUs. Let us consider the availability of intra prediction sources for two of the CTUs. CTU1 is in the top row of a tile, so it cannot use any samples from above to predict the pixels within the CTU. CTUs A and B are in a different tile and cannot be used as prediction sources. CTU C is in the same tile and can be used for prediction. CTU2, on the second row of the tile, has more prediction sources available. For example, CTU D to the left and CTU1 above are both within the same tile and can be used as prediction sources. This means that CTU2 might end up being compressed more efficiently than CTU1 since it has more prediction sources available.

This example illustrates a trade-off that affects slices and tiles. Increasing the number of slices or tiles can be beneficial to processes such as packetisation, error

1 Note that each row/column has to contain an integral number of CTUs. If there are 24 CTUs in each picture row and the picture is split into 5 columns of tiles, then there will be four columns with a width of 5 CTUs each and one column with a width of 4 CTUs.

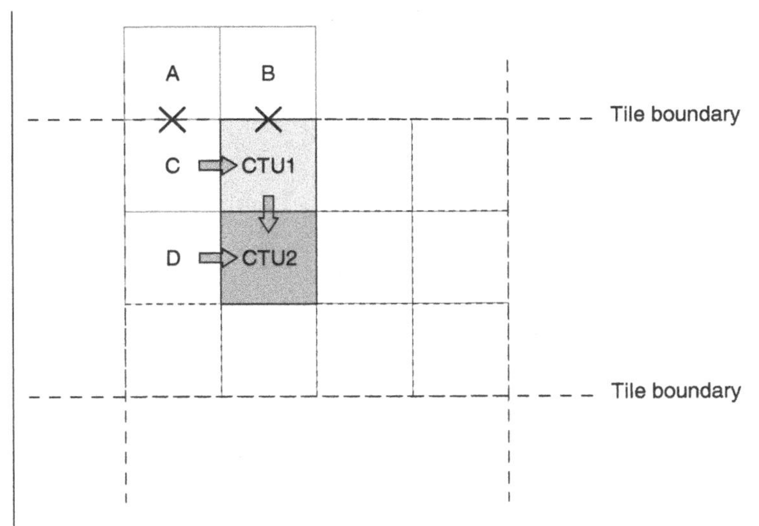

Figure 4.39 Prediction within tiles

handling and/or parallel processing. However, increasing the number of slices or tiles may reduce the opportunities for efficient prediction, which can in turn reduce compression efficiency.

4.5.9 Combining Slices and Tiles in HEVC

In HEVC, slices and tiles may coexist in a coded sequence and within a coded picture. The restriction is that one or both of the following must be true:

a) All CTUs in a slice segment belong to the same tile and/or
b) All CTUs in a tile belong to the same slice segment.

Figure 4.40 shows a valid partitioning in which each tile contains one or more slices. Within the tile, slices are arranged in raster scan order, just as in a complete video frame. In this example, tile 1 (T1) contains two slices, S1 and S2. T2 and T3 each contain one slice, S3 and S4, respectively, and T4 contains three slices, S5, S6 and S7. In this example, all CTUs in a slice and therefore in any slice segments within the slice belong to the same tile. All the CTUs in S1 and S2 belong to T1, all the CTUs in S3 belong to T2 and so on. Condition (a) is true for all the slices.

Figure 4.41 shows another valid example, this time with tiles contained within slices. Slice 1 (S1) contains two tiles, T1 and T2. Note that the CTUs of the slice are coded in raster order within each tile, i.e. top to bottom, left to right. S2 contains tiles T3 and T4. All CTUs in T1 and T2 belong to S1 and all CTUs in T3 and T4 belong to S2. Condition (b) is satisfied for all the tiles.

The partitioning of Figure 4.42 is **not** a valid combination of slices and tiles in an HEVC bitstream. Slice 5 (S5) is the culprit – it crosses two tiles, T3 and T4, so every slice is **not** contained within just one tile. But at the same time, T3 and T4 each contain multiple slices. Not all CTUs in slice 5 belong to the same tile – some are in T3 and

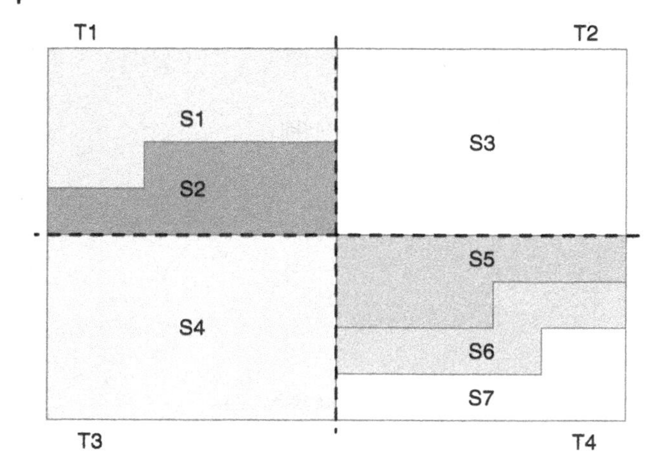

Figure 4.40 HEVC slices contained within tiles

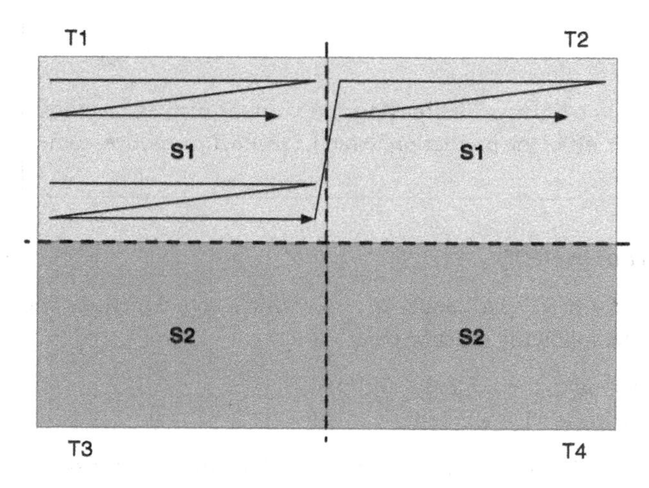

Figure 4.41 HEVC tiles contained within slices

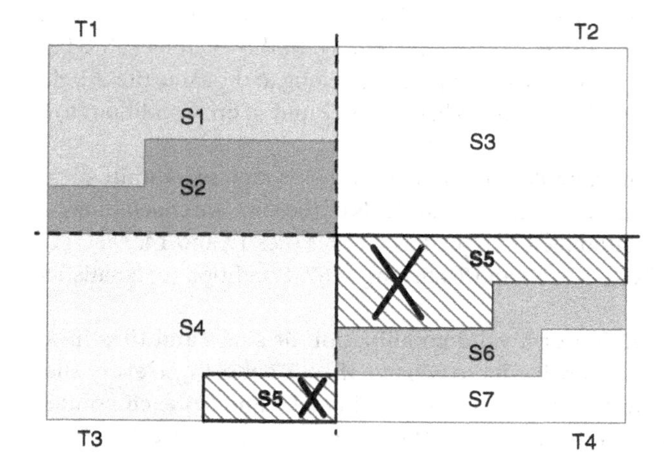

Figure 4.42 NOT a valid HEVC slice/tile structure

some are in T4. Not all CTUs in tiles T3 and T4 belong to the same slice. Neither condition (a) nor condition (b) is satisfied.

4.5.10 HEVC Structures and Parallel Processing

It may be beneficial to parallelise the decoding of a coded picture, for example, using multiple processors or processing cores to speed up the time taken to decode a complete picture. Parallel decoding at the picture level involves allocating certain decoding operations to different processors. Depending on the encoding choices, this may be done in a number of ways when decoding an HEVC sequence.

4.5.10.1 Parallel Decoding Using Tiles

As we saw earlier in Section 4.3.3, tiles within a coded picture may be decoded in parallel. The following features of HEVC tiles make them suitable for parallel processing:

- Tiles are rectangular and may have a regular size. Having a regular and therefore predictable tile size may be helpful since each tile contains the same number of CTUs and should therefore be decoded using approximately the same amount of processing resources.
- Predictions do not cross tile boundaries, so decoding one tile does not necessitate waiting for another tile to be decoded.
- Slices cannot cross tile boundaries, and tiles cannot cross slice boundaries.
- For each tile in a slice segment or some of the tiles, an encoder may optionally insert an entry point marker in the slice segment header. In this scenario, each marker specifies the offset in bytes to the start of a tile in the coded bitstream[2]. This allows each decoding process to start decoding a tile at the appropriate point, without having to wait for the entire picture to be entropy decoded. Figure 4.43 shows the general idea. Once the slice segment header has been decoded, Core 1 can jump ahead and start entropy decoding Tile N at the appropriate byte position in the bitstream. In parallel, Core 2 can start entropy decoding Tile N + 1 at its starting point in the coded bitstream.

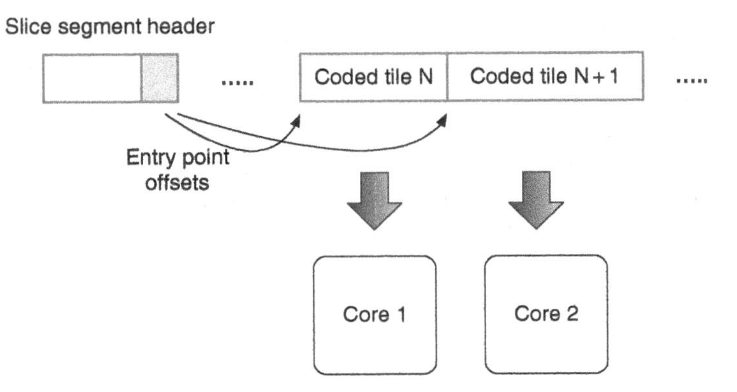

Figure 4.43 Entropy decoding HEVC tiles using Entry Point Offsets

2 The same entry_point_offset mechanism may be used for parallel decoding of CTU rows, so-called wavefront parallel processing.

4.5.10.2 Parallel Decoding Using Slices

Each slice segment in an HEVC bitstream can be decoded somewhat independently, for example, by sending each slice segment to a different processing core for decoding. This may be less straightforward than parallel decoding of tiles, for example for the following reasons:

- Slice segments are not constrained to have a fixed or regular size (CTUs per slice segment), so the processing time per slice may be unpredictable.
- Dependent slice segments cannot be decoded until the associated independent slice segment is decoded since the dependent slice segments require the slice header information of the independent slice segment (see Section 4.5.7).
- Unlike tiles, each slice segment has an associated header, albeit a small one for a dependent slice segment, so using multiple slice segments purely for the purpose of parallel decoding may be inefficient.

4.5.10.3 Wavefront Parallel Processing/Parallelising CTU Rows

An optional HEVC feature sometimes described as WPP enables parallel decoding of rows of CTUs, signalled by entropy_coding_sync_enabled_flag in the PPS. If this flag is 1, then each row of CTUs may be decoded in parallel, with certain conditions. First, the slice segment header contains an entry_point_offset for some or all of the rows of CTUs in the slice segment. Each offset indicates the offset in bytes to the start of the row of CTUs in the coded bitstream. Note that this is the same entry_point_offset mechanism used for tiles. Second, CABAC entropy coding is constrained so that each successive CTU row can be decoded after two CTUs from the previous row have been decoded. This works in the following way:

- When encoding or decoding each row of CTUs, the entropy coding state or context state is stored after the second CTU in the row[3].
- When encoding or decoding each row of CTUs after the first row, the entropy coding state of the first CTU in the current row is initialised with the entropy coding state of the second CTU in the previous row.

This entropy coding synchronisation mechanism enables each row of CTUs to be decoded in parallel. The restriction is that each successive row has to wait for the entropy coding state to be stored for the previous row, so each decoding process can start once two CTUs in the row above have been decoded. In real time, this produces a ripple or wavefront effect as the parallel decoding processes start for each row. Figure 4.44 shows an example. Four rows of CTUs are shown, and the CTUs currently being decoded are highlighted. The first row of CTUs is allocated to Core 1, the second to Core 2 and so on. Core 1 has decoded seven CTUs, Core 2 has decoded five and Core 3 has decoded three. Core 4 has been waiting for Core 3 to complete decoding of the second CTU in its row. This is now complete and Core 4 can start decoding the first CTU in row 4.

---

3 As we will see in Chapter 8, CABAC entropy coding involves a context model or probability model that can depend on values in the upper and upper-right CTUs. This means a decoder cannot start decoding the next row until at least the upper and upper-right CTUs of the previous row have been decoded, so that it can use all the necessary context models.

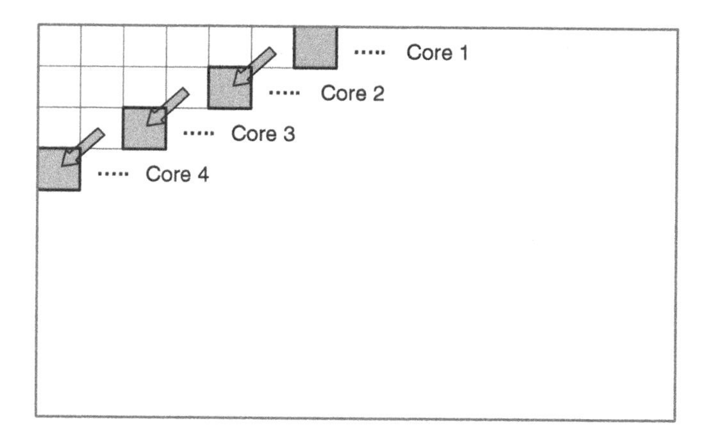

Figure 4.44 Wavefront parallel processing decoding, current CTUs highlighted

4.5.11 Coding Tree Units and Coding Units

A decoder needs to know the size of the basic unit of coding, i.e. the standard unit of coded picture data that it handles as it processes each frame. In previous standards, this was a 16×16 pixel Macroblock. HEVC introduces the CTU with a size of 16×16, 32×32 or 64×64 pixels set in the SPS and fixed for the duration of the sequence (see Section 4.3.1). As we saw earlier, the CTU size places a bound on the computational requirements of a decoder. By setting the CTU size to a maximum of 64×64 pixels, the decoder can be designed with the appropriate amount of memory and processing capacity to handle this amount of data per CTU.

The CTU comprises luma and chroma Coding Tree Blocks (CTBs). For example, a 64×64 pixel CTU corresponds to one CTB of 64×64 luma samples and two CTBs of 32×32 chroma samples, assuming the sampling format is 4:2:0 (Figure 4.45).

Whilst the CTU size is fixed for an entire HEVC video sequence, the CU size can vary from CTU to CTU. The CU corresponds to a square block of pixels and can take any value between minimum and maximum sizes defined in the SPS. Each CU comprises luma and chroma Coding Blocks (CBs). In the example of Figure 4.46, the 32×32 pixel CU (CU2) consists of a 32×32 luma CB and two 16×16 chroma CBs.

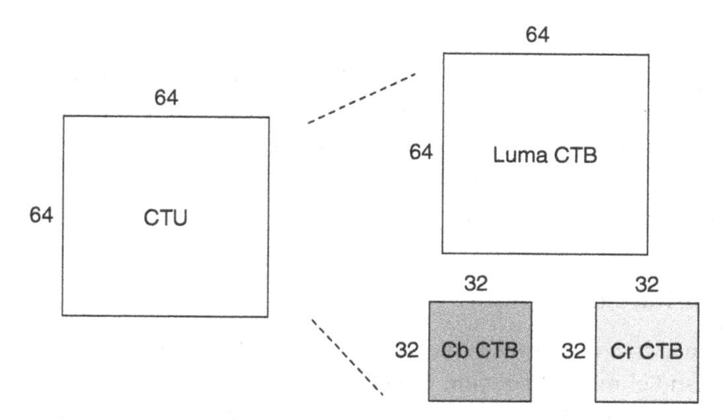

Figure 4.45 HEVC CTU and CTBs, 4:2:0 sampling format

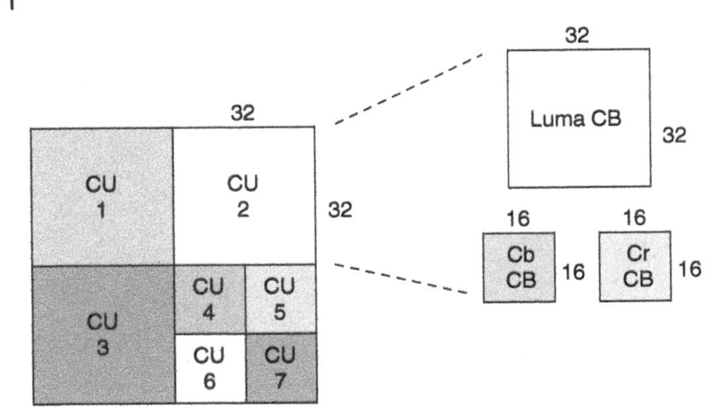

Figure 4.46 HEVC CU and CBs, 4:2:0 sampling format

In an HEVC codec, the CTU is split into CUs using quadtree partitioning. A quadtree splits a block such as a square CTU or CU into four quadrants, with the option to continue to recursively split each quadrant. At each level of the quadtree, a single-bit flag, split_cu_flag, can indicate split or no split for each square. This makes a quadtree an efficient way to adapt the size of coding block or CU while minimising the number of bits required to signal how the block is split.

Consider the example shown earlier in Figure 4.14 and illustrated further in Figure 4.47. A 64×64 CTU is partitioned into CUs using a quadtree. The first level of the quadtree splits the 64×64 into four quadrants, each 32×32 sample. The first three are leaves, forming CUs 1, 2 and 3. The fourth 32×32 quadrant is split again at the next level of the quadtree, into four 16×16 quadrants. These are all leaves, giving a complete partitioning into three 32×32 and four 16×16 CUs. For this particular partitioning, the number of split_cu_flags sent depends on the smallest CU size. For example, if the smallest CU size is 8×8, then a total of 9 split_cu_flags are sent to signal this CTU partitioning, as shown in Table 4.4. Each of the smallest CUs, CU4–CU7, could optionally be split further, so the split_cu_flag must be sent for each. If the smallest CU size is defined to be 16×16, then there is no option to split a 16×16 CU, and so the split_cu_flag is not sent for CUs 4–7 and a total of 5 split_cu_flags are sent for this particular CTU partitioning.

4.5.11.1 Quantisation Groups

A further, optional partitioning of the CTU is into one or more Quantisation Groups. A Quantisation Group is a multiple of CU size and effectively makes it possible for an encoder to group together multiple CUs for the purpose of signalling and calculating QP. Within a Quantisation Group, delta QP, i.e. the change from the previously signalled QP, is sent at most once and so all the CUs share the same predicted QP.

4.5.12 Prediction Units and Transform Units

In HEVC, every CU is partitioned into one or more prediction units (PUs) and one or more TUs. Each PU contains luma and chroma Prediction BLocks (PBs), and each TU contains luma and chroma Transform Blocks (TBs). PU and TU partitionings in HEVC are not required to exactly match each other, as we saw earlier in Figure 4.15.

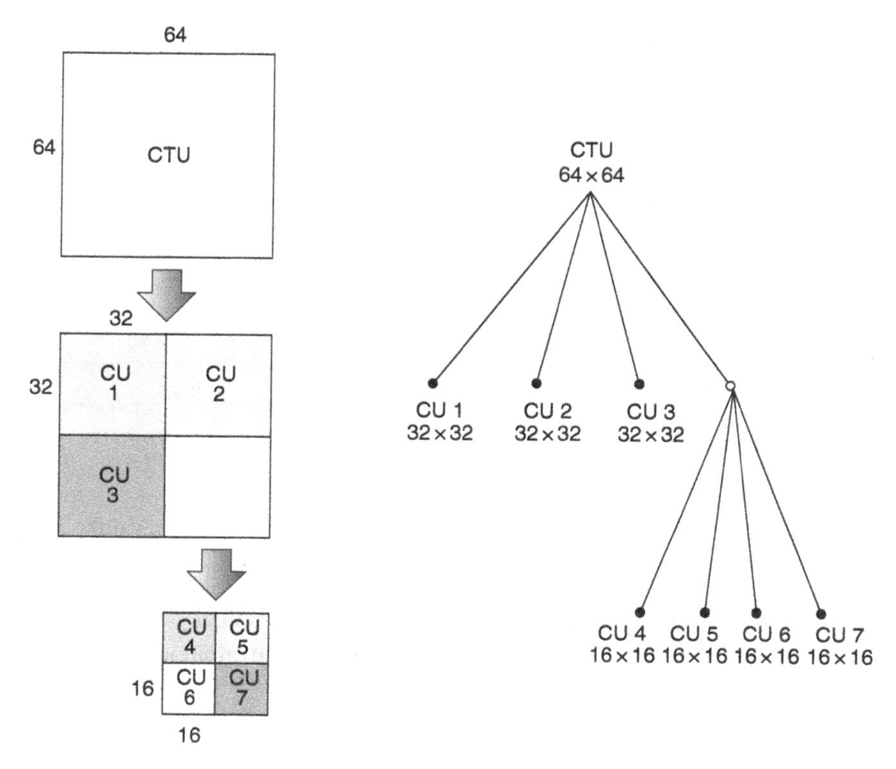

Figure 4.47 CTU/CU partitioning and corresponding quadtree

Table 4.4 Example: split_cu_flag

Level	Split_cu_flag (min CU 8×8)	Split_cu_flag (min CU 16×16)
CTU 64×64	1 (split)	1 (split)
CU1 32×32	0 (no split)	0 (no split)
CU2 32×32	0	0
CU3 32×32	0	0
Final 32×32 region	1 (split)	1 (split)
CU4 16×16	0 (no split)	Not present
CU5 16×16	0 (no split)	Not present
CU6 16×16	0 (no split)	Not present
CU7 16×16	0 (no split)	Not present

4.5.12.1 Prediction Unit Structures

Each HEVC CU is coded using intra or inter prediction. A CU coded in intra-mode can be treated as a single PU, i.e. the PU is the same size as the CU, or split into four square PUs, because the HEVC intra prediction modes operate on square blocks (see Chapter 5). A CU coded in inter-mode can be split into two or four PUs in a number of ways, including square and rectangular PUs, as illustrated in Figure 4.48. For both intra and inter

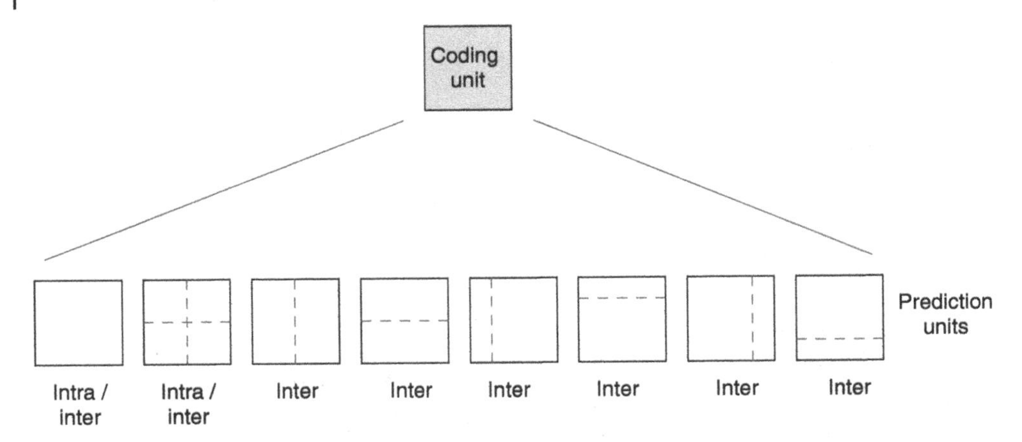

Figure 4.48 HEVC Coding Unit partitioned into prediction units

prediction, a CU is split once at most into PUs. It is not a recursive process, unlike the partitioning into TUs discussed in Section 4.5.12.2.

Each PU is predicted using a set of parameters, an intra prediction mode for an intra-PU, or an inter prediction mode plus further parameters such as motion vectors, reference picture index or indices, and merge index for an inter-PU. We discuss partitioning and prediction of intra-PUs and inter-PUs in Chapters 5 and 6, respectively.

4.5.12.2 Transform Unit, Transform Block and Transform Sub-Block

An HEVC CU is partitioned into one or more TUs using another recursive quadtree structure, known as a transform tree. Starting at the CU, at each level of the transform tree except for the lowest possible level, a flag is sent in the bitstream, split_transform_flag. If split_transform_flag is 0, the tree is not divided further, illustrated as the left-hand branches in Figure 4.49. If split_transform_flag is 1, the tree is split into four transform trees, each with half the original CU horizontal and vertical dimensions, illustrated as the right-hand branches in Figure 4.49. The process continues until either all split_transform_flags are 0 or the tree has reached the lowest level and has been split to the lowest allowable level. The TUs themselves are the 'leaves' of the quadtree. Each TU in a colour video sequence comprises a luma TB and two chroma TBs.

The number of levels allowed in the transform tree depends on the maximum and minimum TU size and the maximum transform tree depth, all of which are specified in the SPS. Possible TU sizes in HEVC range from 32×32 down to 4×4.

Figure 4.50 shows four examples of a 32×32 CU partitioned into TUs, assuming that all TU sizes between 32×32 and 4×4 are enabled:

a) A single 32×32 TU, indicated by split_transform_flag = 0.
b) Four 16×16 TUs.
c) Two 16×16 TUs and eight 8×8 TUs.
d) Two 16×16 TUs, two 8×8 TUs and twenty-four 4×4 TUs.

TBs larger than 4×4 are coded in the HEVC bitstream as a series of transform sub-blocks (TSBs), which are each always 4×4 samples in size. In Chapter 8, we will discuss the scanning and coding of these TSBs.

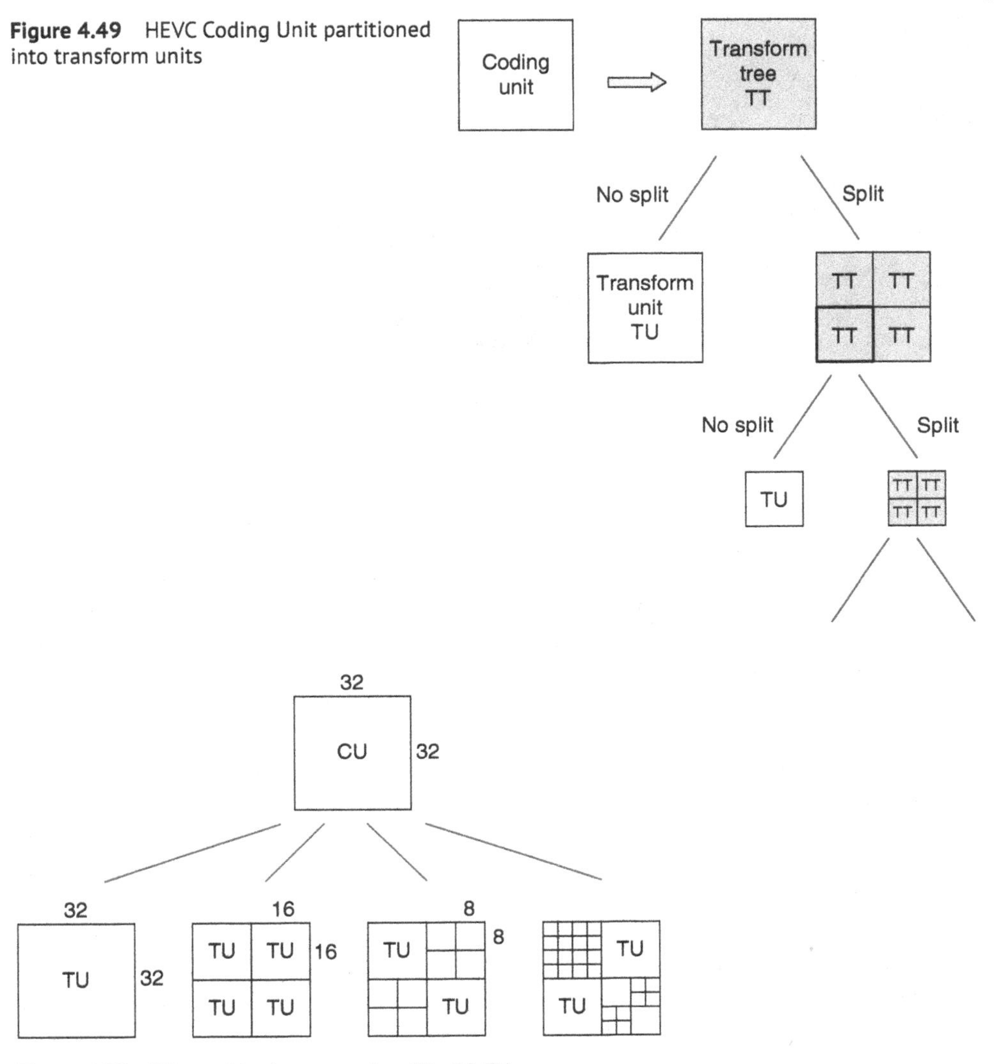

Figure 4.49 HEVC Coding Unit partitioned into transform units

Figure 4.50 TU partitioning examples: 32 × 32 CU

4.5.13 HEVC Structures Example

Consider the structures in an HEVC-encoded bitstream, displayed using the Parabola Explorer bitstream analyser. The partitioning choices in this example are made by the HEVC encoder. The size of each GOP and CTU is fixed throughout the sequence. Partitionings of each CTU are determined by the encoder based on the characteristics of the video frames. For example, the encoder tends to choose more detailed partitionings, i.e. smaller CU, PU and TU blocks, in complex moving areas and less detailed partitionings, i.e. larger blocks, in static or flat areas of the scene.

Figure 4.51 shows a sample frame from the test video sequence *Kristen and Sara*. The original sequence is in 720p format, i.e. 1440×720 pixels per frame. The encoder has

Figure 4.51 Kristen and Sara: sample frame. *Source:* Parabola Research

chosen to use a GoP size of 64, i.e. each GoP consists of 64 coded frames, as illustrated in Figure 4.52. The first frame of each GOP is an I-picture, coded using only intra prediction. Note that the number of bits in each I-picture, shown in the right-hand bar chart of Figure 4.52, is significantly higher than in any inter-coded picture. Of the 63 inter-coded pictures in each GOP, every eighth picture is coded as a P-picture, using inter prediction from one previous I- or P-picture. The remaining pictures are coded as B-pictures using two or more references. If the frame rate is 60 frames per second, this means that an intra-coded picture occurs just less than once per second.

Each picture is partitioned into 64 × 64 CTUs, as shown in Figure 4.53, some of which are further partitioned into 32 × 32, 16 × 16 or 8 × 8 CUs.

Figure 4.54 shows a close-up of a single 64 × 64 CTU. This particular CTU covers a complex moving area around the woman's face and is partitioned into a total of 28 CUs, as shown in Figure 4.55.

The CUs are further partitioned into intra- or inter-PUs. This CTU is part of an inter-predicted picture, which means that the encoder has the choice of intra or inter prediction for each CU. In this CTU, three of the 8 × 8 CUs are predicted in intra-mode. The remaining 25 CUs are predicted in inter-mode, with some of the inter-coded partitions identified in Figure 4.56. Each inter-PU has an associated motion vector, as shown in Figure 4.57.

The TU partitions for this CTU are shown in Figure 4.58. Two of the 8 × 8 CUs and one of the 16 × 16 CUs are subdivided into 4 × 4 TUs. For the remaining 25 CUs, there is no further subdivision and the TU size is the same as the CU size, i.e. 16 × 16 or 8 × 8 samples.

Finally, Figure 4.59 shows the presence or absence of transform coefficients in each coded TU. Some TUs have no coefficients, indicated by no circles – this means that after prediction, transform and quantisation, there are no remaining non-zero transform coefficients in the bitstream. Four of the TUs contain non-zero Y, Cb and Cr coefficients, and the remaining TUs contain non-zero Y coefficients but no coefficients in the Cb or Cr blocks.

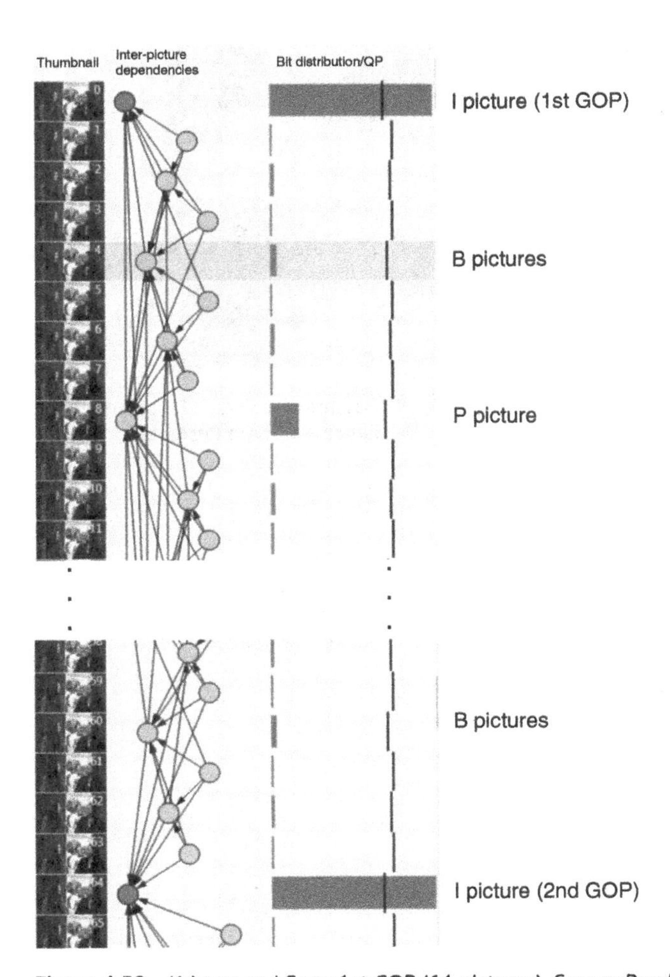

Figure 4.52 Kristen and Sara: 1st GOP (64 pictures). *Source:* Parabola Research

Figure 4.53 Kristen and Sara: picture partitioned into CTUs and CUs. *Source:* Parabola Research

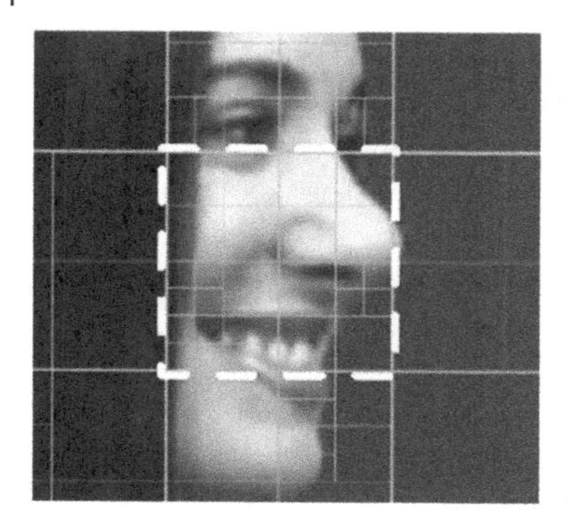

Figure 4.54 Kristen and Sara: close-up with CTU highlighted. *Source:* Parabola Research

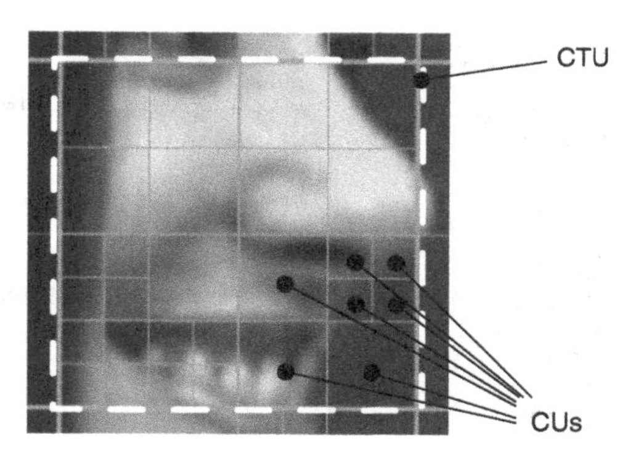

Figure 4.55 Kristen and Sara: CTU partitioned into 28 CUs. *Source:* Parabola Research

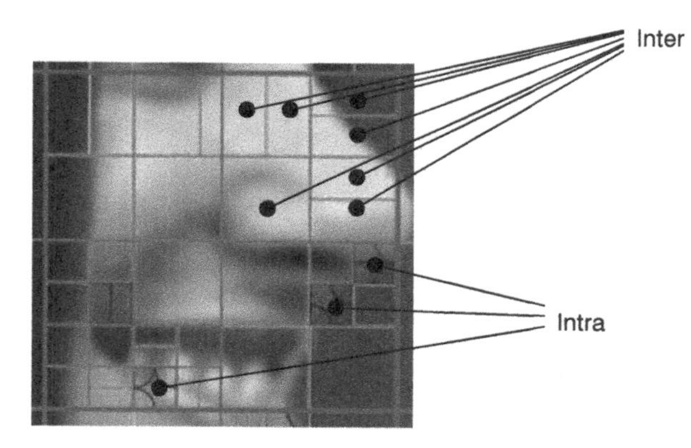

Figure 4.56 Kristen and Sara: CTU showing intra-PUs and inter-PUs. *Source:* Parabola Research

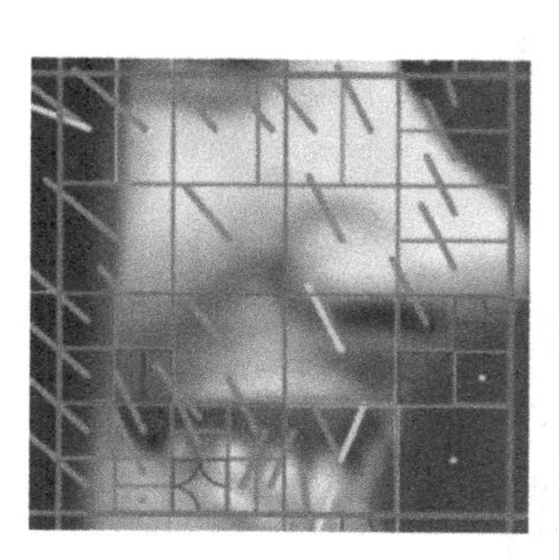

Figure 4.57 Kristen and Sara: Motion vectors. *Source:* Parabola Research

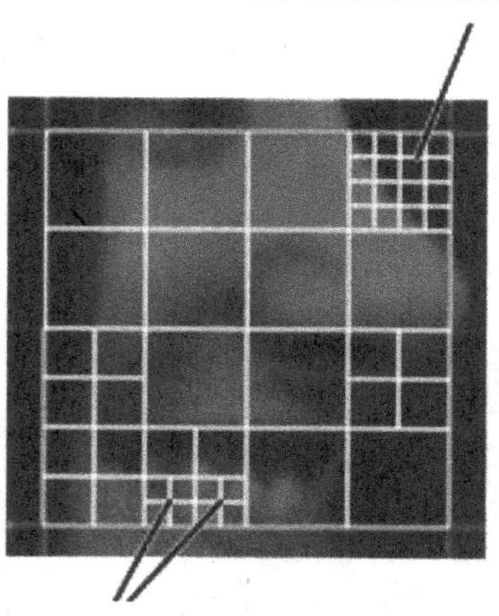

Figure 4.58 Kristen and Sara: TU partitions. *Source:* Parabola Research

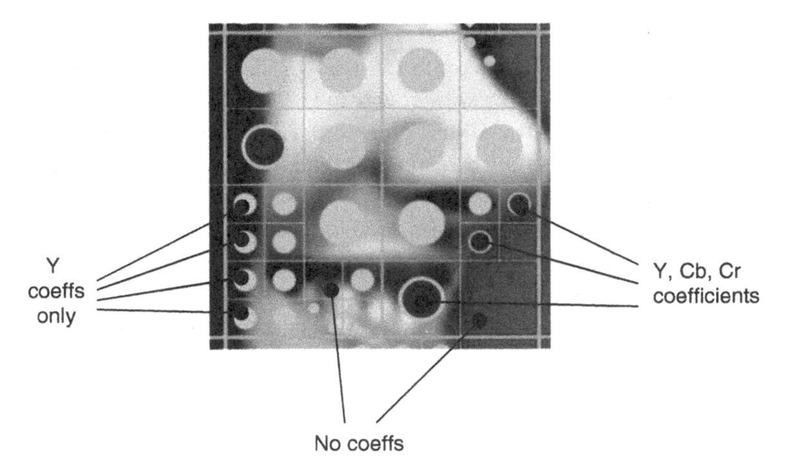

Figure 4.59 Kristen and Sara: TUs showing presence of non-zero coefficients, indicated by circles. *Source:* Parabola Research

4.6 Structures in Versatile Video Coding/H.266

Whilst it is not backwards compatible with HEVC, the newer VVC standard builds on many of the concepts of HEVC. Some of the structures in VVC are natural extensions of those found in HEVC, whilst others are actually simplifications of the HEVC approach.

4.6.1 Sequence Down to Picture

VVC specifies VPSs, SPSs and PPSs in a similar way to HEVC. In addition, the adaptation parameter set (APS) signals picture or slice parameters such as loop filter parameters that may change frequently from one slice or picture to the next, but which would be inefficient to signal with PPSs.

Repeating picture structures such as GOPs are signalled in VVC in a somewhat simpler way to HEVC, but still with the capability to signal the reference picture arrangement for a GOP flexibly and efficiently. Random access and bitstream splicing are supported in a similar way to HEVC, with some simplifications. The concept of Gradual Decoding Refresh is introduced in VVC, in which a decoder can begin decoding at an inter-coded frame rather than an intraframe, such that a correct decoded picture is gradually built up at the decoder over time.

4.6.2 Picture Down to Coding Unit

VVC specifies a picture header that applies to all slices in a coded picture, a concept that existed in older standards such as H.263 but was missing from H.264 and HEVC. Syntax elements in the picture header do not need to be signalled again at the slice header level.

The slice concept is redefined. Slices in VVC comprise either rectangular slices, which are each a rectangular region that is an integral number of tiles or a rectangular subset of a tile, or raster slices each comprising complete tiles in raster order. A new structure, the subpicture, defines a rectangular region of the picture that may contain one or more slices.

The Basic Unit, the CTU, can be up to 128 × 128 samples in size, compared with the maximum 64 × 64 sample CTU in HEVC, and the maximum transform size is 64 × 64. A CTU is partitioned into CUs, first using a quadtree in a similar way to HEVC, then by one or more binary splits and/or ternary splits, as illustrated in Figure 4.60. Each of the leaf nodes of this partitioning process is a CU. CUs can thus be square or rectangular, with a minimum size of 4 × 4 samples.

In most cases, the CU block size is also the block size used for prediction and transformation, so there is no separate PU or TU partitioning. VVC transforms can also be square or rectangular, matching the size and shape of each CU.

In a CTU in an I-picture, luma and chroma may optionally be partitioned separately. The luma and chroma components of the CTU do not have to use the same partitioning into CUs.

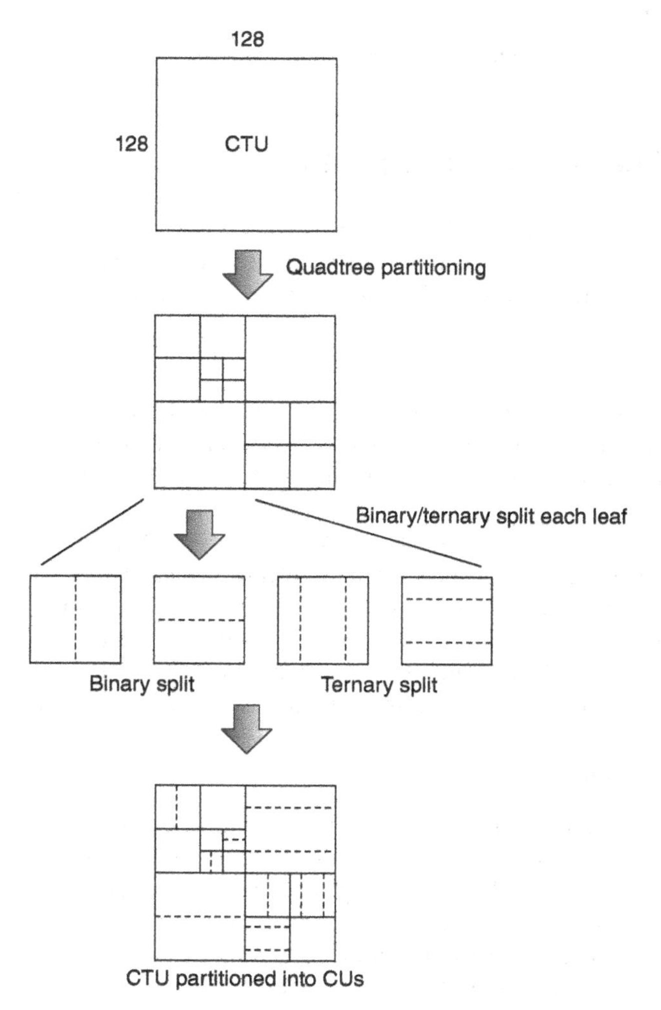

Figure 4.60 VVC Coding Tree Unit partitioning into Coding Units. *Source:* Elecard Streameye

Example:

Figure 4.61 shows part of an HD video frame that was encoded using VVC, with one 128 × 128 CTU highlighted. The zoomed-in area shows how the CTU has been partitioned, first by quadtree splitting into square blocks such as the 16 × 16 region that is further highlighted, then by splitting using ternary and binary splits to form the final partitioning into CUs. The highlighted 16 × 16 region contains a mixture of CU sizes, including 16 × 4, 8 × 4, 8 × 8 and 4 × 4.

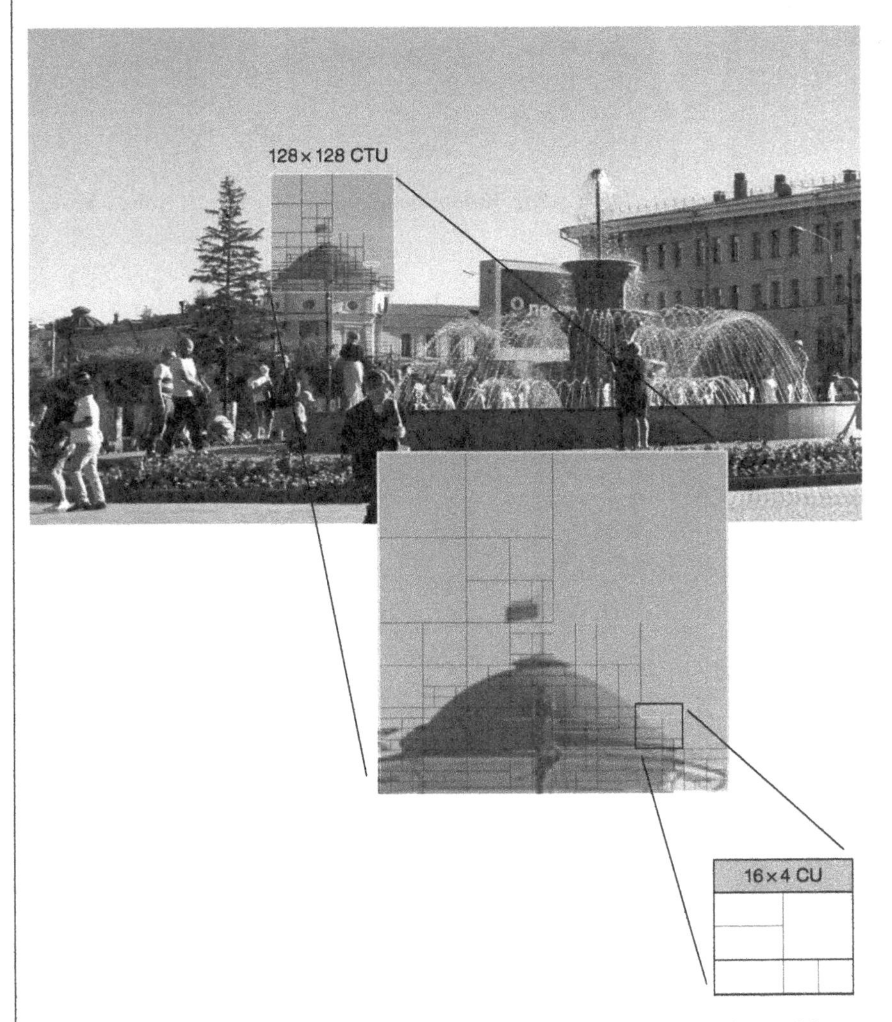

Figure 4.61 Screenshot of VVC partitioning of a 128 × 128 CTU. *Source:* Elecard Streameye

4.7 Conclusion

We have seen in this chapter how a video codec systematically breaks a video sequence down into smaller and smaller data structures, so that it can process image samples in structures that make sense for compression, such as CUs, PUs and TUs. Higher-level structures at the picture level such as slices and tiles help encoders and decoders group together coded data for efficiency in processing or transmission. Higher-level structures still, such as GOPs and parameter sets, organise the coded video pictures into manageable groupings that are useful for practical applications such as video streaming. In the following chapters, we will look at the key processes that are applied to these structures in order to compress video.

Reference

1 ITU-T (2019). *High Efficiency Video Coding*. Recommendation ITU-T H.265, Annex A, Profiles and Levels.

5

Intra Prediction

5.1 Introduction

Intra prediction is the process of creating a prediction block based on pixels in the same frame as the current block. Consider the block of pixels highlighted in Figure 5.1. We want to use pixels outside the block to create a prediction for the pixels inside the block. In this example, we can use some or all of the pixels marked *X*, above and to the left, to create a prediction of the current block. We might not be able to use the pixels marked *Y*, below and to the right. Typically, video encoders and decoders process blocks in a raster order, from left to right and top to bottom, as shown in Figure 5.2. This means that the pixels marked *Y* in Figure 5.1 might not be available when we are processing the current block, but some or all of the pixels marked *X* may be available.

Intra prediction involves predicting a current block from nearby, previously coded pixels in other blocks, such as pixels above and/or to the left of the current block. Some or all of these pixels may be available in previously coded blocks. Figure 5.3 shows three examples of intra prediction types, which are as follows:

- **Flat or DC**: All samples have the same prediction value, for example, the average value of adjacent pixels. The prediction block is a flat region with all samples the same.
- **Planar**: Samples are predicted by fitting a plane function to adjacent pixels. The prediction block is a gradient, with a smooth progression in value depending on the plane function.
- **Directional or Angular**: Samples are predicted by extrapolating adjacent pixels in a particular direction; in this example, extrapolating diagonally down and to the right.

Coding Video: A Practical Guide to HEVC and Beyond, First Edition. Iain E. Richardson.
© 2024 John Wiley & Sons Ltd. Published 2024 by John Wiley & Sons Ltd.
Companion website: www.wiley.com/go/richardson/codingvideo1

Figure 5.1 Pixels surrounding block in the same frame

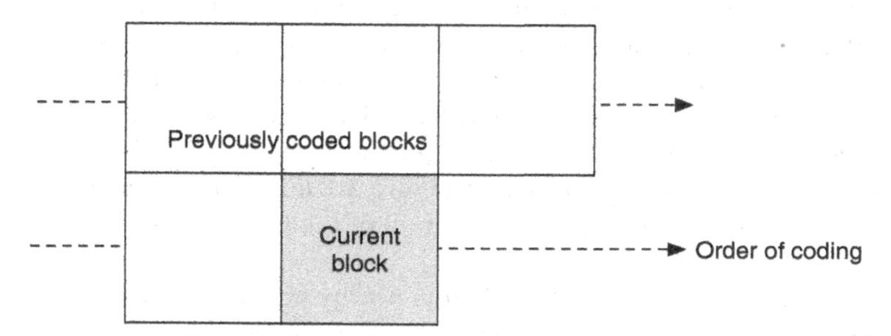

Figure 5.2 Coding blocks in a raster order

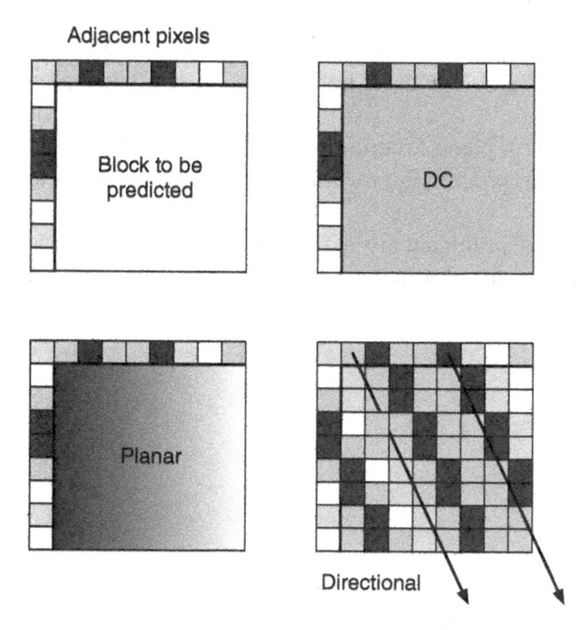

Figure 5.3 Intra prediction types

Example

Figure 5.4 shows a frame split up into square blocks of pixels. Three blocks are highlighted in Figure 5.5. One block contains mostly flat texture, and it might be possible to predict this block efficiently using DC or plane prediction. One block contains a strong angular detail – the edge of a white-painted wall – and might be best predicted with a directional or angular prediction. The third highlighted block contains complex texture. It might be difficult to create an effective intra prediction for this block. In order to intra-code this frame, a video encoder will choose an intra prediction for each block, encode the choice of prediction and the residual and send these in the encoded bitstream. A video decoder will decode the prediction choice and the residual, create the same intra prediction and reconstruct the block.

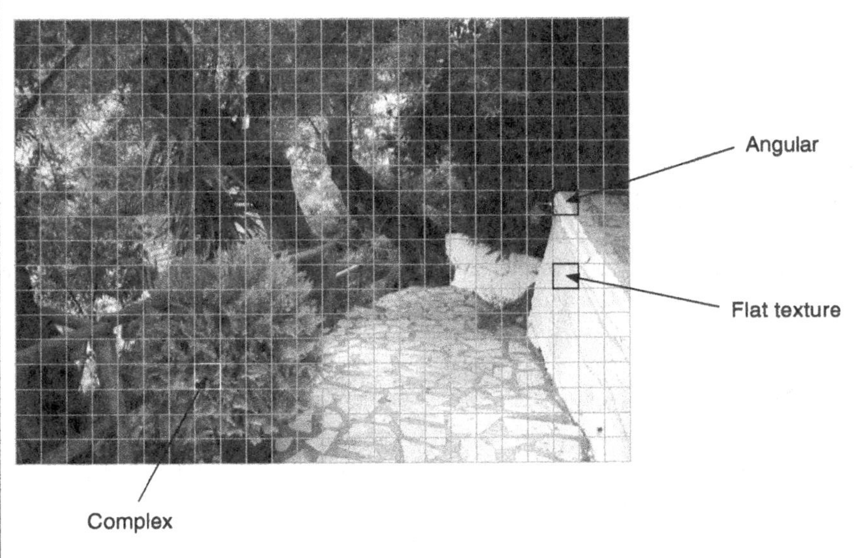

Figure 5.4 Video frame showing three example blocks

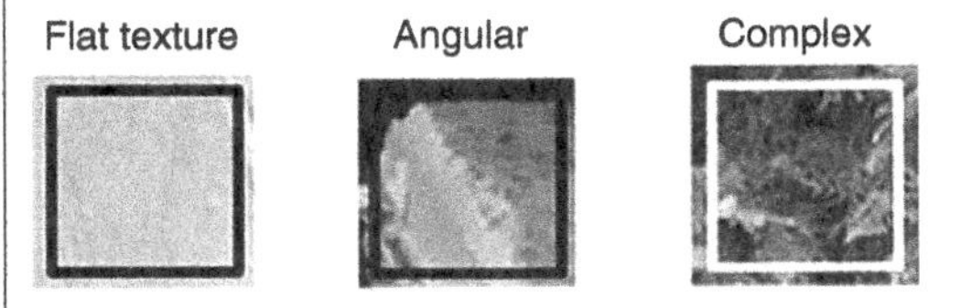

Figure 5.5 Three blocks from video frame

In this chapter, we will consider:

- How does intra prediction work in a video encoder and decoder?
- How should an encoder choose between different intra prediction modes?
- What are the advantages and disadvantages of different intra prediction block sizes?
- How can an encoder signal or communicate intra prediction modes efficiently?
- How is intra prediction implemented in standards-based codecs, including HEVC and VVC?

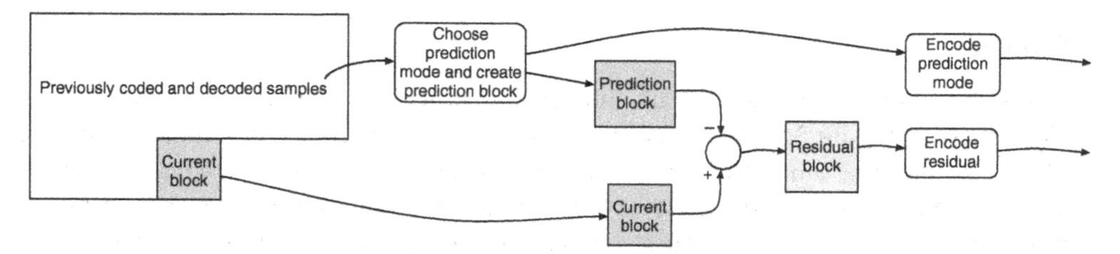

Figure 5.6 The intra prediction process at the encoder

5.2 The Intra Prediction Process

The process of intra prediction in a video encoder is illustrated in Figure 5.6. The encoder selects a prediction mode and creates a prediction block based on previously coded samples, such as the samples above and to the left of the current block. The prediction block is subtracted from the current block to create a block of residual samples. The residual block is encoded and sent in the compressed bitstream along with the encoded prediction mode.

At the decoder, the residual block is decoded. The decoder uses the prediction mode to create an identical prediction block to the one that the encoder created from the previously decoded samples in the same video frame. The decoder adds the prediction block to the residual block and reconstructs the decoded current block.

5.3 Intra Prediction Modes

For an N×N sample block in a video frame, we want to form a prediction from neighbouring, previously coded samples. In general, some or all of the samples adjacent to the current block may be available as prediction sources. These include:

- The sample at the top-left corner of the block,
- N samples immediately above the block,
- N samples above and to the right of the block,
- N samples immediately to the left of the block, and
- N samples below and to the left of the block.

If these samples have already been coded by the time the encoder processes the current block, then they are available for prediction. Figure 5.7 shows a current N×N block for which all the possible samples listed above are available for prediction. Because of the position of this block in the coding sequence, all the upper, top-left and left samples are part of blocks that have already been coded and are therefore available at both the encoder and the decoder. For the block shown in Figure 5.8, the only samples that have already been coded are at the top left, immediately above and immediately to the left of the block. The current block in Figure 5.9 is next to a slice boundary, so the pixels above are not available for prediction. In this case, only the pixels immediately to the left are available.

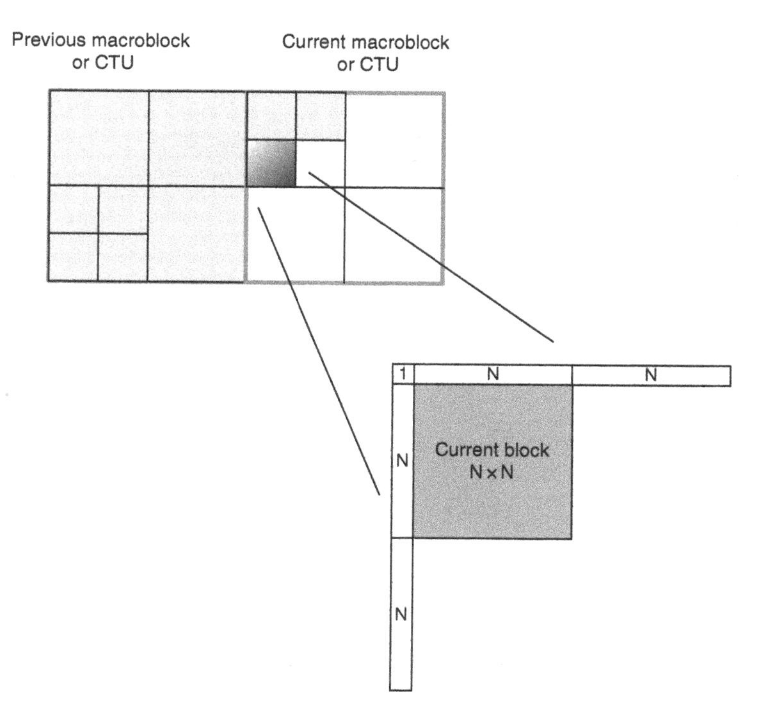

Figure 5.7 Current block (1) – all upper and left samples available for prediction

Figure 5.8 Current block (2) – only immediate above and left samples available for prediction

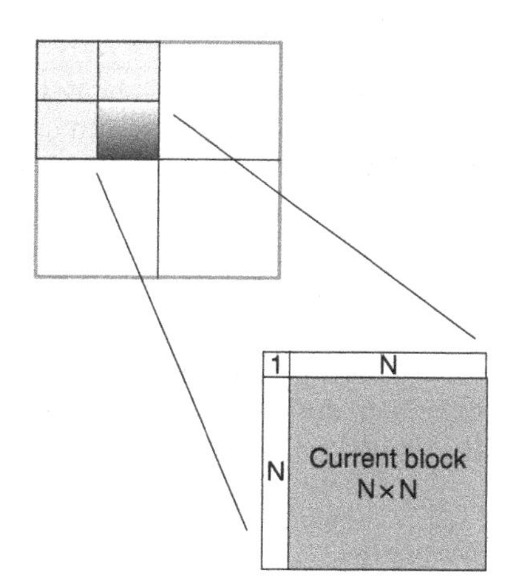

Figure 5.9 Current block (3) – only left-hand samples available for prediction

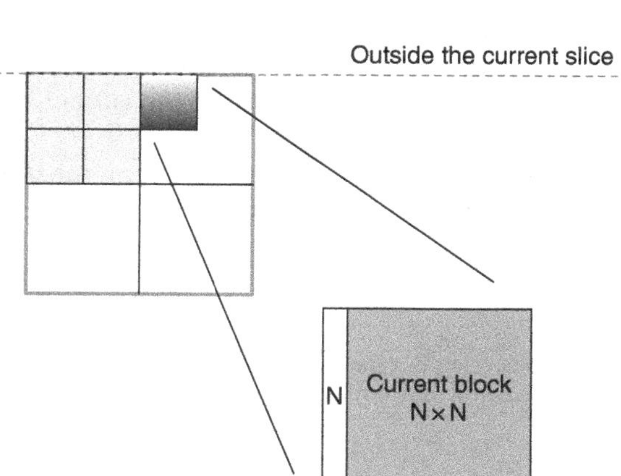

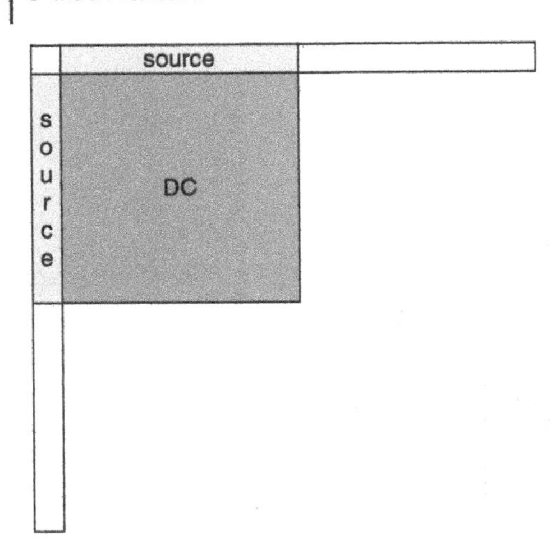

Figure 5.10 Intra prediction: DC

The encoder chooses a prediction for the current block based on a set of intra prediction modes defined in the coding standard and on the available samples above and to the left of the block[1].

5.3.1 DC Prediction

The DC prediction mode creates a prediction by calculating an average, i.e. a mean, of the samples immediately above and immediately to the left of the current block – the samples marked 'source' in Figure 5.10. This average value is copied into all samples of the current block. This mode works well when the current area of the frame is relatively flat and there is not much variation in the texture of the current block.

If the top samples are not available, for example, if this is a block in the top row of the frame, then the prediction may be created by averaging only the left-hand samples. Similarly, if the left samples are not available, the prediction may be created by averaging only the upper samples. If neither left nor upper samples are available, for example, for the first, top-left block of an intra-coded picture, then the prediction may be set to a constant value.

Example:

Figure 5.11 shows a 16 × 16 macroblock on the left and a 4 × 4 block on the right, within the macroblock. We want to create an intra prediction for the current block. All of the samples above and to the left of the current block have already been coded and are available for prediction.

Figure 5.12 shows the DC prediction for this block. The four upper and four left samples are averaged, and the result is applied to every sample in the block. In this

1 Depending on the standard, the encoder may have the option of copying or substituting available samples when certain samples are unavailable.

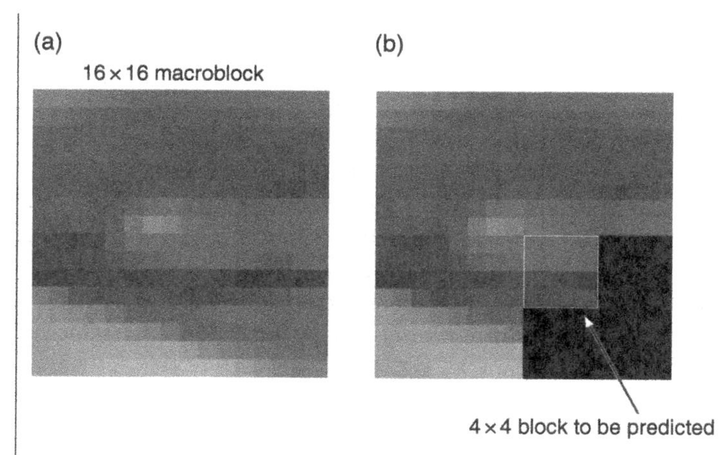

Figure 5.11 Macroblock (a) and current block (b)

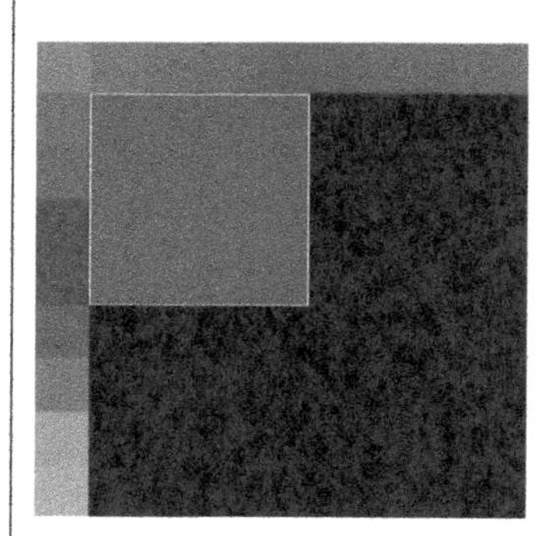

Figure 5.12 DC prediction

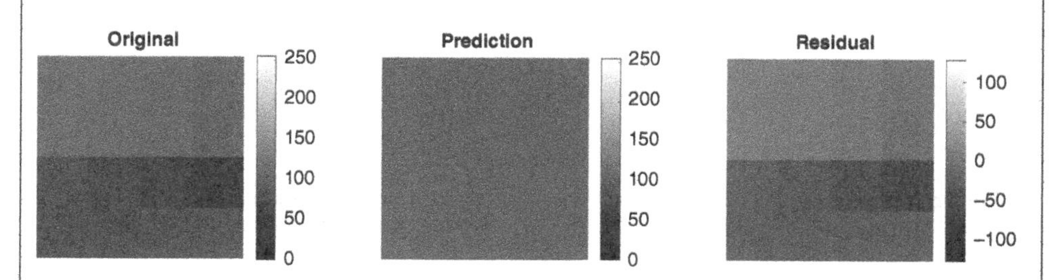

Figure 5.13 Original block, DC prediction and residual

example, the DC prediction is not a particularly good choice, as there is quite a lot of variation in brightness within the current block. Comparing the original block, the DC prediction and the residual, i.e. the difference between the original and prediction, we see that there is still a lot of information in the residual, as shown in Figure 5.13.

5.3.2 Planar Prediction

Planar prediction calculates a gradient function based on certain adjacent samples, for example, the four source samples illustrated in Figure 5.14. Figure 5.15 shows a 16×16 block on the left that is predicted using planar prediction. A plane function is fitted to the samples immediately above and to the left of the block, creating the prediction shown on the right. In this case, the best fit is a gradient from light grey on the left to darker grey on the right. The original block is relatively smooth, so this is a good prediction.

In the second example of Figure 5.16, the original block is more detailed. The best fit is a gradient from dark at the top left to light at the lower right. Plane prediction is less effective for detailed blocks such as in this second example.

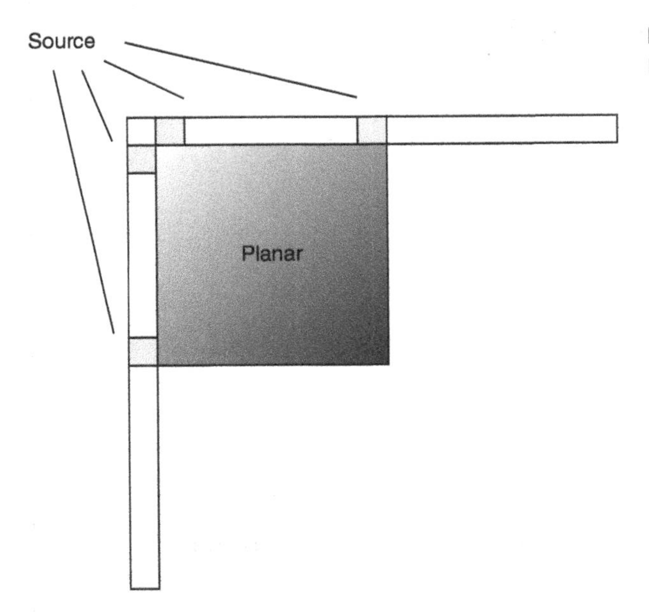

Figure 5.14 Intra prediction: planar

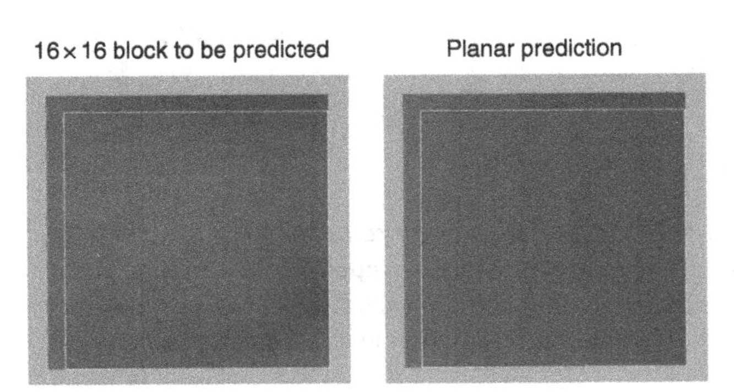

Figure 5.15 Plane prediction example (1)

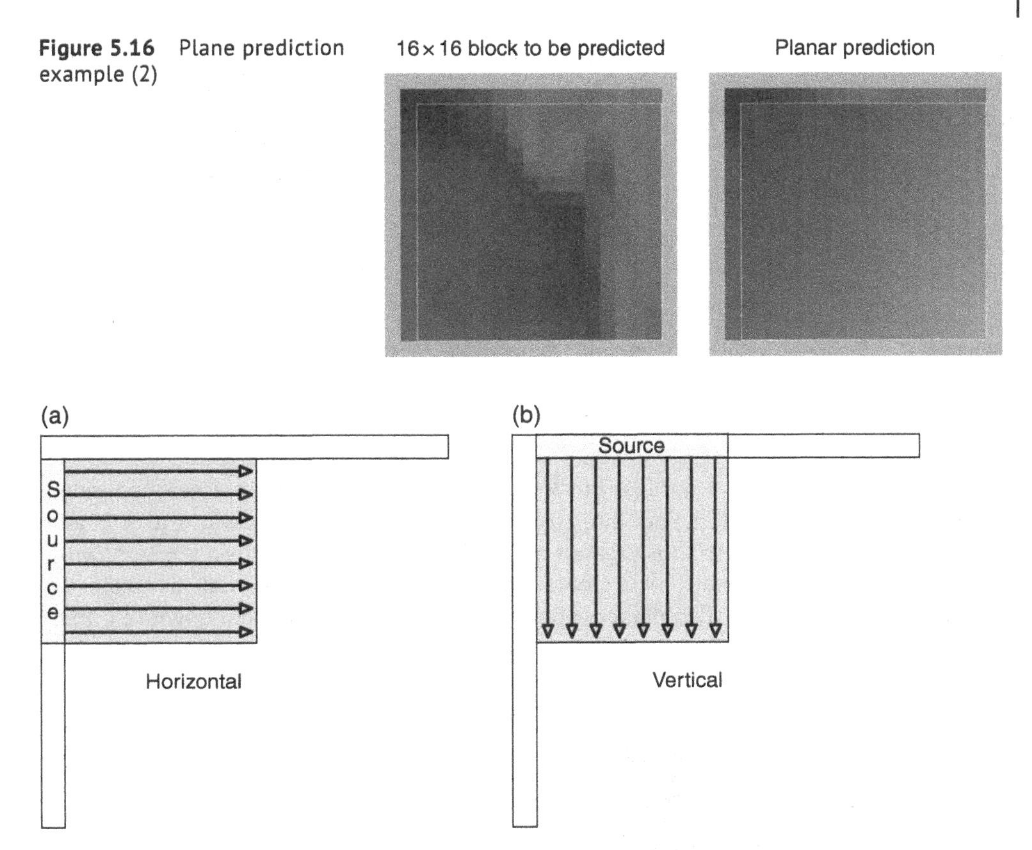

Figure 5.16 Plane prediction example (2)

Figure 5.17 Horizontal (a) and vertical (b) intra prediction modes

5.3.3 Directional Prediction

Directional or angular prediction creates an intra prediction by extrapolating or copying from samples next to the block. Each type or mode of prediction relates to a specific direction. Figure 5.17 shows horizontal prediction in which the left-hand neighbouring samples are copied across to the right and vertical prediction in which the top neighbouring samples are copied downwards.

Three examples of diagonal prediction modes are shown in Figure 5.18, each with a prediction direction that is 45° to the horizontal or vertical. In the first, diagonal (1), the source samples to the left and lower left of the current block are extrapolated at a 45° angle upwards and to the right. In the second mode, diagonal (2), source samples above and to the left of the current block are extrapolated downwards and to the right. In the third mode, diagonal (3), source samples above and to the upper right are extrapolated down and to the left[2].

2 It is not usually possible to create a prediction with a direction from lower right to upper left because the lower right samples have not been decoded yet and so are not available for prediction.

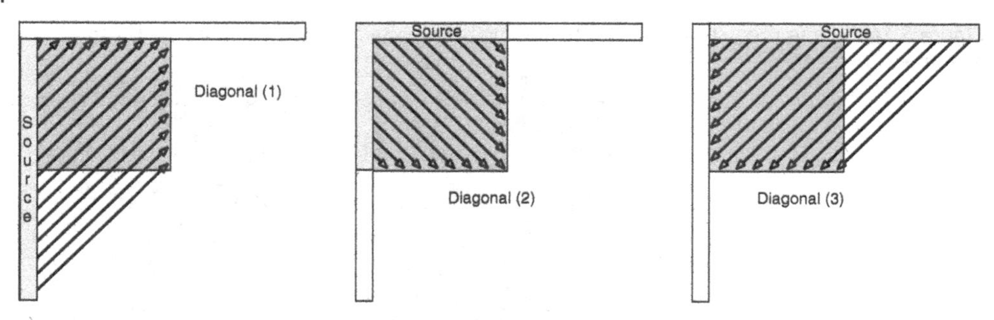

Figure 5.18 Diagonal prediction modes, 45° to horizontal/vertical

Example 1

Figure 5.19 shows a 4×4 original block at the top, with the available neighbouring pixels above and to the left. Four directional predictions are shown below the original block. The vertical prediction is created by copying each of the four top pixels downwards. The horizontal prediction is created by copying each of the four left-hand pixels horizontally to the right. Two diagonal predictions are shown, diagonal down/left and diagonal down/right. Out of these four predictions, the best matches to the original block are probably (1) horizontal or (2) diagonal down/right.

Original block

Figure 5.19 4×4 block and four directional intra predictions

Vertical prediction

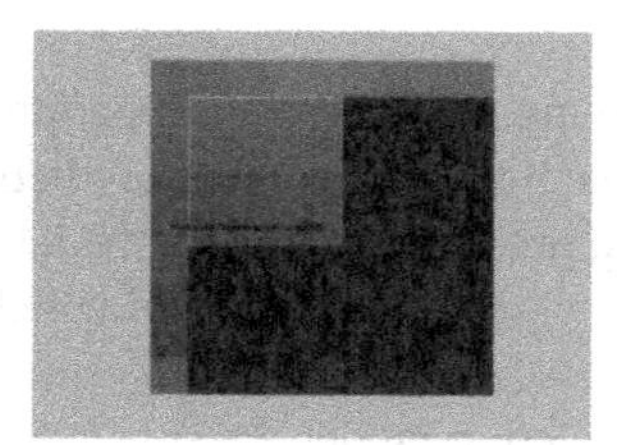

Horizontal prediction

Diagonal (down/left)

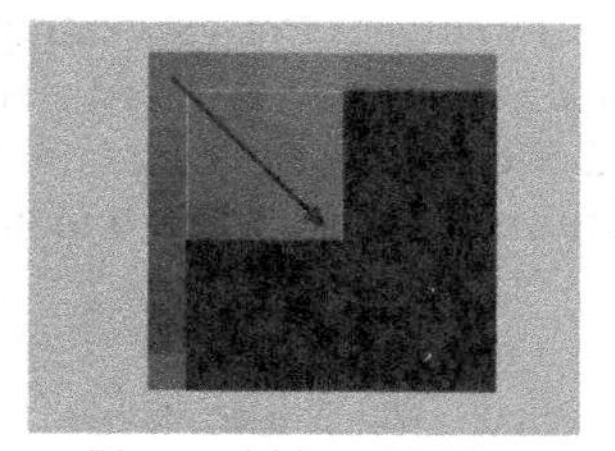

Diagonal (down/right)

Example 2

Figure 5.20 shows another example, this time with the numerical values of the samples shown. At the left is a block of samples to be predicted, with the top and left boundary samples also shown. The upper and left samples belong to blocks that have already been coded and are therefore available for prediction at both the encoder and the decoder. A horizontal prediction is created by copying each of the left-hand samples across, i.e. the values 107, 96, 93 and 80. Subtracting the prediction from the original block gives the residual shown on the right.

For the same original block, a diagonal down/right prediction gives a better result, as Figure 5.21 shows. The diagonal down/right prediction is formed by copying the diagonal top/left neighbours down and to the right. The values in the residual block are closer to zero. An encoder might attempt to predict the block with each of the available prediction modes and pick the mode that minimises the energy in the residual block, as long as this mode does not take too many bits to signal.

The set of available prediction modes and the method of constructing each prediction may be defined in a video coding standard or specification. For example, HEVC/H.265 defines a total of 33 angular or directional prediction modes, as we will see later.

Boundary samples

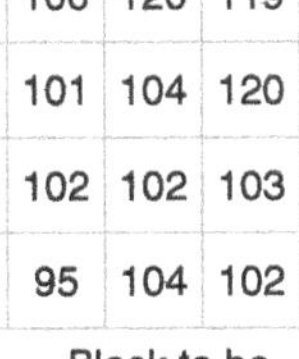

101	105	124	119	118
107	102	106	120	119
96	106	101	104	120
93	100	102	102	103
80	88	95	104	102

Block to be predicted

Horizontal prediction

101	105	124	119	118
107	107	107	107	107
96	96	96	96	96
93	93	93	93	93
80	80	80	80	80

Horizontal residual

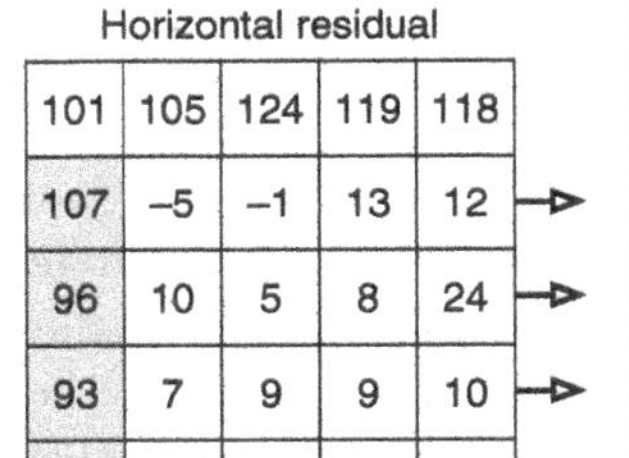

101	105	124	119	118
107	–5	–1	13	12
96	10	5	8	24
93	7	9	9	10
80	8	15	24	22

Figure 5.20 Horizontal prediction example

Boundary samples

101	105	124	119	118
107	102	106	120	119
96	106	101	104	120
93	100	102	102	103
80	88	95	104	102

Block to be predicted

Diagonal down-right

101	105	124	119	118
107	101	105	124	119
96	107	101	105	124
93	96	107	101	105
80	93	96	107	101

Diagonal residual

101	105	124	119	118
107	1	1	–4	0
96	–1	0	–1	–4
93	4	–5	1	–2
80	–5	–1	–3	1

Figure 5.21 Diagonal prediction example

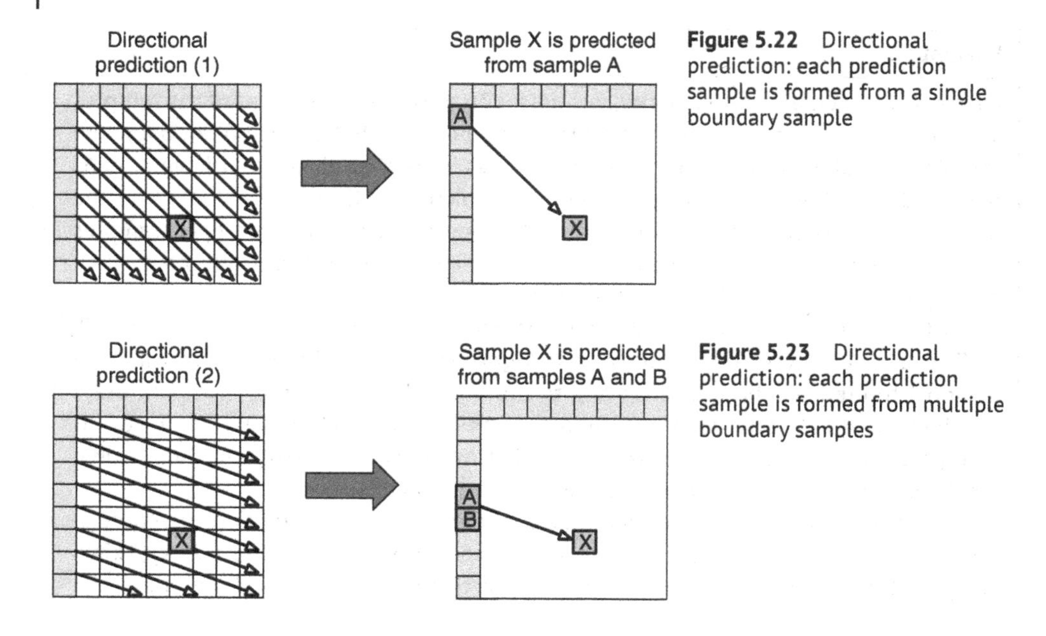

Figure 5.22 Directional prediction: each prediction sample is formed from a single boundary sample

Figure 5.23 Directional prediction: each prediction sample is formed from multiple boundary samples

5.3.4 Sample Filtering

Depending on the angle of directional prediction, each sample in the predicted block may be formed from a single boundary sample or multiple boundary samples. Consider sample *X* in the 8×8 block shown in Figure 5.22. Neighbouring sample A is extrapolated along the 45° direction shown to create predicted sample *X*. In the example of Figure 5.23, the prediction direction is not 45°. The source for prediction sample *X* is somewhere between neighbouring samples A and B, so the encoder and decoder create a prediction for *X* by filtering source samples A and B, by averaging their values. Filtering or smoothing two or more boundary samples may give a more accurate directional prediction.

5.4 Prediction Block Sizes

When creating an intra prediction, an encoder may be capable of choosing between a number of block sizes according to the limits set out in the standard. For example, an H.264 encoder can choose between 4×4, 8×8 and 16×16 intra prediction block sizes. Consider the 16×16 block shown in Figure 5.24. Should the encoder choose a single 16×16 intra prediction, four 8×8 predictions or sixteen 4×4 predictions? The 16×16 prediction, the largest prediction block size in this example, takes the fewest bits to signal. The encoder only needs to send a single parameter or syntax element to specify the prediction mode, such as vertical, horizontal, plane or DC, of the single block A. However, the original block is quite detailed, and it will be difficult to create an accurate prediction for the whole block. Alternatively, the encoder can choose 8×8 prediction block sizes. In this case, it needs to send four syntax elements to specify the modes of each 8×8 prediction block A, B, C and D. However, the predictions may be more accurate, especially since each of the four

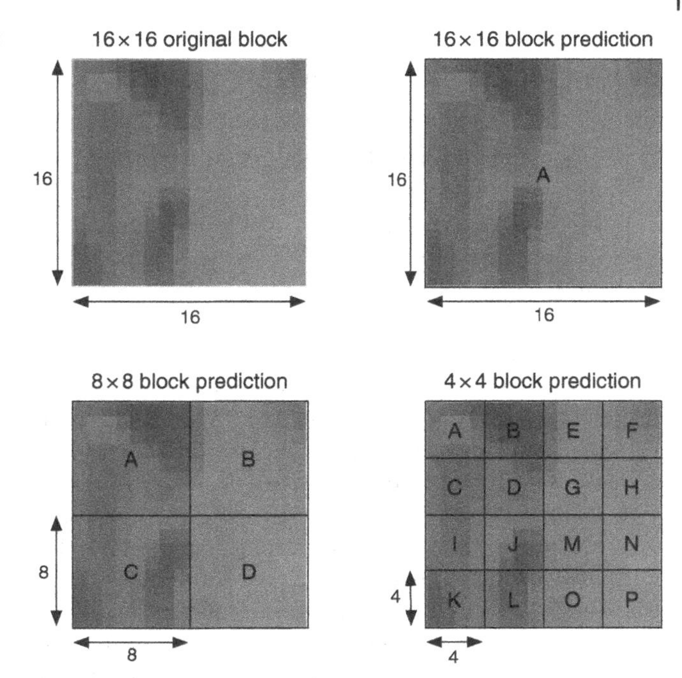

Figure 5.24 16×16 block and prediction size options

Figure 5.25 Example of intra prediction block size choices

quarters has a distinctly different texture. Finally, the encoder can choose 4×4 prediction block sizes. This will probably give the most accurate prediction for each 4×4 pixel block, but the encoder now has to send syntax elements to inform the decoder of the prediction modes of all sixteen 4×4 blocks A to P.

In general, a larger block size is best for regions of the video scene containing smooth texture or simple, strong details, since a large block size will give a reasonably accurate intra prediction and will cost the fewest bits to signal to the decoder, as illustrated in Figure 5.25. A small block size may be better for regions with complex details. Smaller prediction blocks generally give a more accurate prediction but require more bits to signal the prediction modes of each block in a video frame, since there are more prediction blocks per macroblock or CTU and therefore more prediction modes to be signalled.

5.5 Signalling Intra Prediction Choices

For every prediction block, the encoder must communicate the choice of prediction mode. This typically involves encoding a parameter in the bitstream and sending it along with each coded block residual. If the encoder selects a smaller prediction block size in an effort to improve prediction accuracy, the number of prediction modes communicated per frame of video increases. Figure 5.26 shows an example in which four intra-predicted blocks A, B, C and D are encoded into a video bitstream. Each intra prediction mode and each coded residual occupy space in the bitstream. The encoder should choose an intra prediction for each block that reduces the size of the coded residual. At the same time, the encoder should code the prediction mode using as few bits as possible.

H.264 introduced the concept of coding the most probable intra prediction mode. As with many other aspects of video coding, it is a reasonable assumption that the mode of the current block is likely to be the same as, or similar to, the mode of previously coded block(s) in the same area of the frame. Consider the current block in Figure 5.27. Two of the four surrounding blocks are coded using intra prediction mode 1 and two are coded using intra prediction mode 2. It is probable that the mode of the current block will be 1 or 2, illustrated by the graph showing probability (P) of each mode for the current block.

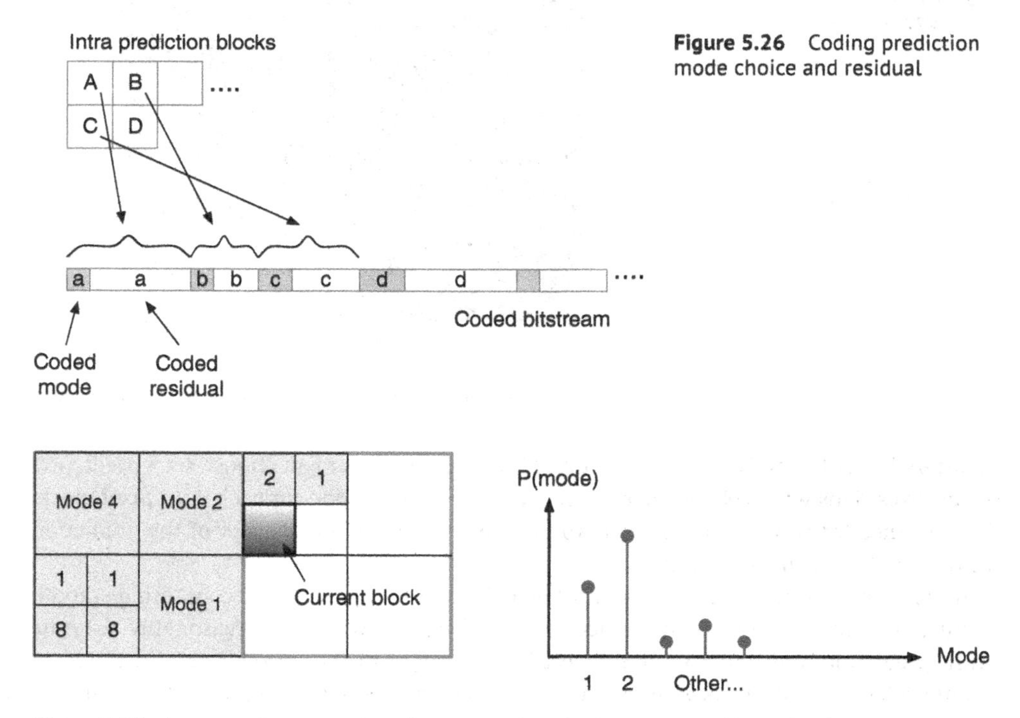

Figure 5.26 Coding prediction mode choice and residual

Figure 5.27 Intra prediction modes of neighbouring blocks and probability of current block mode

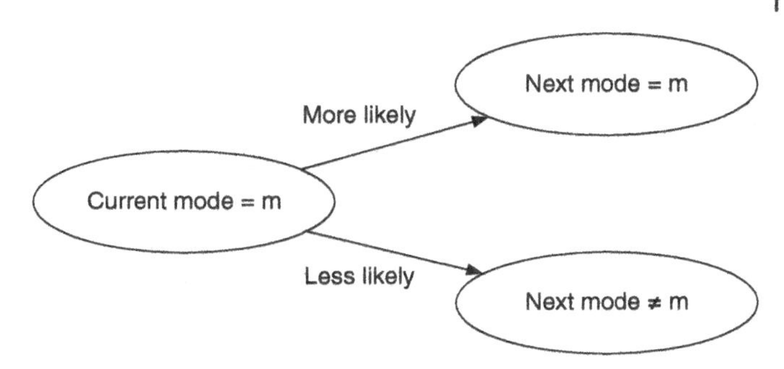

Figure 5.28 Most probable mode, H.264 intra prediction

In an H.264 codec, the most probable mode (MPM) is taken to be the same mode as the last block that was coded using intra prediction. The MPM takes the fewest bits to code (Figure 5.28). If the actual choice of mode for this block is different, then we need to signal the actual prediction mode.

5.6 Choosing a Prediction

For each block of a video frame, the encoder has to choose a prediction mode. This can be a computationally demanding problem when there are many different modes and multiple block sizes to choose from. If enough computing resources are available, if encoding speed is not an issue and/or if the encoder has plenty of computational power, then the encoder can do the following:

```
For each possible prediction block size {
        For each prediction mode {
                Calculate the number of bits to code the residual
                Calculate the number of bits to code the
                  prediction mode
        }
}
```

The encoder chooses the combination of block size(s) and modes that produce the minimum number of encoded bits.

Example:

A current N × N sample block can be intra-predicted as a single block or as four N/2 × N/2 sample blocks, as shown in Figure 5.29. The encoder tests every possible prediction mode for the N × N sample block size and notes the minimum number of bits required to code the mode and the residual. The encoder tests every prediction mode for each of the four N/2 × N/2 sample blocks and records the minimum number of bits required to code each mode and residual. The final choice is the combination of block size and prediction mode(s) that gives the smallest total number of bits. In this case, the smallest overall total is obtained by predicting the block in N × N mode, the left-hand option in this example.

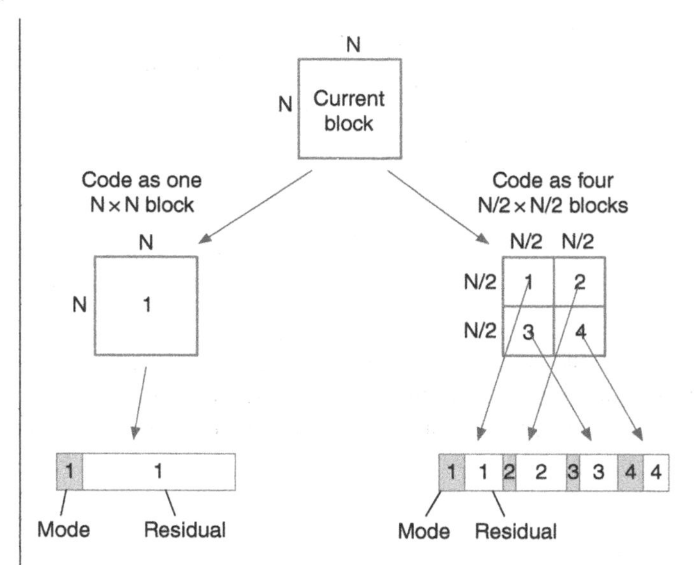

Figure 5.29 Intra prediction mode selection example

Depending on the number of available prediction modes, it may not be possible for the encoder to fully test every possible combination of block size and mode. In practice, a video encoder may simplify the process of choosing a prediction mode by selecting a subset of available modes to test.

5.7 HEVC Intra Prediction

H.265/HEVC intra prediction [1] creates a prediction for the current prediction block (PB) from neighbouring samples to the left and above the current PB.

An intra prediction block is a square block of luma or chroma samples that can range in size from the CTU size, a maximum of 64×64 samples, down to 4×4 samples. As described in Chapter 4, the CTU is partitioned into coding units (CUs) that are made up of coding blocks (CBs). Each CB in an intra-coded CU is partitioned into PBs according to the following rules, as illustrated in Figure 5.30:

i) If the CB is larger than the smallest allowable CB size, the PB is the same size as the CB.
ii) If the CB is the smallest allowable CB size, which is defined in the sequence parameter set as 8×8 or larger, the PB may be either (1) the same size as the CB or (2) split into four quadrants of 4×4 samples or larger. The choice of whether to split the CB is signalled with a flag in the bitstream.

Figure 5.7 shows a block of size N×N samples with upper and left samples available for prediction. In the example of Figure 5.7, an intra prediction could be constructed from any of the 2N samples above, the 2N samples to the left and/or the sample above and to the left, all of which have been coded and are available for prediction. Depending on the block's position in the CTU and the picture, some or all of these samples may not be coded yet and therefore not available for prediction.

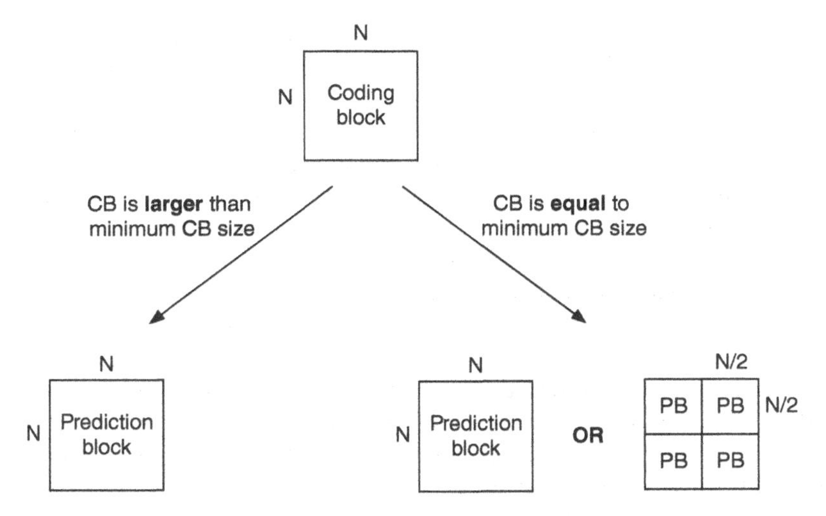

Figure 5.30 HEVC intra prediction block size choices

5.7.1 HEVC Intra Prediction Modes

HEVC supports 35 intra prediction modes. Mode 0 is planar prediction, mode 1 is DC prediction and modes 2–34 are directional or angular prediction modes, as illustrated in Figure 5.31.

Planar prediction, mode 0, fits a plane surface to the block. The surface is created by interpolating between four pixels, two above and two to the left of the current block,

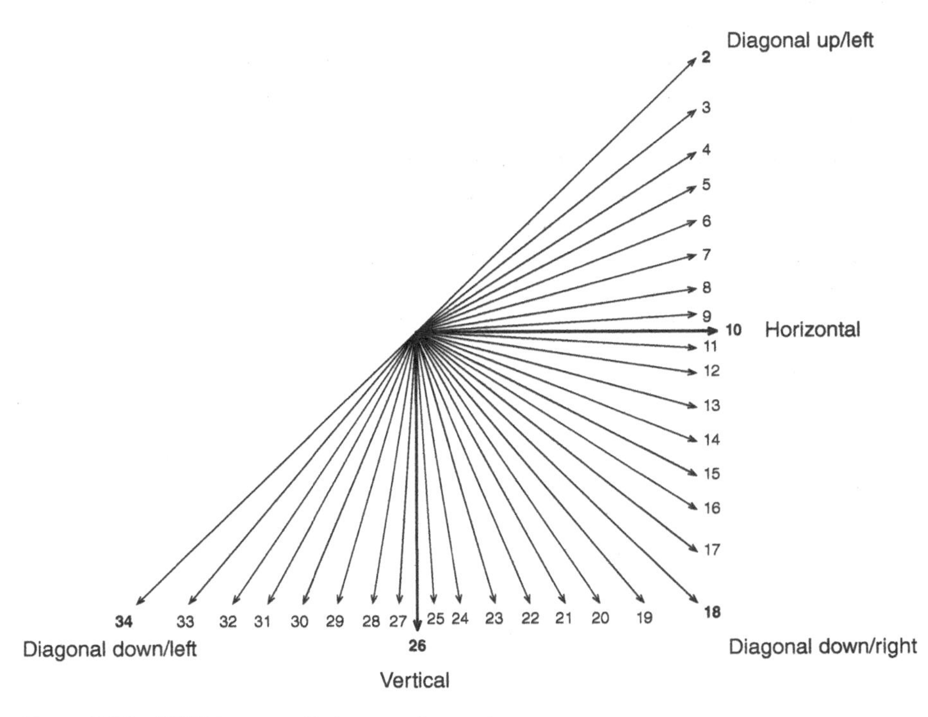

Figure 5.31 HEVC intra prediction angular modes

similar to Figure 5.14 earlier. DC prediction, mode 1, calculates an average of the above and left pixels and applies this average value across the entire block, as illustrated in Figure 5.10 earlier.

HEVC's 33 directional or angular prediction modes can be applied to intra-predicted PUs up to 32×32 in size. Each mode creates a prediction by extrapolating across the block from neighbouring samples above and/or to the left, in the direction shown above. Modes 10 and 26 use horizontal extrapolation from samples immediately to the left or vertical extrapolation from samples immediately above, respectively.

The remaining modes are diagonal predictions. Mode 2 extrapolates from samples to the left of the block, upwards and to the right at a 45° angle, similar to Figure 5.18, left-hand diagram. Mode 18 extrapolates from above and left samples, downwards and to the right at a 45° angle, and mode 34 extrapolates from upper samples, downwards and to the left at a 45° angle. The other diagonal modes extrapolate at angular intervals between modes 2 and 34. The available directions are chosen to be closely spaced around the horizontal and vertical modes, i.e. around modes 10 and 26, and more widely spaced around the 45° diagonals, modes 2, 18 and 34. This reflects the fact that video content often contains more horizontal and vertical features than diagonal features.

5.7.2 HEVC Intra Prediction Filtering

When predicting a particular sample using an angular prediction mode, the prediction source may lie exactly on a boundary sample position, in which case that sample is used as the prediction source. For example, sample *X* is predicted using mode 18, as shown at the left of Figure 5.32. The prediction source for *X* is the single sample A.

However, in many cases, the prediction source lies between boundary samples, for example, when *X* is predicted using mode 15, shown at the right of Figure 5.32. The prediction for *X* is constructed by interpolating between A and B, using bilinear interpolation with 1/32 sample accuracy.

For larger prediction block sizes, 8×8 or larger, boundary samples are smoothed using a three-tap linear filter for certain prediction modes. The intra prediction is constructed by extrapolating from the smoothed boundary samples, see Table 5.1.

At slice or tile boundaries, some of the neighbouring samples may not be available. The encoder may copy available neighbouring samples into the unavailable sample positions, making it possible to use a wider number of the possible prediction modes in these situations.

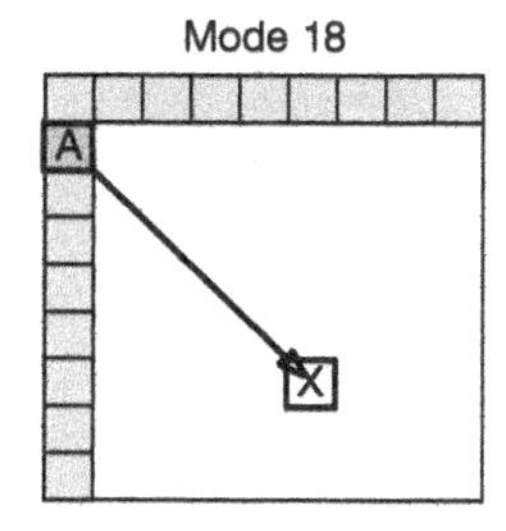

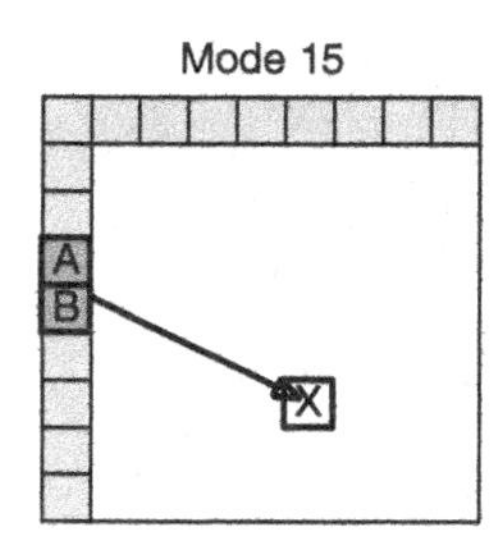

Figure 5.32 HEVC intra prediction filtering

Table 5.1 Intra prediction reference sample smoothing

Intra prediction block size	Reference sample smoothing
4×4	None
8×8	Only for diagonal modes (2, 18, 34)
16×16	All angular modes except (9, 10, 11) and (25, 26, 27)
32×32	All angular modes except (10, 26)

The boundaries of predicted sample blocks are filtered to smooth discontinuities across block boundaries in certain situations, when the block size is 16×16 or smaller and when certain prediction modes are selected.

5.7.3 Signalling HEVC Intra Prediction Modes

HEVC codes the intra prediction mode of each PB based on the observation that neighbouring blocks tend to have the same or similar intra prediction modes. The encoder signals the decoder to use either a MPM from a list of three likely candidates or to use a specific mode that is not on the candidate list.

5.7.3.1 HEVC Luma Prediction Mode Signalling

A list of three candidate prediction modes, the MPM list, is constructed, depending on the intra prediction modes of neighbouring, previously coded blocks to the left and above (see Figure 5.33). The list of candidates is determined as shown in Figure 5.34. Any unavailable neighbours, e.g. blocks that were coded using inter prediction rather than intra prediction, are set to DC mode. If both neighbours have the same directional prediction mode, then the MPM candidates are set to that mode, plus or minus one. Other cases are handled as shown in Figure 5.34. The end result is that each of the three candidate modes is set to either the mode of left-hand block A (+/−1 in some cases), the mode of upper block B, or planar, DC or vertical modes.

For each luma intra-PB, a flag in the bitstream, prev_intra_luma_pred_flag, indicates whether the decoder should use one of the MPMs, i.e. candidate mode 0, 1 or 2, or use a specific intra-mode signalled in the bitstream, as illustrated in Figure 5.35.

For each chroma PB, the encoder signals that the chroma PB should either use one of the planar, DC, horizontal or vertical modes or use the same intra prediction mode as the corresponding luma PB.

5.7.3.2 HEVC Chroma Prediction Mode Signalling

The chroma components of a CU have a limited number of choices of intra prediction mode. Both components can inherit the luma prediction mode, or they can both be coded using modes 0 (planar), 1, (DC), 10 (horizontal), 26 (vertical) or in certain cases 34 (diagonal-down).

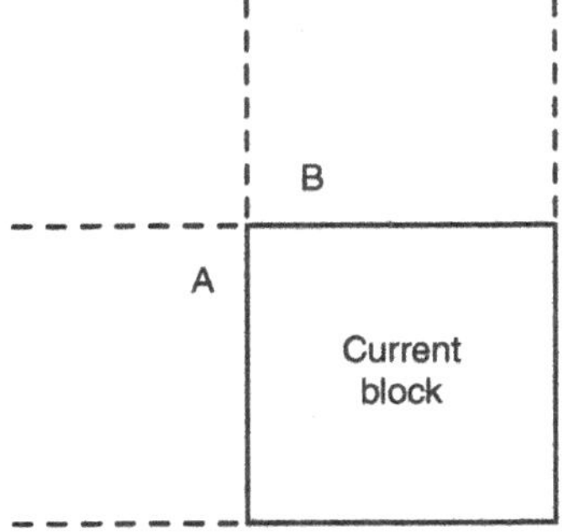

Figure 5.33 Current, left (A) and above (B) blocks

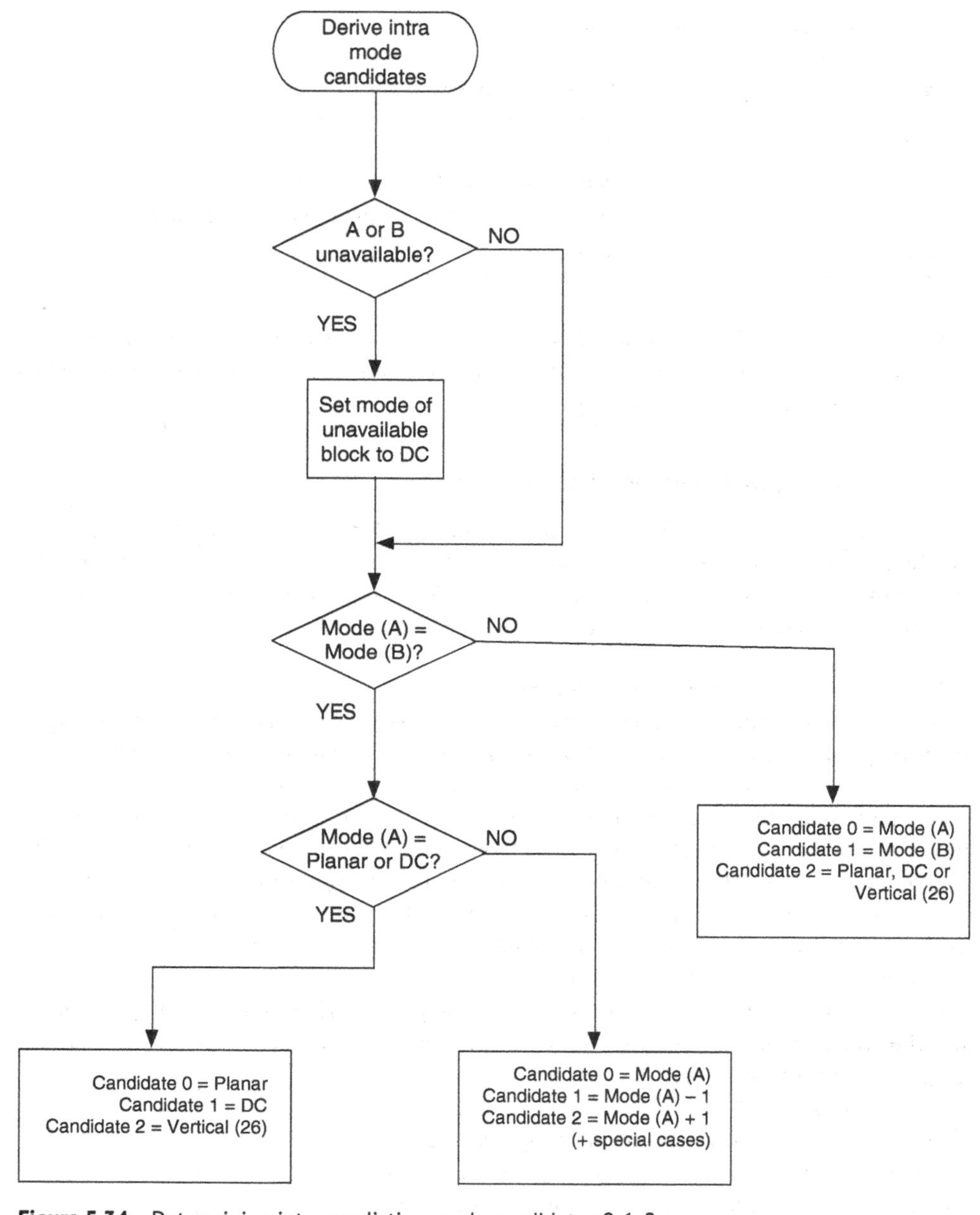

Figure 5.34 Determining intra prediction mode candidates 0, 1, 2

5.7.4 HEVC Intra Prediction Examples

Figure 5.36 shows a region of an intra-coded frame with the CU grid superimposed on the left and the encoder's chosen intra prediction modes on the right. In the right-hand figure, directional prediction modes are represented by straight lines, angled in the direction of prediction from left and/or above; planar prediction is represented by two-quarter circles;

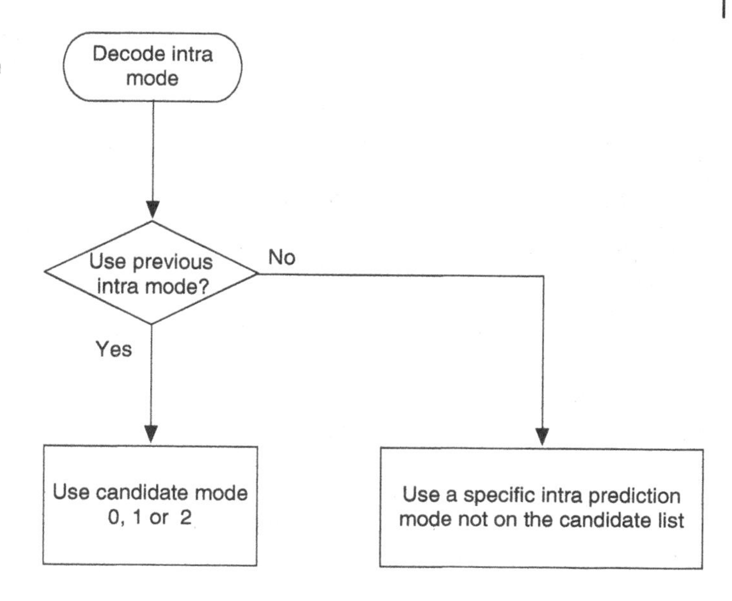

Figure 5.35 Use a most probable mode, or signal a specific mode

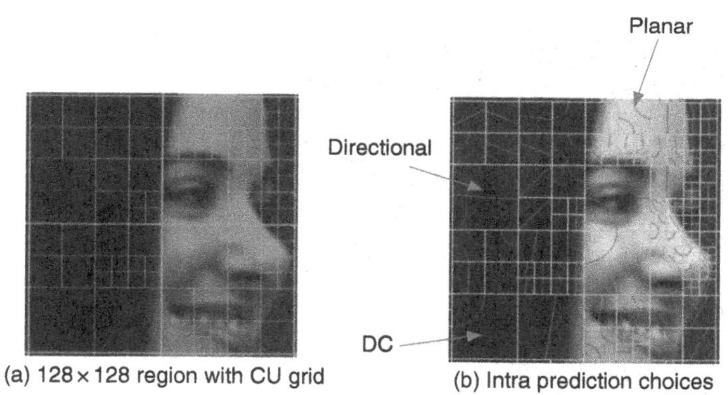

Figure 5.36 HEVC intra prediction: (a) region and (b) prediction choices. *Source:* Source of original image: Parabola Research

and DC prediction is represented by one-quarter circle. Some blocks have a mix of symbols, indicating different prediction choices for luma and chroma. For example, the lower left 32 × 32 CU uses DC prediction for luma and directional prediction for chroma.

The resulting intra predictions are shown in Figure 5.37, without and with the CU grid shown. Each block prediction is formed using the decoded and reconstructed samples available above and to the left of the block. Many of the block predictions are reasonably accurate, with some obvious exceptions such as the 32 × 32 planar prediction around the eye[3]. A more accurate prediction could have been formed by splitting this 32 × 32 CU into smaller CUs/PUs and predicting each separately. However, this would have increased the number of bits required to code the partition choice and prediction choice for each of the smaller blocks.

3 Note that these prediction samples are not intended to be displayed to the viewer – they are merely numerical predictions that attempt to reduce the number of bits that need to be encoded.

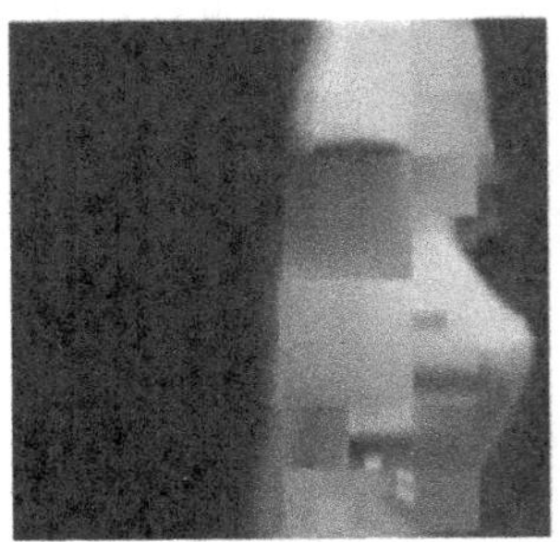

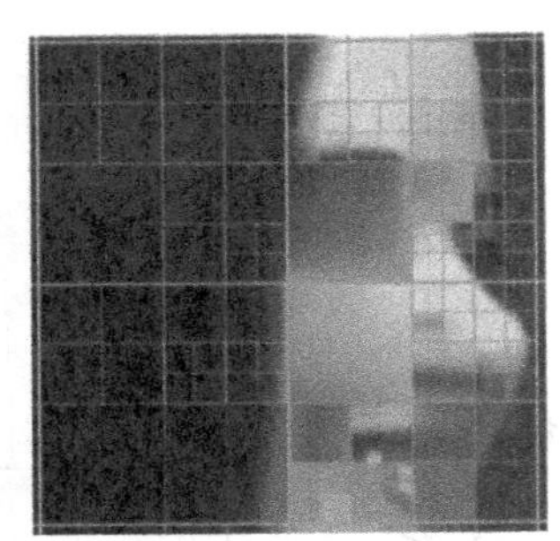

(a) Intra predicted blocks (b) Intra predicted blocks + grid

Figure 5.37 HEVC intra prediction: (a) intra-predicted blocks (b) intra-predicted blocks with PU grid shown

The number of coded bits per CU/PU is illustrated in Figure 5.38. Higher bit counts are represented by brighter blocks and lower bit counts by darker blocks. The highest numbers of coded bits occur in regions that are complex and/or where the encoder has been unable to accurately predict the block.

We will consider a single 16×16 block in the lower right CTU, highlighted in Figure 5.39, with a strong horizontal feature, i.e. the nostril. When the encoder or

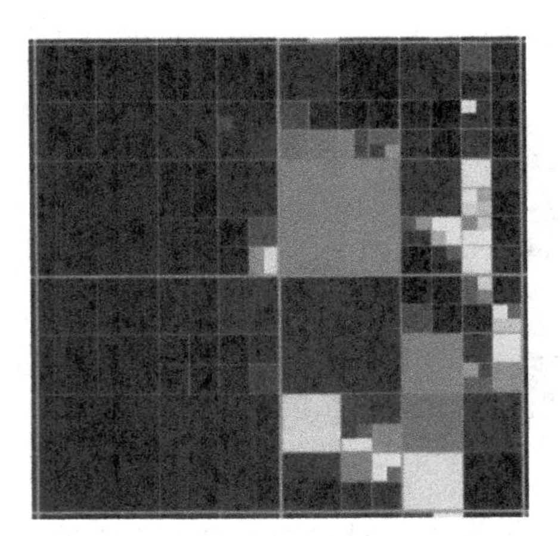

Figure 5.38 Coded bit count (per block)

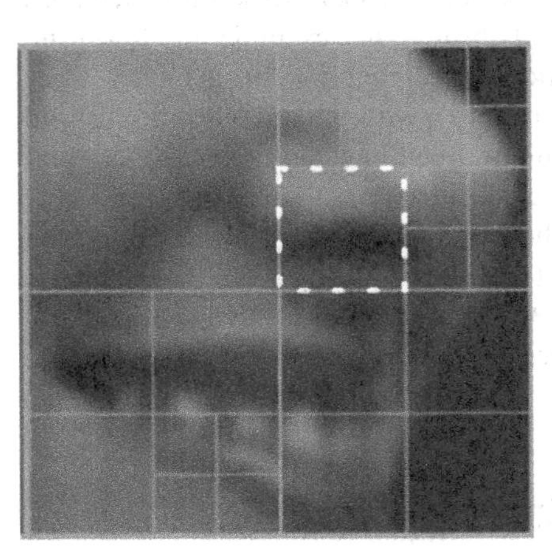

Figure 5.39 16×16 block to be predicted using HEVC intra prediction

decoder is ready to process this block, the neighbouring blocks shown in Figure 5.40 have already been encoded or decoded and are available for prediction. The encoder evaluates each of the available prediction modes. Note that directional modes 2–9 are not available because the lower-left samples have not yet been encoded. The available modes are planar, DC and directional modes, i.e. modes 10–34.

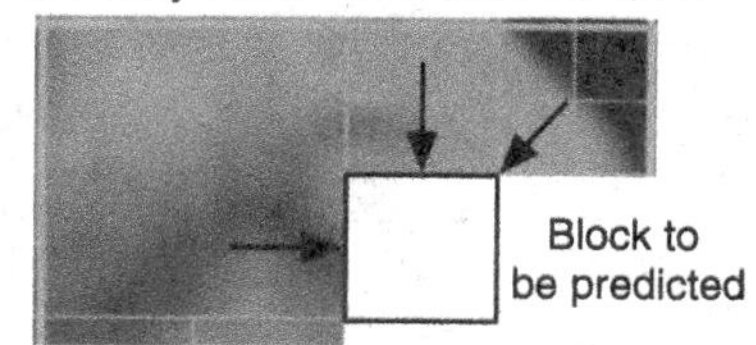

Figure 5.40 16 × 16 block and previously coded samples

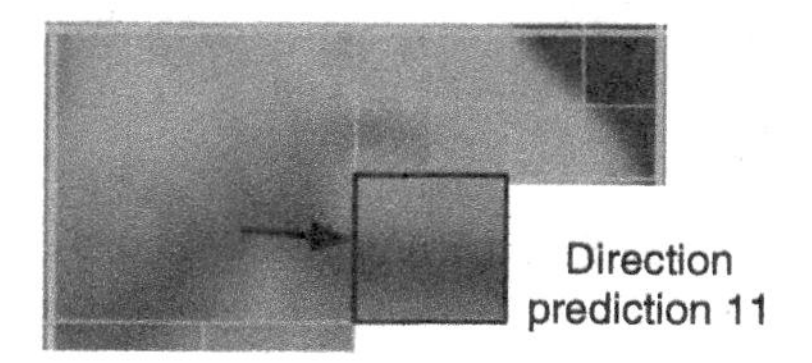

Figure 5.41 Chosen intra prediction mode

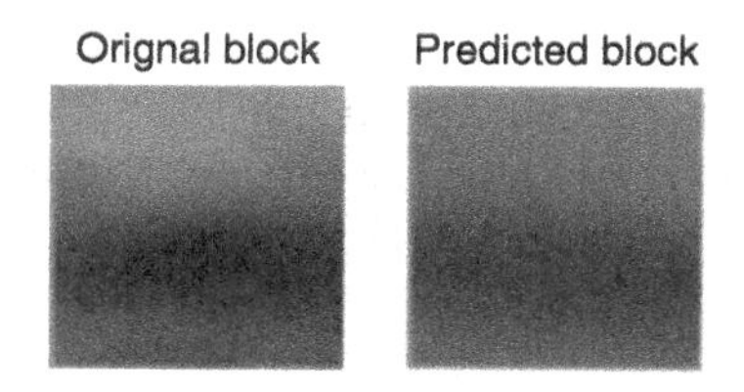

Figure 5.42 Original + intra-predicted block

The encoder tests all of the available modes, for example, by creating each possible prediction, subtracting it from the block and measuring the energy in the residual, and chooses directional prediction mode 11. The encoder creates the prediction by extrapolating to the right and down from the left-hand block, as shown in Figure 5.41. As Figure 5.42 shows, this is a reasonably accurate prediction of the original 16 × 16 block of samples.

The complete intra prediction and reconstruction process for this block is as follows:

The encoder:

- Determines which neighbouring blocks have been coded and are available for prediction, in this case, the immediate left neighbour and the neighbours above and to the right.
- Evaluates the available prediction modes.
- Chooses a prediction mode to minimise the overall cost of encoding the mode and the residual.
- Forms the predicted block, in this case by extrapolating previously coded samples from the left of the current block.
- Encodes the choice of prediction mode and the residual or difference between the original and predicted blocks.
- Decodes the residual.
- Adds the predicted block to the residual and stores it to form future predictions.

The decoder:

- Receives and decodes the choice of prediction mode and the residual.
- Forms the predicted block, identical to the encoder's predicted block, by extrapolating previously coded samples.
- Adds the predicted block to the residual and stores it for eventual display/output and to form future predictions.

Figure 5.43 shows the chosen of intra predictions for a complete 720p-resolution coded frame. At this scale, the intra-predicted blocks form a recognisable image, even though this

Figure 5.43 Intra predictions for a complete frame. *Source:* Source of original image: Parabola Research

is not intended for display to the viewer. Areas of strong texture are reasonably accurately predicted. More complex areas, such as faces, are harder to predict accurately using intra prediction.

Another intra prediction example is shown in Figure 5.44. The source sequence is *Jockey*, 3840×2160 pixels per frame. This is a small portion of an intra-coded picture, with one CTU highlighted. The selected CTU size is 64×64 and is predicted with a single intra-PU

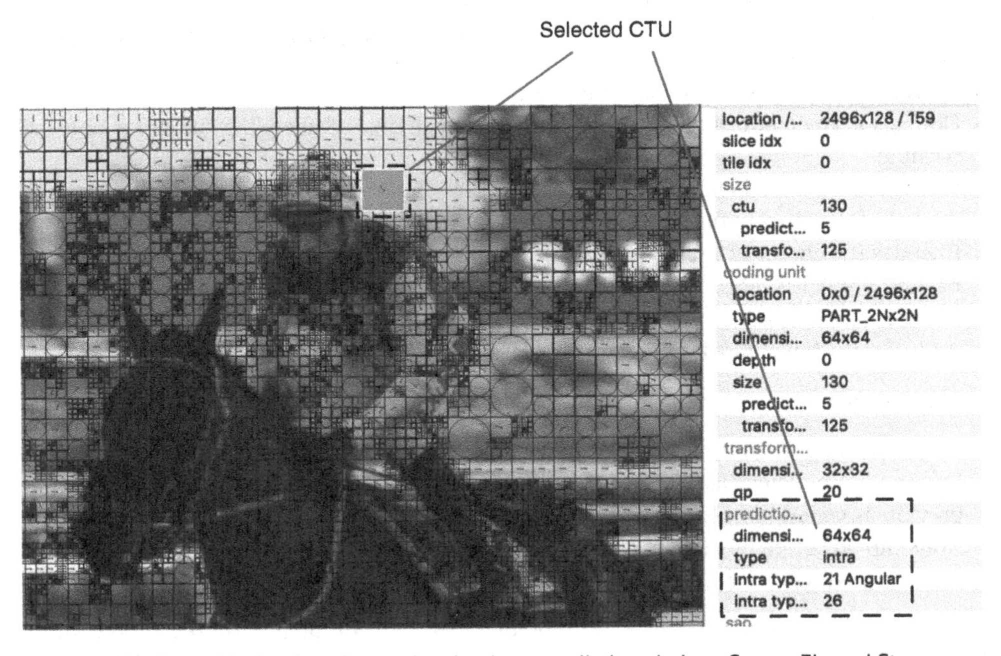

Figure 5.44 Part of Jockey intraframe showing intra prediction choices. *Source:* Elecard Streameye

Figure 5.45 Complete intra-coded frame from Jockey (3840 × 2160 pixels). *Source:* Elecard Streameye

of size 64 × 64 using intra-mode 21 for luma prediction and mode 26 for chroma prediction. Mode 21 is a diagonal-directional prediction mode. The chroma component is predicted using vertical prediction. Note that detailed areas such as the horse rider himself tend to be coded with smaller intra prediction modes. Figure 5.45 shows the complete intra-coded picture.

5.8 VVC Intra Prediction

Intra prediction in the VVC standard extends many of the concepts of HEVC intra prediction, providing more intra prediction options at the cost of increased computation. Enhancements include the following.

5.8.1 Intra Prediction Modes

VVC adds 32 further directional modes to HEVC's 33, giving a total of 65 directional modes. The extra modes are shown as dotted arrows between the solid arrows of the HEVC modes in Figure 5.46. A square block is predicted by extrapolating neighbouring samples according to one of the prediction directions, such as the example in Figure 5.47. Some of the angular modes are modified for the prediction of non-square blocks. Along with DC and planar modes, this gives a total of 67 possible intra prediction modes for each CU.

The mode of each block is coded in a similar way to HEVC, with the list of candidate MPMs extended from HEVC's 3 modes to 6.

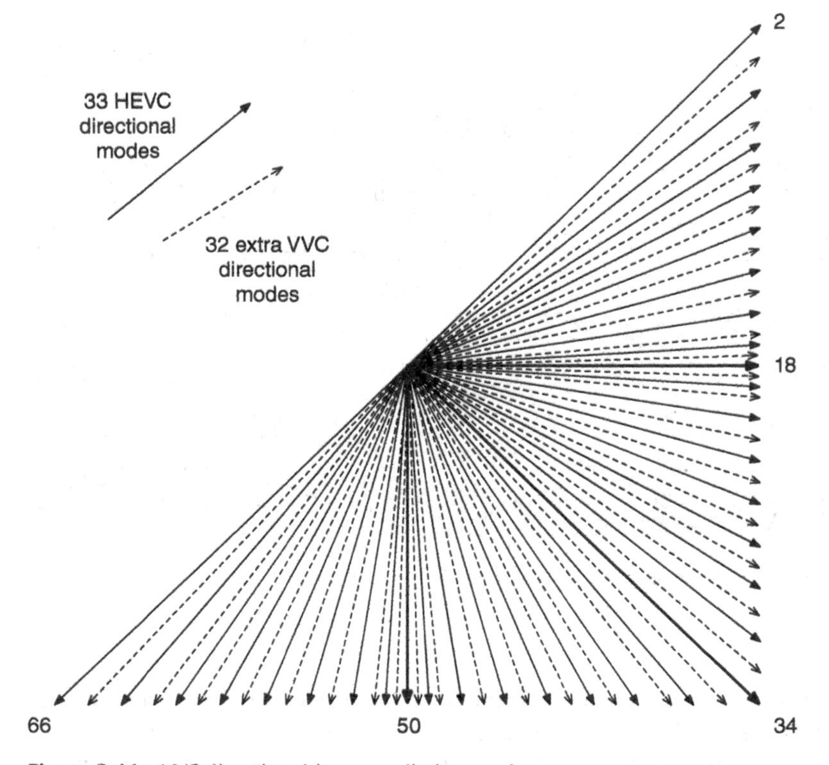

Figure 5.46 VVC directional intra prediction modes

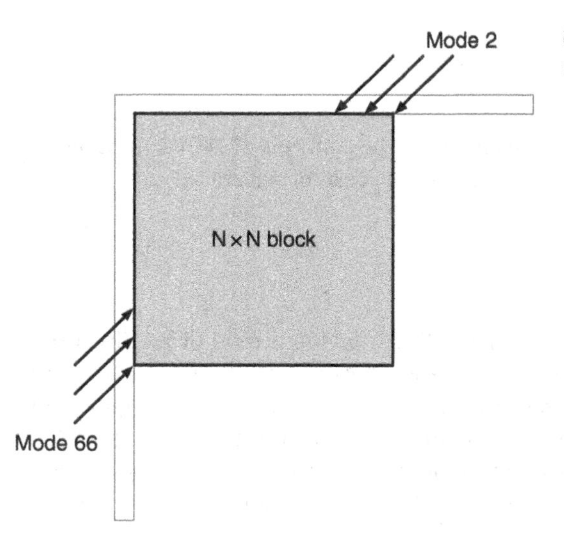

Figure 5.47 VVC directional prediction of square blocks

5.8.2 Intra Prediction of Rectangular Blocks

VVC coding units can be square or rectangular with aspect ratios of 1:1, 1:2, 1:4 or greater. HEVC only supports the intra prediction of square blocks, whereas any of the rectangular VVC coding units can be intra-predicted. When predicting a rectangular block using directional prediction, the set of prediction directions shown in Figure 5.46 may not be ideal. Instead, the directions are remapped, as illustrated in Figure 5.48. This figure shows the modified directional modes for blocks that have an aspect ratio of 2:1, i.e. N × (N/2) samples. In this case, six of the directional modes are removed and replaced with extra, wide-angle modes. A block with this aspect ratio is predicted as shown in Figure 5.49. The remapped modes 2 and 66 have now effectively shifted around to provide more effective coverage of all the directional angles that are likely to appear in a block with these dimensions.

5.8.3 Intra Interpolation Filters

The interpolation filter used in HEVC intra prediction is extended to four taps from two and applied depending on the directional prediction mode.

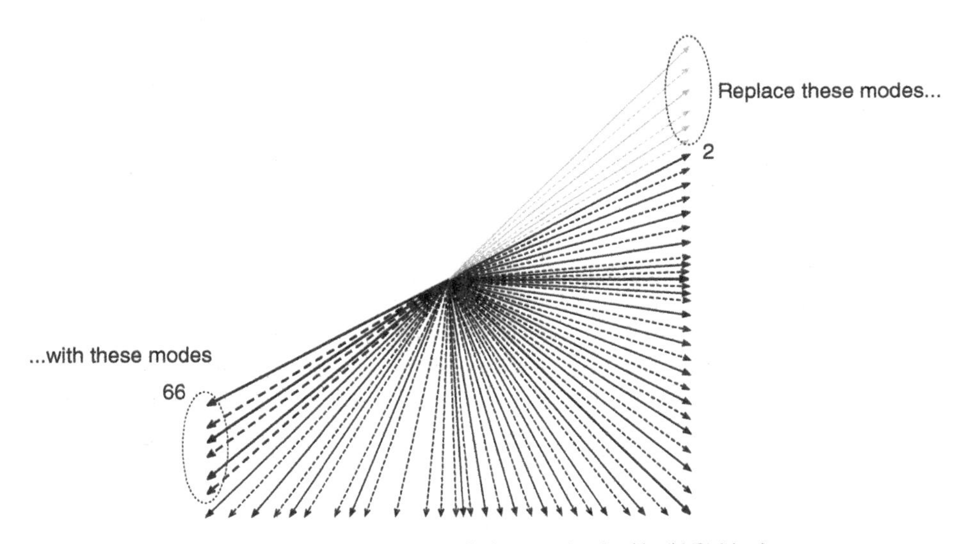

Figure 5.48 Remapped VVC directional prediction modes for N × (N/2) blocks

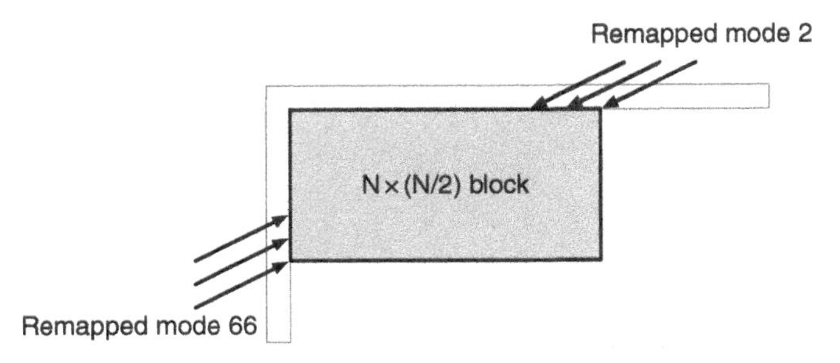

Figure 5.49 VVC directional prediction for N × (N/2) blocks

5.8.4 Cross-Component Prediction

VVC supports 8 prediction modes for chroma blocks, including 3 modes based on chroma prediction from luma.

Luma and chroma components of a block often have structural similarities. VVC introduces cross-component modes for chroma prediction. For these modes, after predicting the luma block, the reconstructed luma block is down-sampled and used to predict the chroma components. The prediction is formed based on the correspondence between edge pixels in the chroma and luma components. A prediction model is generated based on the similarity or difference between the edge samples of the luma and chroma blocks, see Figure 5.50. Based on this model, the reconstructed luma block is used to create an intra prediction from the chroma block.

5.8.5 Extra Reference Lines

Intra prediction in HEVC and the earlier H.264/AVC standards uses reconstructed samples from the lines of samples adjacent to the current block, Line 0 in Figure 5.51. VVC supports the use of two further reference lines: Line 1, which is one sample removed from the current block, and Line 3, which is three samples removed from the current block, as illustrated in the figure. The encoder can signal the selected reference line, and the block is predicted using directional modes only from reconstructed samples in the selected line.

5.8.6 Intra Sub-Partitions

In VVC, intra-predicted luma blocks may be divided into two or four horizontal or vertical sub-blocks prior to carrying out intra prediction. Smaller blocks are divided into two partitions, and larger blocks are divided into four partitions. Figure 5.52 shows the general idea for smaller blocks. An 8×4 luma block shown on the left is partitioned vertically into 4×4 sub-blocks A and B. Partition A is intra-predicted using previously reconstructed top and left samples as usual. Partition A is then reconstructed, and partition B is predicted using samples from the top edge and from partition A. This has the effect of breaking up intra prediction processes for larger blocks and increasing the opportunities for parallelisation.

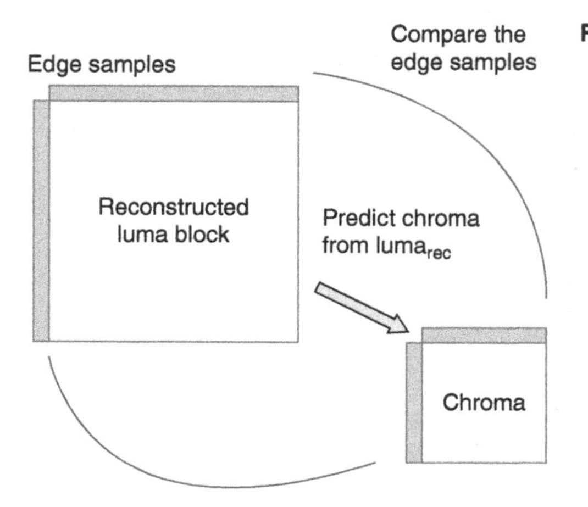

Figure 5.50 VVC chroma prediction from luma

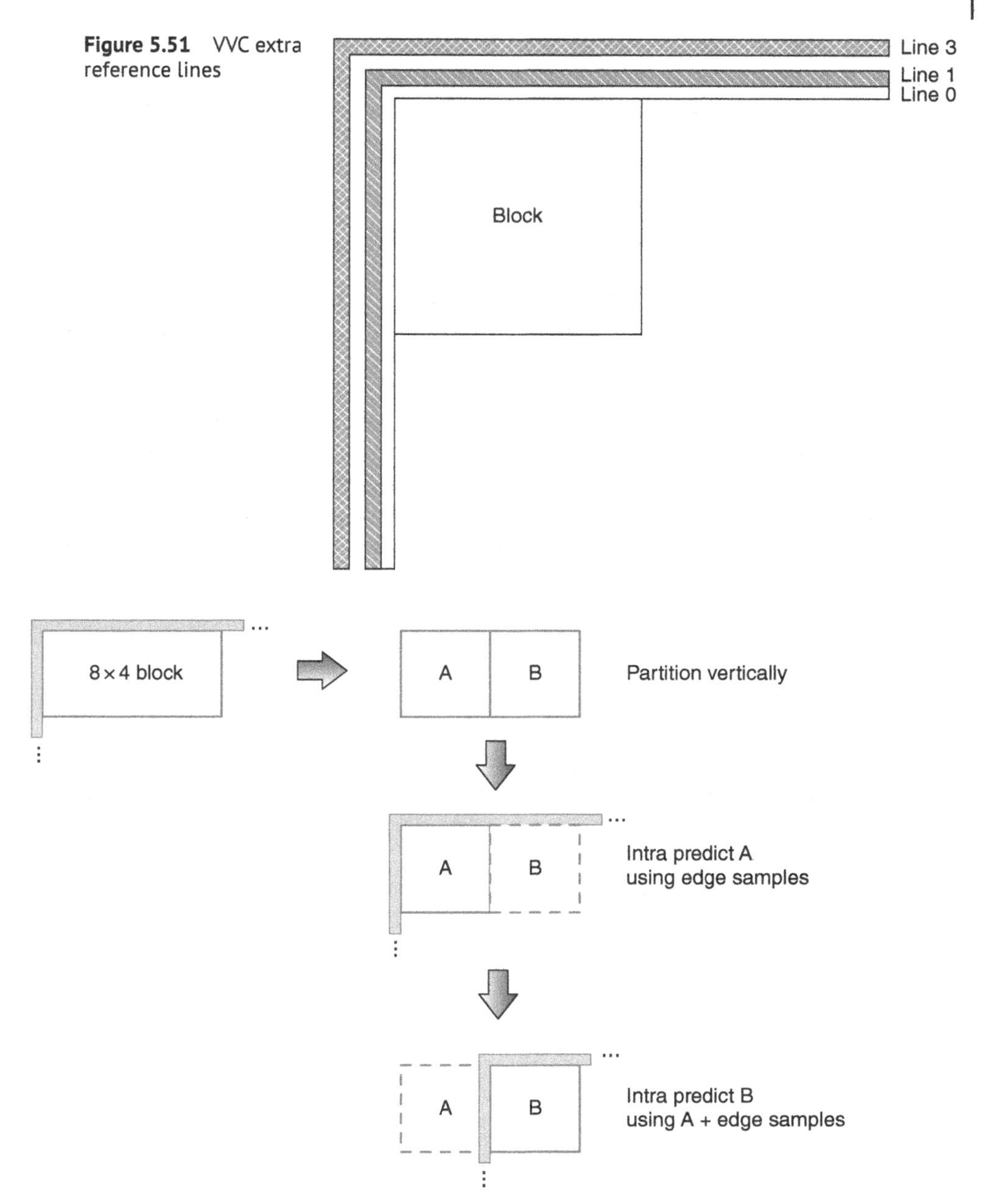

Figure 5.51 VVC extra reference lines

Figure 5.52 VVC intra sub-partitions example

5.8.7 VVC Intra-coding Performance

The authors of Ref. [2] have demonstrated that the VVC Test Model, VTM 6.1, outperforms HEVC's HM16.2 Test Model by an average of around 25% for intra-only coding. This means that for the same visual quality, the intra prediction and coding tools of VVC result in bitstreams that are around 25% smaller than the equivalent bitstreams produced by an

HEVC encoder. Intra-only encoding in VVC can be many times more computationally complex than intra-only encoding in HEVC. Intra-only decoding time is reported to be around 2× the decoding time of HEVC intra-decoding, which suggests that VVC intra-only decoding is around twice as complex as HEVC intra-only decoding.

5.9 Conclusions

In this chapter, we have seen how intra prediction can create a prediction block by using and extrapolating previously coded information in the same frame. This is important for coding the first frame in a video sequence when no previously coded frames exist and for inserting regular anchor frames into a video sequence so that a decoder can access the sequence at random points. Intra prediction is a block prediction mode that can be used in inter-coded frames when other predictions are not efficient and for coding single video frames or still images. For example, Apple iOS supports HEVC's intraframe coding mode as an option for efficient compression and storage of photographs [3]. Intra prediction has its limits, and in Chapter 6, we will see how interframe prediction and coding can further improve prediction and thus video compression efficiency.

References

1 Lainema, J., Bossen, F., Han, W.J. et al. (2012). Intra coding of the HEVC standard. *IEEE Transactions on Circuits and Systems for Video Technology* 22: 1792–1801.

2 Saldanha, M., Sanchez, G., Marcon, C.,et al. (2021). Performance analysis of VVC intra coding. *Journal of Visual Communication and Image Representation.* 79, 103202. doi: https://doi.org/10.1016/j.jvcir.2021.103202.

3 Hannuksela, M.M., Lainema, J., and Vadakital, V.K. (2015). The high efficiency image file format standard [Standards in a Nutshell]. *IEEE Signal Processing Magazine* 32 (4): 150–156.

6

Inter Prediction

6.1 Introduction

Inter prediction is the process of creating a prediction block based on pixels in a different frame from the current video frame. We saw in Chapter 2 that for each block of pixels in the current video frame, it is often possible to find a similar block of pixels in a different video frame. Inter prediction creates a prediction block based on previously coded pixels in a different video frame. This could be the frame just before the current frame i.e. one frame in the past, the frame just after – one frame in the future, or an older previous frame or newer future frame. Figure 6.1 shows a few options for predicting a block in the current frame. We can use the most recent previous frame (1) as our prediction source, or we could use an older frame (2) as the prediction source. We can use a future frame (3) as the source of the prediction. We can make a prediction by combining data from two or more frames, for example, one in the past and one in the future (4).

As Figure 6.2 shows, the encoder decodes or reconstructs each coded frame and creates inter predictions from these previously coded reference pictures. The encoder sends the residual, which is the difference between the prediction and the actual frame data, together with instructions so that the decoder can create an identical prediction.

The decoder stores decoded frames and uses these as reference pictures. It uses the instructions sent by the encoder to create the same prediction as the encoder. This means that the video encoder and video decoder maintain the same set of reference pictures and create identical predictions. The frame or frames chosen as the source of the prediction must be available at the decoder. This means that it is usually necessary to compress and send the prediction source frame(s) before sending the current frame. This in turn means that if the encoder uses future frames as prediction sources, it must reorder the frames in the bitstream, so that the decoder receives these future frames before the current frame.

In this chapter, we will consider:

- The basic concepts of inter prediction.
- Forward, backward and biprediction.
- Multiple reference pictures and sub-pixel interpolation.
- The trade-offs involved in choosing inter prediction block sizes.

Coding Video: A Practical Guide to HEVC and Beyond, First Edition. Iain E. Richardson.
© 2024 John Wiley & Sons Ltd. Published 2024 by John Wiley & Sons Ltd.
Companion website: www.wiley.com/go/richardson/codingvideo1

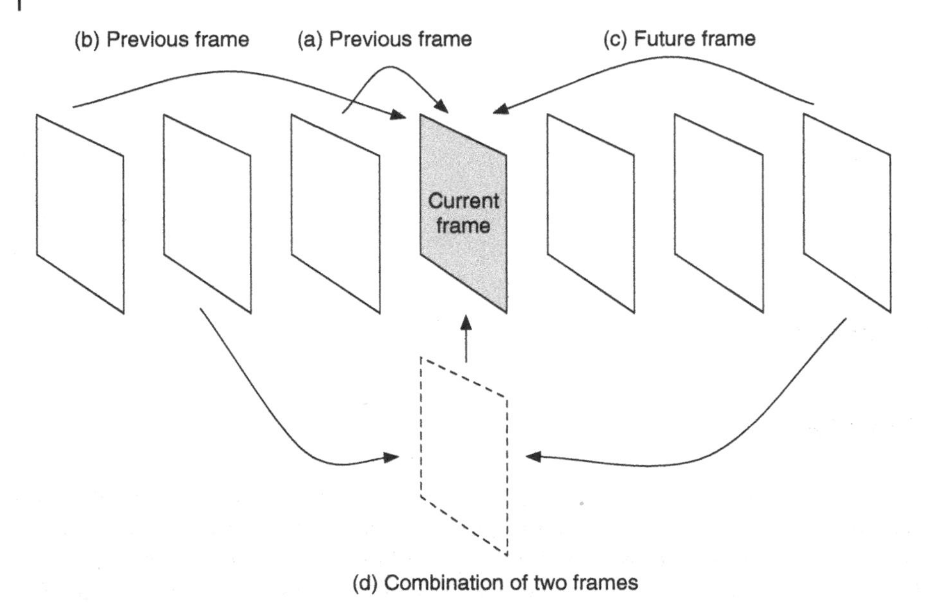

Figure 6.1 Predicting using inter prediction, from (a) a previous frame, (b) an older past frame, (c) a future frame or (d) combinations of frames

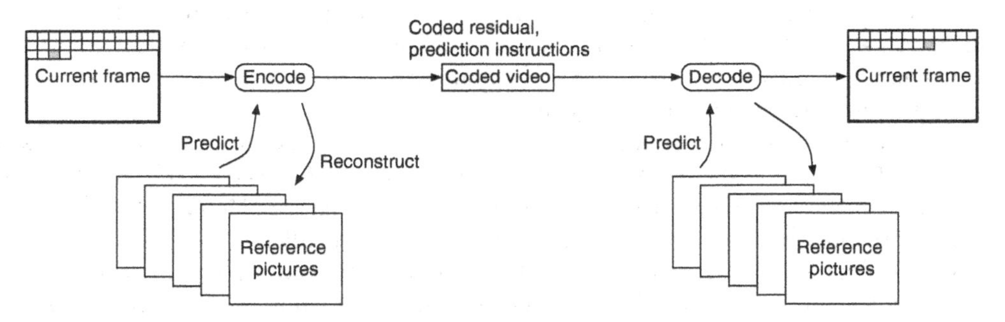

Figure 6.2 Prediction in the encoder and decoder

- Why is it sometimes a better idea to skip the block completely or to switch to intra prediction?
- How inter prediction works in H.265/High Efficiency Video Coding (HEVC) and H.266/VVC.

6.2 Inter Prediction – the Basics

We will start with a simple example: predicting from the previous frame in the video sequence (Figure 6.3). For every block in the current frame (N), the encoder creates a prediction from the previous frame ($N-1$). We can create the prediction from the same position in the previous frame, which is shown as the solid line in Figure 6.4, or from a different position, offset from the original block position, shown as the dotted line in Figure 6.4.

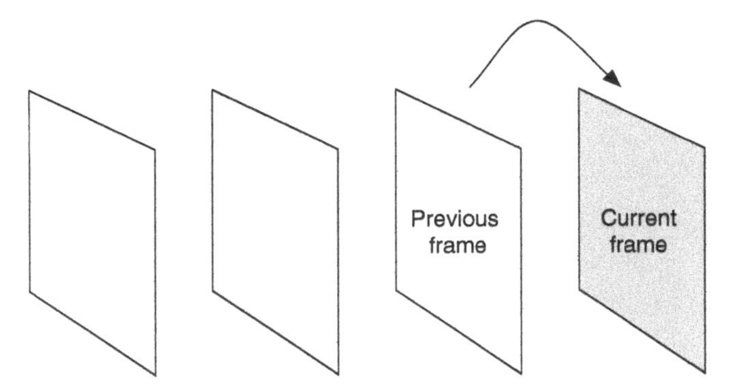

Figure 6.3 Inter prediction from previous frame

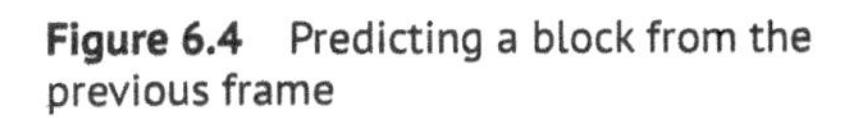

Figure 6.4 Predicting a block from the previous frame

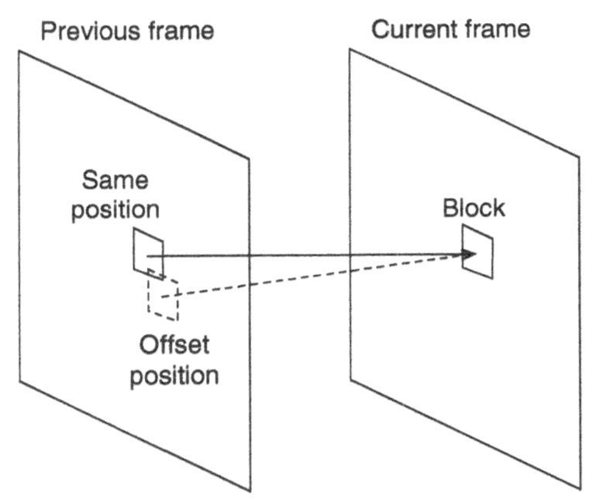

For a particular block in the current frame, shown with a solid white line in Figure 6.5, the block of pixels in the same position in the previous frame, also indicated with a solid white line, is often similar. However, if objects have moved between frames or if the camera itself is moving, we might find a closer match for the block of pixels in a different position in the previous frame, such as the block marked with a dotted white line in Figure 6.5. If we consider the block in the same position to be at location (0,0), then we can specify an offset to different block positions (see Figure 6.6). For example, the dotted area in Figure 6.5 is offset by 5 pixels horizontally, i.e. in the x direction, and by −4 pixels vertically or in the y direction, i.e. an offset of (+5, −4), assuming that pixels are numbered from the lower left. This offset is known as a motion vector.

An encoder carries out the following steps when inter-predicting a square block A in a current frame, with reference to a previously coded frame:

1) Encoder selects an offset to a square region B in a previously coded frame (Figure 6.7). This region in the previously coded frame is the same size as A and is the prediction source.

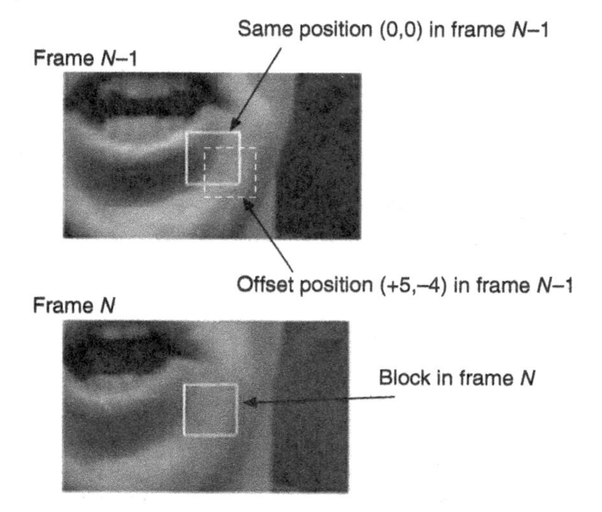

Figure 6.5 Inter prediction example

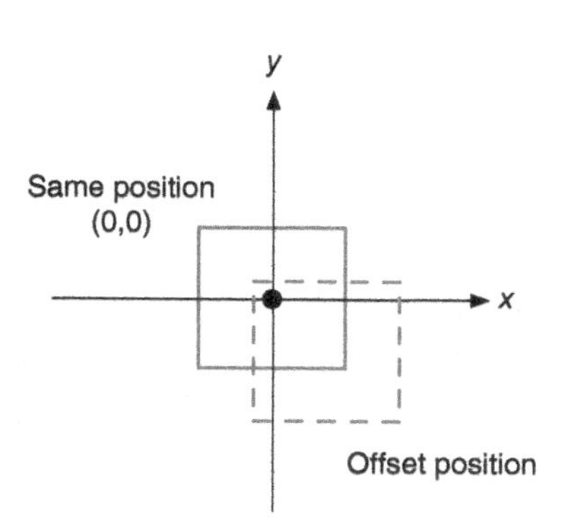

Figure 6.6 Offset to inter prediction source

2) Encoder subtracts each pixel of the prediction source from each pixel of the current block. The difference is the residual block C.
3) Encoder codes and transmits (1) the residual block and (2) the offset – motion vector (see Figure 6.8).

A decoder reconstructs a decoded block using inter prediction with the following steps:

1) Decoder receives and decodes (1) the residual block C and (2) the offset (see Figure 6.9).
2) Decoder creates the prediction source B by identifying the same square region in the previously coded frame using the offset received from the encoder.
3) Decoder adds each pixel of the residual block to each pixel of the prediction source. The result is the decoded block A, as shown in Figure 6.9. This may or may not be identical to the original block A, depending on whether information is discarded during encoding.

Figure 6.7 Block in current frame, offset region in previous frame

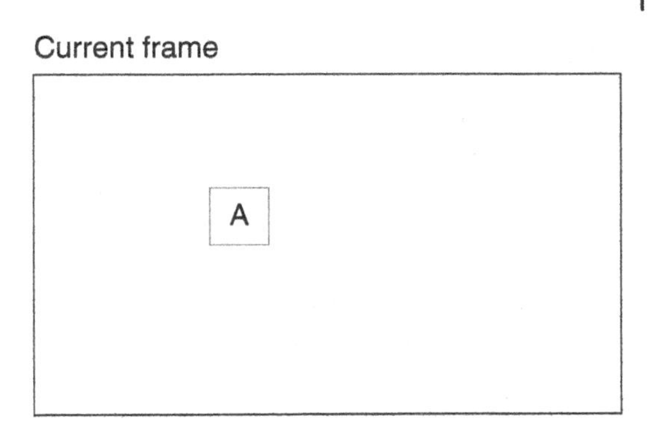

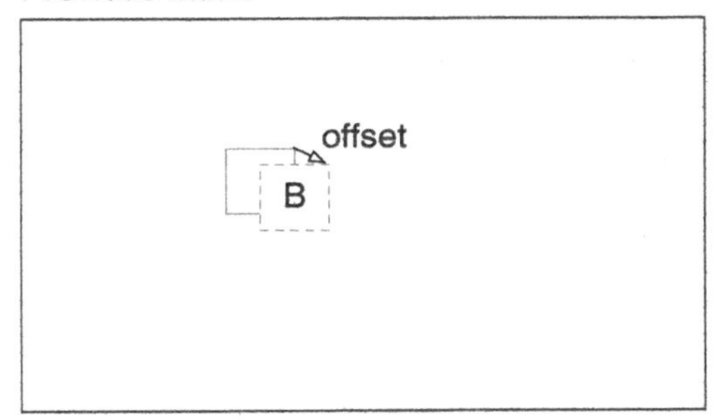

Figure 6.8 Inter prediction: encoding a block

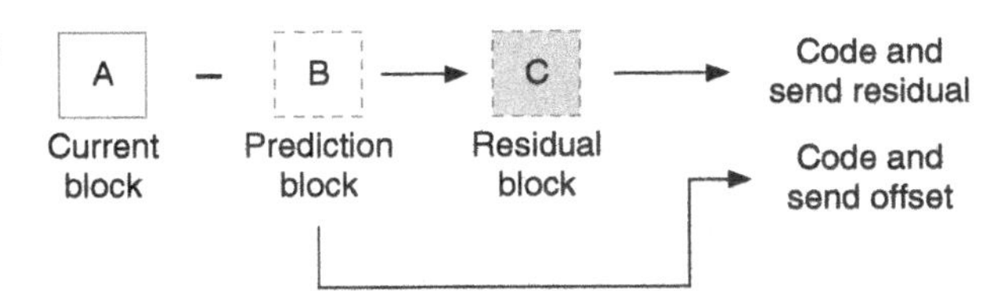

Figure 6.9 Inter prediction: decoding a block

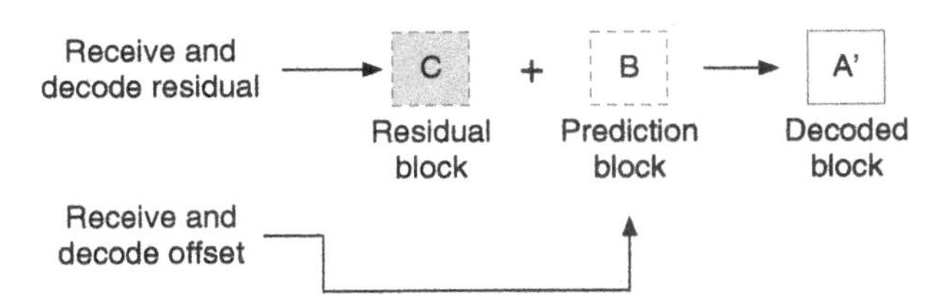

Example

Figure 6.10 shows a 4×4 square block of pixels in a current video frame, with the numerical values of each of the sixteen pixels shown. The encoder creates a prediction from the previous video frame. The region surrounding the position of the original block is shown in Figure 6.11. The encoder chooses a prediction source – in this example, this is the 4×4 area highlighted[1]. It is two pixels to the left, i.e. −2 in the x direction, and two pixels up, i.e. +2 in the y direction. Hence, the offset to the best prediction area is (−2, +2), as shown in Figure 6.12.

The encoder subtracts the pixel values of the prediction block, the area found at position (−2,+2) in the previous frame, to create a residual block shown in Figure 6.13. The residual block is coded and sent to the decoder, together with the offset or motion vector (−2, +2) and any other information needed by the decoder to re-create the prediction.

The decoder uses the offset (−2,+2) to find the same region and create the same prediction. It decodes the residual and adds the prediction and the residual to create a decoded block (Figure 6.14). Note that if the compression process is lossy, as in this example, the decoded residual is not identical to the residual created by the encoder. The result is that some of the decoded pixel values, highlighted in grey, are not identical to the originals.

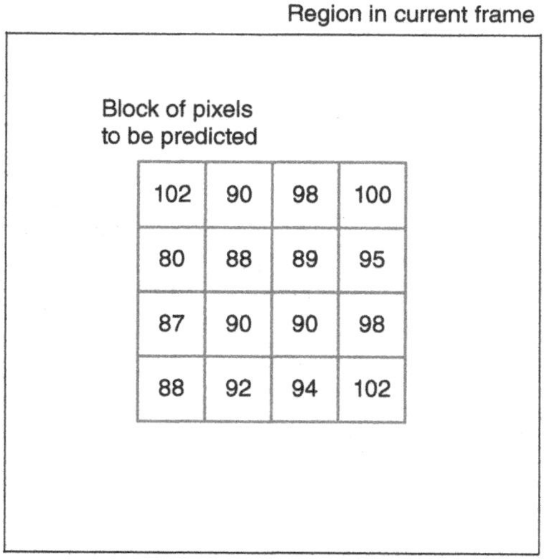

Region in current frame

Block of pixels to be predicted

102	90	98	100
80	88	89	95
87	90	90	98
88	92	94	102

Figure 6.10 4×4 block of pixels in current frame

1 How does the encoder choose a prediction source? As we saw earlier for intra-prediction, the encoder will typically choose a prediction that minimises some combination of (1) the coded residual size, (2) the number of bits required to tell the decoder how to re-create the prediction and (3) computing time.

Figure 6.11 Surrounding region in previous frame

Best match (–2, +2)

Region in previous frame

Y							
100	**92**	**98**	**100**	78	90	47	40
80	**87**	**82**	**96**	80	86	44	40
84	**87**	**90**	**90**	98	87	46	42
83	**88**	**92**	**98**	100	65	46	52
60	62	80	100	65	48	30	50
68	65	72	65	60	52	35	50
40	42	72	66	60	48	40	47
43	42	78	66	62	51	41	47

X

Same position (0,0)

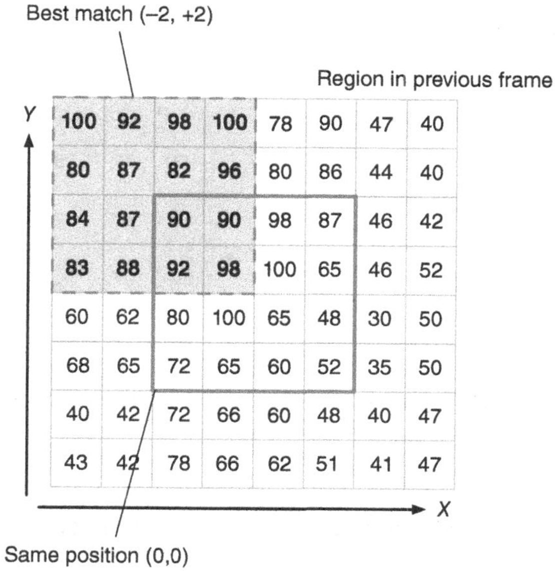

Figure 6.12 Offset or motion vector

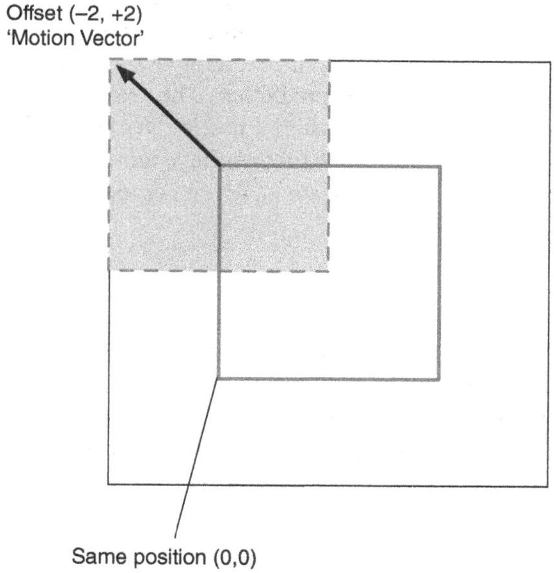

Offset (–2, +2)
'Motion Vector'

Same position (0,0)

Original block

102	90	98	100
80	88	89	95
87	90	90	98
88	92	94	102

–

Prediction block

100	92	98	100
80	87	82	96
84	87	90	90
83	88	92	98

=

Residual block

2	–2	0	0
0	1	7	–1
3	3	0	8
5	4	2	4

Figure 6.13 Original, prediction and residual blocks

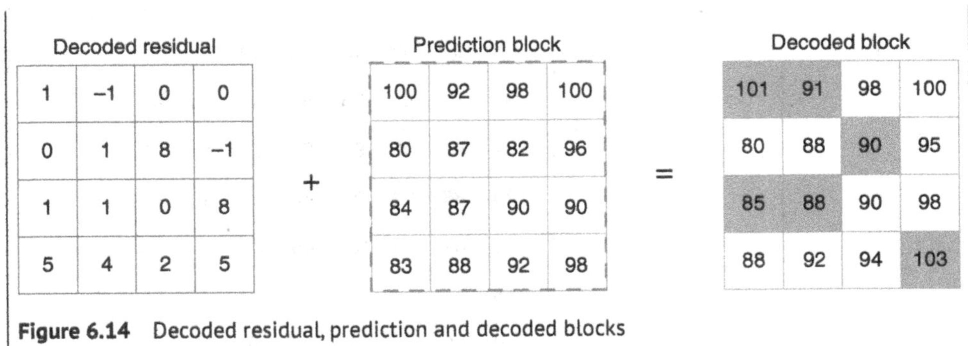

Figure 6.14 Decoded residual, prediction and decoded blocks

6.2.1 Motion Vectors for Inter Prediction: The Terminology Problem

In the previous section, we considered the offset from the current block position to the position of a prediction region in a previously coded frame. The offset from the current block to the position of the prediction block is conventionally described as a motion vector, and the process of predicting a block using inter prediction can be described as motion-compensated prediction. The term *motion vector* can be confusing for at least two reasons. Firstly, when a motion vector points to a prediction block in a previous frame, it points to where objects have moved from. Secondly, motion vectors point to the best match, which may or may not correspond to actual object motion. We will look at both of these aspects.

1) Motion vectors may not point in the direction of motion.

If the prediction source is in a previous frame and if objects are moving from frame to frame, then the offset points to where pixels are moving *from*, not where they are moving to.

Consider the ball shown in Figure 6.15. The ball is moving diagonally up and left, and so it changes its position from frame $N-1$, the previous frame, to frame N, the current frame. In the current frame N, the previous position of the ball is shown as a dotted line. Now consider block A in frame N, as shown in Figure 6.16. This block partly covers the moving ball, and we want to find an inter prediction source for block A in the previous frame $N-1$. The best prediction is where the ball moved from, i.e. down and right. When the object, the ball in this case, is moving up and left, the offset to the best prediction in the previous frame is down and right. So the offset or motion vector points to where the object moved from, which is exactly opposite to the direction of motion.

If we consider the example from the *Foreman* sequence (Figure 6.17), we see the same situation. From frame $N-1$ to frame N, the pixels of the man's face are moving up and to the left. The best prediction for the current block in frame N, outlined with a solid line, is a square region of pixels down and to the right, denoted with a dotted line. The offset or motion vector points to where the pixels moved *from*, i.e. down and to the right.

Note that if we predict from a future frame, then the motion vector points to where the block of pixels is moving *to*, i.e. in the direction of motion.

Figure 6.15 Object moving up and to the left

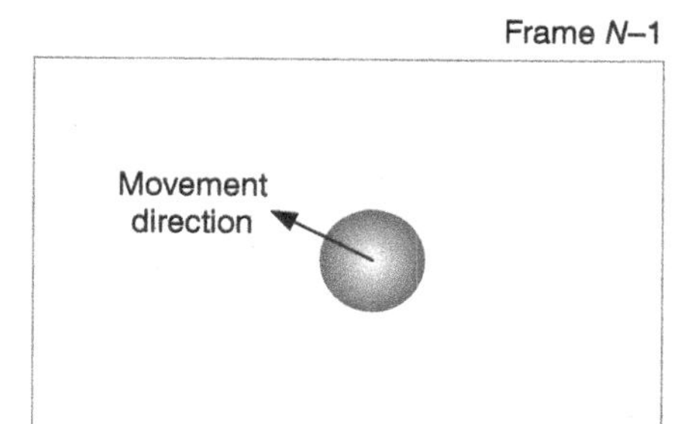

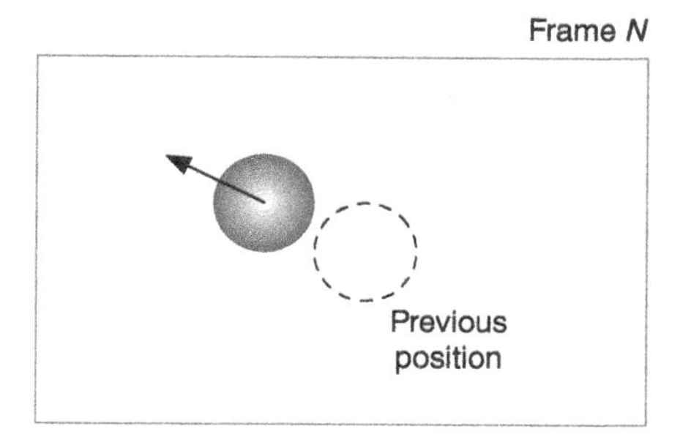

Figure 6.16 Block in frame N with offset to prediction source in frame $N-1$

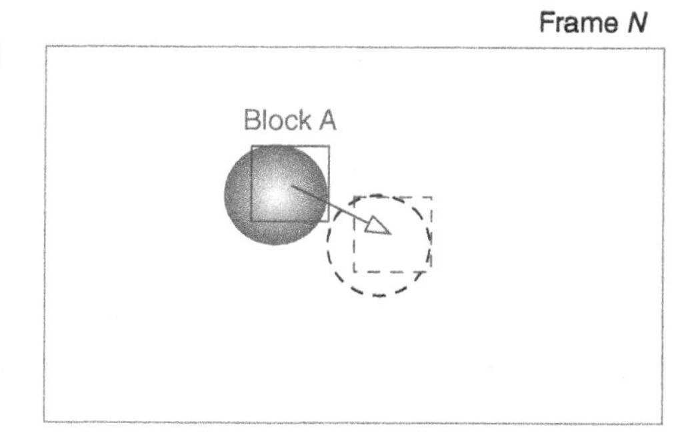

2) Motion vectors do not necessarily correspond to motion.

Consider two frames from a video clip (Figures 6.18 and 6.19). The motion vectors for each block of Frame 2 are superimposed on the frame. In this clip, the camera position is fixed and the boats and the water are moving gently. Some of the motion vectors point in seemingly random directions, particularly in the water at the bottom of the frame and in the sky above the boats, as shown in Figure 6.19.

When the encoder chooses a motion vector for a block, it is choosing the offset to the best prediction, i.e. the prediction that minimises the number of coded bits for the block, including the coded residual and the coded motion vector. Sometimes this offset corresponds to motion, and sometimes it does not. In this example, the water texture changes from frame to frame, so it is difficult to find a good match in the previous frame. The encoder searches for the best match and finds a match in a seemingly random position. This is not a mistake

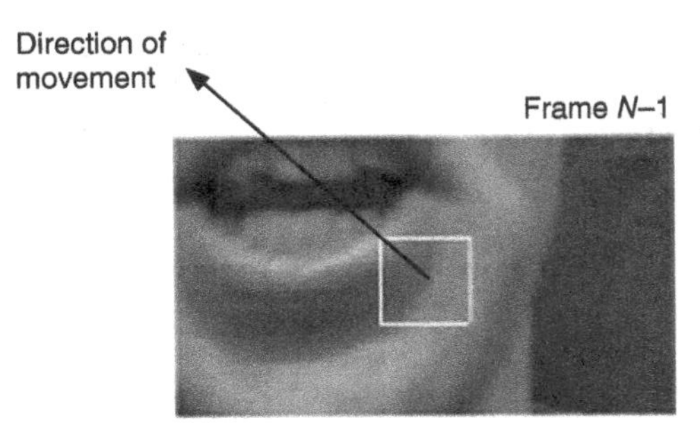

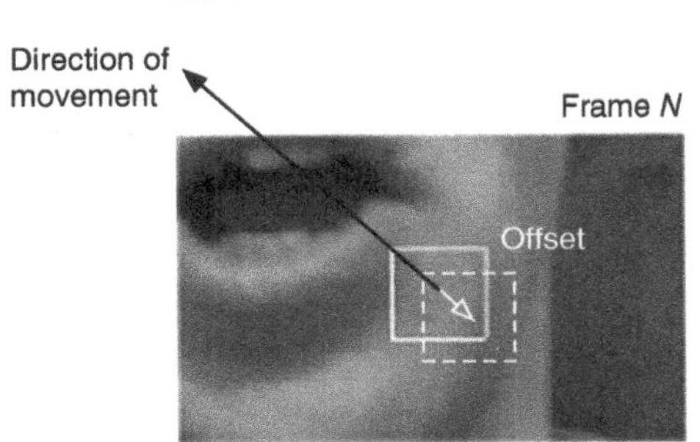

Figure 6.17 Inter prediction: offset and direction of movement

Figure 6.18 Boats at *Gourdon* – Frame 1

Figure 6.19 Boats at *Gourdon* – Frame 2 with motion vectors. Overlay source: Elecard Streameye

or a failure – if the offset or motion vector identifies the prediction that minimises the number of coded bits, then the encoder has done its job successfully, whether or not the motion vector actually corresponds to motion.

6.3 Forward, Backward and Biprediction

In Section 6.2, we saw how a video codec can predict a block in the current frame from a region of the same size as the previous frame. In theory, a video encoder can use any data as a prediction source, as long as it has already been coded by the time the encoder makes the prediction. If the data have already been coded, then they have already been sent to the decoder, and so the decoder can create the same prediction.

Possible sources for creating an inter prediction include:

- The previous frame,
- Older past frames,
- Future frames,
- Combinations of previously coded frames, past and/or future.

6.3.1 Picture or Slice Types

In Chapter 4, we saw that a picture or slice, which itself corresponds to part or all of a picture, can be coded as an I-, P- or B-picture or I-, P- or B-slice, using one of the popular video coding standards.

In an I-picture or I-slice, all the coding units (CUs) or Macroblocks are coded using intra prediction, i.e. there is no option to use inter prediction. In a P-picture or P-slice, CUs or MBs may be coded using inter prediction with one reference picture, or using intra prediction.

In a B-picture or B-slice, CUs or MBs may be coded using inter prediction with two reference pictures, which is known as biprediction. They may alternatively be coded using inter prediction using one reference picture, or intra prediction.

This means that when encoding a P- or B-slice, an encoder can choose a mix of prediction types depending on what is best for the current MB or CU. For example, in a B-slice, one CU may be efficiently predicted by using two prediction sources, i.e. biprediction. The next CU may be difficult to predict from previously coded frames, and the best prediction choice might be to use intra prediction. Thus, the difference between I-, P- and B-slices is in the choices of prediction types that are available.

6.3.2 Forward Prediction from Past Frames

Instead of predicting from the immediately preceding frame $N-1$, the encoder can choose to predict a block from an older frame, e.g. $N-2$, $N-3$, as shown in Figure 6.20. Predicting from an older frame, i.e. a frame that should be displayed before the current video frame, can be described as forward prediction. For many blocks, the best prediction may well be found in the previous frame $N-1$. However, in some cases, a better prediction might be available in an older frame. For example, if there is periodic movement in the scene, such

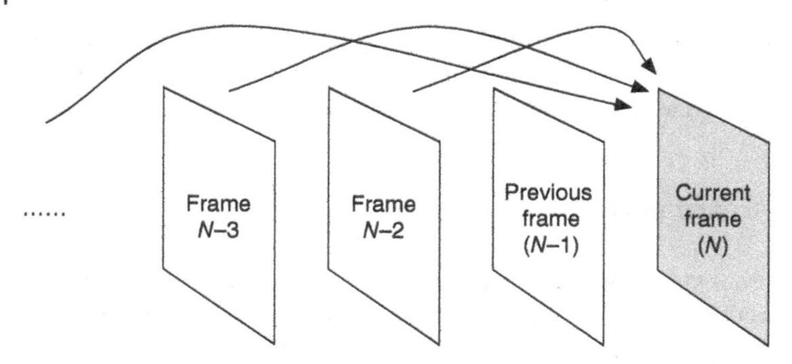

Figure 6.20 Prediction from past frames

as the moving waves in the *Gourdon* clip shown in Figure 6.18, we might find a matching block in an older frame rather than in the previous frame. Alternatively, if the changes from frame to frame are complex and somewhat random, we might simply find a prediction that reduces the residual by looking at an older frame. The frame we choose to predict from is known as a reference frame or reference picture.

The above discussion implies that the more reference frames we consider, the better. However, there are downsides to considering many previous frames as possible reference frames for each block of pixels. First, the encoder has to signal the choice of prediction frame, i.e. $N-1$, $N-2$, etc., which may take more bits than simply assuming prediction from the previous frame. Second, both the encoder and decoder have to store all of the frames that might be used for prediction, which may require more memory to store all of the pixels of the previously coded frames. Third, searching for a prediction match, known as motion estimation, is computationally intensive, and there is often a diminishing return, i.e. a point at which it is not worth the computational effort to search a much older reference frame.

6.3.3 Backward Prediction from Future Frames

Another option is to create an inter prediction from a future frame, i.e. a frame that will be displayed after the current frame. An encoder can predict a block in current frame N from a region in frame $N+1$, $N+2$, $N+3$, as illustrated in Figure 6.21. This may seem counter-intuitive since these frames should be displayed in the future. However, backwards prediction is feasible if the encoder and decoder can change the coding order of video frames so that any future frames used for prediction are sent before the current video frame. This has implications for coding and decoding delays, as we will see later. Once again, the encoder has to communicate to the decoder the choice of prediction source, i.e. which future frame contains the source of the prediction.

Backwards prediction can be particularly useful when areas of a video frame are uncovered. Consider frame N, as shown in the centre of Figure 6.22. Behind the moving jockey, the number 2 is uncovered as he passes. The pixels in this uncovered region do not exist in the previous frame, frame $N-1$. However, they are available in the next frame, frame $N+1$.

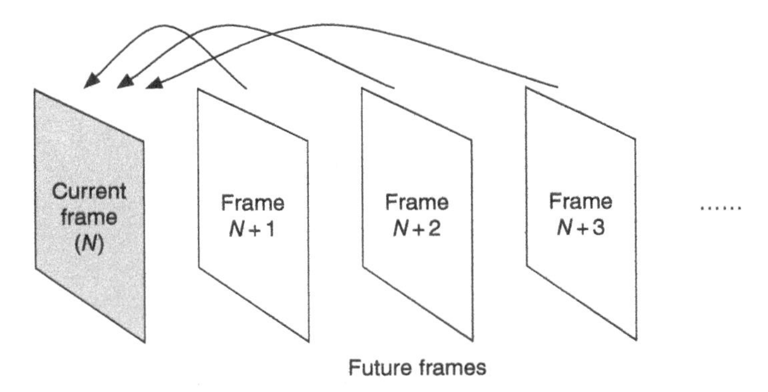

Figure 6.21 Prediction from future frames

Figure 6.22 Uncovered region – predict from future frame

For this region, the encoder can probably find a better prediction by looking at frame $N + 1$ rather than frame $N - 1$.

Consider two blocks, labelled 1 and 2, in a close-up region of Frame N shown in Figure 6.23. Block 1 covers the jockey's shoulder. The same region is clearly visible in the previous frame, frame $N - 1$, and the encoder predicts Block 1 from this region in frame $N - 1$. Block 2 is part of the number 2. The same region is covered in Frame $N - 1$ but available in the future frame

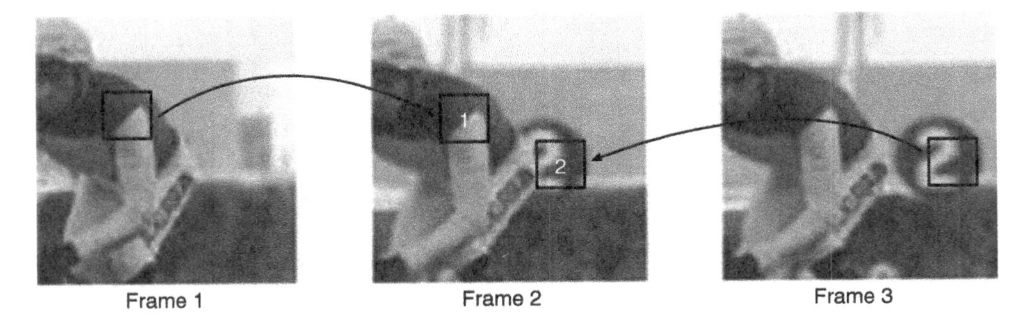

Figure 6.23 Predicting from past or future frames: example

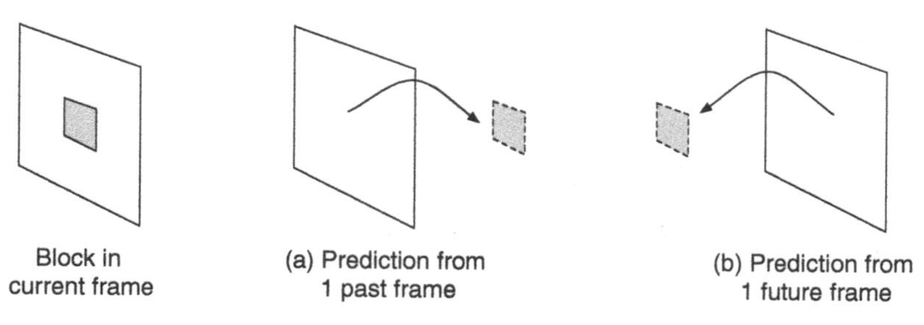

Figure 6.24 Prediction from a past or future frame

$N+1$, and so the encoder predicts Block 2 from the future frame, $N+1$. The reference frame for Block 1 is frame $N-1$, and the reference frame for Block 2 is frame $N+1$.

Figure 6.24 illustrates the two options of creating a prediction from a past frame – forward prediction, and creating a prediction from a future frame – backward prediction.

6.3.4 Biprediction from Two Frames

A third option is to create an inter prediction by combining pixels in two reference frames. Figure 6.25 shows current frame N, which is part-way through a dissolve or fade between two scenes, a crowd scene shown in Frame $N-1$ and a city scene shown in Frame $N+1$. The highlighted block contains elements of both scenes – it is a composite of a girl's shoulder from the crowd scene and a building from the cityscape. It may not be possible to create an accurate prediction from either the past frame $N-1$ or the future frame $N+1$. However, the encoder might be able to create a good prediction by combining or averaging pixels from the past frame and the future frame. This is known as biprediction.

A biprediction block is created by averaging each pixel of two source blocks, each from a different frame, as shown in Figure 6.26. The encoder identifies a block in the first frame, Source 1, and a block in the second frame, Source 2. The biprediction block is formed by averaging each pair of pixels from Source 1 and Source 2. This biprediction block is then subtracted from the original block to create a prediction residual. Biprediction, predicting a block from two reference frames, can often give a better prediction than uni-directional prediction, which predicts from a single reference frame. However, the encoder needs to

Figure 6.25 Biprediction example: create block in frame N as average of past and future regions

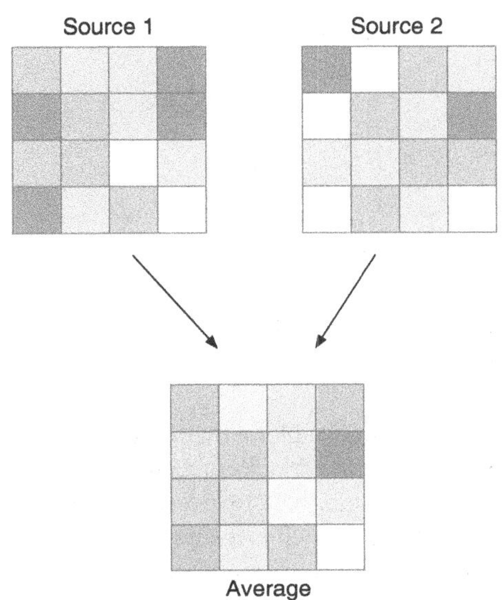

Figure 6.26 Creating a bipredicted block

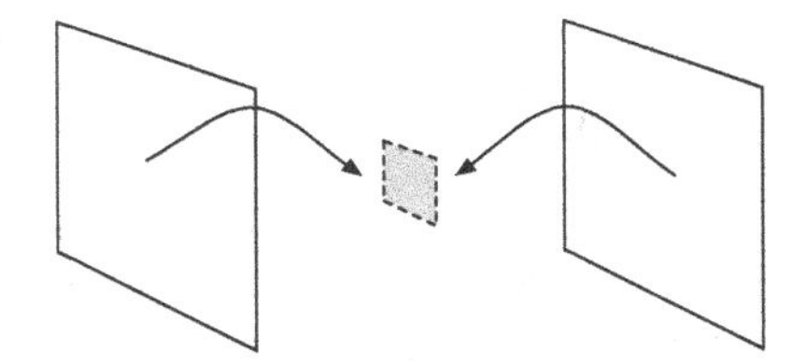

Figure 6.27 Biprediction: one past and one future frame

send offsets to two prediction source blocks, e.g. Source 1 and Source 2 in Figure 6.26, so biprediction may require more bits to signal the choice of prediction source.

Biprediction has been a feature in video coding since the publication of the MPEG-1 and MPEG-2 video coding standards in the early 1990s. In these early standards, the encoder was restricted to creating a biprediction using a specific past frame and a specific future frame (Figure 6.27). Recent standards such as H.264 and HEVC give the encoder more flexibility so that a bipredicted block can be created from one past and one future frame, or from two past frames, or from two future frames, as illustrated in Figure 6.28. In theory, more prediction options are possible, such as creating a prediction by combining data from three or more frames. Again, biprediction has implications for the way in which reference pictures are handled, which we will discuss further in Section 6.7.

6.4 Inter Prediction Block Sizes

In the previous sections, we saw how inter prediction predicts a block within the current frame from a region in a previously coded frame. How should the encoder choose the *size* of this block?

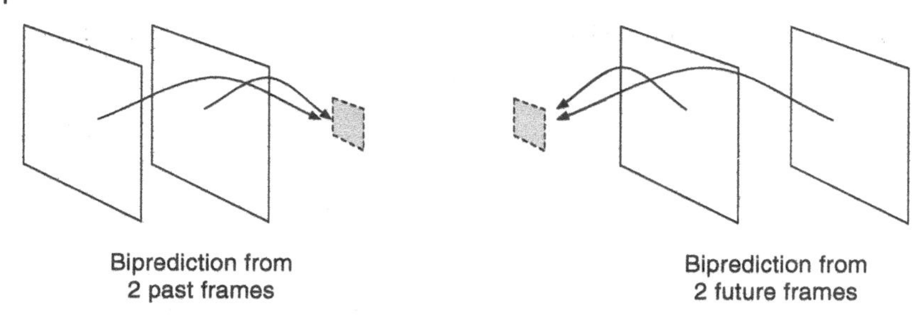

Figure 6.28 Biprediction: two past or two future frames

We saw in Chapter 4 that a video encoder partitions each frame into basic units of coding known as macroblocks or Coding Tree Units. These are square blocks, typically a power of 2 to each side, e.g. 16×16 pixel macroblocks or CTUs of size 16×16, 32×32 or 64×64. A prediction block is a subset of a macroblock or CTU. We could predict the entire macroblock or CTU as one prediction block by choosing a prediction block size equal to the macroblock or CTU size. Alternatively, we could choose a smaller block size, such as a square or rectangular block that is smaller than the macroblock or CTU.

Example 1

Figure 6.29 shows a frame from the *Jockey* sequence partitioned with a larger prediction block size on the left and a smaller block size on the right. The larger block size has the advantage that there will be fewer motion vectors to send, so prediction and coding might be more efficient for areas of the background that are easy to predict. The smaller block size on the right may be better for predicting complicated

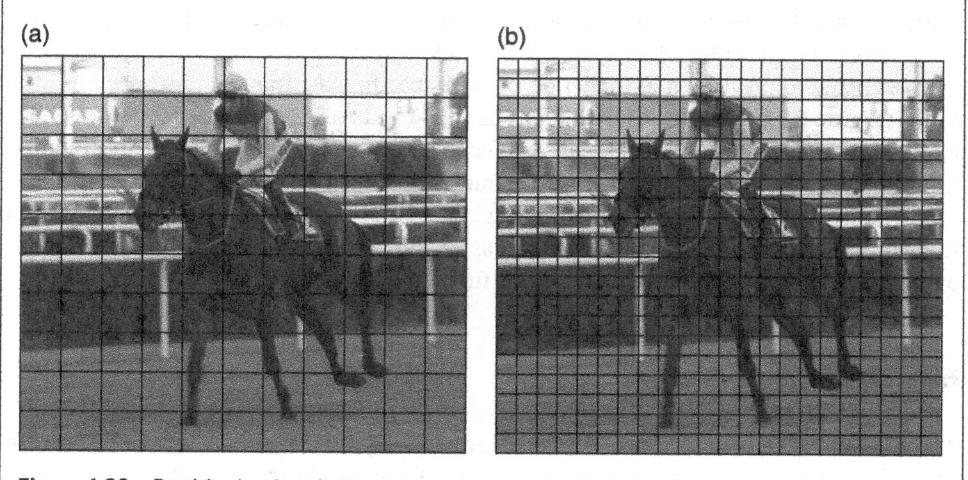

Figure 6.29 Partitioning into larger (a) or smaller (b) prediction block sizes

Figure 6.30 Partitioning using a mixture of prediction block sizes

areas such as the horse and the jockey. As with intra prediction, the best option may be to switch between two or more block sizes, using larger blocks for areas of the background with less movement and smaller blocks for complex moving regions, such as the example in Figure 6.30. Here, for areas of background or areas such as the horse's body where all the pixels are moving roughly in the same direction, the encoder chooses a large prediction block size. For areas containing more complex movement, such as the boundaries of the moving regions, the encoder chooses a small prediction block size.

The encoder communicates extra information for every prediction block. As well as signalling the choice of prediction block size and perhaps further prediction mode information, the encoder must communicate a motion vector for each inter prediction block. As we will see later, there are a number of ways the encoder can reduce the amount of extra overhead information, such as by inferring or by predicting some or all of the overhead. In general, however, smaller prediction block sizes mean more prediction blocks per frame and more information sent to describe the mode and motion vector of each prediction block.

Square prediction blocks are computationally simple to process and require little overhead to signal, especially if their dimensions are powers of 2. For example, if we start with a square basic unit of coding such as a CTU and split up the unit using a quadtree (see Chapter 4), a single bit is enough to indicate a split into four 32×32 prediction blocks or prediction units (PUs). Rectangular prediction blocks may require more computational complexity and more overhead to signal the block size. Successive standards have introduced increasing flexibility into the size and shape of inter prediction blocks, as Figure 6.31 illustrates.

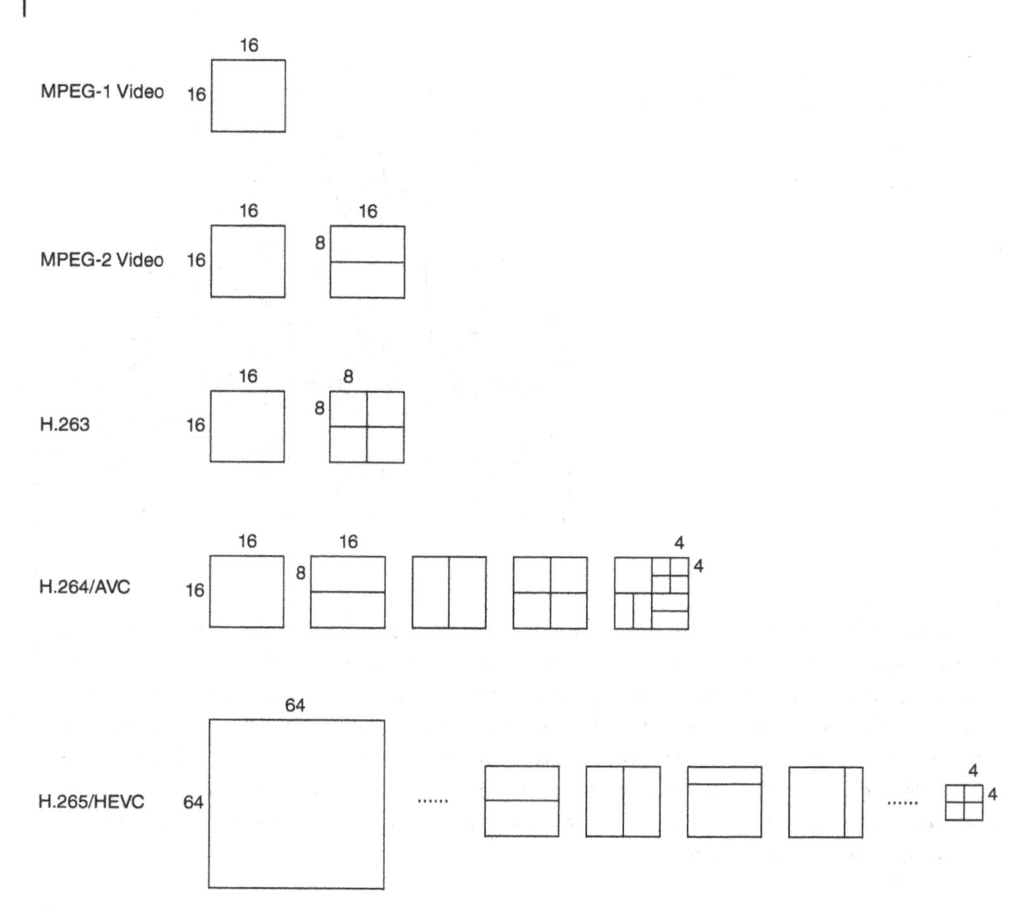

Figure 6.31 Inter prediction block sizes in selected standards

6.5 Motion Vectors

Inter prediction attempts to generate a prediction block or region that is similar to a current block or region. A motion vector is an offset from the position of the current block to the position of the prediction block. As we saw earlier, a motion vector may or may not correspond to the actual motion of the pixels from one video frame to another.

Each motion vector points to the location of the prediction region relative to the current block position. A motion vector has an *x*-component, which indicates the horizontal shift, and a *y* component, which indicates the vertical shift. For example, the motion vector shown in Figure 6.32 has an *x*-component of −2 and a *y*-component of +3, which means that the prediction region is 2 pixels to the left and 3 pixels up from the current block position.

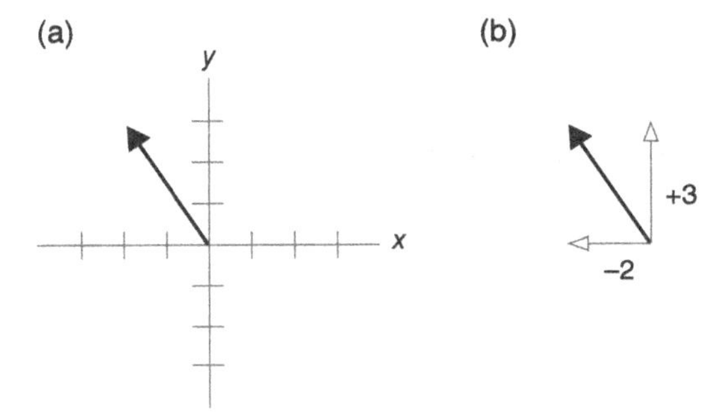

Figure 6.32 Motion vector (a), x and y components (b)

Example 1

Figure 6.33 shows two frames from the *Foreman* sequence, each 352×288 pixels. Each 16×16 block of Frame 2 (see Figure 6.34) is predicted from a 16×16 region in Frame 1.

Figure 6.33 Two frames from the *Foreman* sequence (352×288 pixels)

Figure 6.34 16×16 blocks to be predicted (Frame 2)

The motion vectors of each block are shown in Figure 6.35. There are several interesting aspects to note:

1) There are some larger, random-looking vectors near the top of the image. These larger vectors occur within the area of the white hat. They do not correspond to actual motion but are merely somewhat random best matches because there is no matching texture for the encoder to find in the previous frame.
2) Apart from these random vectors, most of the other vectors are small or zero. The distribution of the x-component of the motion vectors is shown in Figure 6.36. The lower axis shows each of the possible vector values, from −10 to +10, and the left axis shows the number of x components that have each value. Of the 396 blocks in Frame 2, over 200 have zero x vectors and most of the rest are clustered around small positive and negative values.
3) If you consider any block position in the frame, the current vector is usually very similar to neighbouring vectors. Thus, there is a high probability that the current block's motion vector is the same as or similar to neighbouring block motion vectors.

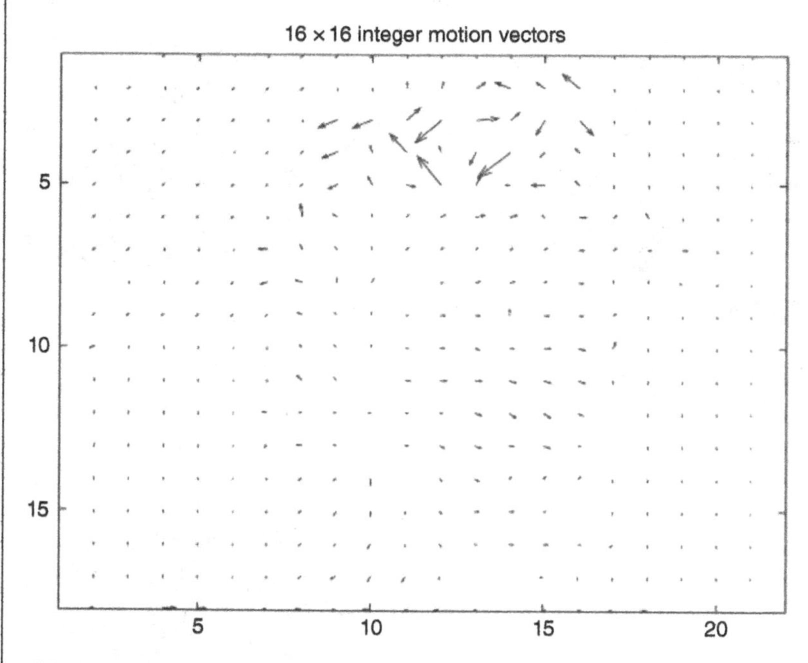

Figure 6.35 Motion vectors for each 16×16 block

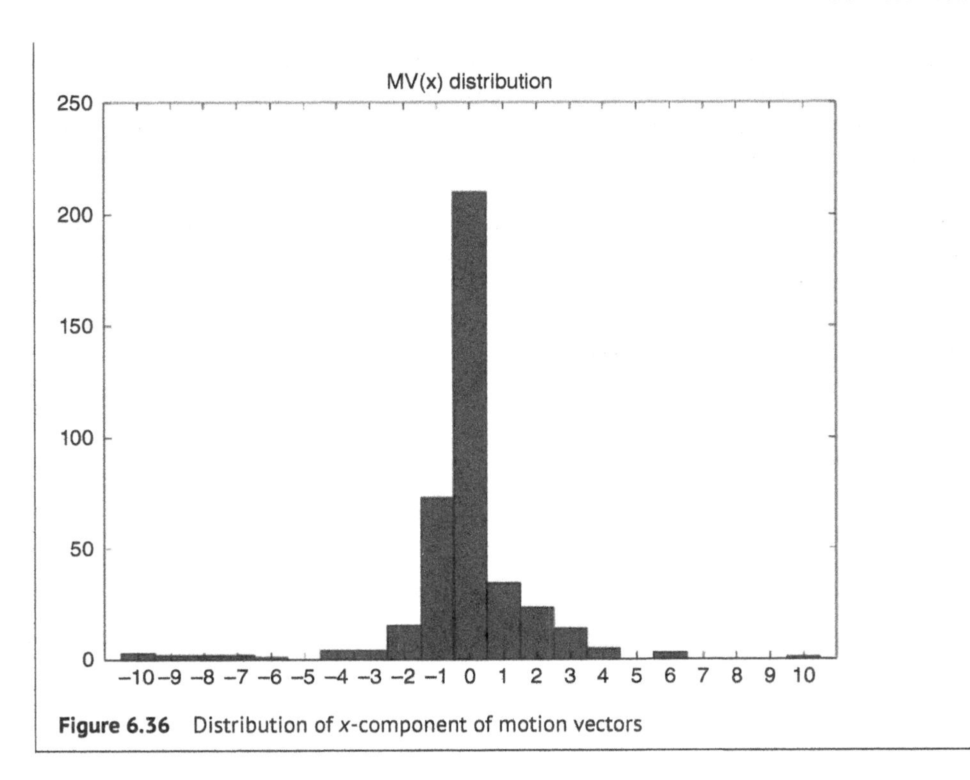

Figure 6.36 Distribution of x-component of motion vectors

Example 2

Another pair of frames from the *Foreman* clip is shown in Figure 6.37. This time, the camera is panning to the right, which means that most of the frame is moving to the left. Notice the right-hand side of the building. The pixels of the building move to the left in Frame 2, as the camera pans to the right.

Figure 6.37 Two frames from *Foreman*, panning camera

We can see that most of the 16×16 block motion vectors for frame 2, as shown in Figure 6.38, are pointing to the right. As we saw earlier, if pixels are moving to the left from the reference frame, Frame 1, to the current frame, Frame 2, then the motion vectors will point to the right.

Because most of the motion vectors are pointing to the right, the distribution of motion vectors has changed, as shown in Figure 6.38. We can see that the dominant motion vector value in the x-direction is around +6. Once again, motion vectors tend to be similar to those of neighbouring blocks.

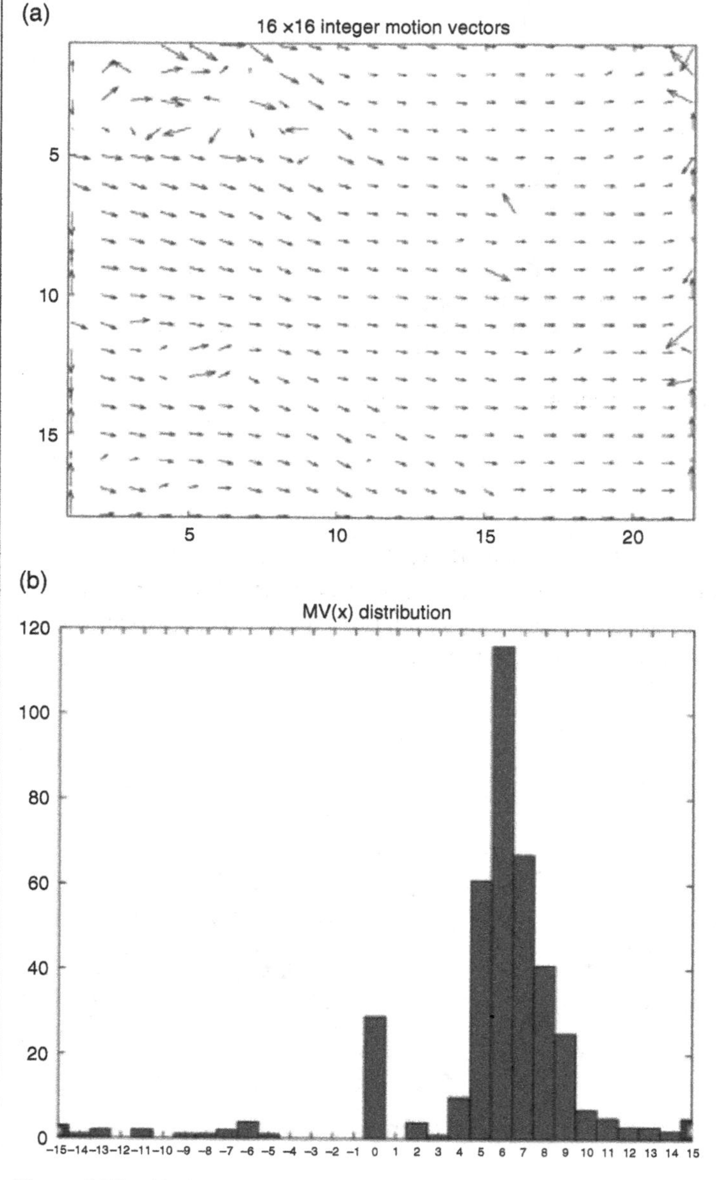

Figure 6.38 Motion vectors (a), distribution of motion vector x-values (b)

6.5.1 How Does the Encoder Choose Each Motion Vector?

In the above examples, the encoder selected the motion vector that gives the best match to the current block. The best match is the prediction block that minimises a particular measure of difference, which in these examples was the mean squared error (MSE) between the current block and the prediction block. MSE can be calculated by:

- Subtracting each pixel of the prediction from each pixel of the current block (d)
- Calculating the square of the difference (d^2)
- Adding up the result for each pair of pixels in the block, 256 values in the case of a 16×16 pixel block
- Dividing by the number of pixels in the block, 256 in this case.

The smaller the difference between pairs of pixels in the prediction block and the current block, the smaller the calculated value of MSE will be. This means that in this example, the motion vector for each block points to the prediction block that gives the smallest MSE.

Inter prediction motion vectors tend to have certain predictable properties, such as:

- Motion vector magnitudes tend to be clustered around 0 if the camera position is fixed, or around a fixed value if the camera is moving.
- Motion vectors in a local area of the image tend to be correlated, so the current block motion vector is likely to be similar to neighbouring block motion vectors.

We will see in Section 6.8 that these properties can be used to minimise the number of bits required to send each motion vector.

6.6 Sub-Pixel Interpolation

So far, we have discussed motion compensation, i.e. inter prediction using motion vectors, where each motion vector points to a position in a reference frame that is offset by a number of pixels. Figure 6.39 shows a current frame containing a ball. The ball exists in the

Figure 6.39 Object moving by an integer number of pixels

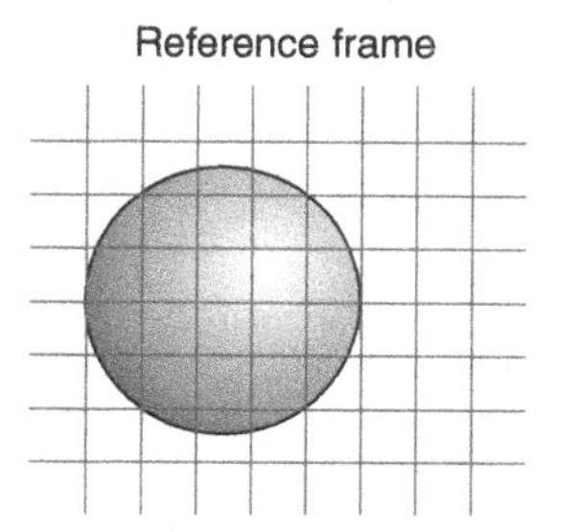

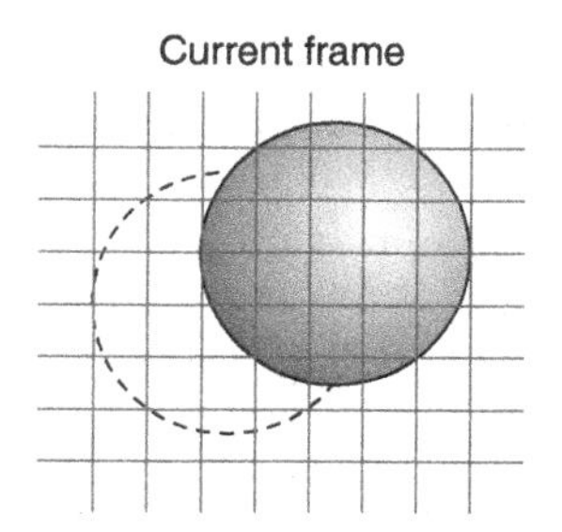

Motion vector: (–2, –1)

reference frame in a position two pixels left and one pixel down. If we want to predict the block containing the ball, we can use a motion vector (−2, −1), and we will create an accurate prediction.

Unfortunately, objects in a scene rarely move in neat, integer steps. In Figure 6.40, the ball in the current frame is in the same position as the previous example. However, the ball in the reference frame can be found 1.5 pixels to the left and 0.5 pixels down, i.e. the ball has moved by a non-integer number of pixels from one frame to the next. In a natural video scene, objects will tend to move by arbitrary amounts, for example, by 0.87 pixels, −14.92 pixels, etc.

We can create a better prediction in the example of Figure 6.40 if we can make it possible to send a non-integer motion vector, in this case (−1.5, +0.5). We can do this by re-sampling the reference frame to a higher resolution, for example, by re-sampling at half-pixel positions as shown in Figure 6.41 and choosing a motion vector that points to pixels in the re-sampled reference frame.

You may have noticed that the above examples are simplifications of the real situation. In reality, the ball would appear much more pixelated than shown. Let us look at a more realistic example. Figure 6.42 shows a section of a current frame with a dark vertical line that is two pixels wide. The same line exists in the reference frame shown on the left of the figure, but because of fractional-pixel movement from one frame to the next, the line appears smeared across three horizontal pixels. If the encoder tries to predict a 4×4 block in the top left of the region, Figure 6.43, right, it will fail to find a good match at an integer-offset position, Figure 6.43, left.

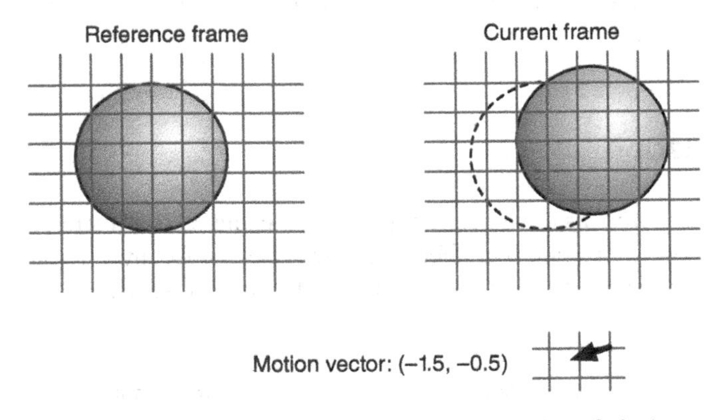

Figure 6.40 Object moving by a non-integer number of pixels

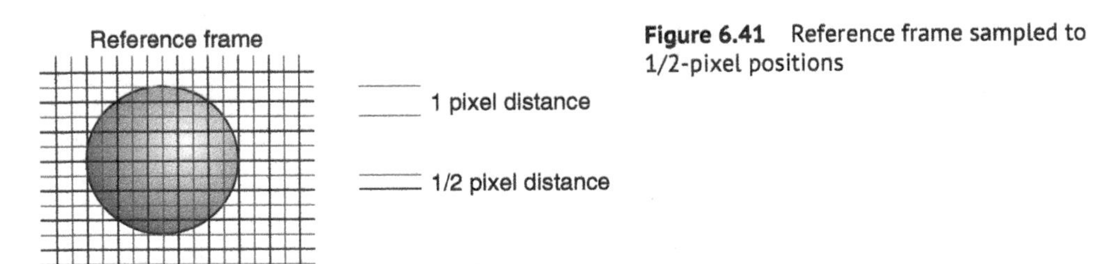

Figure 6.41 Reference frame sampled to 1/2-pixel positions

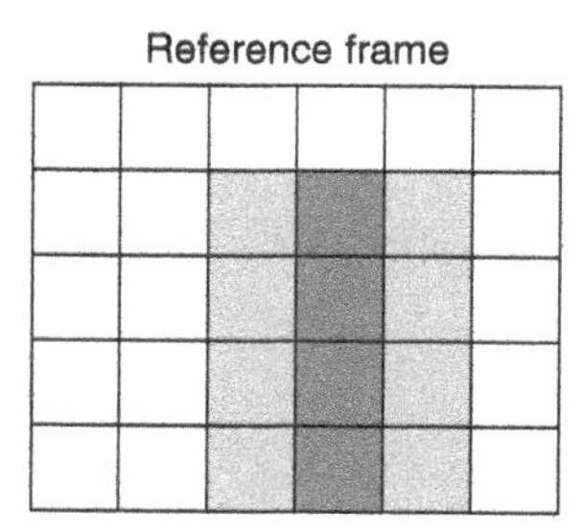

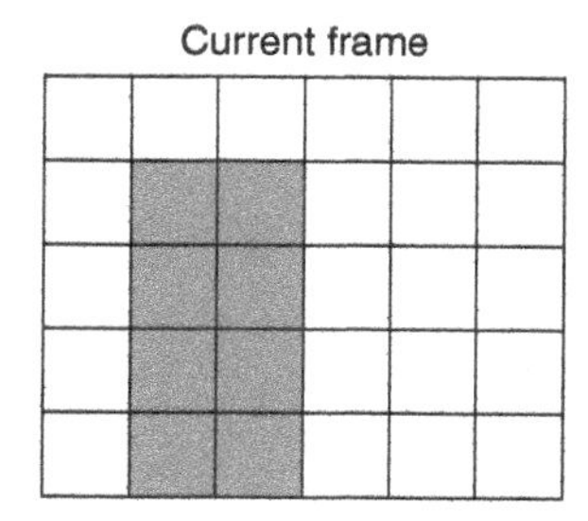

Figure 6.42 Reference frame and current frame, integer pixel positions

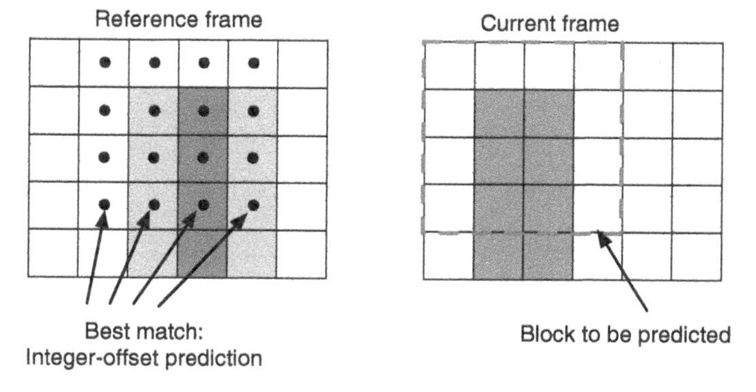

Figure 6.43 Block to be predicted, best match, integer motion vector

Figure 6.44 shows the reference frame area interpolated to half-pixel accuracy. The original, integer pixel positions are marked with white circles. The missing half-pixel positions are interpolated from neighbouring integer pixel positions. We have interpolated to twice the number of samples in the horizontal and vertical positions.

The encoder can now search for a 4×4 block within this expanded, re-sampled region. We can find a better match in the position shown in Figure 6.45. The motion vector that

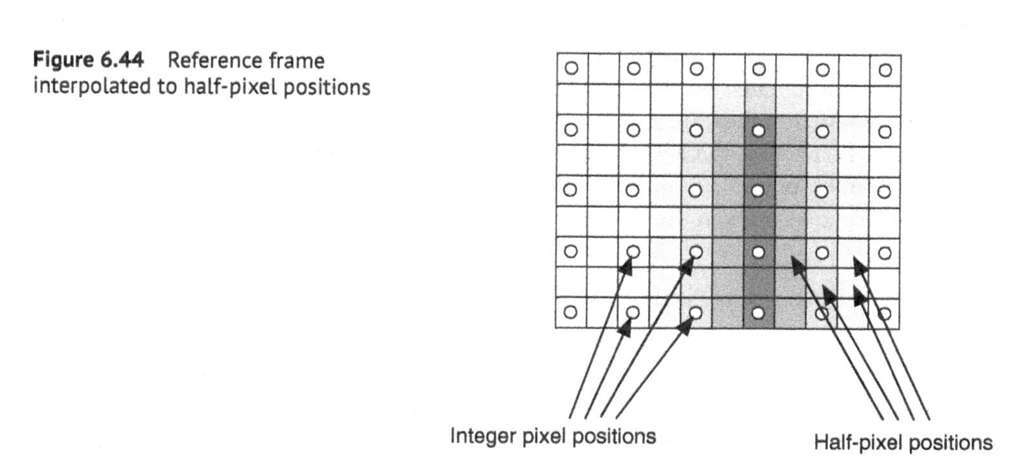

Figure 6.44 Reference frame interpolated to half-pixel positions

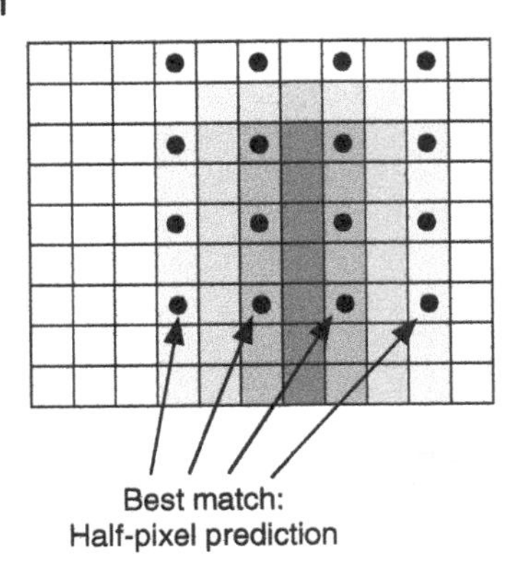

Figure 6.45 Best match in re-sampled reference frame

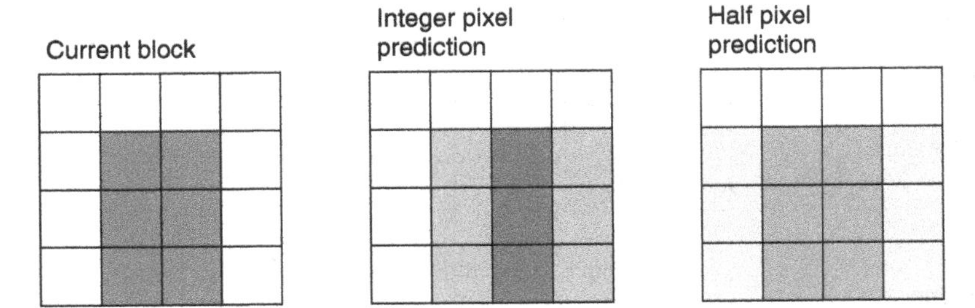

Figure 6.46 Result of integer and half-pixel prediction

points to this block is (+1.5, 0). This is not a perfect match because the re-sampled pixel values are not identical to the current frame pixel values, but it is a better match than the integer position shown in Figure 6.43.

Figure 6.46 shows the current 4×4 block and the results of creating an integer-pixel prediction and a half-pixel prediction. The integer prediction is clearly not a good match for the original block. The half-pixel prediction shows the 'best match' sample values selected from the half-pixel interpolated grid, i.e. the black dots in Figure 6.45. The half-pixel prediction is a better match, but it is not an exact match. The brightness or luma levels of the interpolated pixel positions are not identical to the original pixels, because these interpolated values are created by averaging integer pixel values, as shown in Figure 6.44. In a sense, the actual pixel values we are searching for do not exist in the reference frame. At most, sub-pixel prediction can create interpolated samples that are closer to the pixel values in the current frame, as this example shows.

6.6.1 What Does a Sub-Pixel Interpolated Reference Frame Look Like?

On the left-hand side of Figure 6.47 is a close-up of a Quarter Common Intermediate Format (QCIF) frame, resolution 176×144 pixels, without any interpolation. Interpolating the frame to half-pixel positions, shown as the middle image, creates four times the number of pixels. This interpolated image looks smoother, but there is no new information in it. In

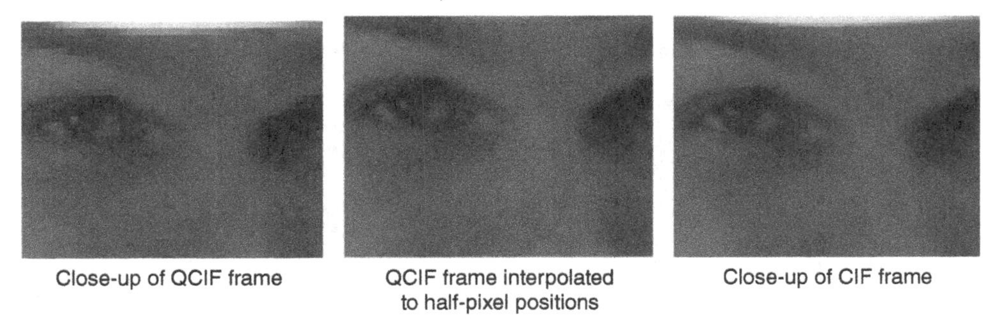

Figure 6.47 QCIF frame close-up, interpolated version, CIF version

contrast, a higher-resolution version of the original frame, shown at the right, has more actual detail. For example, the eye in an actual higher-resolution CIF frame on the right is clearer and more detailed than the eye in the interpolated QCIF frame in the middle, even though there is the same number of pixels in both images, i.e. 352×288 pixels for the whole frame.

6.6.2 Interpolation Filters

Interpolated sub-pixel positions are created by combining neighbouring pixel values, typically using filtering. The way in which neighbouring values are combined or filtered has an effect on the quality of the interpolation. For example, on the left of Figure 6.49 is a 40×40 pixel close-up of a QCIF frame from the *Foreman* sequence, shown in Figure 6.48. For comparison, Figure 6.49 shows the same region from the CIF version of the frame on the right. Note that the QCIF frame is significantly less detailed than the CIF frame, as we would expect.

Figure 6.48 QCIF frame from *Foreman* sequence

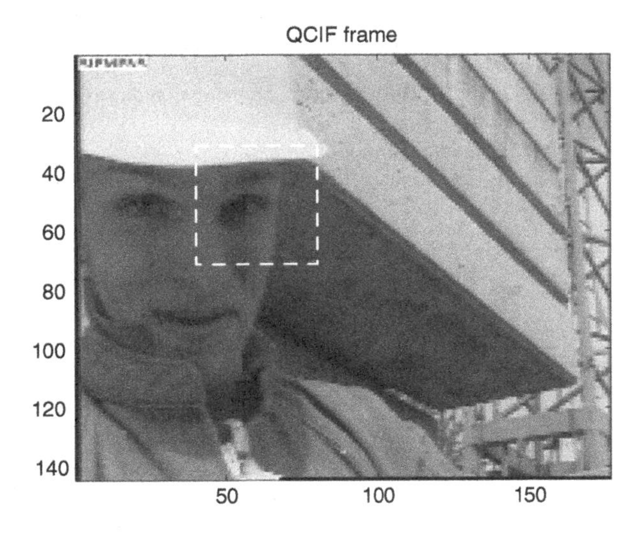

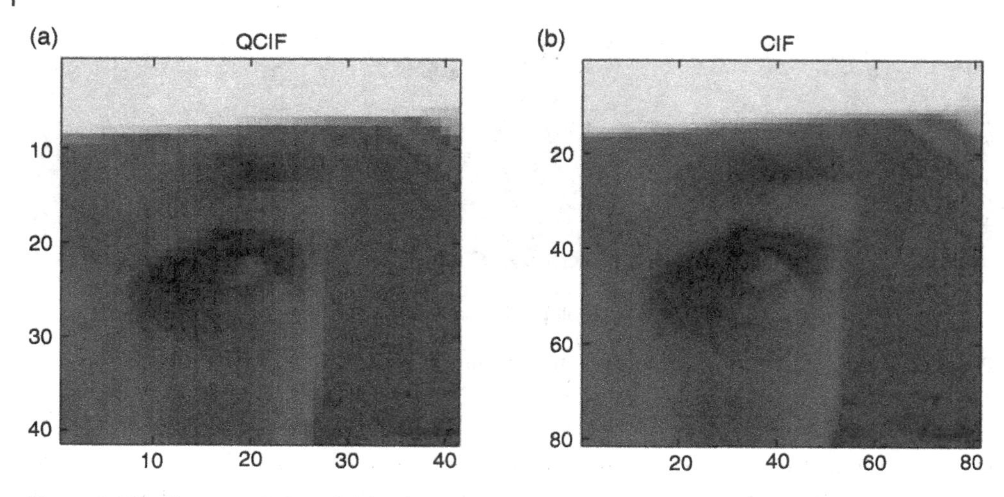

Figure 6.49 Close-up region of QCIF frame (a) and double-resolution CIF frame (b)

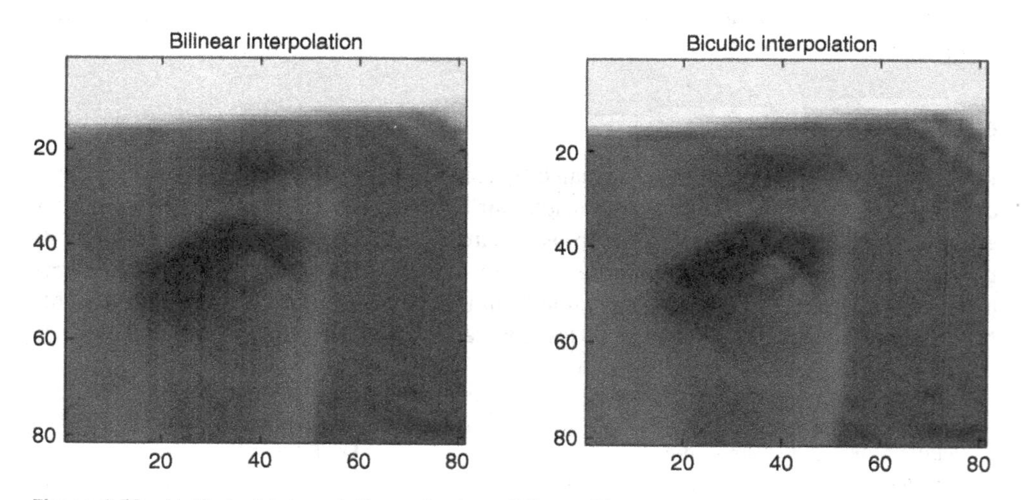

Figure 6.50 Half-pixel interpolation using two different filters

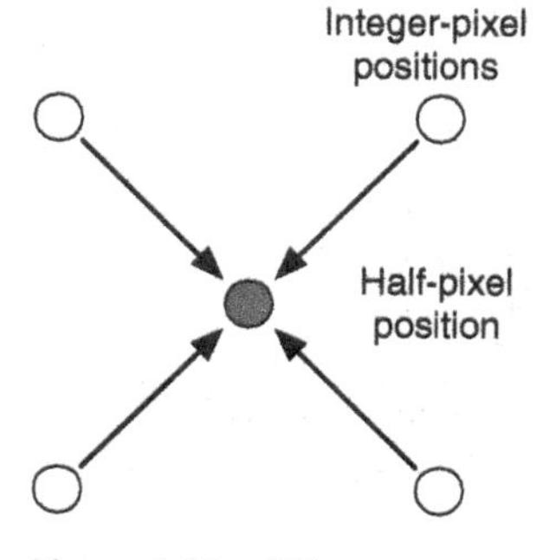

Figure 6.51 Bilinear interpolation

If we interpolate the half-sample positions in the QCIF region, we get an interpolated version that looks like one of the examples in Figure 6.50. Each of the two examples is generated using a different interpolation filter. The left-hand example uses bilinear interpolation, i.e. simple averaging of the neighbouring pixels, to generate each half-pixel location. Each half-pixel position is generated by averaging up to four neighbouring integer-pixel positions, as illustrated in Figure 6.51. The right-hand example in Figure 6.50 uses bicubic interpolation, where each half-pixel position is generated based on up to sixteen neighbouring integer-pixel positions,

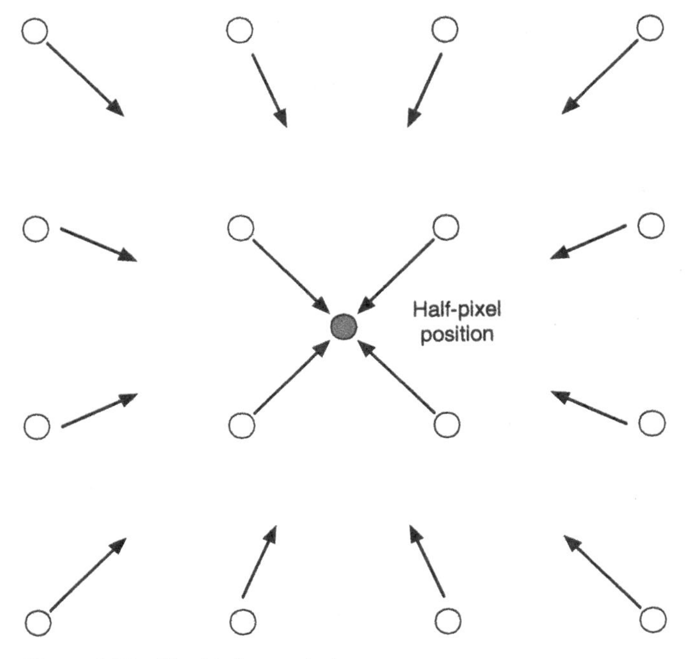

Figure 6.52 Bicubic interpolation

as illustrated in Figure 6.52. Bicubic or higher-order interpolation tends to do a better job of upsampling edges in the original data. Here, the bicubic interpolated example looks sharper than the bilinear interpolated example in Figure 6.50.

6.6.3 Trade-Offs in Sub-Pixel Interpolation

There are two main ways to increase the performance or accuracy of sub-pixel motion-compensated prediction. First, as the example in Figure 6.50 shows, we can use a higher-order interpolation filter. A higher-order filter such as a bicubic interpolation filter can produce a more accurately interpolated reference frame and hopefully a better inter prediction. This comes at a cost of increased computational complexity at both the encoder and decoder, since both encoder and decoder must create the same interpolated reference for prediction.

Second, we can increase the sampling density. Figure 6.50 illustrates the effect of re-sampling the reference frame to a half-pixel density, i.e. interpolating to half-sample positions. We can interpolate to quarter-sample positions (see Figure 6.53), eighth-sample positions or indeed to an arbitrary sampling density. Figure 6.54 shows the same close-up region of the *Foreman* frame interpolated to quarter-pixel resolution and eighth-pixel resolution. There is little apparent difference between these versions, apart from the larger number of interpolated pixel positions, since interpolation does not add any new information. However, higher-density interpolation makes it possible for motion vectors to point to a greater number of positions between integer pixels.

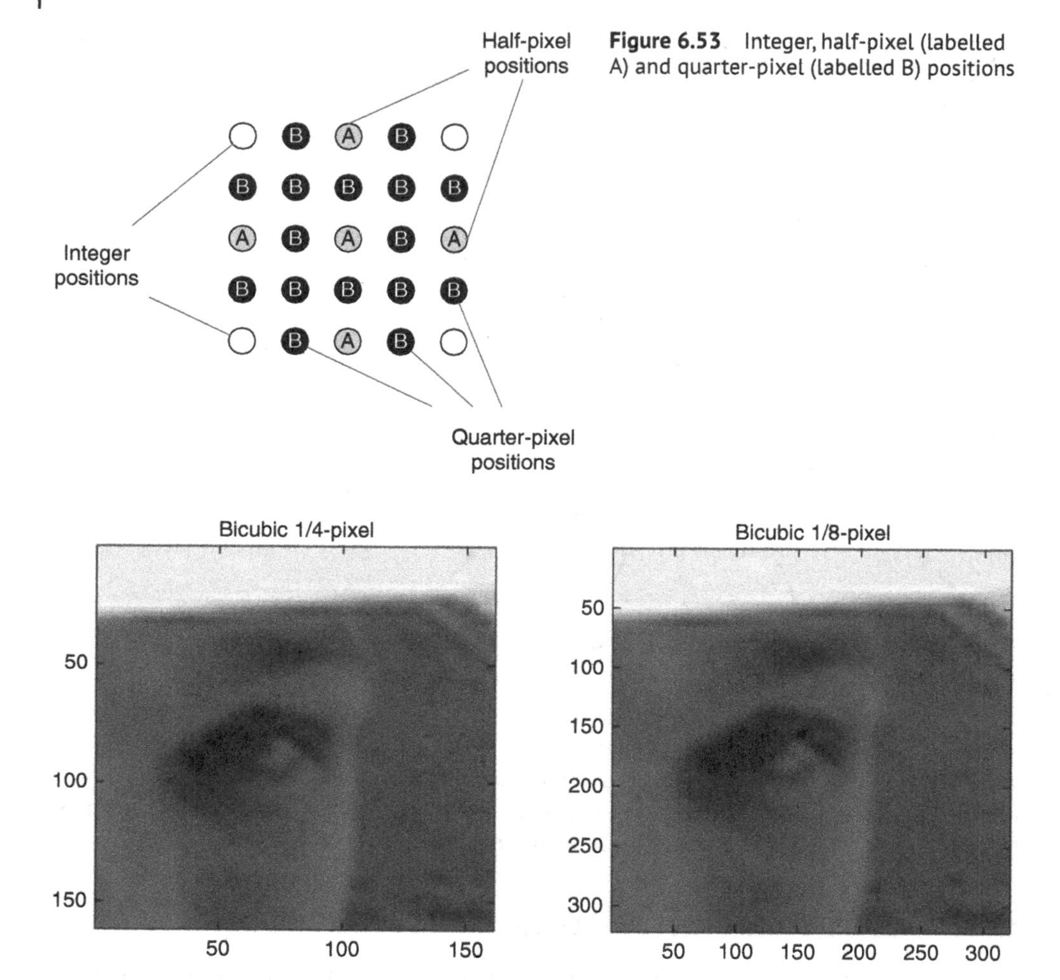

Figure 6.53 Integer, half-pixel (labelled A) and quarter-pixel (labelled B) positions

Figure 6.54 Quarter-pixel and eighth-pixel bicubic interpolation, 'Foreman'

For example, if the reference frame is interpolated to quarter-pixel resolution, then motion vectors can point to the apparent motion of integer, half-pixel and quarter-pixel steps, as illustrated in Figure 6.55. In this figure, the integer pixel positions are shown as circles. Starting from the origin (0,0), which represents no motion, we can create motion vectors such as (+0.75, +0.5) and (−0.75,−1.0). If the reference frame is interpolated to eighth-pixel resolution, as shown in Figure 6.56, we can choose motion vectors such as (−0.125, +1.125) and (+0.625, −0.5). As we increase the resolution of the interpolated frame, our chosen motion vectors can more closely approximate actual interframe motion, which, of course, will take arbitrary values.

There are two potential disadvantages of increased interpolation density. First, increased interpolation density requires more computational processing. Both encoder and decoder must generate and store the necessary sub-pixel values, which requires computation and memory. Increasing the interpolation density by a factor of two, for example, from quarter- to eighth-pixel interpolation, requires four times as many interpolated points to be

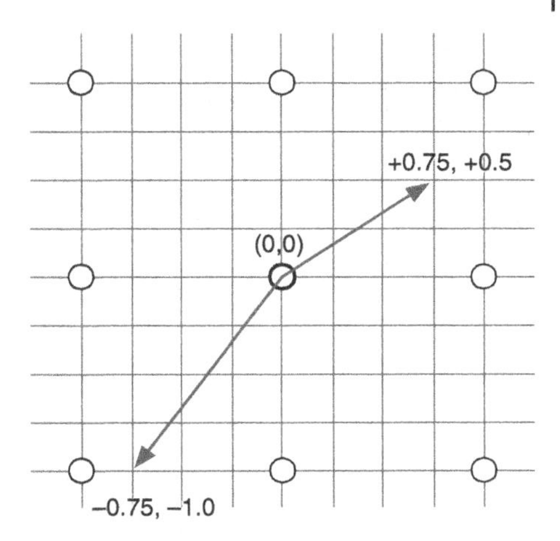

Figure 6.55 Motion vectors with quarter-pixel interpolation

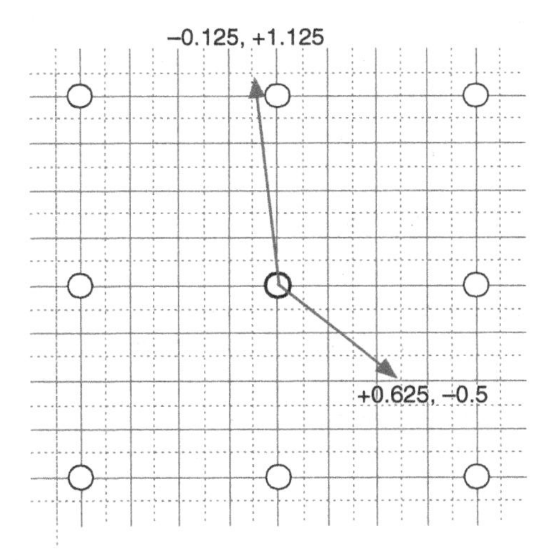

Figure 6.56 Motion vectors with eighth-pixel interpolation

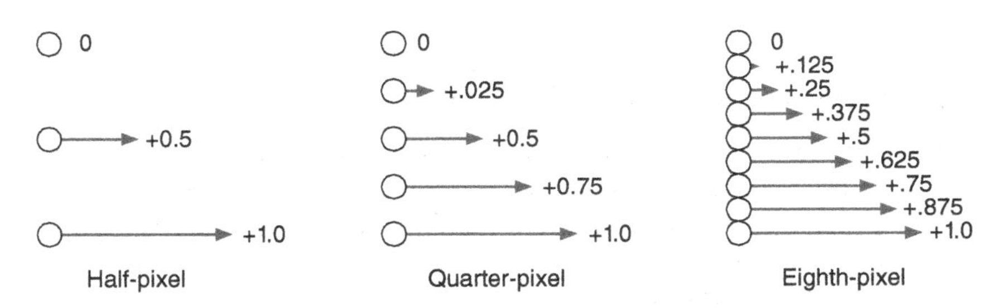

Figure 6.57 Half, quarter and eighth-pixel vectors in the range (0,1)

calculated and stored. Second, higher-precision motion vectors typically require more bits to encode and transmit. Figure 6.57 shows the number of possible horizontal vectors between 0 and 1 for half-pixel, quarter-pixel and eighth-pixel interpolation, respectively. Regardless of the method of encoding and sending vectors, a motion vector in the range (0,1) will require more bits to encode as the interpolation resolution increases.

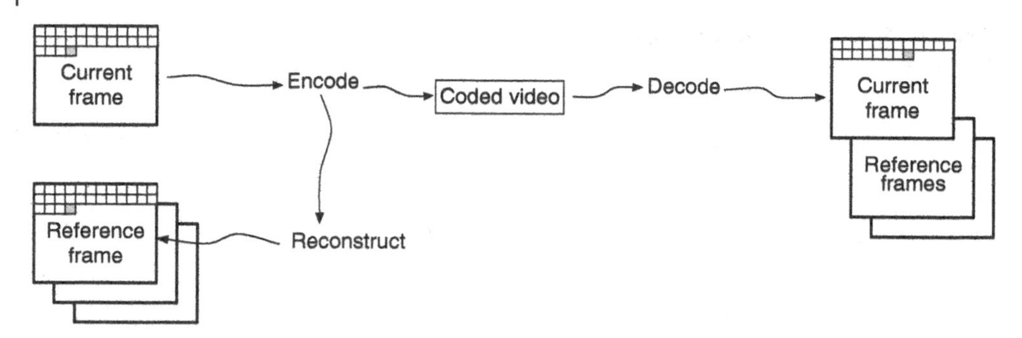

Figure 6.58 Encoding, decoding and reference frames

At a certain point, the cost of the extra bits required to send motion vectors outweighs the benefit of greater motion-compensated prediction accuracy. It turns out that this point depends in part on the design of interpolation filters. With more sophisticated interpolation filters, a higher interpolation depth can provide an overall compression gain [1].

6.7 Reference Pictures

6.7.1 Reference Pictures for Inter Prediction

We saw in Section 6.2 that a block in the current video frame is inter-predicted from a region, typically of the same size, in a previously coded frame known as a reference frame or reference picture. Both the encoder and decoder need to be able to construct the same prediction for every block, which means that both need to have the same set of reference frames available. As the encoder processes each block of the current frame, it encodes the block into the coded video bitstream and also decodes the block to create a reconstructed block, which it uses to build up a reference frame (see Figure 6.58). This mimics the behaviour of the decoder, which also decodes each block to build up a current decoded frame for output and optionally, for use as a reference frame. In this way, both the encoder and decoder construct identical sets of reference frames.

Example 1: One Reference Picture, Forward Prediction Only

The sequence of coded pictures shown in Figure 6.59 starts with an intra-picture, picture 0, labelled I for intra. Because this is the first coded picture, all blocks in picture 0 are coded using intra prediction only. Picture 0 is coded and sent to the decoder. Both encoder and decoder store a decoded version of picture 0 as a reference picture for inter prediction. The encoder encodes picture 1 as an inter-coded picture, labelled P, using picture 0 as a prediction reference. This means that all the inter-predicted blocks in picture 1 are predicted from picture 0. Picture 1 is sent to the decoder, which also uses picture 0 as a reference picture, as illustrated in Table 6.1. In this example, the encoder and decoder are restricted to using a single reference picture, so picture 1 replaces picture 0 as the reference picture. Picture 2 is encoded and sent, using picture 1 as a reference for any inter predictions, and so on.

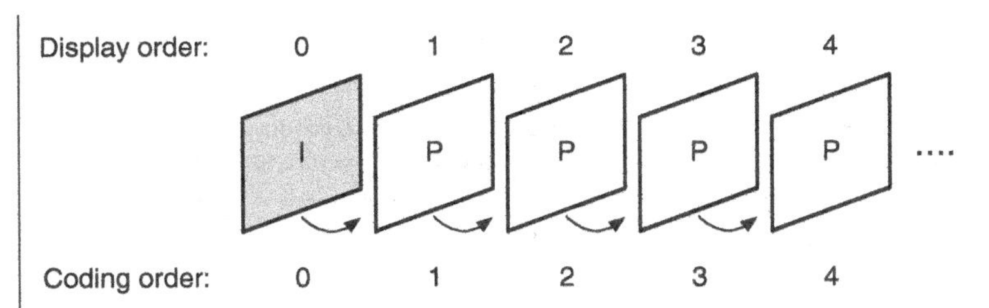

Figure 6.59 Single reference frame, forward prediction

Table 6.1 Reference pictures (example 1)

Encoder: current picture	Encoder: reference pictures	Decoder: current picture	Decoder: reference pictures
0 (I)	—	0 (I)	—
1 (P)	0	1 (P)	0
2 (P)	1	2 (P)	1
3 (P)	2	3 (P)	2
....		....	

Example 2: Up to Four Reference Pictures, Low Delay

A similar example with an increased number of reference frames is shown in Figure 6.60. Both encoder and decoder can store and use up to four reference pictures. Notice that each picture is predicted only from past pictures. This can be described as a low-delay configuration, since the decoder can always decode and display each picture as soon as it has been received.

Pictures 0 and 1 are coded as before. When the encoder comes to Frame 2, it can use both pictures 0 and 1 as reference pictures for inter prediction. Each block can be predicted from picture 0, from picture 1 or from a combination of pictures 0 and 1, using biprediction (see Section 6.3). Frame 3 can be coded using pictures 0, 1 and 2 as references, and so on. The limit on active reference pictures is 4. In order to use picture 4 as

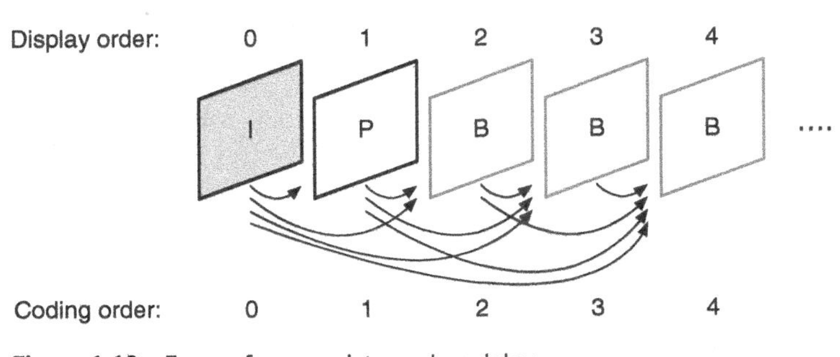

Figure 6.60 Four reference pictures, low delay

Table 6.2 Reference pictures (example 2)

Coding order	Encoder: current picture	Encoder: reference pictures	Decoder: current picture	Decoder: reference pictures
0	0 (I)	—	0 (I)	—
1	1 (P)	0	1 (P)	0
2	2 (B)	0, 1	2 (B)	0, 1
3	3 (B)	0, 1, 2	3 (B)	0, 1, 2
4	4 (B)	0, 1, 2, 3	4 (B)	0, 1, 2, 3
5	5 (B)	0, 2, 3, 4	5 (B)	0, 2, 3, 4
....	....		....	

a reference, the encoder has to remove one of the other pictures, 0, 1, 2 or 3, from its buffer of reference pictures. This decision, i.e. which reference pictures remain available for prediction, has to be communicated to the decoder so that both encoder and decoder have identical reference picture buffers. When picture 5 is encoded, the encoder keeps decoded intra-picture 0 in the reference buffer and removes the oldest inter-picture (1) from the buffer, as shown in Table 6.2.

Consider two inter-predicted blocks A and B in picture 4, as illustrated in Figure 6.61. By the time picture 4 is decoded, both encoder and decoder have pictures 0, 1, 2 and 3 available as reference pictures. For each block, the encoder can choose one or two prediction sources from any of these reference pictures. The encoder chooses to predict block A from a nearby region in picture 3. The encoder predicts block B using biprediction from pictures 2 and 0. The encoder needs to send the following information to the decoder:

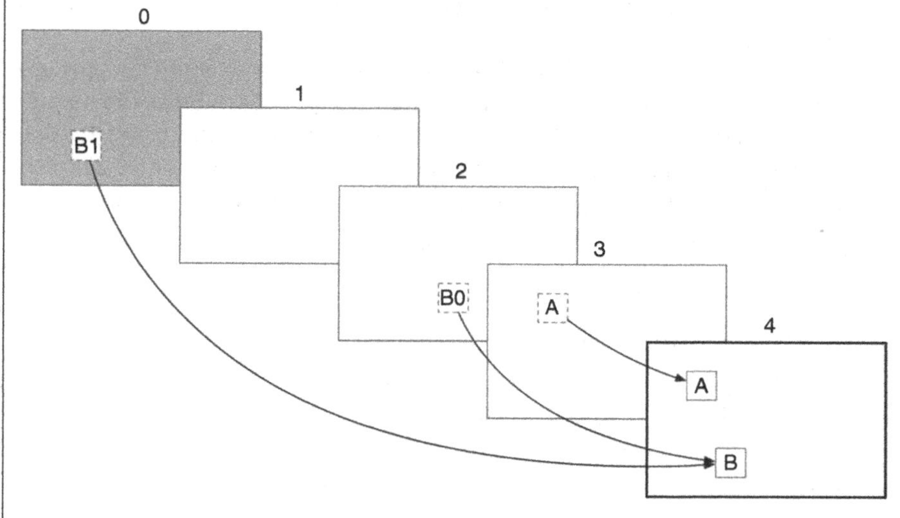

Figure 6.61 Inter prediction choices (four reference pictures)

Block A: pointer to reference picture 3, one motion vector, coded residual.
Block B: pointer to reference picture 2, motion vector for reference picture 2, pointer to reference picture 0, motion vector for reference picture 0, coded residual.

With this information, the decoder can create the same predictions for block A and block B and can successfully decode the blocks.

Example 3: Two Reference Pictures, Higher Delay

In this example, as shown in Figure 6.62, unlike the previous examples, the coding order and display order are not the same. Coded pictures are deliberately sent out of order so that all the necessary reference pictures are available before decoding each picture. The configuration in this example can be described as a higher delay because the decoder has to wait a few frames before displaying decoded frames. As we will see, this type of configuration can give better compression performance at the expense of an increase in decoding delay.

The encoder codes and sends intra-coded picture 0 as before. The next picture to be coded and sent is picture 4, which is predicted using picture 0 as a reference. Picture 4 can be decoded as soon as it is received, but it cannot be displayed because the decoder has not received pictures 1, 2 or 3.

Picture 2 is coded and sent next, using past picture 0 and future picture 4 as its two references. Each block in picture 2 can be predicted from picture 0, picture 4 or both using biprediction. Picture 2 can be decoded as soon as it is received, since the decoder has pictures 0 and 4 available, but it cannot be displayed yet, as Table 6.3 illustrates. Picture 1 is coded next, using past picture 0 and future picture 2 as references. Once picture 1 has been decoded, the decoder can output or display pictures 1 and 2. Picture 3 is coded using pictures 2 and 4 as its references. After decoding picture 3, the decoder can output pictures 3 and 4. At this point, the decoder has caught up and completed processing of all of the first 5 pictures, 0, 1, 2, 3 and 4.

The coding structure carries on, with the encoder coding picture 8 using forward prediction from picture 4 followed by pictures 6, 5 and 7, as shown in Figure 6.62.

Consider three blocks in picture 2, shown in Figure 6.63. Picture 2 is coded after picture 0 and picture 4, as shown in Figure 6.62. Only pictures 0 and 4 are available for prediction. For each block in picture 2, the encoder can choose inter prediction from

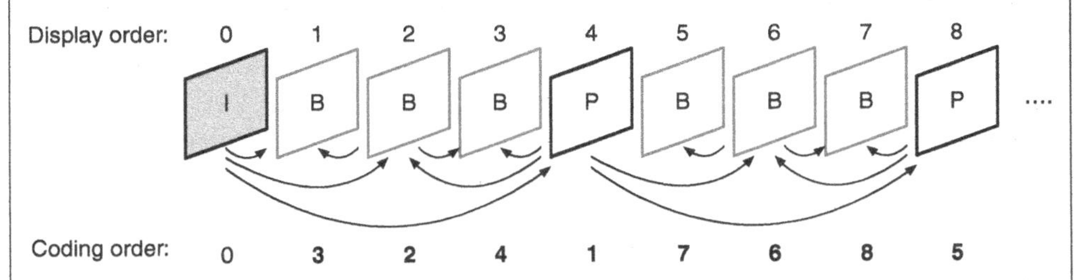

Figure 6.62 Two reference pictures, higher delay

Table 6.3 Reference pictures (example 3)

Coding order	Encoder: current picture	Encoder: reference pictures	Decoder: current picture	Decoder: reference pictures	Decoder: output pictures
0	0 (I)	—	0 (I)	—	0
1	4 (P)	0	4 (P)	0	
2	2 (B)	0, 4	2 (B)	0, 4	
3	1 (B)	0, 2	1 (B)	0, 2	1, 2
4	3 (B)	2, 4	3 (B)	2, 4	3, 4
5	8 (P)	4	8 (P)	4	
6	6 (B)	4, 8	6 (B)	4, 8	
7	5 (B)	4, 6	5 (B)	4, 6	5, 6
8	7 (B)	6, 8	7 (B)	6, 8	7, 8
9	12 (P)	8	12	8	
....	......		...		

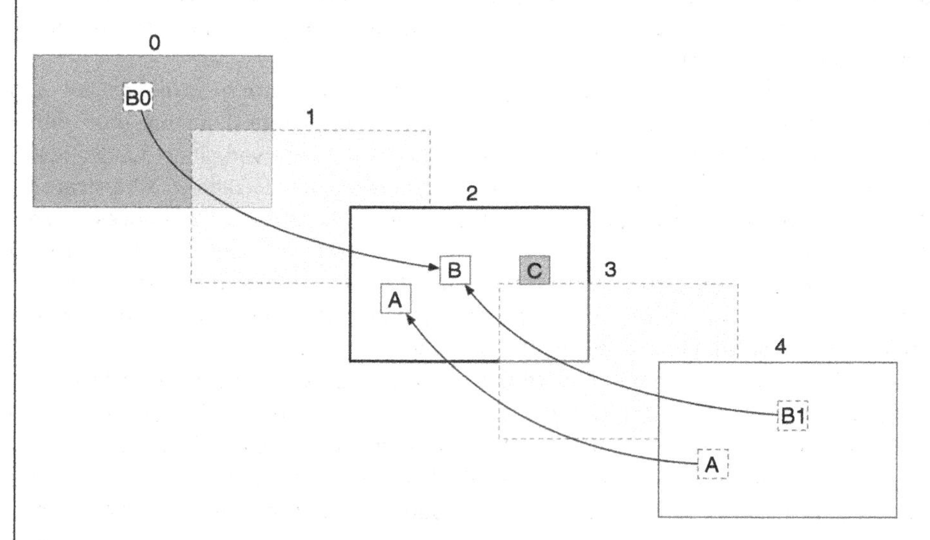

Figure 6.63 Prediction choices: two reference pictures, forward and backward prediction

picture 0, inter prediction from picture 4, biprediction from both 0 and 4, or intra prediction.

Block A is inter-predicted from picture 4, so the encoder sends a pointer to picture 4, a motion vector and a coded residual for this block.

Block B is bipredicted from pictures 0 and 4, and the encoder sends pointers to pictures 0 and 4, motion vectors to identify the source region in pictures 0 and 4, i.e. two motion vectors, and a coded residual.

Block C is intra-predicted so the encoder sends intra prediction mode information and the coded residual.

6.7.2 Reference Picture Structures

The examples in Figures 6.59, 6.60 and 6.62 illustrate different ways of arranging reference pictures and coded pictures. Coded and reference pictures can be organised in a number of ways, depending on the requirements of the application, for example:

- **To minimise end-to-end delay**: In an application such as video calling or video conferencing, it is important that the delay between capturing a frame at the source and displaying it at the destination is as low as possible. A short delay gives a more natural conversational experience.
- **To maximise compression efficiency**: For example, biprediction with forward and backward reference frames tends to be more efficient than forward-only prediction, which in turn is more efficient than intra-only prediction. A highly efficient sequence may consist mostly of B-pictures with as few I- and P-pictures as possible.
- **To provide random access capability**: In many applications, it is important that a viewer can start decoding part-way through a sequence with a relatively short delay. This typically requires the use of intra-coded pictures at regular intervals. When a user joins a sequence part-way through, the decoder finds the nearest I-picture and starts decoding from there.
- **To provide scalability, i.e. the ability to extract a lower-fidelity version of the video clip**: For example, pictures that are not used as prediction references themselves can be discarded without affecting the remaining coded pictures.
- **To reduce or limit the storage requirements of the decoder**: For example, the coding structure of Figure 6.59 only requires the decoder to store a single reference frame. Once each frame has been decoded, the decoder can overwrite the reference buffer and re-use the same memory. In contrast, Figure 6.60 requires the decoder to store four reference frames, since blocks in later pictures can be predicted from any one of the four stored references.

Some of these aims may be mutually exclusive. For example, high compression efficiency might be achieved at the expense of significant delay and memory requirements at the decoder and vice versa.

6.7.2.1 Low-Delay Picture Structures

Figures 6.59 and 6.60 are examples of low-delay picture structures. In each case, the sequence starts with an I-picture, since there are no previously coded pictures to predict from. Further I-pictures may be inserted at intervals to enable a decoder to start decoding at a later point and/or to re-synchronise if there is a transmission error. If the encoder and decoder are communicating in both directions, for example, in a video calling application, then it may not be necessary to insert further I-pictures unless the decoder signals that there has been a transmission problem.

Each of the remaining pictures is coded using P- or B-prediction, i.e. each block is inter-predicted from one or two reference pictures. All blocks are predicted from past pictures – pictures that are older in display order. This means that the decoder can display each frame as soon as it is decoded, which minimises the delay at the decoder. Hence, the contribution of the video codec to end-to-end delay need not be more than a one-frame delay.

6.7.2.2 Random Access Picture Structures

Figure 6.62 is an example of a random access picture structure, which can achieve better compression efficiency than a low-delay structure, assuming an equal number of I-pictures, at the expense of an increase in decoding delay. The sequence starts with an I-picture. For random access capability, it is necessary to insert I-pictures, typically at regular intervals. The repeating structure of reference pictures is sometimes described as a Group of Pictures (GoP). If I-pictures occur once every second, say, then a decoder attempting to start decoding at an arbitrary point has to wait at most one second before it receives and decodes a fresh intra-coded picture.

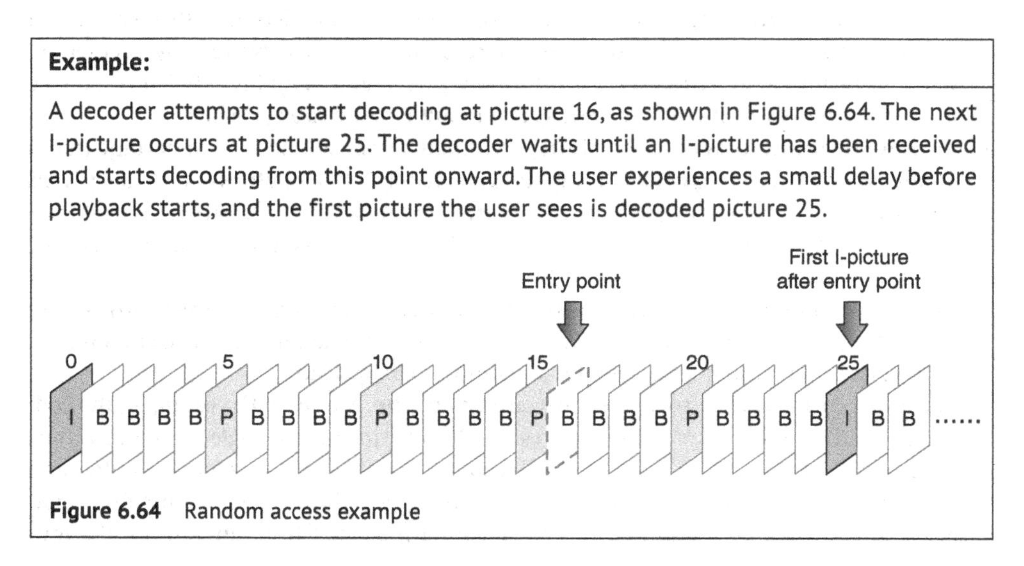

Example:

A decoder attempts to start decoding at picture 16, as shown in Figure 6.64. The next I-picture occurs at picture 25. The decoder waits until an I-picture has been received and starts decoding from this point onward. The user experiences a small delay before playback starts, and the first picture the user sees is decoded picture 25.

Figure 6.64 Random access example

Hierarchical Picture Structures Figure 6.65 shows another random access picture structure. In this example, I-pictures are sent every 16 pictures, and all of the other pictures are coded as B-pictures. The coded pictures are arranged in a hierarchy, which we can see more clearly in Figure 6.66. The lowest level or layer of the hierarchy, layer 0, includes pictures 0, 8 and 16. The single B-picture in this layer, picture 8, is predicted using only pictures 0 and 16 as reference pictures. At the next layer, layer 1, pictures 4 and 12 are predicted using only layer 0 pictures as reference pictures, using pictures 0, 8 and 16 as references. Layer 2 pictures are predicted using lower-layer pictures, using pictures 0, 4, 8, 12 and 16 as references.

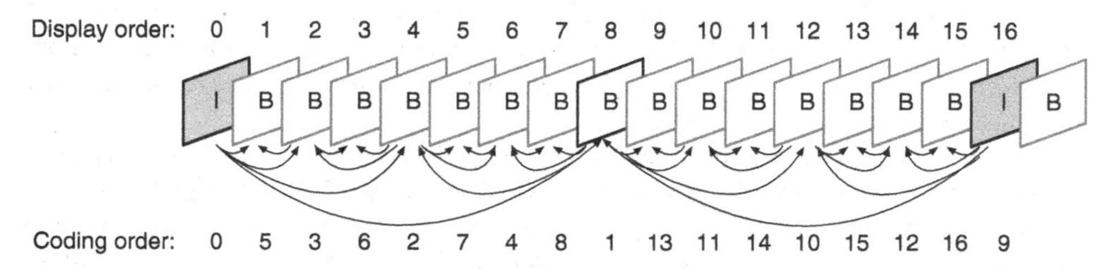

Figure 6.65 Hierarchical Group of Pictures structure (view 1)

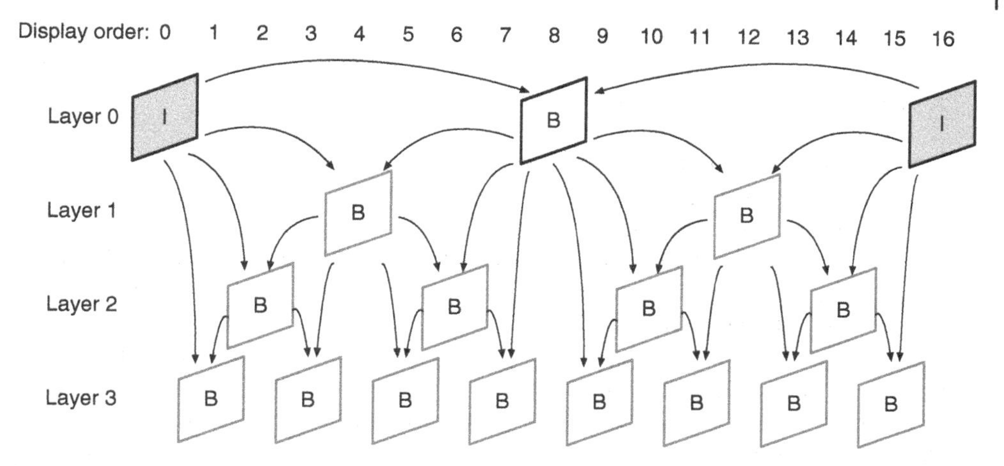

Figure 6.66 Hierarchical Group of Pictures structure (view 2)

In this type of hierarchical picture structure, pictures at each layer are predicted using only lower-layer pictures as references. This has two potential benefits:

1) It is easy to strip out higher layers to create a lower frame rate video clip, for example, to reduce the bitrate of the coded video sequence. For example, layer 3 can be removed from Figure 6.66 to leave a sequence containing 8 pictures per GoP. Because none of the layer 0, 1 or 2 pictures use layer 3 for prediction, the remaining pictures can be successfully decoded.
2) A hierarchical GOP structure may have better compression efficiency than a non-hierarchical structure [2].

6.7.3 Reference Picture Buffers

We have seen how each coded picture is reconstructed at the encoder and decoded at the decoder and is then potentially available for use as a reference picture. Both encoder and decoder maintain identical reference picture buffers or decoded picture buffers (DPBs), each containing a set of previously coded pictures. Inter-predicted blocks in the current picture can be predicted from one or more of these reference pictures.

Practical considerations include:

- Maintaining a limited number of stored reference pictures at the encoder and the decoder. The maximum amount of data is limited by the memory capacity, typically determined by the Level capabilities of the decoder, specified in a video coding standard such as H.265. The total number of reference pictures has to stay within this maximum limit.
- Adding new pictures to the reference picture buffer at the encoder and decoder.
- Removing older pictures when they are no longer needed for prediction and to ensure the maximum number of reference pictures is not exceeded.

- Organising or ordering reference pictures in the reference picture buffer. This may include marking reference pictures as short- or long-term (see below) and organising list(s) of indices that point to reference pictures in the buffer.

6.7.3.1 Adding Pictures to the Reference Picture Buffer

In H.264, each reconstructed or decoded picture is marked as (1) used for reference, short-term, (2) used for reference, long-term or (3) not used for reference, as illustrated in Figure 6.67. Short-term reference pictures are typically held for a limited amount of time and will be automatically removed from the reference picture buffer as new pictures are decoded and added. Long-term reference pictures, as the name suggests, persist in the reference picture buffer until they are explicitly removed. A decoded picture that is unused for reference may be displayed but is not stored in the reference picture buffer. For example, the B-pictures at Layer 3 in Figure 6.66 are not used to predict any other pictures, so they would be marked as unused for reference and not held in the reference picture buffer.

6.7.3.2 Removing Pictures from the Reference Picture Buffer

In the scenario shown in Figure 6.67, when a new picture is added to the short-term buffer, it may be necessary to remove an older picture in order to make room. The decoded picture buffer or reference picture buffer contains a limited number of pictures, determined by the coding standard and/or by information sent in the coded bitstream. For example, consider a situation where the reference picture buffer contains a maximum of 8 pictures, three of which are marked as long-term, as shown in Figure 6.67. Five pictures are currently marked as short-term. When a new short-term picture is coded, one of the short-term pictures has to be removed or overwritten to make space for the new short-term picture.

6.7.3.3 Ordering Reference Pictures

The reference picture buffer contains a set of previously coded pictures. For each inter-coded block, the encoder needs to indicate one or more of these reference pictures as prediction sources. In H.264 and HEVC, the encoder and decoder construct one or two ordered lists of pointers to reference pictures. A PU in a B-slice can be predicted from a reference picture in List 0, a reference picture in List 1, or from two reference pictures using biprediction, one each from List 0 and List 1. For example, block A in Figure 6.63 is predicted from

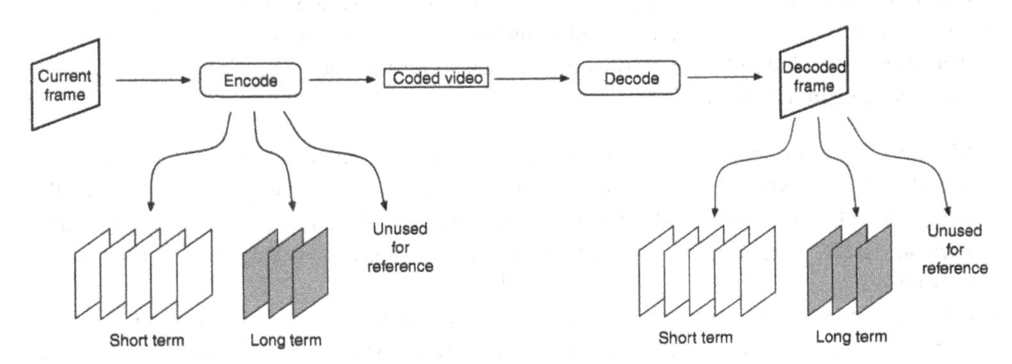

Figure 6.67 Decoded pictures: short-term, long-term, unused for reference, H.264

picture 4, which is contained in List 1, and Block B is bipredicted from picture 0 in List 0 and picture 4 in List 1.

These lists have a default order, which may be altered by sending commands in the bitstream. Examples of default orders are as follows:

List 0:

1) All the short-term reference pictures preceding the current picture in display order, in descending display order, most recent first.
2) All the short-term reference pictures after the current picture in display order, in ascending display order.
3) All the long-term reference pictures, in order of long-term index.

List 1:

1) All the short-term reference pictures after the current picture in display order, in ascending display order.
2) All the short-term reference pictures preceding the current picture in display order, in descending display order, most recent first.
3) All the long-term reference pictures, in order of long-term index.

Example:

In Figure 6.68, current picture 5 has five reference pictures: 0, 2, 4, 6 and 8. Picture 0 is marked as Long-Term, and pictures 2, 4, 6 and 8 are Short-Term. The default ordering of these reference pictures is as follows:

List 0:

Past short-term reference pictures, most recent first: 4, 2
Future short-term reference pictures, ascending: 6, 8
Long-term reference picture: 0

List 1:

Future short-term reference pictures, ascending: 6, 8
Past short-term reference pictures, most recent first: 4, 2
Long-term reference picture: 0

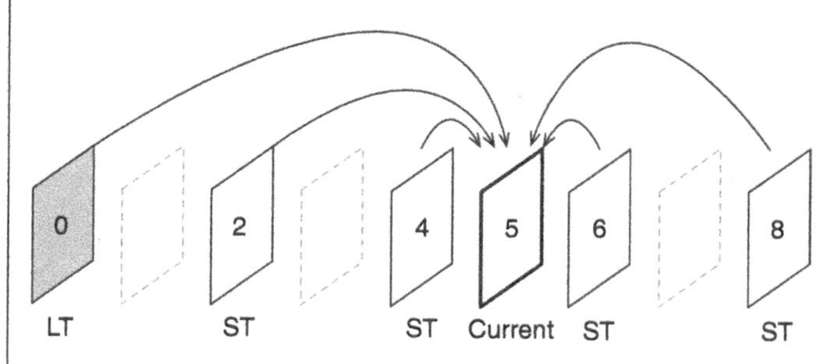

Figure 6.68 Current picture with five reference pictures

Figure 6.69 shows the ordering of List 0 and List 1 for this picture, i.e. current picture 5. If the encoder sends an index to List 0 Ref. [2], this indicates picture 6. If the encoder sends an index to List 1 Ref. [1], this indicates picture 8.

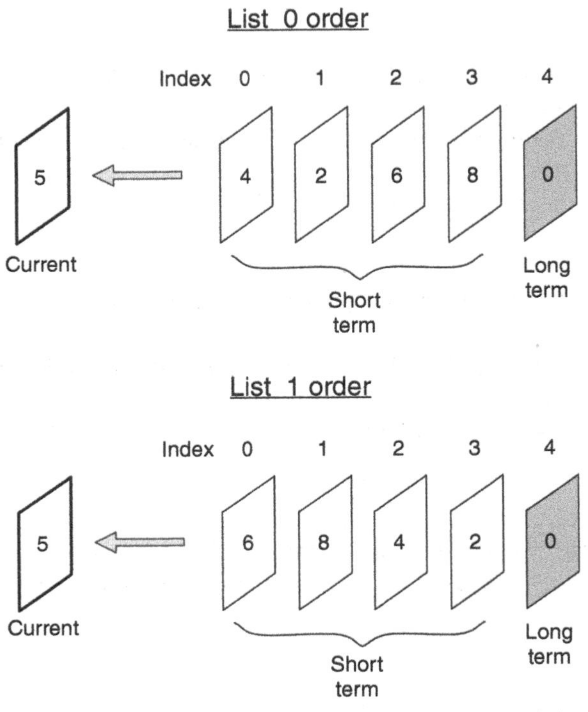

Figure 6.69 List 0 and List 1 reference picture ordering

6.7.3.4 Communicating Reference Picture Structures

We have seen a number of different reference picture structures including low delay, e.g. Figure 6.60, and random access, e.g. Figures 6.64 and 6.65. The encoder needs to communicate each reference picture structure to the decoder so that the decoder knows which reference pictures are available for a given inter-coded picture.

In HEVC, this can be done using reference picture sets (RPSs), which will be discussed later in Section 6.12.7.

6.8 Signalling Inter Prediction Choices

6.8.1 Inter Prediction Parameters

Figure 6.70 shows a rectangular block in a current picture, bipredicted from a past and a future reference picture. In order to decode this block, the decoder needs to know the following prediction parameters:

- **Motion vector 0**: The motion vector (offset) pointing to the first reference block.
- **Motion vector 1**: The motion vector pointing to the second reference block.

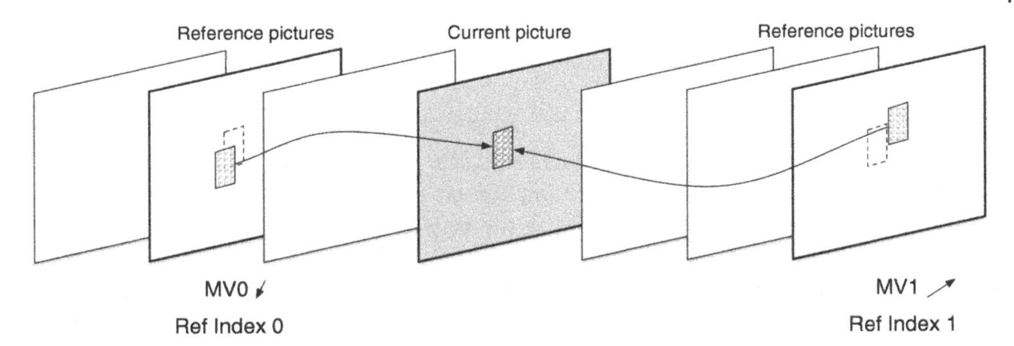

Figure 6.70 Inter prediction parameters

- **Reference index 0**: An identifier for the reference picture containing the first reference block.
- **Reference index 1**: An identifier for the reference picture containing the second reference block.

If the current block is predicted from a single reference picture, using P-prediction, then the decoder only needs MV0 and Ref Index 0. If the current block is predicted from two reference pictures using biprediction, then the decoder needs to know MV0, MV1, Ref Index 0 and Ref Index 1.

It is important to communicate these parameters to the decoder as efficiently as possible, using the smallest possible number of bits. The prediction parameters for each block can be communicated, as shown in Figure 6.71, with the parameters split into base parameters and delta parameters.

The base parameters, such as a motion vector predictor, are either inferred or selected:

- **Inferred**: Base parameters are automatically generated at the decoder, based on previously decoded block parameters. No information is sent in the bitstream.
- **Selected**: Base parameters are selected at the decoder, from a set or list of parameters generated from previously decoded blocks. A selection index, identifying the chosen base parameters, is sent in the bitstream.

The delta parameters are either sent or not sent:

- **Sent**: A delta value, i.e. the difference between the base parameters and the actual prediction parameters, is encoded and sent in the bitstream. The decoder can add this delta value to the base parameters to create the block prediction parameters.

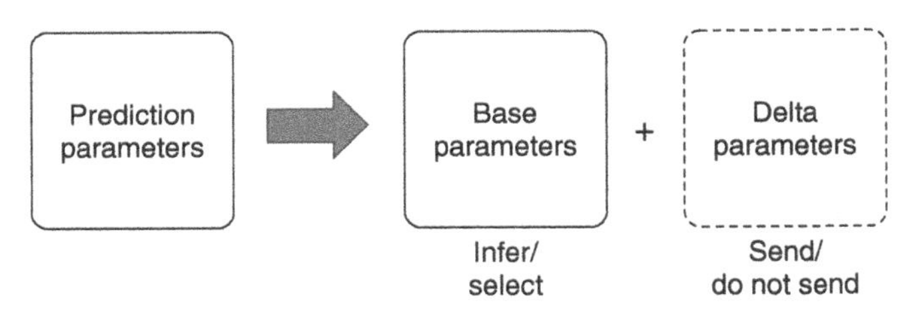

Figure 6.71 Communicating prediction parameters

- **Not sent**: No delta value is sent, so the current block prediction parameters are set to be identical to the base parameters.

Example:

The current block X is inter-predicted using a single motion vector. The encoder has calculated that a vector of (+3, +1.5) will give the smallest residual for this block. We will look at how we might calculate a base vector (motion vector predictor) and a delta vector for this block, based on previously coded neighbouring blocks, as shown in Figure 6.72.

Assuming that blocks are coded in a raster scan order, i.e. left to right, top to bottom, then by the time we are ready to code block X, the neighbouring blocks A, B, C and D may already have been coded. If these blocks have already been coded, then the encoder and the decoder can use their parameters during the coding of block X.

Calculating a base motion vector:

We will look at three methods of calculating a base vector.

a) **Median (A, B, C)**: The base vector is defined as the median of the X-component and the median of the Y component of blocks A, B and C.
Median (Ax, Bx, Cx) = median (+3, +4, −1) = +3
Median (Ay, By, Cy) = median (+1.5, +2.5, +2) = +2

Hence, the base vector is (+3, +2)

b) **Most common vector amongst (A, B, C, D)**: The base vector is defined as the mode of (A, B, C, D). The most common vector (not necessarily the most common X and Y values) is (+4, +2)

c) **Select a candidate from (A, B, C, D)**: Choose a candidate and send a parameter or index that identifies the candidate. The encoder chooses a candidate that minimises the delta between the base vector (the candidate) and the actual vector X. In this case, the best candidate is B and the base vector is (+3, +2.5).

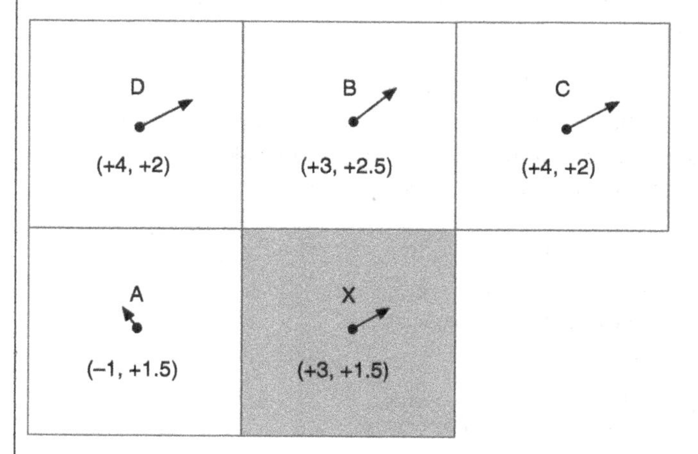

Figure 6.72 Motion vector X and neighbours

Coding or not coding the delta vector:

Consider two options for the delta vector, i.e. the difference between the actual vector X and the base vector: (i) code and send the delta or (ii) do not send it at all.

i) Code and send the delta vector:
   Option (a) above: Base vector is (+3, +2), delta vector is (0, −0.5)
   The decoder calculates the same base vector (+3, +2), receives the delta vector and adds it to the base vector to restore the original X vector (+3, +1.5).
   Option (b): Base vector is (+4, +2), delta vector is (−1, −0.5)
   Option (c): Base vector is (+3, +2.5), delta vector is (0, −1)
ii) Do not send a delta vector:
   In this case, the decoder is stuck with the base vector, i.e. it has to use the base vector to predict X.

Option (i), code and send the delta vector, allows the decoder to use the correct or optimal choice of motion vector (+3, +1.5). This may result in a smaller residual, i.e. fewer residual bits, but at the expense of more bits to code the delta.

Option (ii), do not send a delta vector, results in the decoder using a possibly sub-optimal motion vector[2], e.g. (+3, +2). This may result in a larger residual and more residual bits but saves the bits which would have been sent to code the delta vector.

6.8.2 Signalling the Base Motion Vector Parameters

A decoder can infer the base motion vector automatically based on previously coded information, select it from a list of candidates or generate it by scaling previously coded information.

a) Generate the base motion vector automatically from previously coded block information.

In this first case, a base motion vector is generated automatically, i.e. no information is sent to indicate a choice of base motion vector. For example, H.264 median vector prediction, as illustrated on the left of Figure 6.73, works in a similar way to the median prediction example above. A base motion vector or motion vector prediction is generated automatically as the median of three neighbouring block motion vectors (A, B, C). For certain block sizes in H.264, a base motion vector is generated by copying from a left-hand block or an upper block, as illustrated on the right of Figure 6.73.

b) Generate a list of candidates and select the base motion vector from the list.

In this second case, the encoder generates a list of candidates or possible base motion vectors and sends a parameter to the decoder, which indicates a selected candidate. For example, the VP8 and VP9 video codecs generate a list of candidate base vectors, illustrated

2 Of course, the encoder knows how it chooses to code the motion vector. If the encoder chooses not to send a delta vector, it will use the actual motion vector received by the decoder (e.g. +3 and +2) to create its own prediction and generate a residual.

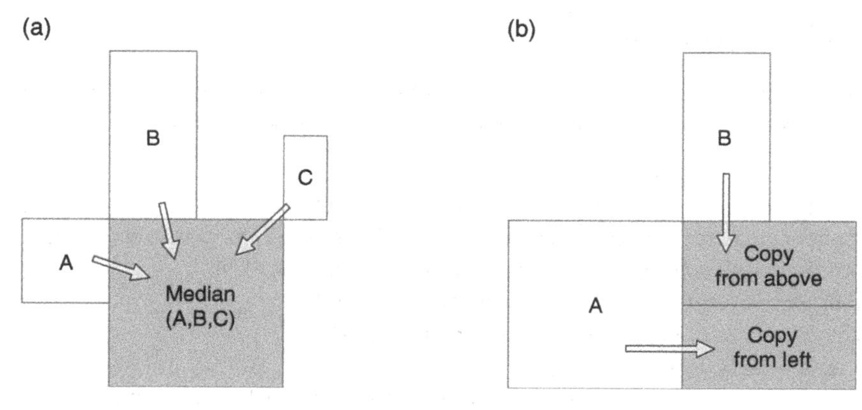

Figure 6.73 Median vector prediction (a), copy prediction (b)

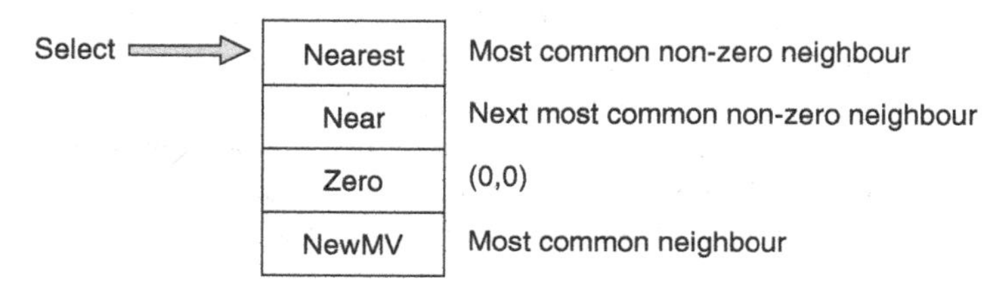

Figure 6.74 VP8/VP9 candidate motion vectors

in Figure 6.74. The encoder determines an ordered list of neighbouring block vectors, from the most common to the least common. If two or more neighbouring blocks have identical vectors, then this vector may become the most common or nearest non-zero neighbour. The encoder sends a parameter, which indicates one of the following:

i) **Nearest**: Most common non-zero motion vector from neighbouring, previously coded blocks.
ii) **Near**: Next most common non-zero motion vector.
iii) **Zero**: The (0,0) motion vector.
iv) **NewMV**: Most common motion vector from neighbouring blocks.

The NewMV choice is followed by a coded delta vector; the other choices are not followed by a delta.

H.265/HEVC generates a candidate list that may include spatial neighbours at the lower left, left, upper left, above and upper right in the same picture and/or temporal neighbours at the lower right or same location in the reference picture, as illustrated in Figure 6.75. The encoder sends a parameter that selects one of these candidates: a merge index in merge mode or a candidate index in advanced motion vector prediction (AMVP) mode.

c) Generate a base motion vector by scaling a previously coded motion vector.

This involves identifying a previously coded motion vector and scaling this vector to generate a base motion vector. For example, H.264's Temporal Direct mode works as follows. First, a motion vector, MV, is identified for a block in a reference picture, i.e. a block that

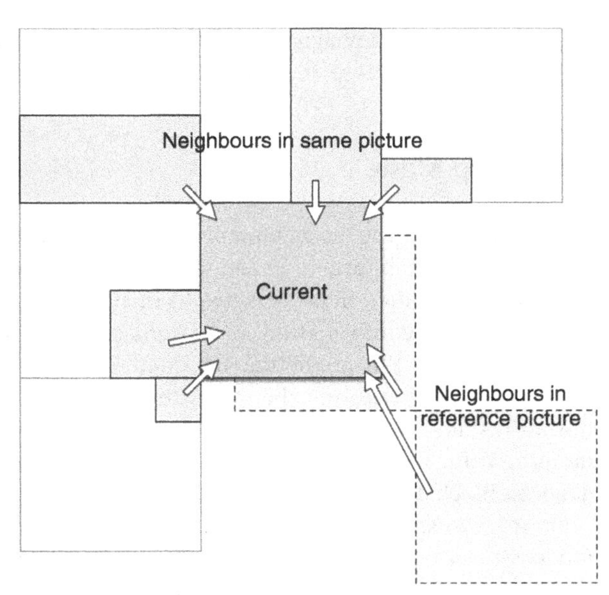

Figure 6.75 HEVC candidate motion vectors

has already been coded and sent. Second, two scaled motion vectors, List 0 and List 1 vectors as discussed in Section 6.7.3, are generated by scaling this motion vector, MV. The scaling depends on the temporal distance between the current frame and the List 0 or List 1 reference pictures.

6.8.3 Signalling the Delta Motion Vector Parameters

The options here are to send or not to send a delta motion vector.

a) Do not send any delta parameters:

In H.265/HEVC's merge mode, H.264's Direct Modes and VP9's Near, Nearest and ZeroMV modes, no delta parameter is sent. Instead, the decoder uses the base motion vector as is. For example, if an HEVC PU is coded using merge mode, the decoder identifies a base motion vector from a neighbouring block indicated by the merge index, and this becomes the motion vector for the current PU.

b) Code and send the delta parameters:

In H.265/HEVC's AMVP mode or VP9's NewMV mode, a delta motion vector is calculated and encoded in the video bitstream. The decoder constructs the base motion vector or motion vector prediction and adds the delta motion vector to generate a motion vector for the current block.

6.8.4 Signalling Reference Picture Choices

In H.264, HEVC and VP9, each inter prediction block has one or two motion vectors. Unless the block's inter prediction parameters are inherited, as in HEVC's merge mode, the encoder sends an index signalling the choice of reference picture for each motion vector. In the HEVC case, this is an index into a list of candidate reference pictures, List 0 or List 1.

6.9 Skip Mode

Skip mode, one of the most common modes in many practical coding scenarios, is a mode where no residual information and no prediction information is sent.

If the inter prediction process results in an accurate prediction, the residual block will contain very little information. After transform and quantisation, the block may contain all-zero data. If the prediction parameters are inferred rather than sent, e.g. using HEVC's merge mode, and the quantised residual block(s) are all-zero, then no prediction or residual data needs to be sent. The decoder simply infers a prediction mode by merging from a neighbouring block, generates a prediction block and this becomes the decoded block data.

Figure 6.76 shows a frame coded as a B-picture. Many of the CUs are coded in skip mode. One part of the picture is highlighted, and the skipped CUs are marked with S. In this example, much of the picture consists of skipped CUs, i.e. no data are sent for these CUs.

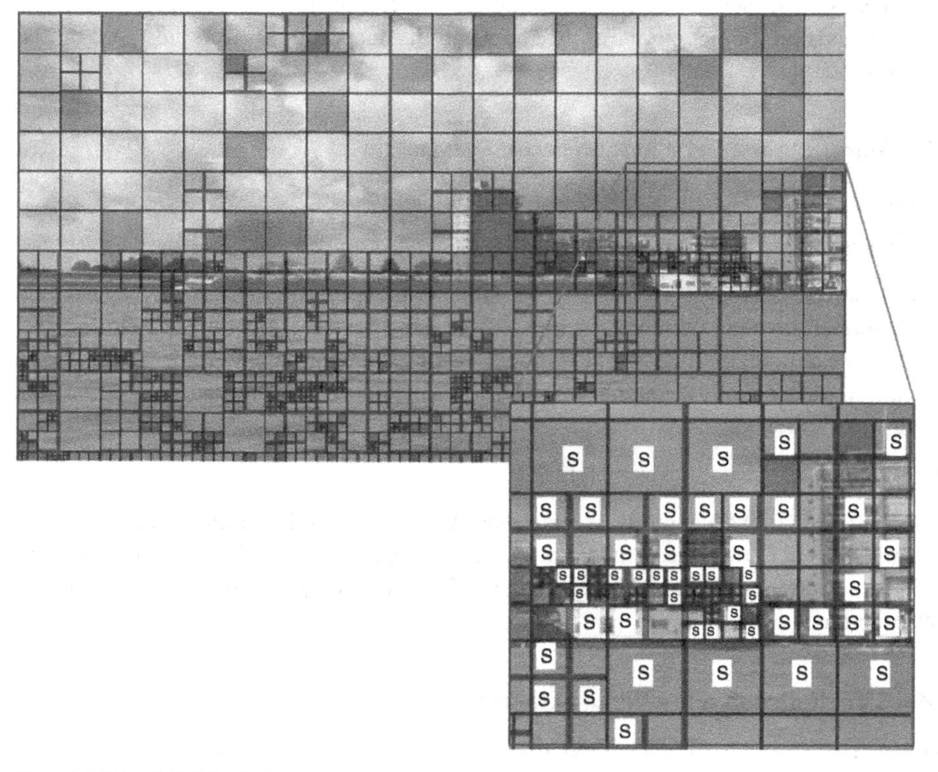

Figure 6.76 HEVC B-picture showing skip CUs (S). *Source:* Solveig Multimedia/Zond265

6.10 Loop Filter

Figure 6.77 Close-up of original video frame

Figure 6.78 Close-up of decoded frame with no filtering

Figure 6.79 Close-up of decoded frame with deblocking and Sample Adaptive Offset filtering

The encoder and decoder must use the same prediction source in order to create an identical prediction. This means that both encoder and decoder must form a prediction from decoded, rather than original, video information, as illustrated earlier in Figure 6.2. Video coding is typically a lossy process, which means that the decoded frame is not identical to the original video frame. The coding process introduces distortion such as blocking and ringing effects. For example, Figure 6.77 shows a close-up region from an original, uncompressed video frame. When the frame is encoded and decoded at a relatively high compression ratio, the result is a somewhat distorted decoded frame, as shown in Figure 6.78, with distortions or artefacts such as visible block edges and ripple-like distortions. This can be a problem for prediction. If the prediction source, for example, a reference picture such as Figure 6.78, is significantly distorted, it may not be possible to find a good prediction match for the current block. In this example, the reference frame contains obvious blocking and ringing distortion. Because of this, the best available motion-compensated prediction for the current block is distorted. This means that the residual after prediction will contain significant energy, even if the motion-compensated prediction is as accurate as possible.

One solution to this problem is to filter the reference frame to reduce distortion before forming the prediction. If the filtering is successful, the filtered reference frame contains less distortion, and it is therefore easier to find an accurate match for the current block. Both encoder and decoder must carry out identical filtering operations. This means that the filter process forms part of the encoding loop, and so this type of filter is often described as a loop filter. Figure 6.79 shows the same decoded picture with filtering applied – in this case, HEVC's deblocking and Sample Adaptive Offset filters – that help to reduce blocking and ringing distortions. If we try to predict a later frame from this filtered reference picture, we are likely to find a better match for the new frame.

Loop filtering algorithms are discussed in detail in Chapter 9.

6.11 When Inter Prediction Does Not Find a Good Match

Inter prediction relies on an assumption that we can find a good match for a block in the current frame, by looking at a previously coded frame. For example, if we predict a block in frame *N* from the previous frame, $N-1$, inter prediction will work effectively if we can find a block of the same size within a search region, perhaps nearby the location of the block in frame *N*. Consider the factors that may make inter prediction work badly. What circumstances might lead to the encoder failing to find a good prediction for the current block, i.e. failing to find a similar block in a reference frame?

a) Non-translational motion:

Video codecs such as H.264 and HEVC assume a translational motion model. A block in the current frame is predicted from a block of the same size, shape and orientation in a reference picture, offset by an *X*/*Y* motion vector. Of course, objects and features in real-world video scenes do not only move in translational steps. Figure 6.80 shows a number of ways in which objects can move from one frame to the next. At the top is translational motion – our square block in the current frame can be found in the same size and shape in the reference frame, offset by an *X* and *Y* amount. Next is rotation – our block has rotated relative to its position in the reference frame. The block may be skewed, or it may have increased or decreased in scale. The block may have deformed, for example, if it corresponds to the surface of a beach ball that is being deflated. And of course, real-world objects can and do move using combinations of these and other changes.

More complex motion models can capture some or all of these types of motions. For example, an affine motion model can represent motion using more parameters rather than the simple two parameters of *X*/*Y* translation.

b) Motion smaller than the block size:

The choice of block or region size for inter prediction is a compromise. A smaller block size may be capable of more accurately reflecting complex, real-world motion. However, each inter-predicted block requires some overhead to be coded, such as *X*/*Y* motion vectors, prediction mode and the size of the block itself. As we saw earlier, smaller block sizes can offer more accurate predictions at the expense of greater overhead. Current standards set a lower limit on block size, typically around 4×4 pixels. Because of the trade-off involved in choosing an inter prediction block size, the codec will often encounter situations where object motion is at a scale smaller than the block size. Figure 6.81 shows two examples. In the first, the current block contains two overlapping circular objects. These are moving in different directions, and so it will not be possible to find a single block in the reference picture that contains the same visual information. This is a very common problem, particularly around the edges of objects that are moving relative to a fixed background. In the second example, the current block contains three small objects that are each moving in different directions. Again, it will not be possible to find a single block in the reference frame that contains all three objects. Both of these problems might be solved or mitigated by choosing a smaller block size, but this would have the associated trade-off mentioned above.

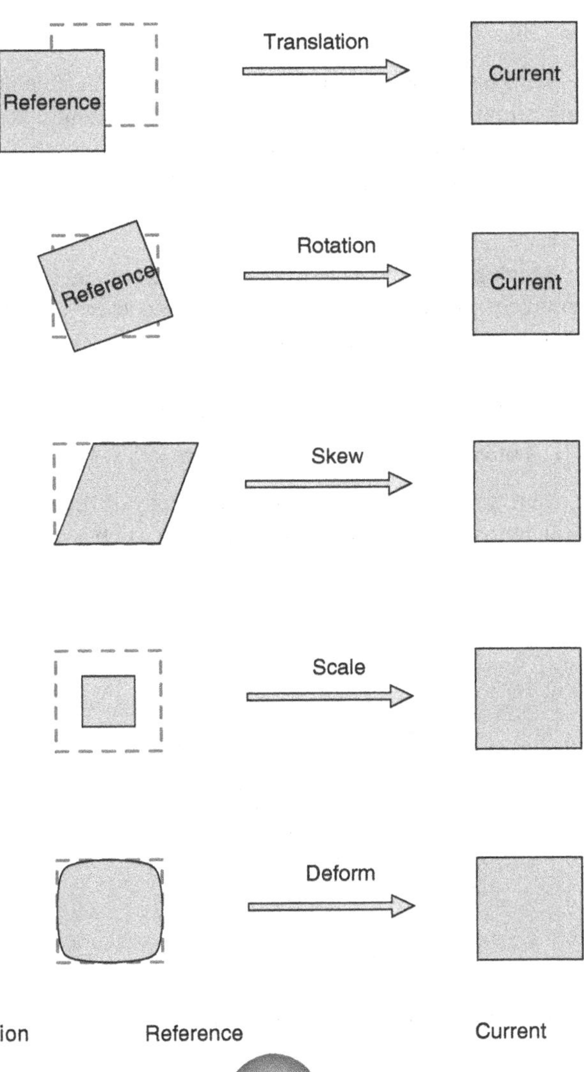

Figure 6.80 Translational and non-translational block prediction

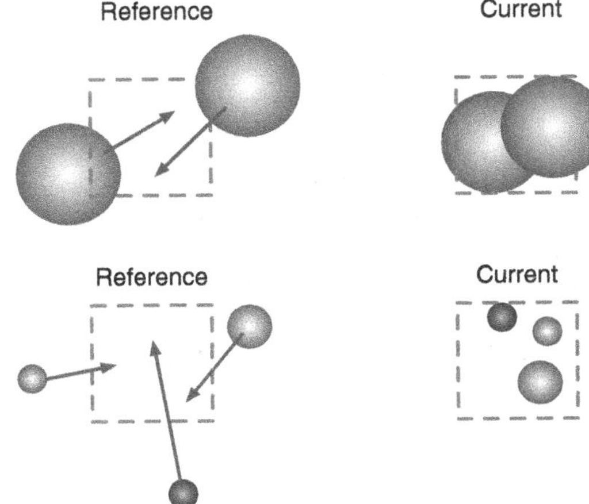

Figure 6.81 Sub-block motion

c) Illumination changes:

Another reason that the encoder may not be able to find an accurate match is changing illumination. In the example of Figure 6.82, a light is switched on, brightening most of the areas of the frame. Despite the fact that nothing has moved between the two frames, the pixels of the highlighted block have clearly changed due to the lighting change. Of course, prediction from a future frame, where the light is on, may help if the reference picture structure makes this possible.

Moving objects will often change the illumination in the scene as they interact with the light sources (see Figure 6.83). In this example, there is nothing moving in the highlighted block, but the book at the top left is moving and casting a shadow that changes the pixel values within the block. Illumination compensation, in which the encoder signals a change in illumination as part of the prediction process, may provide better prediction in these types of scenarios.

d) Random motion, complex motion, fades, etc.:

Certain types of motion are notoriously difficult for a block-based video codec to handle efficiently. Motion that appears random, such as water ripples (see Figure 6.84), smoke and explosions, makes it difficult to find a genuine match in a reference picture. Fades and

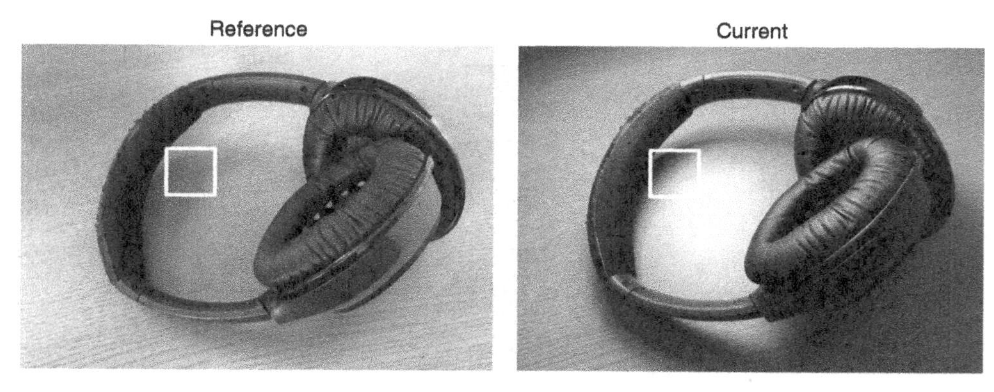

Figure 6.82 Scene with changing illumination (light going on)

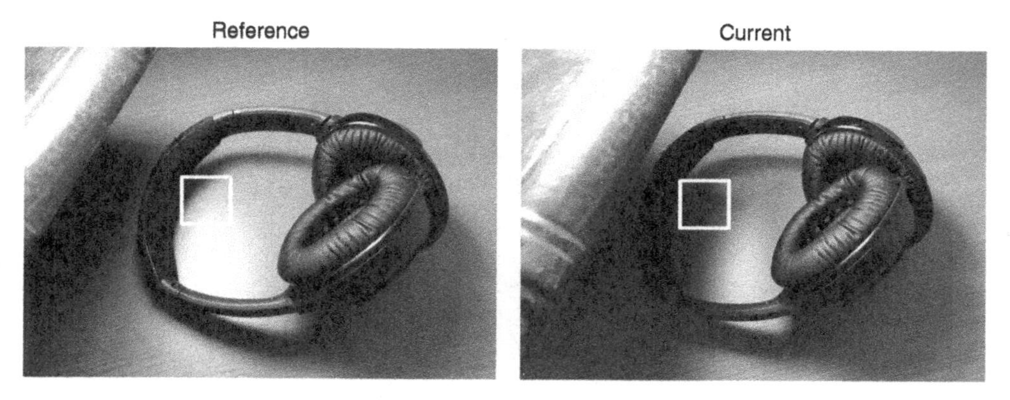

Figure 6.83 Scene with changing illumination (moving objects)

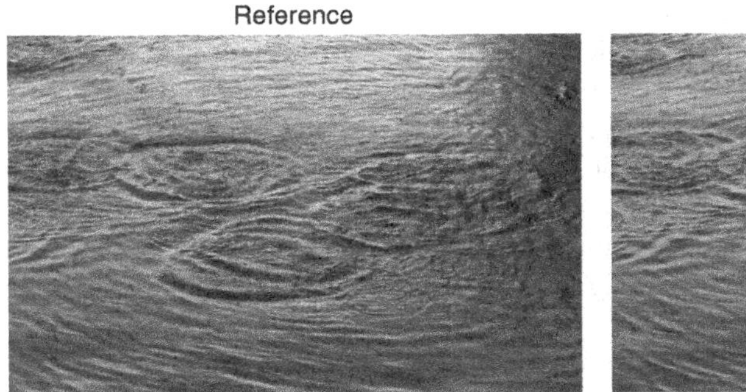

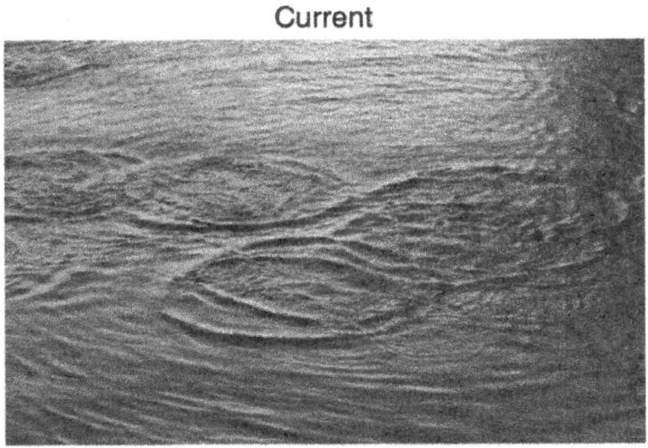

Figure 6.84 Scene with semi-random motion

other editing effects, such as the cross-fade shown earlier in Figure 6.25, can also be difficult to handle using motion-compensated inter prediction. Techniques such as weighted prediction, a form of biprediction where one reference may have a greater or lesser weight than the other, can be useful for fades.

A video encoder has to encode any clip it encounters, including, of course, video scenes containing all of the types of motion and changes mentioned above. If the encoder conforms to one of the present standards such as H.264 or HEVC, it is limited to prediction using translational motion. If the encoder cannot find a genuine match for a block using motion-compensated prediction, it may find a suitable inter prediction from a neighbouring block that just happens to be similar to the current block, as illustrated in Figure 6.19, or it may revert to intra prediction, as we will see later.

6.12 HEVC Inter Prediction

6.12.1 Overview

HEVC achieves efficient inter prediction through a combination of techniques, including:

- The ability to adapt the PU size from 64×64 down to 4×4 samples, with varying rectangular block shapes.
- Flexible choices of prediction type, direction and reference.
- Interpolation to sub-pixel positions for accurate motion compensation.
- The ability to merge a number of PUs, sharing one set of prediction parameters across multiple PUs, which makes it possible to efficiently predict motion for irregular-shaped moving regions.
- Adaptive deblocking and smoothing filters to reduce artefacts in the reconstructed picture with the aim of improving further prediction (see Chapter 9).
- The ability to select between intra- and inter prediction at a range of block sizes, depending on the video content.
- Candidate lists for predicting or copying motion vectors from previous blocks, known as merge mode or AMVP modes.

We will go through the details later, but first here is an example to show the general concepts.

6.12.1.1 HEVC Inter Prediction Example

Figure 6.85 shows a close-up of part of a P-picture coded using HEVC. This region contains four 64×64 sample CTUs labelled 1–4, each of which is partitioned into PUs. The figure shows motion vectors superimposed on the inter-coded PUs.

Figure 6.86 shows the prediction mode of each of the PUs in CTU3, which was at the lower left. Because this is within a slice in a P-picture, all the PUs must be coded using inter prediction with one reference or using intra prediction. Most of the PUs are inter-predicted using one reference, with the PU size ranging from 32×32 down to 4×8 luma samples. The

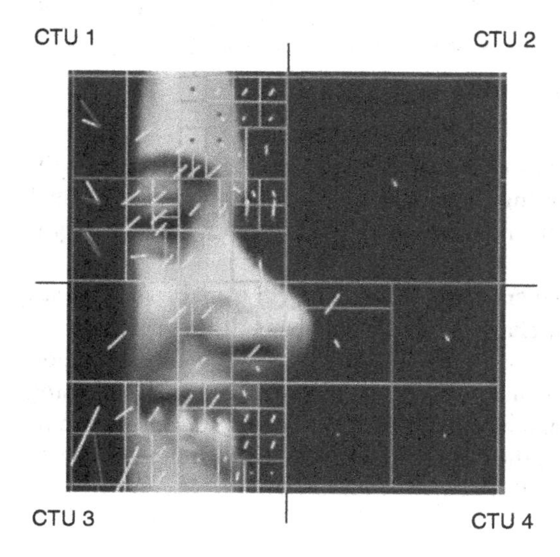

Figure 6.85 Region of a coded picture containing four CTUs

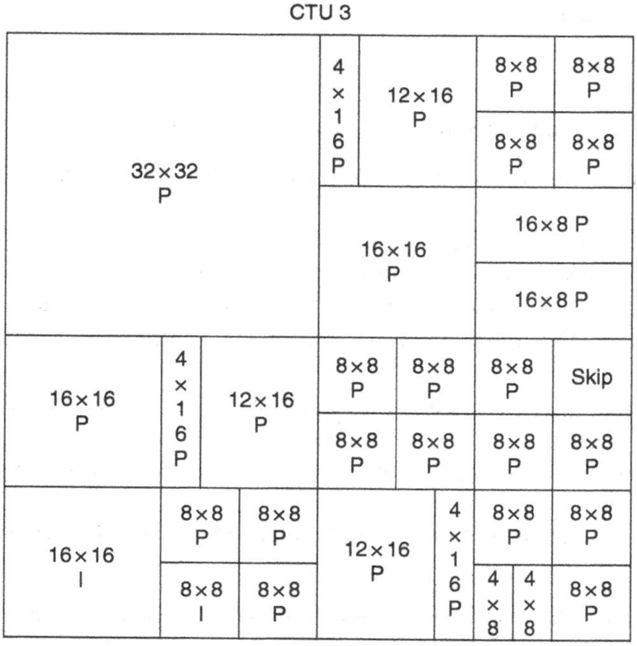

CTU 3

32×32 P
4×16 P
12×16 P
8×8 P
8×8 P
8×8 P
8×8 P
16×16 P
16×8 P
16×8 P
16×16 P
4×16 P
12×16 P
8×8 P
8×8 P
8×8 P
Skip
8×8 P
8×8 P
8×8 P
8×8 P
16×16 I
8×8 P
8×8 P
8×8 I
8×8 P
12×16 P
4×16 P
8×8 P
8×8 P
4×8
4×8
8×8 P

Figure 6.86 CTU3 showing prediction block types

encoder has chosen to encode one PU in skip mode and two using intra prediction.

This encoder uses larger PUs for regions where the motion and/or detail are relatively uncomplicated. Smaller PUs are selected for complex and/or boundary regions, such as the edge of the woman's face, to give a more accurate prediction of movement.

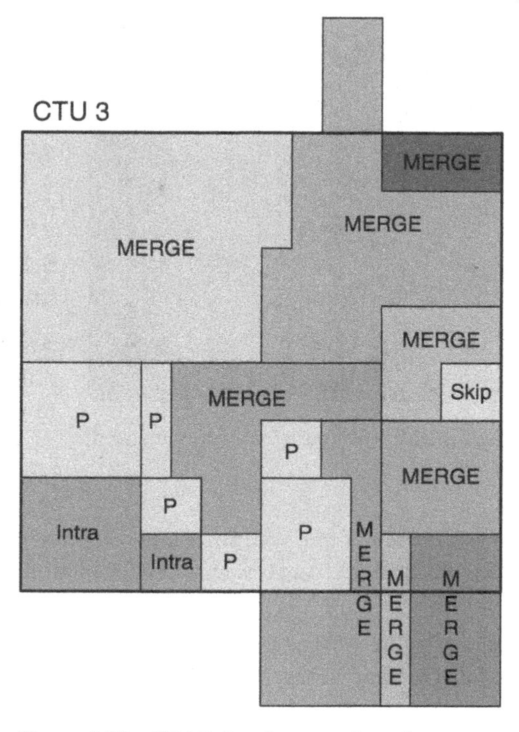

Figure 6.87 CTU 3 showing merging of motion regions

A disadvantage of using a large number of small PUs is that the overhead required to code motion parameters is greater. This can be mitigated by merging adjacent PUs such that the same motion parameters are sent once and then applied to a group of PUs. In Figure 6.87, the PUs in each shaded region of CTU 3 are merged. This means that the motion parameters, i.e. the motion vector and choice of the reference picture, are sent only once for each merged region, with the first of the merged PUs in coding order. Instead of sending motion parameters for subsequent PUs in the merged region, the encoder indicates that each PU inherits its parameters from an adjacent PU. As Figure 6.87 shows, merged regions may have irregular shapes and may cross over CTU boundaries. This makes it possible to efficiently predict irregular-shaped regions with similar motion parameters.

The result is an efficient interframe prediction that can be represented with a small number of bits. Figure 6.88 shows the prediction formed by the encoder, a very close match to the actual reconstructed image. The number of bits required to code each block, including

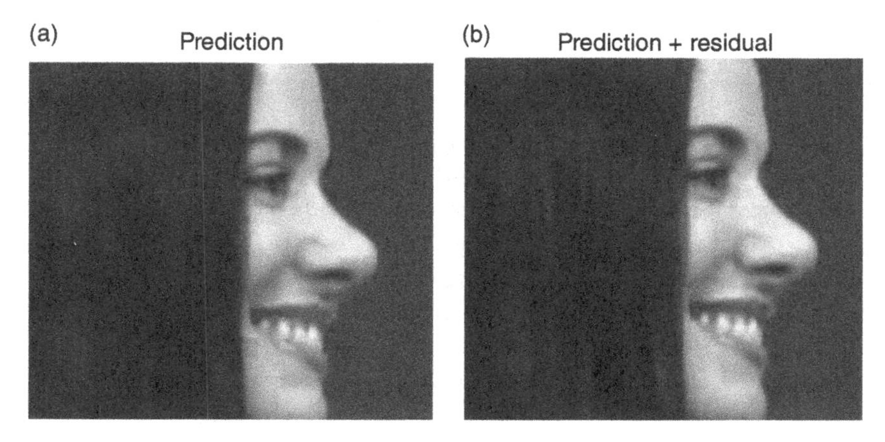

Figure 6.88 Inter-predicted region (a) and reconstructed region (b)

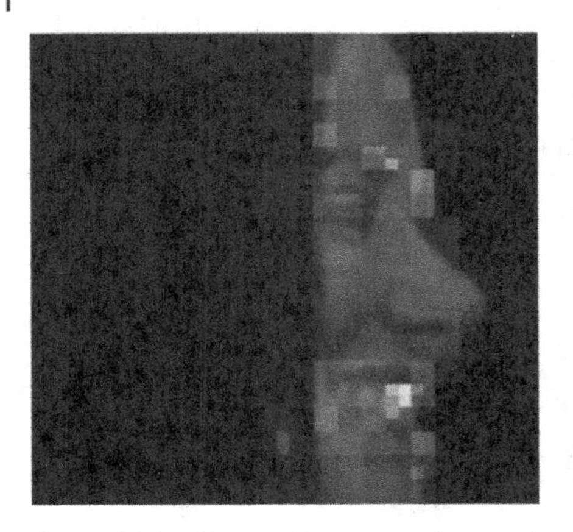

Figure 6.89 Bits per block (superimposed on image)

the prediction parameters and the residual, is shown in Figure 6.89, where brighter blocks contain more coded bits. The bit count tends to be highest around the boundaries of complex moving regions such as the woman's mouth.

6.12.2 HEVC Inter prediction Unit Partitioning

An HEVC CU can be partitioned in a number of different ways for inter prediction, as illustrated in Figure 6.91. Each CU is partitioned into one, two or four PUs using one of four symmetric partition modes, 2N×2N, N×2N, 2N×N and N×N, or one of four asymmetric partition modes. Each of the resulting PUs is separately predicted using inter prediction from one or two reference pictures from reference picture lists 0 and 1.

6.12.2.1 Slice Types and PU Types

Each CU may be partitioned in intra-mode or inter-mode if the current slice is a P- or B-slice, see Table 6.4.

6.12.2.2 Symmetric PU Partitions

An inter-coded CU of size 2N×2N, where $N = 4, 8, 16$ or 32, can be partitioned into one of four symmetric partition types, as illustrated in Figure 6.91:

a) 2N×2N, i.e. the CU is partitioned into a single PU of the same size as the CU,
b) N×2N, where the CU is partitioned into two PUs,

Table 6.4 Slice types and PU types

Slice type	PU type(s)
I	CU partitioned into intra-PUs only
P	CU partitioned into intra- or inter-PUs. Each inter-PU may be predicted from one List 0 reference picture, with one motion vector (MV)
B	CU partitioned into intra- or inter-PUs. Each inter-PU may be predicted from: a) One List 0 reference picture, with one MV b) One List 1 reference picture, with one MV c) One List 0 AND one List 1 reference picture, with a total of two MVs, i.e. biprediction Option (c) is not allowed for the smallest possible PU partition sizes of 8×4, 4×8 or 4×4

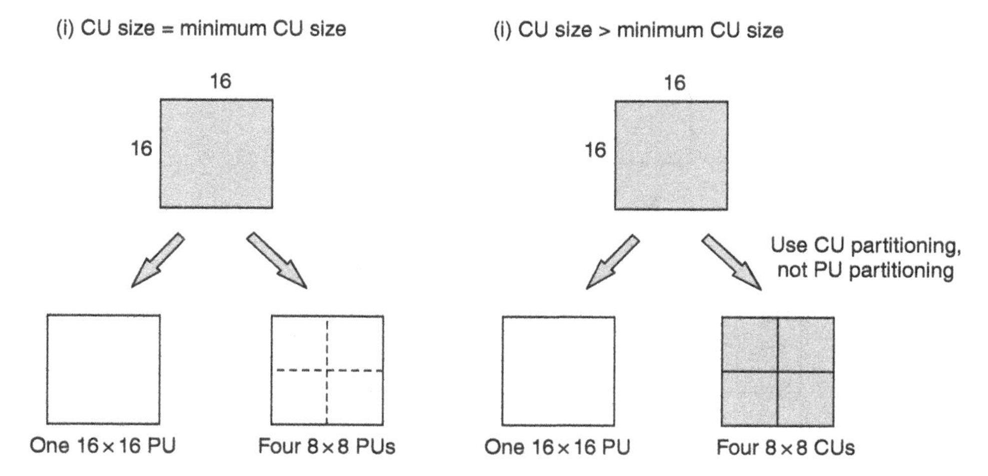

Figure 6.90 Symmetric partitions of an inter-CU: examples

c) 2N×N, where the CU is partitioned into two PUs, or
d) N×N, where the CU is partitioned into four PUs.

The final symmetric partition, into four N×N PUs, is only allowed in HEVC if the CU size is already the minimum size available, since the encoder could achieve the same result simply by selecting a smaller CU size.

Example (i): Current CU size is 16×16, which is the minimum available CU size, determined from a Sequence Parameter Set parameter. Partitioning into four 8×8 PUs is allowed.

Example (ii): Current CU size is 16×16, but the minimum available CU size is 8×8. Partitioning into four 8×8 PUs is not allowed because the encoder can simply select a CU size of 8×8 to achieve the same result (see Figure 6.90).

6.12.2.3 Asymmetric PU Partitions

When enabled by a Sequence Parameter Set flag, the encoder can choose to split a CU into further four asymmetric partition types, as shown in Figure 6.91:

a) N/2×2N (L): A narrow PU, one-quarter of the width of the CU, on the left and a wider PU, 3/4 of the width of the CU, on the right.
b) N/2×2N (R): A wider PU on the left and a narrow PU on the right of the CU.
c) 2N×N/2 (U): A narrow PU at the top and a wider PU at the bottom of the CU.
d) 2N×N/2 (D): A wider PU at the top and a narrow PU at the bottom of the CU.

These asymmetric partitionings give the encoder flexibility in matching PU size and shape to the characteristics of moving objects in the video scene. The potential disadvantages of asymmetric partitioning include (1) higher computational costs due to the encoder searching for more partition types to find the best match and (2) higher signalling costs due to the increased range of partition types. It is up to the encoder whether or not to enable the use of asymmetric partitions. Asymmetric partitions are not available for all CU sizes.

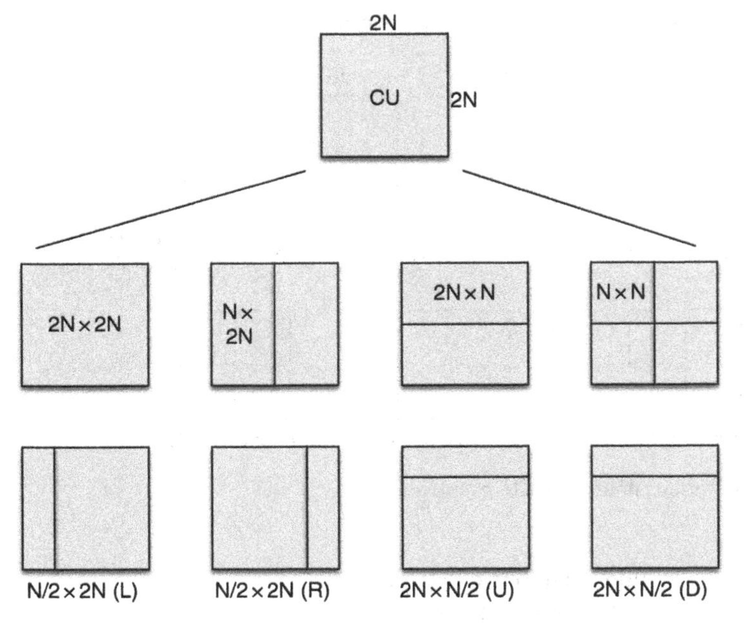

Figure 6.91 CU partitionings into inter-PUs: symmetric and asymmetric

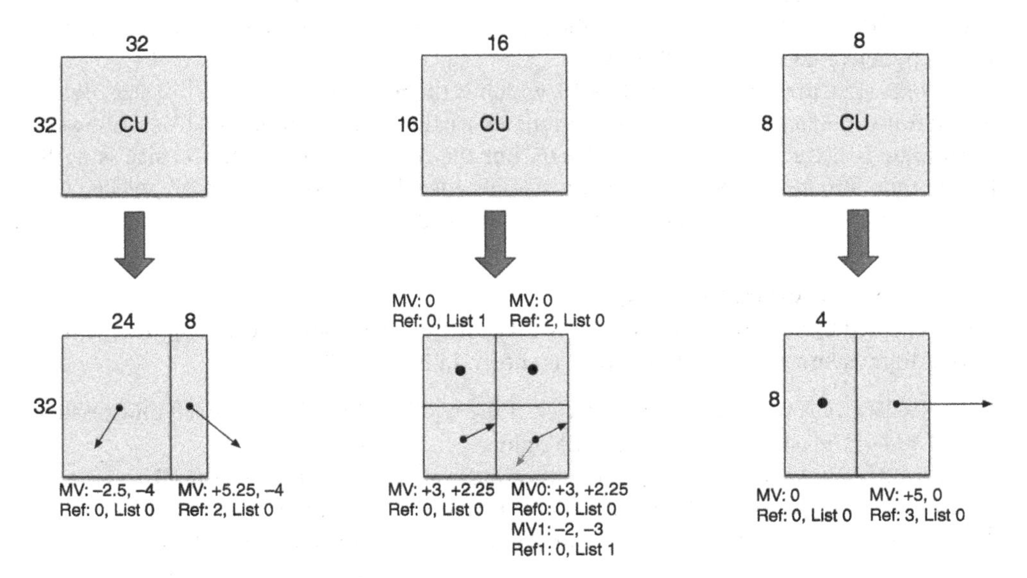

Figure 6.92 Examples of inter-PU partitionings

6.12.2.4 CU Partitionings into Inter-PUs: Examples

Example 1, Figure 6.92:

A 32×32 CU in a P-slice is partitioned into 24×32 and 8×32 PUs, each with one MV and one reference picture from reference picture List 0.

Example 2, Figure 6.92:
A 16 × 16 CU in a B-slice is partitioned into four 8 × 8 partitions:

i) Zero MV, reference picture 0 from reference picture List 1
ii) Zero MV, reference picture 2 from reference picture List 0
iii) One MV, reference picture 0 from reference picture List 0
iv) Two MVs and two reference pictures from List 0 and List 1, using biprediction.

Note that the smallest CU size must be 16 × 16 for this partitioning to be valid.

Example 3, Figure 6.92:
An 8 × 8 CU in a B-slice is partitioned into two 8 × 4 partitions, each with one MV and one reference picture from List 0. Note that biprediction is not allowed because the PU size is smaller than 8 × 8.

Example 4:

Figure 6.93 shows a frame from the *Jockey* sequence that has been encoded using HEVC. The CTU size is 64 × 64 pixels. The encoder has chosen 64 × 64 PUs for areas such as parts of the background where a single inter prediction is suitable for the whole region. Around the jockey and the horse, the encoder tends to choose smaller block sizes, such as the 8 × 4 block highlighted. Figure 6.94 is a close-up of a single 64 × 64 pixel CTU from the same frame, showing the motion vectors for each block. Some PUs are bipredicted, i.e. they have two motion vectors, each pointing to a different reference frame. Others are predicted using a single motion vector.

Figure 6.93 Section of frame showing HEVC inter prediction unit block sizes. *Source:* Elecard Streameye Analyser

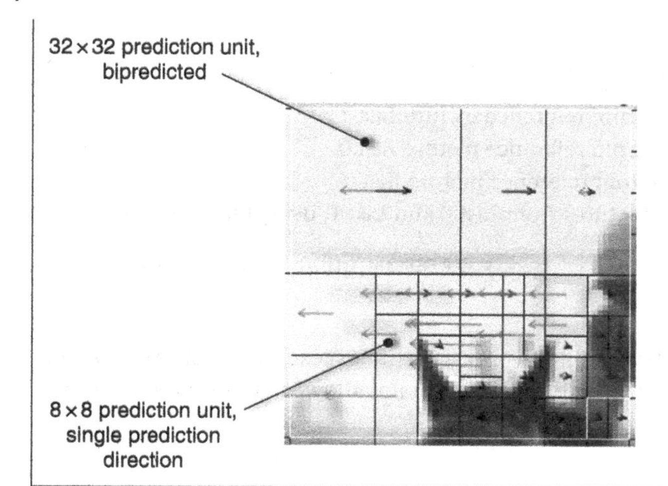

Figure 6.94 Close-up of Coding Tree Unit showing prediction units and motion vectors. *Source:* Elecard

6.12.3 HEVC Inter-PU Coding

An inter-PU can be coded in one of the following ways:

Skip: Send only a Skip flag and a merge index.
Merge: Send a merge index and coded blocks.
AMVP: Send a Predictor Index, a delta motion vector, a reference index and optionally coded coefficient blocks.

Figure 6.95 is an overview of the inter-PU coding process in an HEVC codec. The general idea is to minimise the amount of information that needs to be sent for each PU. If possible, the overhead information, indicating the motion vector and reference index, is copied from a previously coded neighbouring PU and the current PU is coded in skip or merge mode. If unique overhead information is sent, the motion vector is predicted from a previously coded neighbouring PU known as a candidate PU. This can be beneficial, as neighbouring PUs often have similar or identical motion vectors. In simplified form, the process is as follows:

1) Identify a set of candidate PUs, each of which is a spatially adjacent PU in the same picture or a co-located or neighbouring PU in the chosen reference picture.
2) Select one candidate from the list.
3) Send an index or flag that identifies the candidate PU.
4) Code and send any remaining information.

In merge mode, the list contains up to five candidates. For each PU, the encoder sends the candidate index and either a skip indication, which means there is no more data to be coded, or coded residual data.

In AMVP mode, the list contains just two candidates. For each PU, the encoder sends a flag indicating which candidate to use, plus a delta motion vector and reference picture index, plus the coded residual.

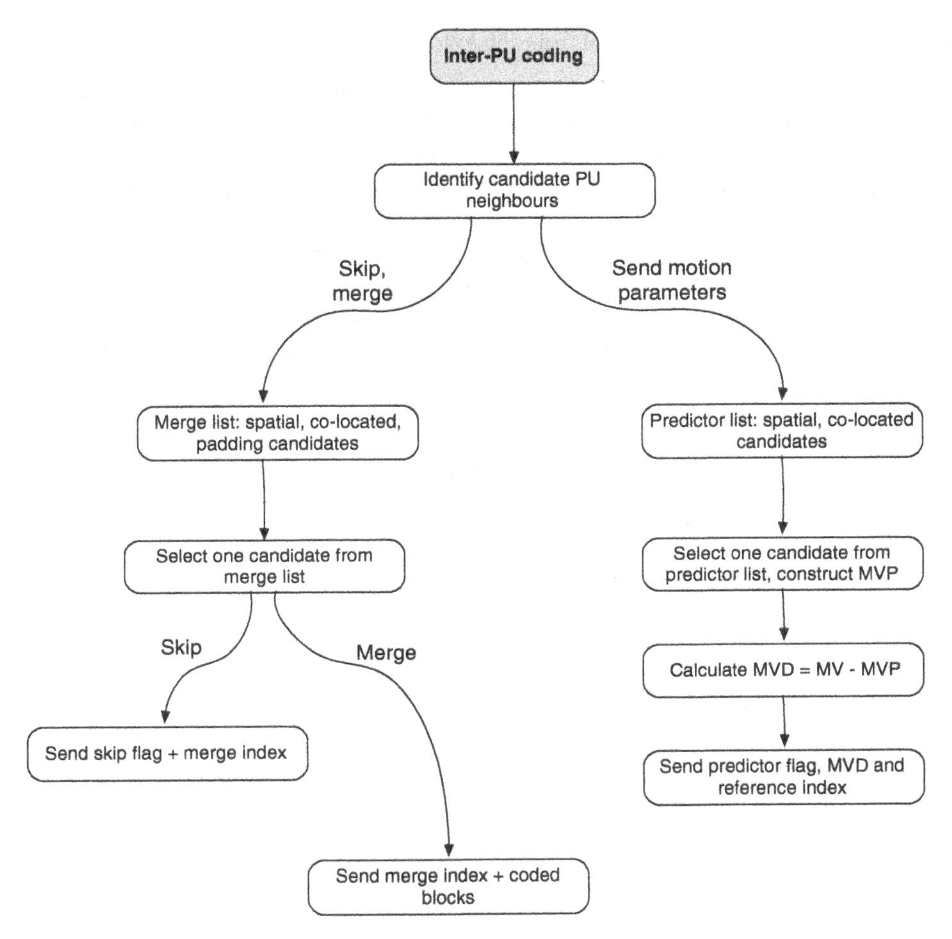

Figure 6.95 HEVC inter-PU coding overview

6.12.3.1 Skip or Merge Coding

Sending a merge index allows a single set of prediction parameters, i.e. motion vector and choice of reference picture, to be applied to multiple adjacent regions in an HEVC picture, as illustrated in Figure 6.87. A PU coded in Skip or merge mode inherits its motion parameters from a spatially or temporally adjacent PU. This means that the motion vector and choice of reference picture are sent only once for each merged region, with the first of the merged PUs in coding order. Instead of sending motion parameters for subsequent PUs in the merged region, the encoder indicates that each PU inherits its parameters from an adjacent PU. As Figure 6.87 shows, merged regions may be of irregular shape and may cross over CTU boundaries. This makes it possible to efficiently predict irregular-shaped regions with similar motion parameters.

The number of merge candidates MaxNumMergeCand is signalled in the slice header. The encoder generates a list of MaxNumMergeCand available merge candidates and signals the choice of candidate by encoding a merge index in the bitstream.

The list of candidates is constructed from the following:

1) Up to four spatially adjacent candidates, selected from a list of up to five neighbouring blocks, such as the neighbouring blocks shown in Figure 6.96. There may be more than one block immediately above or immediately to the left of the current PU, as shown in Figure 6.96, and so the blocks that contain the positions A_1, B_1, B_0, A_0 and B_2, in that order are chosen as the potential candidates (see Figure 6.97). Each candidate is only included if the corresponding block is available for prediction, i.e. already coded, in the same slice and tile and not coded in intra-mode, and if the candidate vector and reference index are not identical to a previous candidate in the list. If all five potential candidates are available, only the first four are included, since there can be a maximum of four spatial candidates.

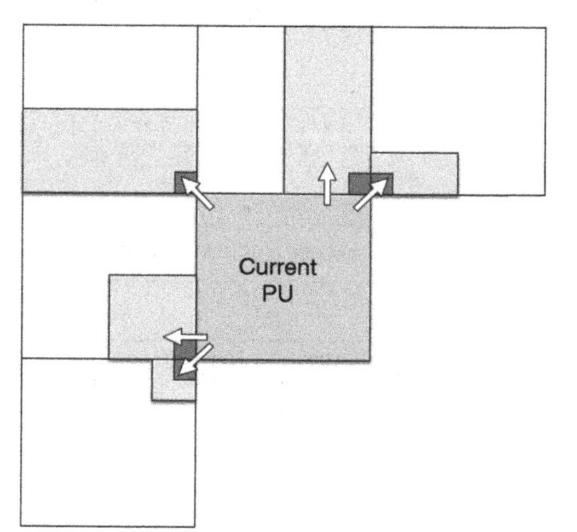

Figure 6.96 Merge mode: spatially adjacent candidates

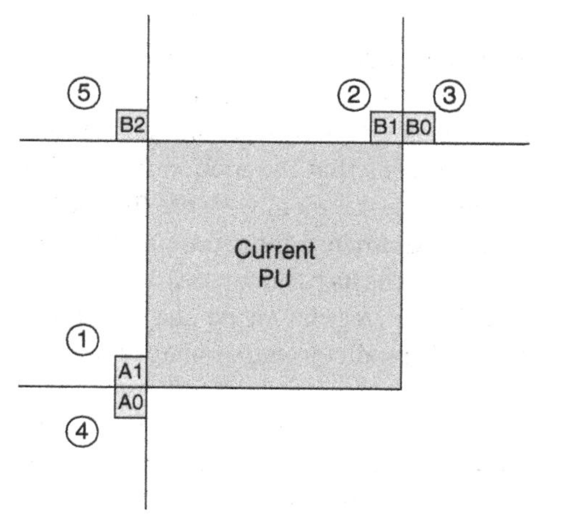

Figure 6.97 Merge mode: evaluating spatial candidates

2) One co-located or temporal candidate, a block below and to the right or in the same position in the reference picture for the current PU (Figure 6.98). The blocks containing the positions T_1 and T_2 are the options for the co-located candidate (see Figure 6.99). If T_1 is available, described in the HEVC standard as ColBr, Collocated Bottom Right, it is chosen as the co-located candidate; otherwise, T_2 is chosen, described as ColCtr, Collocated Centre.
3) Optionally, padding candidates, which are existing candidates or zero motion vectors, to ensure there are always *MaxNumMergeCand* candidates.

Figure 6.100 shows two examples. In the first example, the list of candidates comprises MaxNumMergeCand-1 spatial candidates and one temporal or co-located candidate. In the second example, the list comprises spatial candidates, one temporal or co-located candidate and one padding candidate.

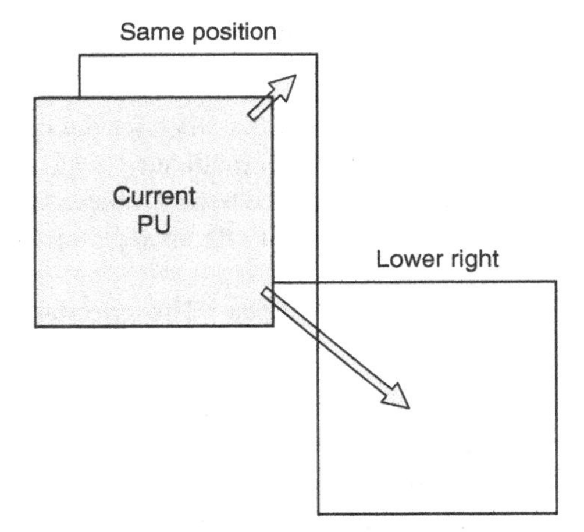

Figure 6.98 Merge mode: co-located candidates in reference picture

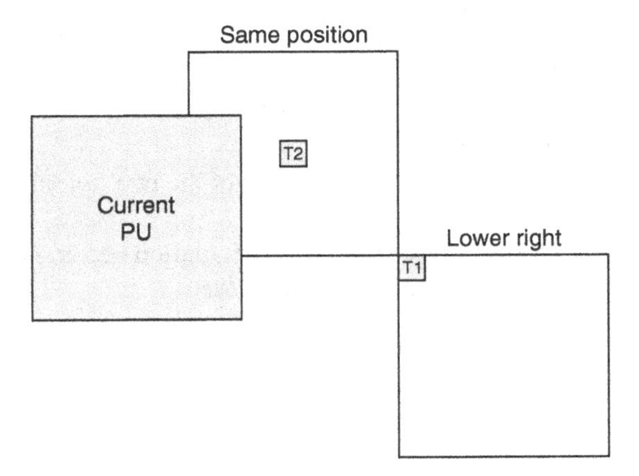

Figure 6.99 Merge mode: evaluating co-located candidates

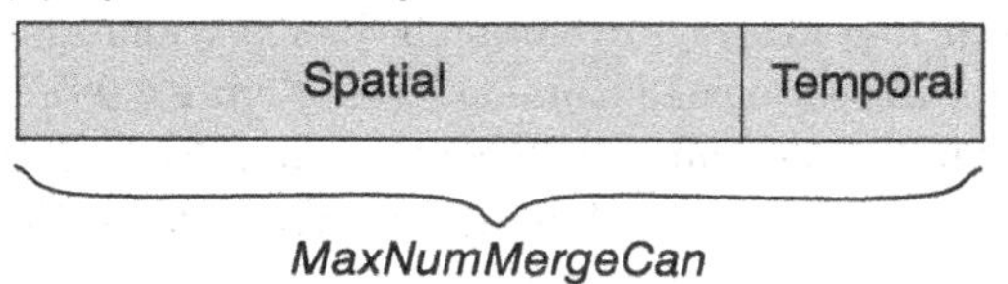

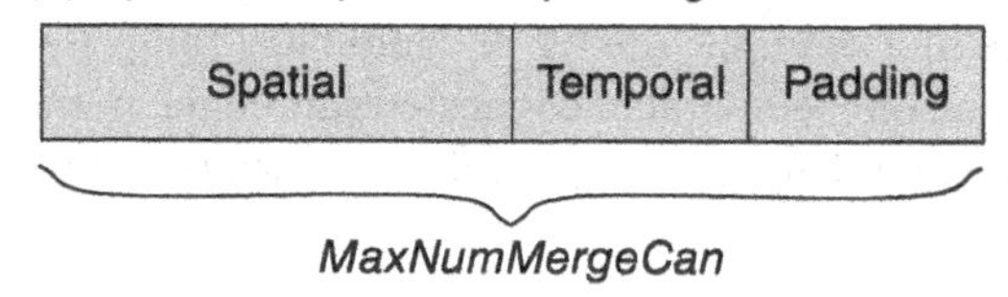

Figure 6.100 Merge candidate list examples, (a) spatial and temporal, (b) spatial, temporal and padding.

6.12.3.2 Inter-PU Coding: Advanced Motion Vector Prediction

If an inter-PU is coded in AMVP mode, the encoder carries out the following for each motion vector:

1) Constructs a prediction MVP for the PU motion vector, using a predictor,
2) Signals the choice of predictor,
3) Signals the difference between the actual MV and the predictor, MVD and
4) Signals the choice of reference picture, REF_IDX.

Constructing the Prediction The encoder generates a list of two candidates from the following:

1) **Predictor a**: A_0 if available, i.e. already coded, in the same slice and tile and not coded in intra-mode, else A_1 if available (see Figure 6.97).
2) **Predictor b**: B_0 if available, else B_1 if available, else B_2 if available (see Figure 6.97).
3) **Predictor c**: T_1 if available, else T_2 if available (see Figure 6.99).

Example 1: A_1 and T_1 are available, all other predictors are unavailable. The choice of predictors is (a, c), i.e. (A_1, T_1).

Example 2: All the predictors are available. The choice of predictors is (a,b), i.e. (A_0, B_0). Predictor c is ignored.

Example 3: A_0, A_1, B_2 and T_2 are available. The choice of predictors is (a,b), i.e. (A_0, B_2). Predictor c is ignored.

The encoder selects from one of the two candidate predictors.

Signalling the Predictor For each motion vector, flag *mvp_lX_flag* is sent to indicate which of the two MVP predictors is chosen.

Signalling the Motion Vector Difference For each motion vector, MVD, the actual motion vector minus the predictor MVP, is signalled in the bitstream.

Signalling the Choice of Reference Picture For each motion vector, REF_IDX signals the choice of reference picture from the available list of references.

Example:

Figure 6.101 shows a B-picture from a 720p clip that was coded using HEVC. The inter prediction blocks are superimposed as a grid. Figure 6.102 shows a close-up of the highlighted region, 256 pixels wide and 192 pixels high. This region is split into a total of 66 PUs. For 41 of the 66 blocks, the encoder has chosen not to send delta prediction parameters. This means that the prediction parameters are inherited as base parameters from neighbouring blocks, using merge mode. For the remaining 15 of the 66 blocks, marked with P, the encoder has sent delta parameters using AMVP mode. Hence, for the majority of the blocks in this region, no prediction parameters are actually sent in the coded bitstream and instead the parameters are inferred by the decoder.

Figure 6.101 Frame from *Barge* clip, showing HEVC inter prediction blocks. *Source:* Elecard Streameye

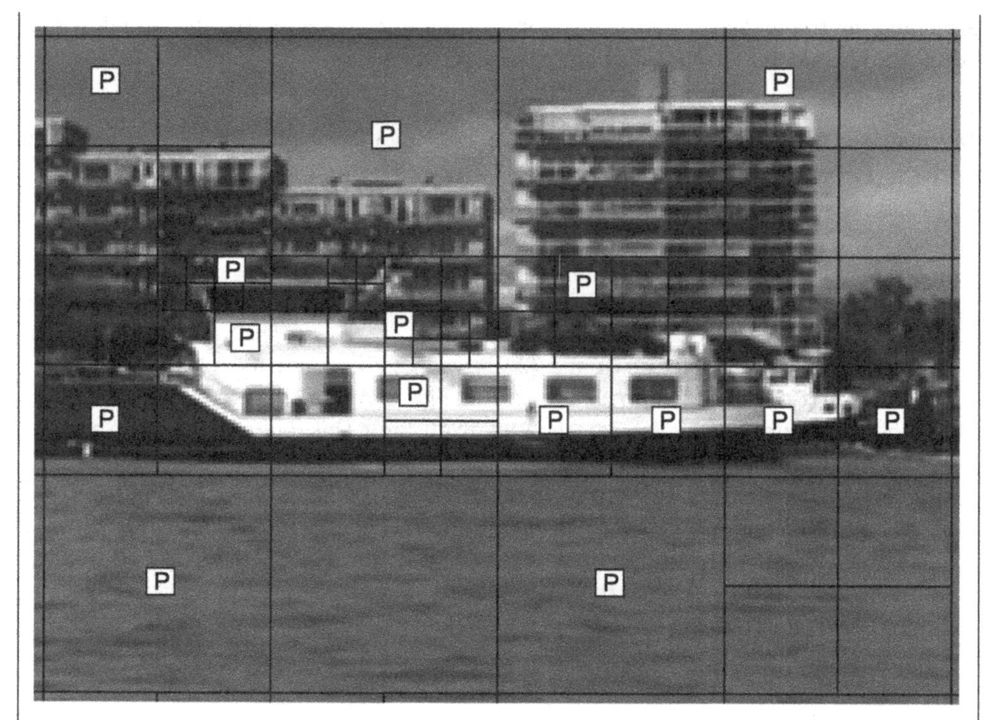

Figure 6.102 256 × 192 pixel region showing blocks with transmitted delta prediction parameters (P) and non-transmitted parameters (no P)

6.12.3.3 HEVC Inter-PU Coding: Examples

a) Coding PUs: AMVP and merge modes

The encoder decides to encode PU1, as shown in Figure 6.103, in AMVP mode. Left-hand neighbours A0 and A1 are available as prediction sources; upper neighbours B0, B1 and B2 are intra-coded and are not available. Blocks T1 and T2 in the reference picture are available.

Predictor a = A0, the first available left-hand predictor.
Predictor b is not available, since B0, B1 and B2 are not available.
Predictor c = T1, the first available co-located predictor.

The list of possible predictors is therefore (A0, T1). The encoder chooses A0 as the predictor and signals:

```
merge_flag = 0, i.e. no merge
mvp_flag = 0, i.e. choose the first predictor, A0
ref_idx choice
mvd = motion vector(current block) - motion vector(block A0)
```

After coding PU1, the encoder decides to code the next prediction unit PU2, as shown in Figure 6.104, in merge mode. Left-hand candidate A0 is not available because it has not yet

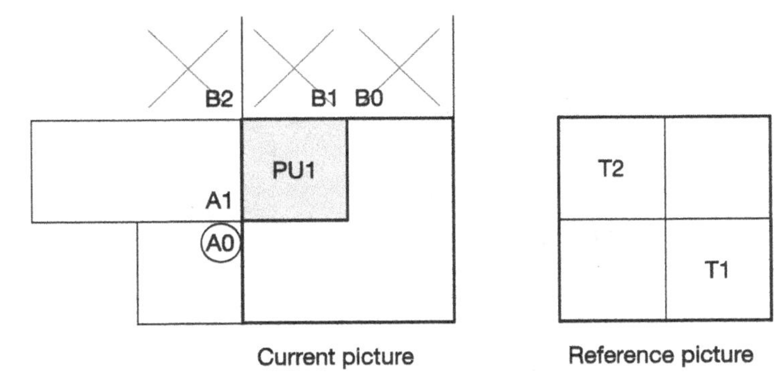

Figure 6.103 Coding PU1, AMVP mode

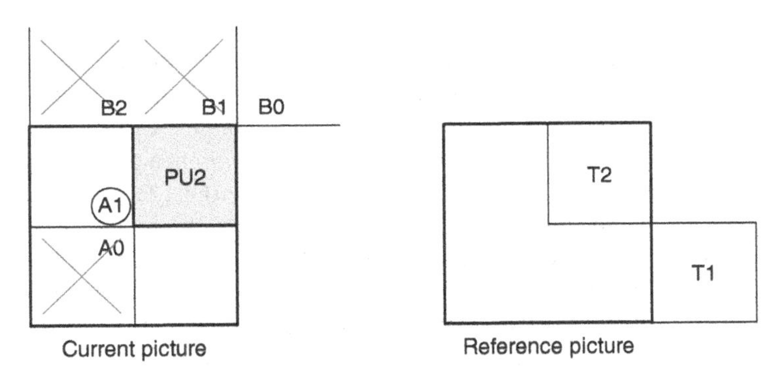

Figure 6.104 Coding PU2, merge mode

been coded. Left-hand candidate A1 is available as it was coded as PU1, and B0 is available. Blocks T1 and T2 in the reference picture are available.

Max_merge_can, the number of merge candidates = 4, and so the list of available merge candidates is:

(A1, B0, T1, Pad), where Pad is a zero motion vector.

The encoder chooses candidate A1 and signals:

```
merge_flag = 1
merge_idx = 0, i.e. choose the first merge candidate
```

The decoder copies the motion vector and reference index from A1, i.e. from PU1.

b) Merging to create irregular regions

HEVC's merge mode can effectively model motion across irregular-shaped regions. We will look at one of the merge regions in Figure 6.87 in more detail.

The merge chain starts with an 8 × 16 PU in CTU1 (see Figure 6.85), which is coded in AMVP mode, which means an explicit signalling of a motion vector and a choice of

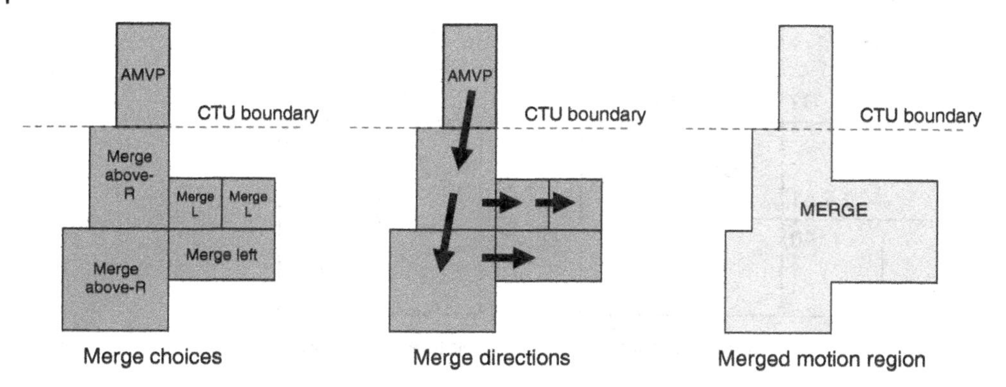

Figure 6.105 HEVC merged motion region example

reference picture. The 12×16 PU below this initial PU, which happens to be across the boundary in CTU3, is coded as Merge, Above-Right (see Figure 6.105), left-hand diagram. This means that this 12×16 PU inherits the motion parameters of the initial PU, which means it is coded with the same motion vector and choice of reference picture, as indicated by the arrow in the centre diagram. The 16×16 PU below the 12×16 PU is also coded as Merge, Above-Right, so it inherits the same parameters. Three more PUs are coded as Merge, Left, so they each inherit the parameters. The end result is that the irregular-shaped region shown in Figure 6.105 at the right-hand side is coded with a single set of motion parameters, namely a single motion vector and choice of reference picture. This illustrates how merge mode enables various irregular-shaped regions to be coded with the same motion parameters.

6.12.4 HEVC Inter Prediction Sources

In a P-slice, each inter-coded PU is predicted from one reference picture, chosen from a list of available reference pictures, List 0. The encoder sends:

- Coded prediction information, indicating merge or prediction
- Reference picture index, which is an index into List 0
- Coded residual information.

In a B-slice, each inter-coded PU may be predicted from one or two reference pictures, chosen from two lists, List 0 and List 1, either one reference picture from List 0, or one reference picture from List 1, or two reference pictures, one each from List 0 and List 1, using biprediction.

Figure 6.106 shows four inter-coded PUs in a current picture, which contain a single B-slice. PU1 and PU2 are each predicted from a single picture that precedes the current picture in display order, a past picture. PU3 is predicted from a single picture that is later than the current picture in display order, a future picture. PU4 is bipredicted from a past picture and a future picture.

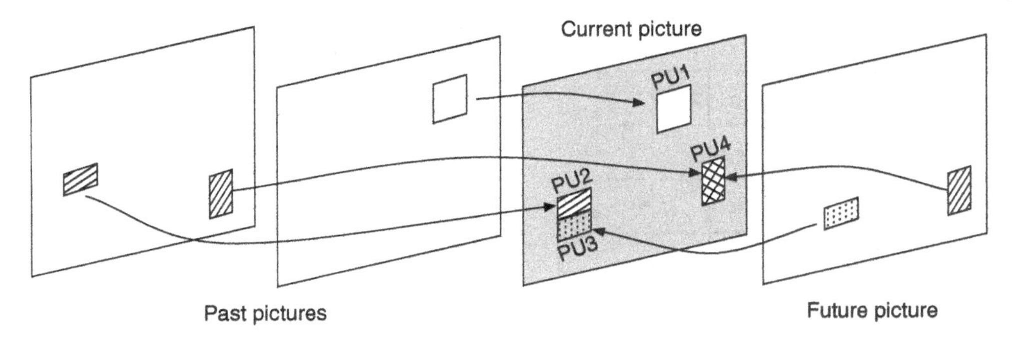

Figure 6.106 Prediction sources for inter-PUs

6.12.5 HEVC Biprediction and Weighted Prediction

PUs in an HEVC B-slice may be coded using biprediction, in which the PU is predicted from two reference pictures. The prediction is formed by averaging samples from each reference picture. The relative contribution of each of the two reference pictures may be adjusted by an optional weighting factor, signalled in the slice segment header. The weighting factor may be used to increase or reduce the contribution of each reference picture. In the example of Figure 6.25, such a weighting factor might be beneficial to control the amount of prediction from each reference as the fade progresses.

In HEVC, weighted prediction can be applied separately to luma and chroma components and is also available in P slices to adjust the relative contribution of luma and chroma in forming the inter prediction.

6.12.6 HEVC Sub-Pixel Interpolation

HEVC uses a set of seven- or eight-tap filters to interpolate the luma component of the reference picture to quarter-pixel resolution. The chroma component is interpolated to eighth-pixel resolution using a set of four-tap filters.

6.12.6.1 Luma Sample Interpolation

Positions A_{ij} in Figure 6.107 are the integer-position luma samples in a reference picture. For every integer sample A_{ij}, there are fifteen quarter-position samples, a...k, n, p, q, r, that the encoder may use for inter prediction[3]. These sample positions are interpolated as described below.

i) Positions a, b, c:

Each quarter-sample position a, b and c is interpolated by filtering integer-position samples in a horizontal direction, as illustrated in Figure 6.108. Position a is interpolated as a

3 Why does the HEVC standard label the quarter-position samples in this way? To avoid any confusion when reading labels such as l, m and o. The light-grey letters are quarter-position samples for other integer samples.

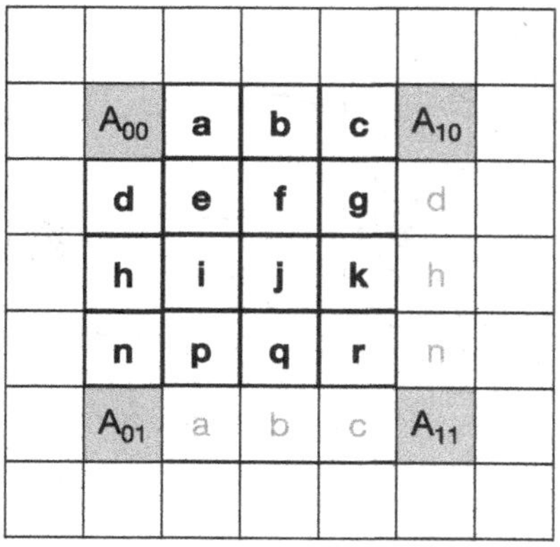

A00
a
b
c
A10
d
e
f
g
d
h
i
j
k
h
n
p
q
r
n
A01
a
b
c
A11

Figure 6.107 HEVC luma interpolation positions

Integer luma positions
A-30
A-20
A-10
A00
a
b
c
A10
A20
A30
A40
-1
+4
-10
+58
+17
-5
+1
+40
+40
-11
-11
+4
+4
-1
-1

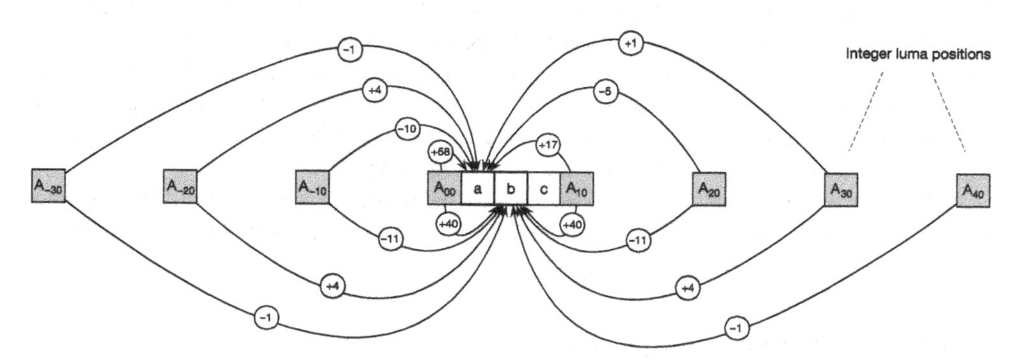

Figure 6.108 HEVC luma sample interpolation filters

weighted combination of seven integer-position samples, four to the left and three to the right:

$$a = (-1*A_{-3,0} + 4*A_{-2,0} - 10*A_{-1,0} + 58*A_{0,0} + 17*A_{1,0} - 5*A_{2,0} + 1*A_{3,0})/64$$

Note that the closest integer-sample positions, $A_{0,0}$ followed by $A_{1,0}$, have the strongest weighting.

Position b is interpolated as a weighted combination of eight integer-position samples:

$$b = (-1*A_{-3,0} + 4*A_{-2,0} - 11*A_{-1,0} + 40*A_{0,0} + 40*A_{1,0} - 11*A_{2,0} + 4*A_{3,0} - 1*A_{4,0})/64$$

Position c is interpolated as a weighted combination of $A_{-2,0}$ to $A_{4,0}$, using a filter that is a mirror image of the filter for a, not shown in Figure 6.108.

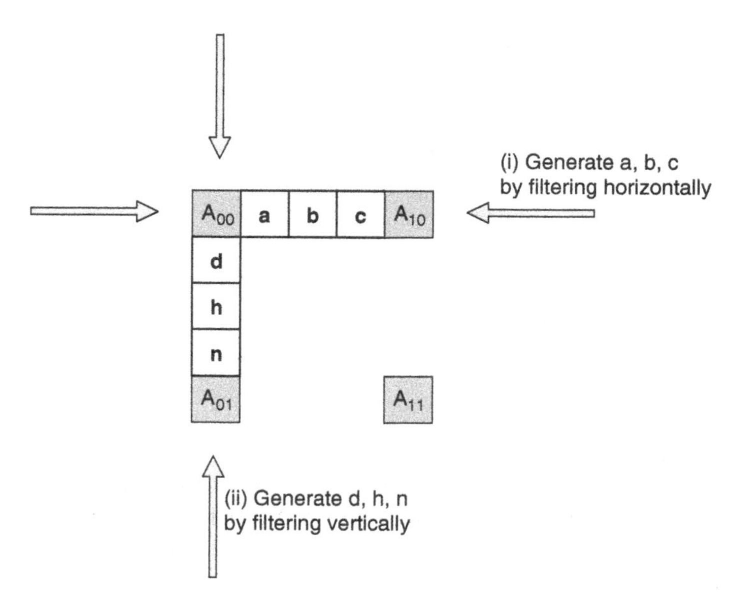

Figure 6.109 HEVC luma sample interpolation, positions a–d, h, n

ii) Positions d, h, n:

Positions d, h and n have vertical integer-position neighbours and are interpolated by filtering these neighbours in a vertical direction. Position d is interpolated as a weighted combination of seven integer-position samples, four above and three below, using the same filter equation as position a, applied vertically. Positions h and n are interpolated in the same way as b and c, applying the filters vertically (Figure 6.109).

iii) Positions e, f, g, i, j, k, p, q, r:

These positions are not aligned horizontally or vertically with integer-position samples, so they are interpolated by filtering previously calculated values a, b and c in a vertical direction. Positions e, f and g are interpolated as a weighted combination of seven values of a, b or c, respectively, in the same way as position d. Positions i, j and k are interpolated the same way as position h, using a weighted combination of eight values of a, b and c, respectively. Positions p, q and r are interpolated the same way as position n, using a weighted combination of seven values of a, b and c, respectively (see Figure 6.110).

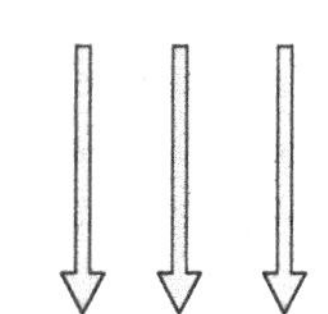

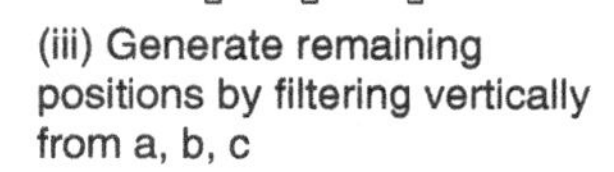

Figure 6.110 HEVC luma sample interpolation, remaining positions

6.12.6.2 Chroma Sample Interpolation

Assuming a sampling format of 4:2:0, there are twice as many chroma samples as luma samples, both horizontally and

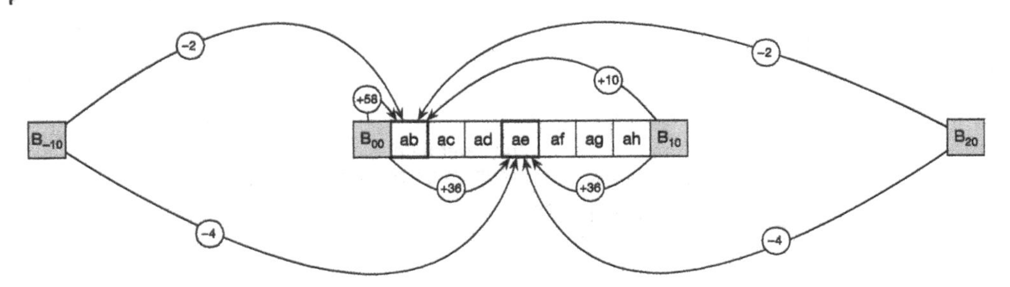

Figure 6.111 HEVC Chroma sample interpolation filters

vertically. This means that a luma motion vector with quarter-sample precision corresponds to a chroma motion vector with eighth-sample precision. There are therefore 63 eighth-sample chroma positions for every integer-sample position B_{ij}. Each of these eighth-sample positions is interpolated using a four-tap filter, i.e. a filter that is computationally simpler than the luma filters described above.

Position ab, shown in Figure 6.111, is interpolated from the four nearest integer-sample positions B_{ij} horizontally, using a four-tap filter:

$$ab = (-2*B_{-1,0} + 58*B_{0,0} + 10*B_{1,0} -2*B_{2,0})/64$$

Notice that the nearest position $B_{0,0}$ is given the highest weighting. Positions ac, ad, ae, af, ag and ah are each interpolated from the same four integer-sample positions $B_{-1,0}$, $B_{0,0}$, $B_{1,0}$ and $B_{2,0}$ using a set of four-tap filters. For example, the filter for position ae is:

$$ae = (-4*B_{-1,0} + 36*B_{0,0} + 36*B_{1,0} -4*B_{2,0})/64$$

Because ae is the same distance from $B_{0,0}$ and $B_{1,0}$, the weights of the inner taps are equal (+36).

In a similar way to the luma samples, the remaining chroma positions are interpolated by filtering vertically from either the integer-sample positions B_{ij} or the previously calculated ab, ac, ad....ah positions, using the same set of four-tap filters.

Example: HEVC Luma Sample Interpolation

Figure 6.112 shows a section of a CIF image on the left. Note the pixelation. The same section has been interpolated to quarter-pixel resolution, i.e. for every integer pixel position, 15 quarter-pixel positions have been created using the HEVC luma interpolation filters to produce the right-hand image. As we saw earlier in the chapter, sub-pixel interpolation cannot create new information, but the seven/eight-tap filters have done a respectable job of upsampling the image without excessive smoothing of edges and without removing detail.

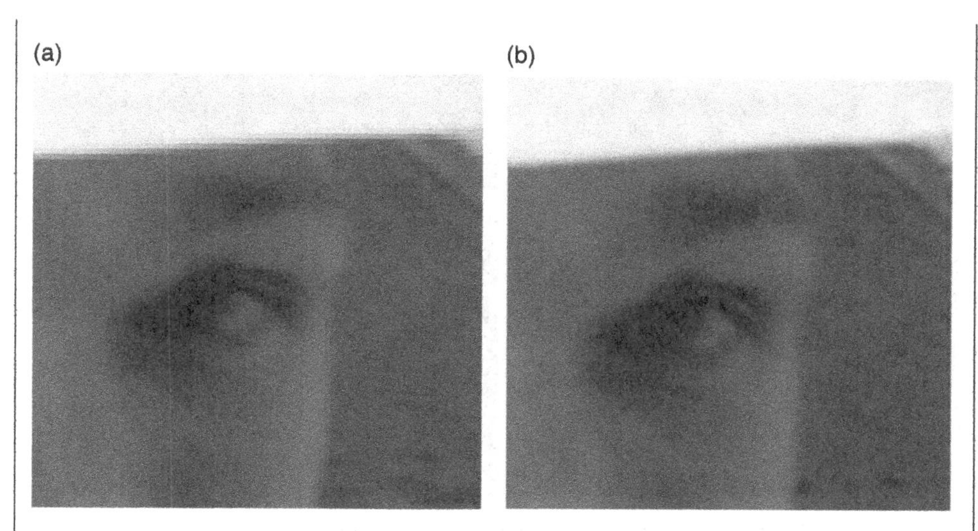

Figure 6.112 Section of original image (a), HEVC luma interpolation (b)

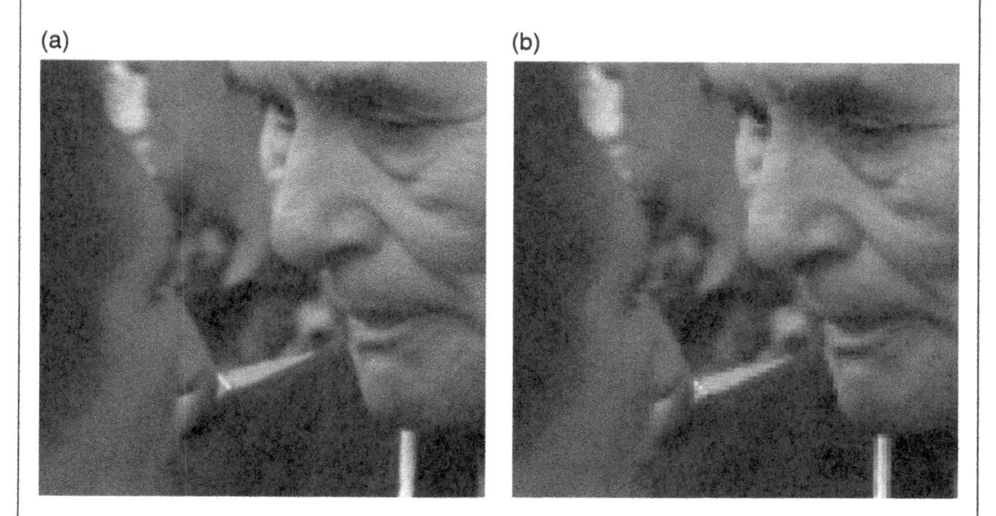

Figure 6.113 Full-resolution original (a), decimated x4 (b)

Example: An original, full-resolution image, such as the left-hand image of Figure 6.113, is decimated by a factor of 4 in the horizontal and the vertical directions, to produce Figure 6.113, right. In other words, only every fourth pixel horizontally and vertically is retained. We will look at the effect of interpolating the subsampled image to quarter-pixel resolution, which is effectively the same as upsampling the low-resolution version to the original resolution.

First, consider bilinear interpolation. Each sub-pixel position is interpolated by averaging the neighbouring full-pixel neighbour, weighting by position. The result, shown

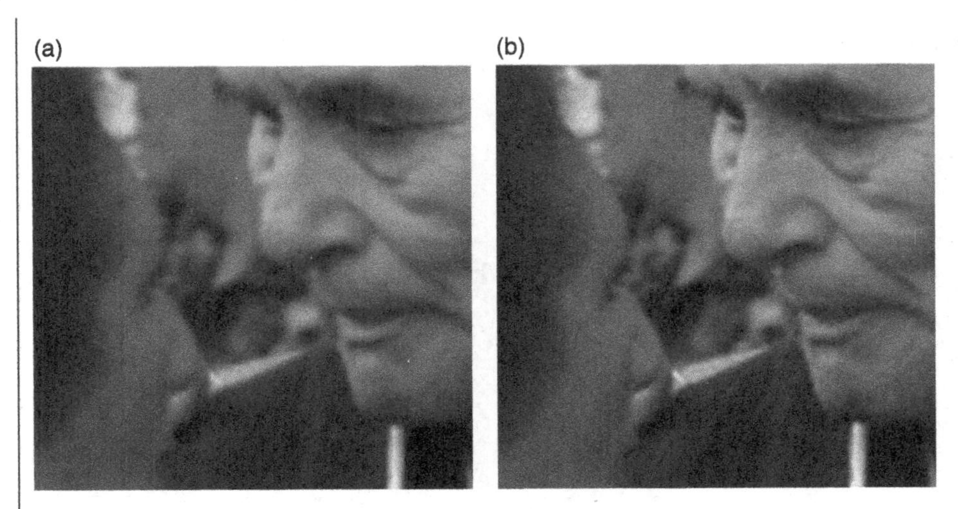

Figure 6.114 Quarter-pixel interpolated using bilinear interpolation (a), HEVC luma interpolation (b)

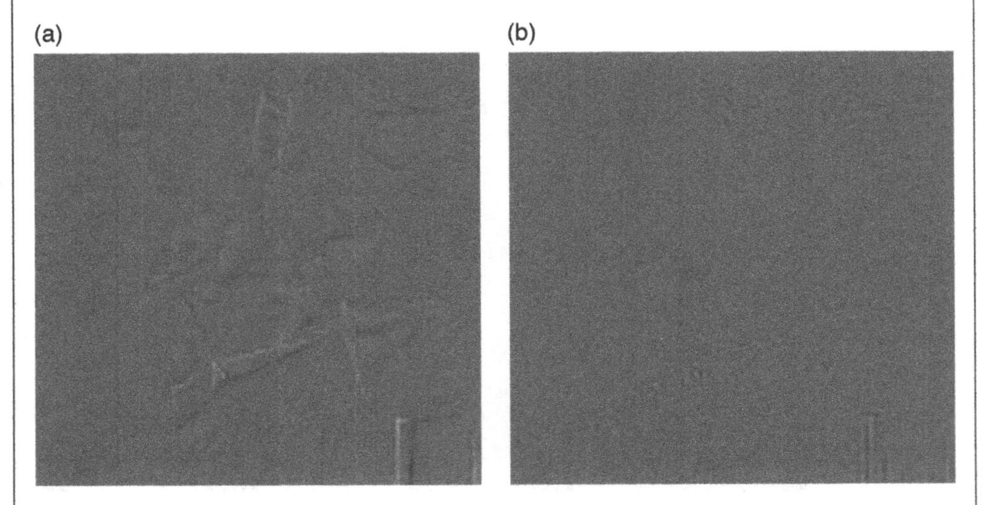

Figure 6.115 Difference between full-resolution and interpolated, bilinear (a) and HEVC (b)

at the left of Figure 6.114, is a blurred version of the original. Image edges tend to lose definition. Using HEVC luma interpolation, shown at the right of Figure 6.114, we see a cleaner image with sharper edges, closer to the original full-resolution image. By using more full-pixel samples to create each interpolated position, the HEVC interpolation does a better job of preserving edges. When we consider the difference between the original full-resolution image and the bilinear interpolated version, on the left-hand side of Figure 6.115, we can see differences around the edges, whereas the HEVC luma interpolated version is much closer to the original, as illustrated by the difference image at the right of Figure 6.115.

6.12.7 HEVC Reference Pictures

A RPS identifies a number of reference pictures to be used for the current picture, relative to the position or Picture Order Count of the current picture. Figure 6.116 shows an RPS that could be used to communicate the Short-Term reference pictures available for coding picture 5 in Figure 6.68. Short-term reference pictures are identified relative to the current picture position. In this example, the RPS is (−3, −1, +1 and +3). HEVC RPSs are discussed in Chapter 4.

6.13 Inter Prediction in VVC

The VVC standard introduces a number of enhancements to the inter prediction processes in the earlier HEVC standard [3], including the following.

6.13.1 Inter Prediction Partitions

As we saw in Chapter 4, VVC CUs can be square or rectangular with aspect ratios including 1:1, 1:2 and 1:4 (see Figure 6.117). In addition, a square N×N inter-coded CU may be split into two triangular partitions, labelled 1 and 2 in Figure 6.118. Each triangular partition is coded with a separate inter prediction merge index.

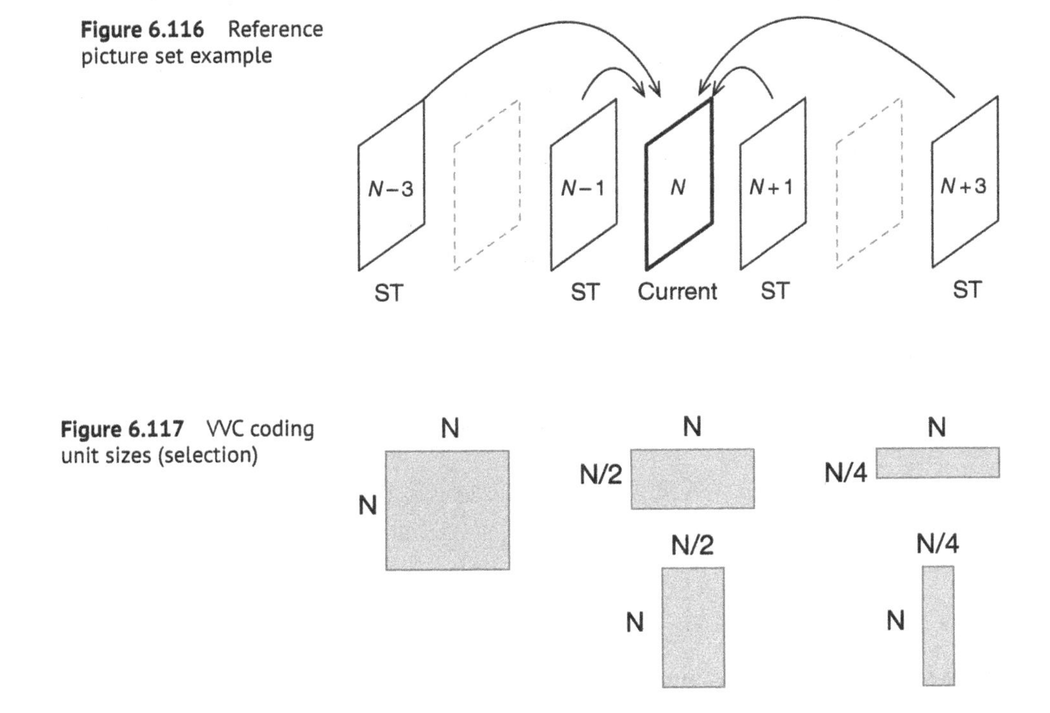

Figure 6.116 Reference picture set example

Figure 6.117 VVC coding unit sizes (selection)

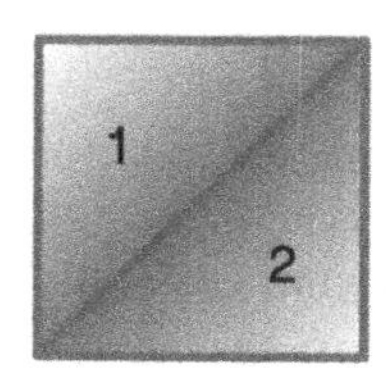

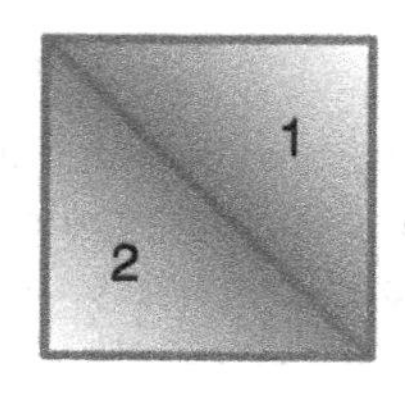

Figure 6.118 VVC triangular partitions

6.13.2 Extended Merge Mode

VVC extends the merge mode concept of HEVC, so that the candidate list, up to 6 per CU, may include:

- Spatial neighbour motion vectors,
- Temporal co-located neighbour motion vectors,
- Motion vectors from a history buffer, which records recently used motion vectors in the current row of CTUs,
- The average of a pair of merge candidates and/or
- Zero motion vectors, if necessary to pad out the list of merge candidates.

A small offset may be added to a merge candidate, i.e. merge mode with MVD.

6.13.3 Affine Motion-Compensated Prediction

In this mode, two or three control point motion vectors are sent per CU, to describe the motion field of the CU using four or six parameters, respectively. The motion vector at the centre of each 4×4 sub-block is derived from the control point motion vectors. Figure 6.119 shows an example of a 32×16 CU with two control point vectors v0 and v1, which describe the motion at the top two corners of the CU. The encoder and decoder extrapolate from these control points to determine the motion vectors of each 4×4 block. The 4×4 block vectors are calculated with 1/16-sample precision.

Adding a third control point, v2, provides more control over the vectors of the 4×4 blocks, as illustrated in Figure 6.120.

Note that the 4×4 subblocks themselves are motion-compensated in the usual way, using translational motion compensation with the derived 4×4 motion vector.

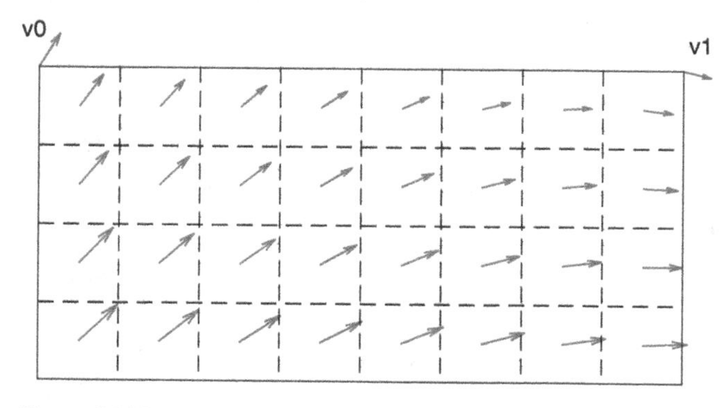

Figure 6.119 Four-parameter affine motion vectors

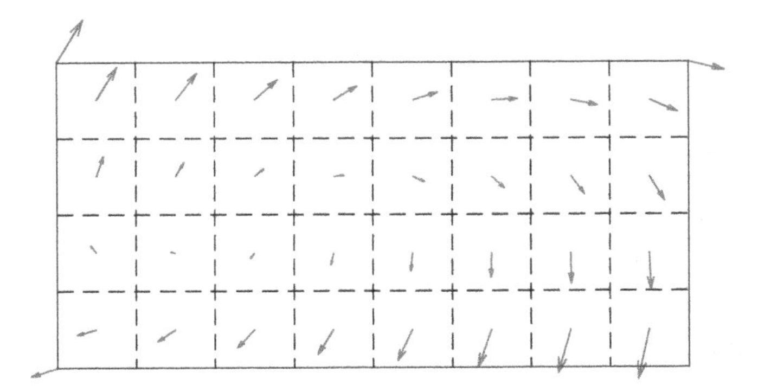

Figure 6.120 Six-parameter affine motion vectors

The control point motion vectors v0, v1 and optionally v2 may be coded using a special version of merge mode or AVMP.

The block-based nature of the resulting motion field is modified by adjusting each prediction sample based on an optical flow equation, which has the effect of smoothing the discontinuities between the motion vectors of each 4×4 block.

6.13.4 Biprediction Enhancements

Enhancements to biprediction, compared with HEVC, include:

- **Symmetric MVD**: Derive a forward and backward pair of List 0 and List 1 reference pictures, a pair at an equal temporal distance forward and backwards. Send a single motion vector difference relative to the derived List 0 reference frame, copy it and reverse it to apply to the derived List 1 reference frame.
- **Biprediction with weighted average**: Signal or derive by merging a weight factor to be applied to the two biprediction references. This enables more or less weight to be given to the List 0 and List 1 references.
- **Biprediction refined with optical flow**: The bidirectional prediction is refined using an optical flow equation, which has the effect of smoothing the motion field and reducing prediction discontinuities at block edges.
- **Biprediction refined at decoder**: The decoder carries out an extra motion estimation search which has the effect of minimising the difference between the List 0 and List 1 predictions.

6.13.5 Sub-Pixel Interpolated Motion Compensation

The resolution of motion vectors and hence of sub-pixel interpolation can be signalled at the CU level. For an inter-CU not coded in affine mode, luma motion vectors can have the resolutions of ¼ sample, ½ sample, integer sample or 4-integer sample. A 4-integer sample means that motion vector x and y components must each be a multiple of 4 in length.

For ¼ sample resolution, each position is interpolated using an eight-tap filter. For ½ sample resolution, each sub-pixel position is interpolated using a six-tap filter.

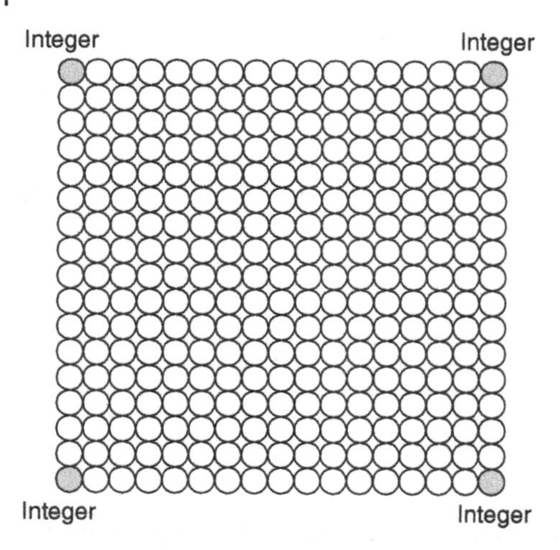

Figure 6.121 Integer pixel samples, dark grey, and 1/16-pixel samples, white

For an affine motion-compensated CU, the 4×4 sub-block motion vectors are calculated with 1/16-sample resolution. The control point motion vectors may be sent with 1/16 sample, ¼ sample or integer-sample resolution. Figure 6.121 illustrates the 1/16-pixel positions that must be interpolated between integer pixel positions for 1/16-pixel resolution.

6.14 Conclusions

In the last two chapters, we have considered intra and inter prediction, the process of creating and signalling a prediction block. Each successive video coding standard has added to the sophistication of these processes, resulting in more and more sophisticated prediction methods, often at the expense of increasing processing complexity. The result should be a prediction block that is a close match to the current block of samples, which is very similar to the current block.

A video encoder selects a prediction type and source for each block, forms the prediction and subtracts it from the current block. Unless the prediction is perfect, the result is a residual block, which as we will see is typically a set of small positive and negative values. In Chapter 7, we will look at the next stage in the video encoding process, which is to transform each residual block into a frequency domain and quantise the resulting frequency coefficients.

References

1 Wedi, T. and Musmann, H.G. (2003). Motion- and aliasing-compensated prediction for hybrid video coding. *IEEE Transactions on Circuits and Systems for Video Technology* 13 (7): 577–586.

2 Schwarz, H., Marpe, D. and Wiegand, T. (2006). Analysis of hierarchical B pictures and MCTF. *2006 IEEE International Conference on Multimedia and Expo*. pp. 1929–1932.

3 Chien, W.J., Zhang, L., Winken, M., et al. (2021). Motion vector coding and block merging in the versatile video coding standard. *IEEE Transactions on Circuits and Systems for Video Technology* 31 (10): 3848–3861. doi: https://doi.org/10.1109/tcsvt.2021.3101212.

7

Transform and Quantisation

7.1 Introduction

We have seen how a video encoder processes a video frame, predicting each block using intra prediction or motion-compensated inter prediction. The encoder subtracts the prediction from the original block to leave a residual block. The next stage is to process the residual block using a block transform and quantisation (see Figure 7.1). After these two processes, the block can be encoded into a compressed bitstream. When the bitstream is decoded, each block is rescaled and inverse-transformed to create a decoded residual.

A block transform has the effect of concentrating the important visual information in a block into a small number of significant values known as transform coefficients. Quantisation is used to remove less significant values, leaving a small number of visually significant coefficients to represent the block. Figure 7.2 shows an example. In the 8×8 residual block on the left, the pixel information is distributed across the block. After transformation, the coefficient block concentrates most of the information into a few significant coefficients at the top left. After quantisation, most of the quantised coefficients are set to zero. The quantised coefficient block can typically be coded into a much smaller number of bits than the residual block.

The amount of quantisation, i.e. the amount of information that is removed, can be adjusted to control the trade-off between compression and quality. More quantisation gives more compression but reduces image quality when the block is decoded, whereas less quantisation gives less compression but increases image quality.

In this chapter, we will consider:

- What transform should we choose?
- How does the transform work?
- How should we quantise the transformed data?
- How do we reverse the process to recover the residual data at the decoder?
- How do we specify the transform, quantisation and inverse processes so that video encoders and decoders can interoperate?
- How should we put this into practice in software or hardware?
- How is this all done in the standards? We will focus on H.265/High Efficiency Video Coding (HEVC) for specific examples.

Coding Video: A Practical Guide to HEVC and Beyond, First Edition. Iain E. Richardson.
© 2024 John Wiley & Sons Ltd. Published 2024 by John Wiley & Sons Ltd.
Companion website: www.wiley.com/go/richardson/codingvideo1

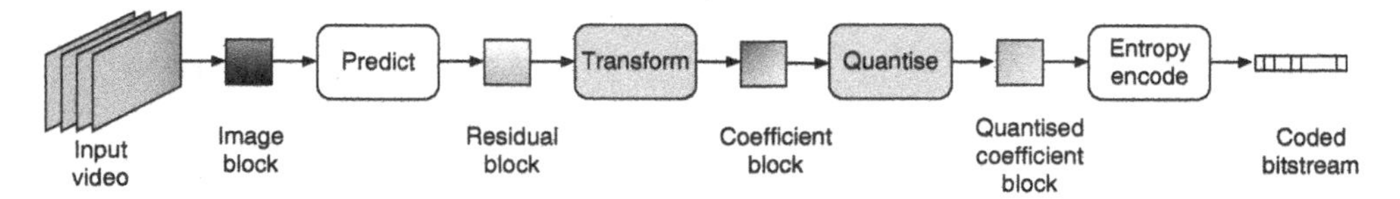

Figure 7.1 Video encoder

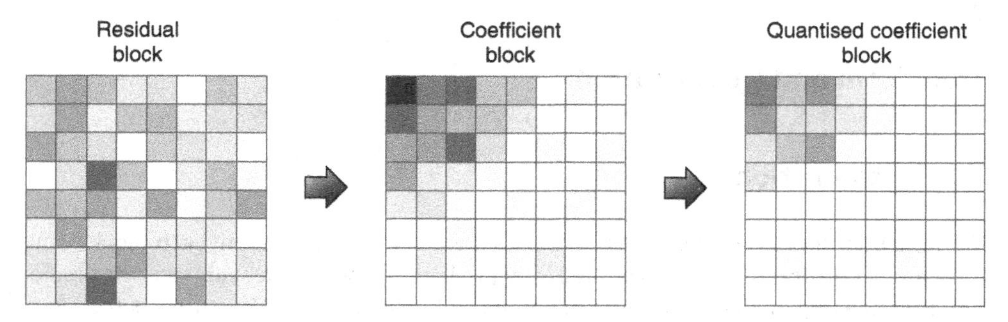

Figure 7.2 Residual, coefficient and quantised blocks

7.1.1 Transform: An Overview

The transform stage converts pixels or samples into a different representation that is suitable for compression. Many video codecs use block transforms. A block transform takes a block of pixels or residual samples and converts it into a block of coefficients (see Figures 7.2 and 7.3). Figure 7.3 shows a block of pixel values on the left, each stored as a number that indicates the brightness or luma of the pixel. The lightest pixel in this example has a luma value of 194, and the darkest pixel has a luma value of 89. A transform converts the pixels into a block of coefficients, each of which is a number that indicates the proportion or weight of a frequency component. In this example, the highest-magnitude coefficient has a value of +392 and the lowest-magnitude coefficient has a value of +2. Note that in the coefficient block, the coefficients at the top left tend to have much higher magnitudes than the coefficients at the lower right. This is one of the useful properties of certain block transforms since it is easier to efficiently encode high-magnitude coefficients when they are grouped together.

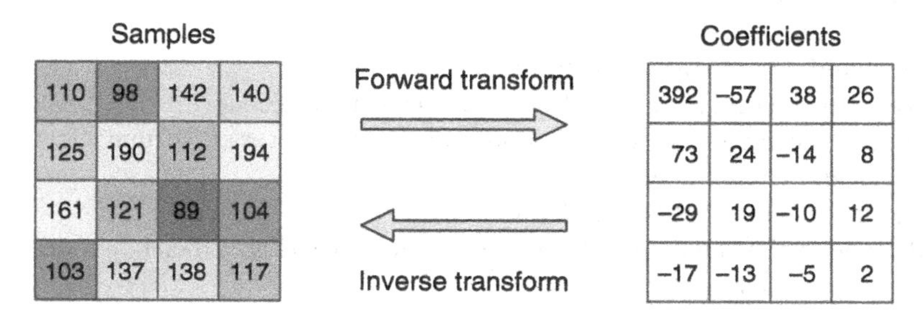

Figure 7.3 4 × 4 block transform example

A block transform may be reversible, which means that applying a forward transform followed by an inverse transform will recreate the original data. By itself, a block transform does not compress the data.

7.1.2 Quantisation: An Overview

Quantisation reduces the information contained in the block of coefficients and makes it easier to compress. However, it is a non-reversible or lossy process, which means that the information removed cannot be restored. Consider the example of Figure 7.4. In this example, each coefficient is quantised by dividing it by the integer 16 and rounding towards zero (see Table 7.1).

The result is that (1) larger-valued coefficients are reduced in magnitude and precision and (2) small-valued coefficients are set to zero. In the example of Figure 7.4, coefficients smaller than ± 16, in this case seven of the 16 coefficients, are set to zero, because of the choice of divisor (16). The block of quantised coefficients has properties that make it suitable for efficient compression. First, many of the coefficients are zero. Second, non-zero coefficients are mostly grouped together in the top left of the block. As we will see later, these two properties lend themselves to an efficient compressed representation of the quantised coefficient block.

The decoder cannot recover the original coefficients, but it can create an approximation to the original block by rescaling the coefficients (see Figure 7.5). In this example, each quantised coefficient is multiplied by the same factor of 16. The rescaled coefficients have

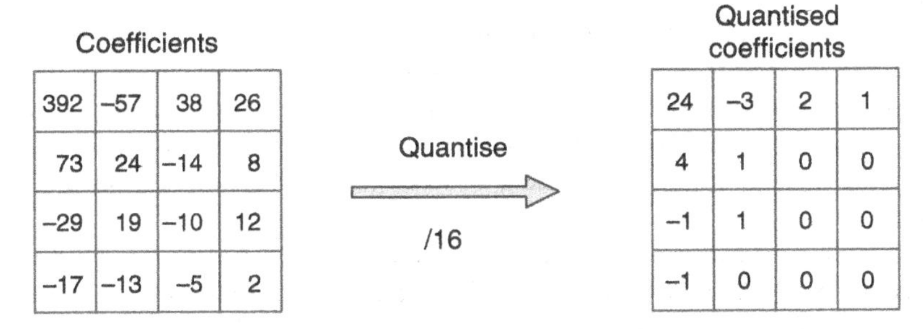

Figure 7.4 Quantisation

Table 7.1 Coefficient quantisation examples

Coefficient	/16	Round towards zero
392	24.5	24
−57	−3.5625	−3
38	2.375	2
−14	−0.875	0
...	...	...

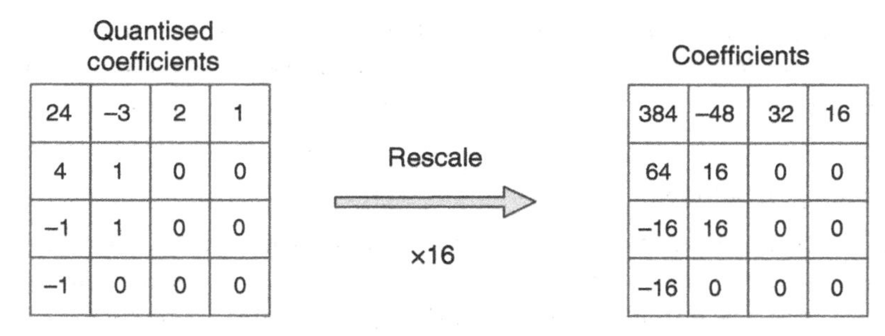

Figure 7.5 Rescaling

similar magnitudes to the original coefficients, e.g. +384 instead of +392 and −48 instead of −57. Comparing the original coefficients with the rescaled coefficients, the rescaled values are not quite the same as the originals and the small-valued coefficients remain at zero.

7.2 Residual Blocks

In a video encoder, the output of the prediction process is a residual frame or picture. Each block in the residual picture has been predicted from neighbouring samples in the same frame or from previously coded frames. The better the prediction, the less information remains in the residual picture, and so a typical residual picture consists of blocks of small positive and negative samples and lots of zeros. Figure 7.6 shows a video frame, and Figure 7.7 shows the residual picture after inter-prediction of each block of the frame. Much of the residual picture is mid-grey, which represents near-zero values. A residual sample shown as grey is zero, or close to zero. In some areas, such as the edges of moving objects, the prediction has not worked so well, and there are larger positive and negative values.

We will consider two blocks within the original frame and the residual.

Block 1: a block of 8×8 pixels in the background is highlighted in the original frame (see Figure 7.6) and in the residual frame (see Figure 7.7). The original block is mid-grey and has luma values ranging from 108 to 114. The residual block contains only small values from −3 to +3 with a number of zeros. The inter prediction process has created an accurate prediction and left us with a near-zero residual.

Block 2: a block of 8×8 pixels around the corner of the man's eye is highlighted in the original and residual frames. The original block has luma sample values ranging from 72, which is dark grey, to 133, which is mid-grey. The residual block has values in the range of −46 to +44. This block is in a complex moving part of the picture, and inter prediction has not found an accurate match. As a result, there is still quite a lot of information in the residual block.

Residual blocks tend to have at least two characteristics that are challenging for compression. First, the information tends to be spread across the block, rather than concentrated in

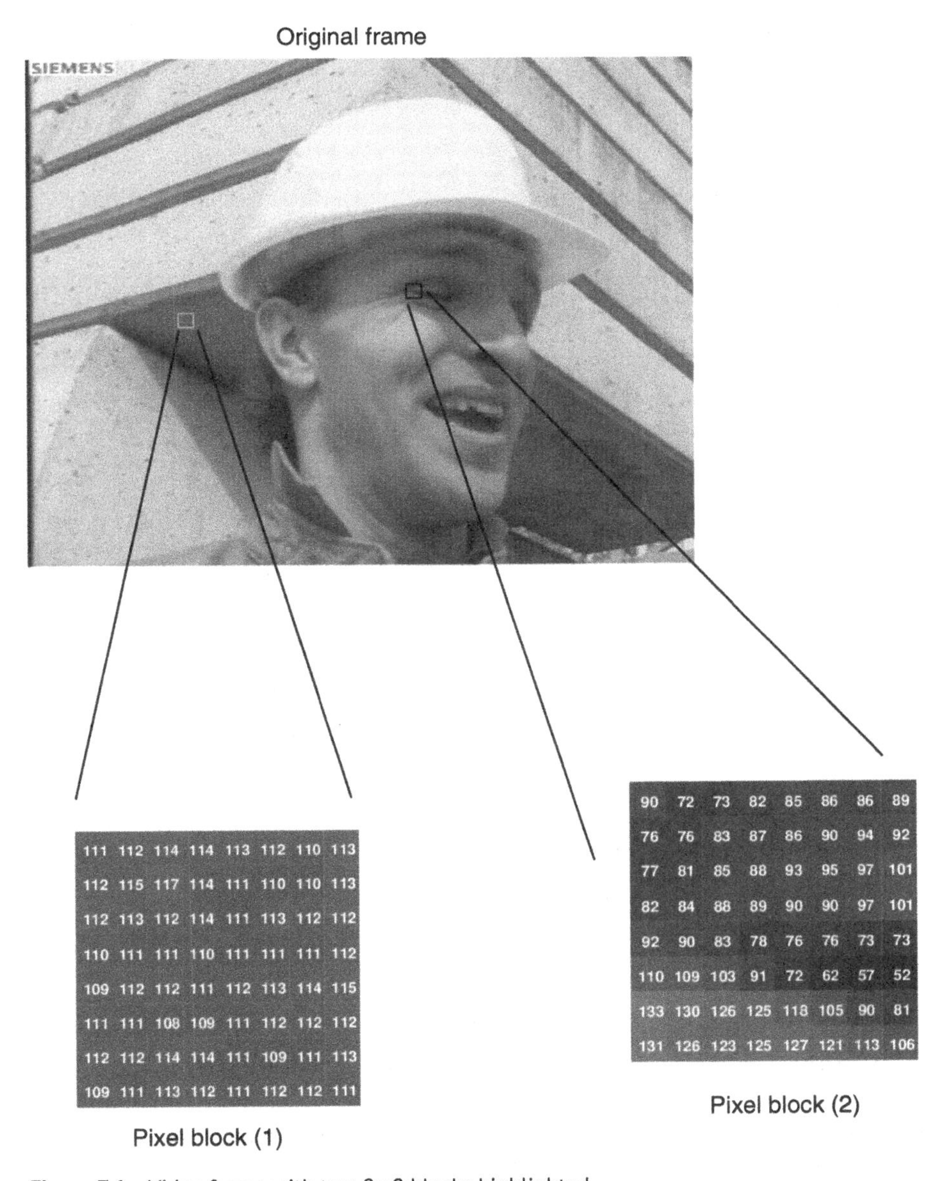

Figure 7.6 Video frame with two 8 × 8 blocks highlighted

one place. Second, all of the samples are of roughly equal significance, as they all play an important part in reconstructing the final image block.

Ideally, we would like the information to be concentrated into a small number of important values, so that we can keep these values and discard the rest, thereby achieving compression. This is what the transform and quantisation stages aim to achieve.

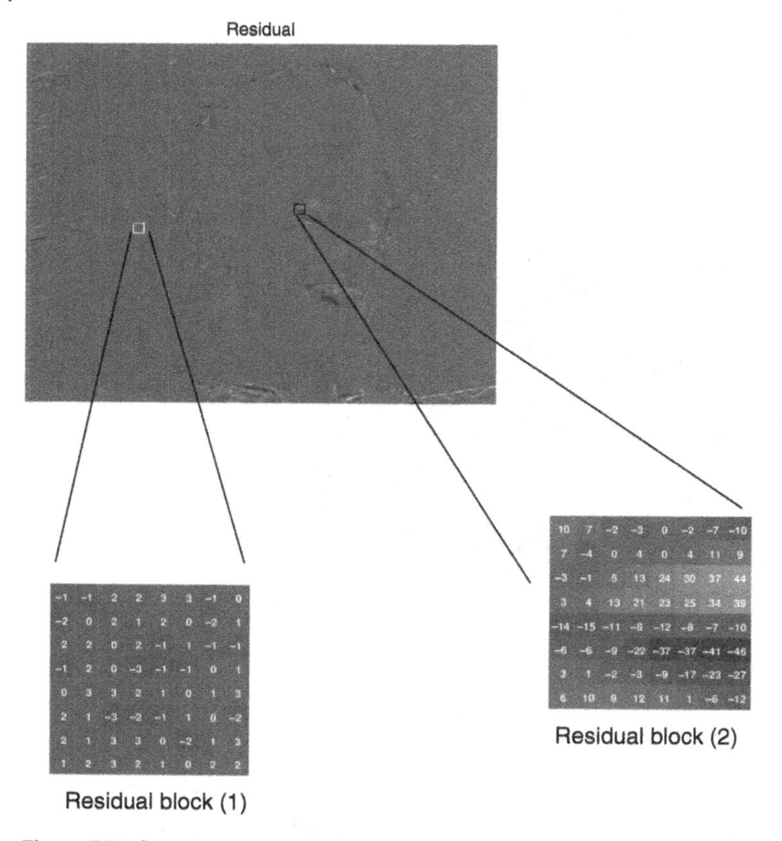

Figure 7.7 Residual frame with two 8 × 8 blocks highlighted

7.3 Block Transforms

7.3.1 What Is the Purpose of a Block Transform?

A block transform converts residual blocks into transform coefficient blocks that are easier to compress efficiently. Preferably, the transform will compact the energy or information in a block into a small number of significant coefficients. As we saw earlier in Figure 7.3, after applying a block transform, the transformed block represents the same information, but most of the data in the block are typically concentrated into a small group of large valued coefficients. The transformed block can be quantised to retain these significant coefficients, thus retaining most of the important visual information, whilst removing small-valued coefficients that do not contribute much to the image.

A block transform takes advantage of certain features of photographic and video images and certain characteristics of human vision, including the following:

- A typical image or video frame is made up of strong features such as objects, edges and textures, with smaller, more subtle variations of texture.
- Dominant features in images, such as lines and textures, are often important to the overall visual effect.

- Our vision system responds to these features, and we tend to notice them more than we do small, subtle variations in texture.
- Dominant features in an image tend to appear as strong spatial frequencies, which I explain below.
- We can remove weak or small-valued spatial frequencies without a significant loss of image quality.

Example

Figure 7.8 shows two versions of an image. The original image at the top contains a lot of detail in the waves in the foreground, plus dominant features such as the outline of the boat and the horizon line. The lower version has been heavily filtered to remove much of the fine detail. Despite the fact that much of the information has been removed by filtering, the dominant features are still visible and the filtered picture is clearly recognisable as a picture of a boat sailing out of a harbour. This demonstrates that we take more notice of dominant, strong features when we are viewing a scene. These dominant features often translate into dominant spatial frequencies in a transformed version of the image data, as we will see later.

Filtered image

Figure 7.8 Original and filtered image

7.3.2 Spatial Frequencies and Basis Patterns

A block within an image can be represented as a sum of basis patterns, each of which corresponds to a spatial frequency component in the block. Each type of block transform has its own set of basis patterns. Figure 7.9 shows the 16 basis patterns for a 4×4 discrete cosine transform (DCT), which we will examine in detail later. The top-left pattern is known as the DC pattern and represents a flat, uniform average brightness of an image block. As we move along the top row to the right, we have three patterns that correspond to increasing horizontal frequency. Moving down the left column, we have patterns that correspond to increasing vertical frequency, and the remaining patterns are combinations of horizontal and vertical variations.

The basis patterns can be thought of as prototypes or building blocks for actual image blocks. Any 4×4 image block can be created by combining these 16 basis patterns with different weights – by scaling each basis pattern and combining them together. Figure 7.10 shows how this works. Each basis pattern with coordinates (i,j) is scaled by a weight W_{ij}, which has the effect of making the pattern brighter or darker. The weighted patterns are added together to create an image block. Changing the weights W_{ij} mixes more or less of

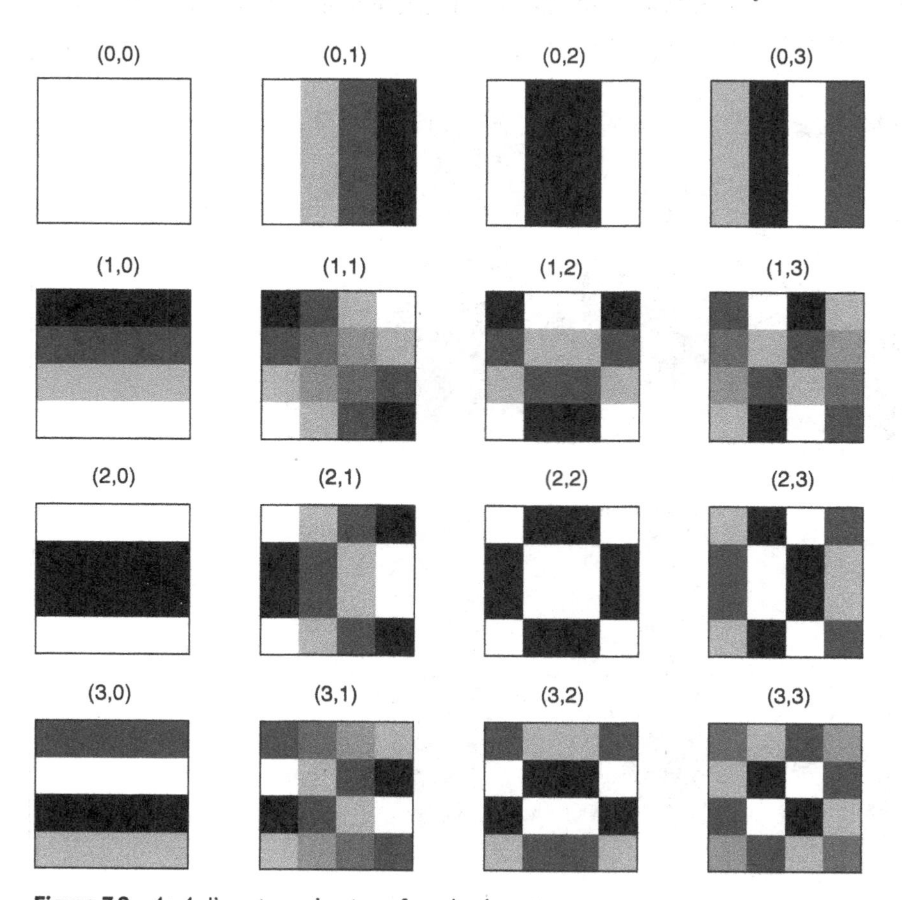

Figure 7.9 4×4 discrete cosine transform basis patterns

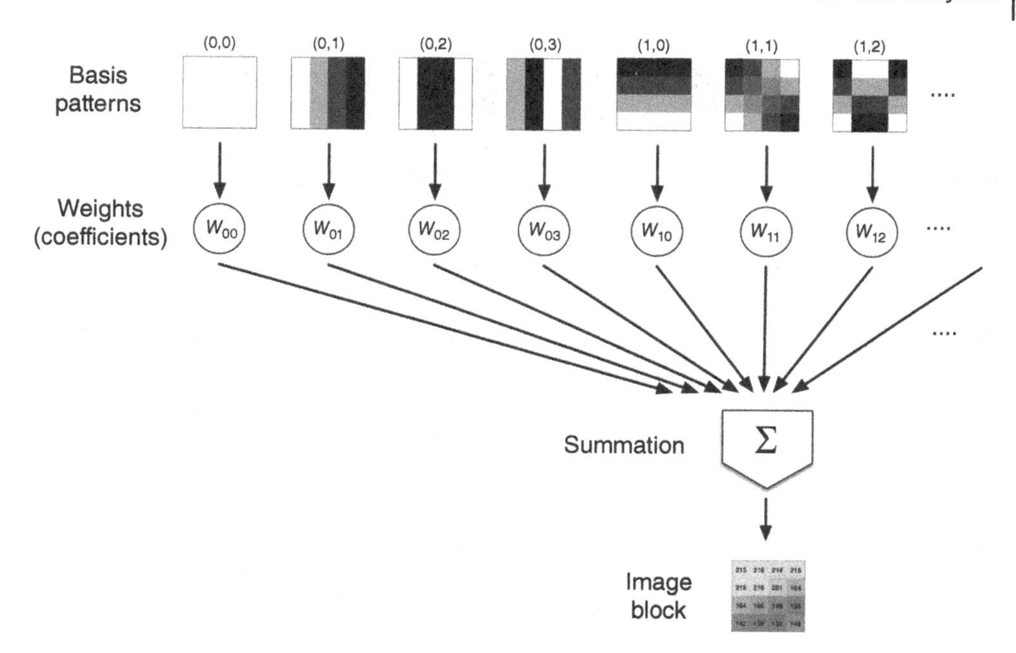

Figure 7.10 An image block is a weighted sum of basis patterns

each pattern into the final image block. By adjusting the set of weights, we can construct any possible 4×4 image block using the same 16 basis patterns.

The set of weights or scales of each basis pattern are the transform coefficients. This means that the forward transform calculates the set of weighting values or coefficients that, when applied to the same set of basis patterns, construct the original image block.

The same process can be applied with different block sizes. Figure 7.11 shows the basis patterns for the 8×8 DCT. Each basis pattern is an 8×8 sample block, and this time, there are 64 basis patterns instead of the 16 basis patterns for the 4×4 DCT. The process works in a similar way. The top-left block represents the average or DC value of the block. Moving to the right, we have increasing horizontal frequencies; moving down, we have increasing vertical frequencies; and the remaining patterns are combinations of horizontal and vertical frequencies. In the same way that any 4×4 block can be constructed as a weighted sum of the 4×4 DCT basis patterns, any 8×8 luma block can be constructed by combining these 64 basis patterns of the 8×8 DCT with different weights or coefficients.

7.3.3 What Does the Transform Do?

The forward transform takes as its input a block of image or residual samples and outputs the coefficients or weights of each of the basis patterns (see Figure 7.12). The transform finds the weights for each basis pattern that would exactly reproduce the image block.

The inverse transform takes a block of coefficients and calculates the corresponding image block. In other words, the inverse transform synthesises an image block based on the coefficients or weights of each of the $N \times N$ basis patterns.

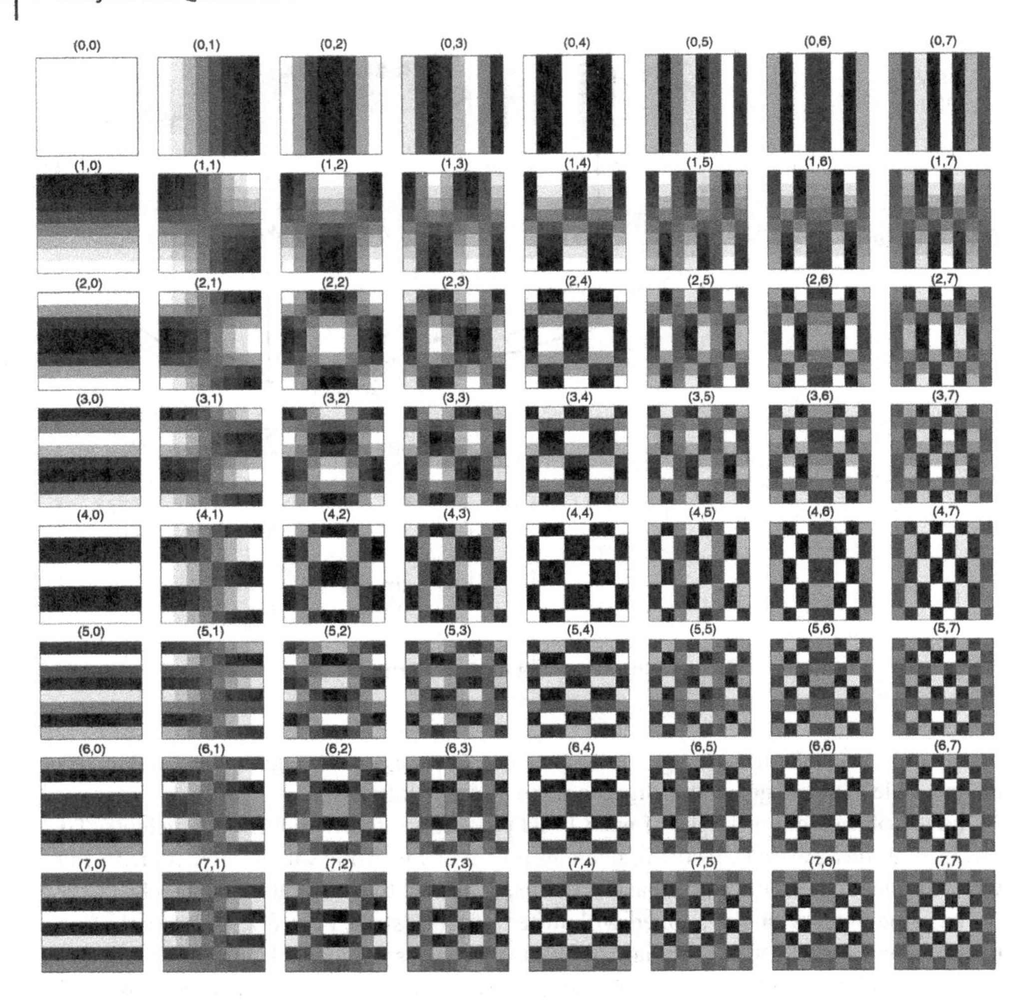

Figure 7.11 8 × 8 DCT basis patterns

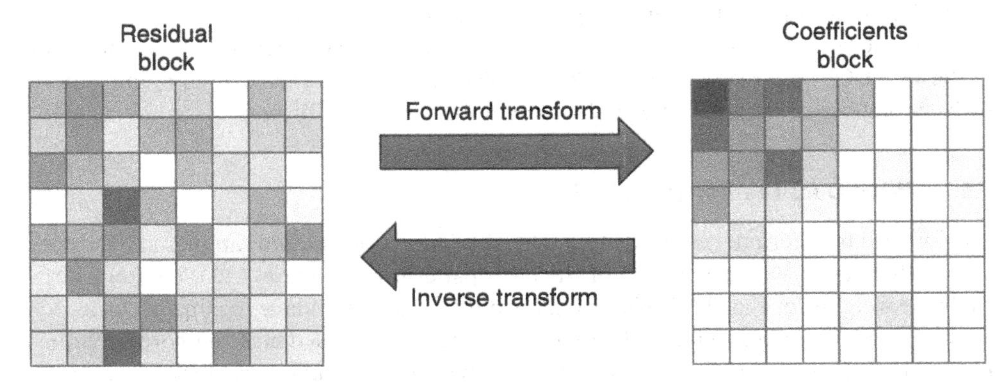

Figure 7.12 Forward and inverse transform

7.3.4 Types of Block Transform

A block transform should decorrelate and compact the information in a block of residual image samples. This means that for a typical image block, the transform should concentrate the spread-out image information into as few significant or large valued coefficients as possible, which are preferably closely grouped together in the transformed block. Figure 7.12 shows the desired outcome, with the block energy concentrated into a cluster of large coefficients in the top left of the transformed block. We will look at several transforms that can achieve this result.

7.3.4.1 The Karhunen–Loeve Transform

The optimal transform for a block of samples depends on the actual information in the block. For a given set of input data, we can derive a theoretically optimal transform, according to the desirable properties of decorrelation and compaction. The Karhunen–Loeve Transform (KLT) can be optimal in certain scenarios.

The KLT is a block transform with coefficients derived from a set of input data such as an image or video frame. There are two problems that make the KLT impractical for many image and video coding applications. First, the KLT is not amenable to fast, computationally efficient implementation. Real-time implementation would require significantly higher computational resources than other alternative transforms. Second, the coefficients of the KLT have to be calculated based on an input data set. For a practical video coding application, these coefficients would have to somehow be communicated from the encoder to the decoder so that the decoder could reverse the transform. If this is not done frequently enough, the KLT coding efficiency will suffer as the transform becomes increasingly sub-optimal for each image block. If this is done frequently, the overhead of communicating the inverse transform to the decoder will reduce the overall compression efficiency.

So far, the KLT has not found widespread practical use, but it continues to be studied as an alternative to transforms such as the DCT [1].

7.3.5 The Discrete Cosine Transform

A very popular and widely used block transform is the DCT, which we will examine in detail below. Developed in 1973 by Rao and Yip and described in what must be one of the most cited papers in the field of image and video compression [2], the DCT has properties that make it attractive for image compression. It approaches the decorrelation and compaction performance of the KLT for typical image blocks, though not necessarily for residual blocks. For a particular block size, we need only a single transform design, as the transform does not need to adapt or change depending on the data to be transformed.

The DCT of a two-dimensional block of pixels or samples is separable, which means it can be carried out using a one-dimensional transform applied twice – once to the rows and once to the columns of the block. The DCT can be implemented in many different ways, including in efficient software and hardware applications.

These last two points are important for fast, computationally efficient implementations. This in turn is important as the block transform has to be carried out for every block of every frame of a video sequence, many thousands of times per second in a real-time video codec.

7.3.6 The Hadamard Transform

The Discrete Hadamard/Walsh–Hadamard transform (DHT) has basis patterns constructed from binary values rather than fractional or multi-valued integer functions as with the DCT or Discrete Sine Transform (DST). In these basis patterns, each sample position can have one of only two values, unlike the multi-valued patterns of the DCT.

A block transform relies on correlating basis patterns with the features of an image or residual block. If the block contains a range of values, then a multi-valued basis pattern is perhaps more likely to correlate with the content of the block. Hence, a DHT may be sub-optimal for typical real-world image blocks containing pixels with a wide range of values. However, a transform such as the DHT with two-valued basis patterns may be effective in certain cases, such as:

- **Small block sizes**: When the block size is just 2×2 samples, a 2×2 DCT is identical to a 2×2 DHT.
- **Certain residual blocks:** If the prediction stage has been very effective, the residual block will contain mostly small positive and negative values and zeros (see block (1) in Figure 7.7). A simple transform such as the DHT may be just as effective as a more complex transform at correlating with such small-magnitude, dispersed values.
- **Graphic images or flat/simple features:** Image blocks within graphic areas, such as an overlay in a TV image, do not necessarily have the typical range of values of a conventional photographic image. These blocks may correlate well with the simple, two-valued basis patterns of the DHT.

7.3.7 Discrete Sine Transform

Unlike the DCT, which is constructed from cosine functions, the DST is based on sinusoidal basis functions. Whilst the DCT can be close to optimal for original image blocks, certain residual blocks may correlate more closely with the DST basis patterns.

7.3.8 The Discrete Cosine Transform in Detail

The DCT is a widely used transform for video compression. Versions of the DCT have been used in numerous video coding standards, including H.261, MPEG-1 and MPEG-2 video, H.263 and H.264/AVC. In practical video codecs, the DCT is applied to square blocks with a power-of-two dimension, i.e. 4×4, 8×8, 16×16, etc.

The DCT can be computed as a matrix multiplication that converts an $N \times N$ block of samples $\mathbf{X}$ into a block of coefficients $\mathbf{Y}$, where $\mathbf{X}$ and $\mathbf{Y}$ have the same dimensions:

2D forward DCT, matrix form

$$\mathbf{Y} = \mathbf{A}\mathbf{X}\mathbf{A}^{\mathrm{T}} \tag{7.1}$$

A is an $N \times N$ transform matrix. For the DCT-II form of the transform [2], the elements of **A** are:

DCT-II matrix elements

$$A_{ij} = C_i \cos\frac{(2j+1)i\pi}{2N} \text{ where } C_i = \sqrt{\frac{1}{N}}(i=0),\ C_i = \sqrt{\frac{2}{N}}(i>0) \tag{7.2}$$

Alternatively, the DCT can be viewed as a sum of products:

2D forward DCT-II

$$Y_{xy} = C_x C_y \sum_{i=0}^{N-1}\sum_{j=0}^{N-1} X_{ij} \cos\frac{(2j+1)y\pi}{2N}\cos\frac{(2i+1)x\pi}{2N} \tag{7.3}$$

Developing the 4×4 DCT

We will examine the properties of the 4×4 DCT. Expanding Eq. (7.3) for each position (i,j), the transform matrix **A** for a 4×4 DCT is:

$$A = \begin{bmatrix} \frac{1}{2}\cos(0) & \frac{1}{2}\cos(0) & \frac{1}{2}\cos(0) & \frac{1}{2}\cos(0) \\ \sqrt{\frac{1}{2}}\cos\left(\frac{\pi}{8}\right) & \sqrt{\frac{1}{2}}\cos\left(\frac{3\pi}{8}\right) & \sqrt{\frac{1}{2}}\cos\left(\frac{5\pi}{8}\right) & \sqrt{\frac{1}{2}}\cos\left(\frac{7\pi}{8}\right) \\ \sqrt{\frac{1}{2}}\cos\left(\frac{2\pi}{8}\right) & \sqrt{\frac{1}{2}}\cos\left(\frac{6\pi}{8}\right) & \sqrt{\frac{1}{2}}\cos\left(\frac{10\pi}{8}\right) & \sqrt{\frac{1}{2}}\cos\left(\frac{14\pi}{8}\right) \\ \sqrt{\frac{1}{2}}\cos\left(\frac{3\pi}{8}\right) & \sqrt{\frac{1}{2}}\cos\left(\frac{9\pi}{8}\right) & \sqrt{\frac{1}{2}}\cos\left(\frac{15\pi}{8}\right) & \sqrt{\frac{1}{2}}\cos\left(\frac{21\pi}{8}\right) \end{bmatrix} \tag{7.4}$$

The cosine function, $\cos\left(\frac{a\pi}{8}\right)$, is symmetrical and repeats after 2π radians or 360 degrees. Some of the elements of A can be simplified because of this symmetry (see Table 7.2 and Figure 7.13).

Applying these relationships simplifies A to:

$$A = \begin{bmatrix} \frac{1}{2} & \frac{1}{2} & \frac{1}{2} & \frac{1}{2} \\ \sqrt{\frac{1}{2}}\cos\left(\frac{\pi}{8}\right) & \sqrt{\frac{1}{2}}\cos\left(\frac{3\pi}{8}\right) & -\sqrt{\frac{1}{2}}\cos\left(\frac{3\pi}{8}\right) & -\sqrt{\frac{1}{2}}\cos\left(\frac{\pi}{8}\right) \\ \frac{1}{2} & -\frac{1}{2} & -\frac{1}{2} & \frac{1}{2} \\ \sqrt{\frac{1}{2}}\cos\left(\frac{3\pi}{8}\right) & -\sqrt{\frac{1}{2}}\cos\left(\frac{\pi}{8}\right) & \sqrt{\frac{1}{2}}\cos\left(\frac{\pi}{8}\right) & -\sqrt{\frac{1}{2}}\cos\left(\frac{3\pi}{8}\right) \end{bmatrix} \tag{7.5}$$

or

$$A = \begin{bmatrix} a & a & a & a \\ b & c & -c & -b \\ a & -a & -a & a \\ c & -b & b & -c \end{bmatrix} \text{ where } \begin{aligned} a &= \frac{1}{2} \\ b &= \sqrt{\frac{1}{2}}\cos\left(\frac{\pi}{8}\right) \\ c &= \sqrt{\frac{1}{2}}\cos\left(\frac{3\pi}{8}\right) \end{aligned} \tag{7.6}$$

Table 7.2 Cosine expressions

Cosine expression	Simplifies to
$\frac{1}{2}\cos(0)$	$\frac{1}{2}$
$\cos\left(\frac{5\pi}{8}\right)$	$-\cos\left(\frac{3\pi}{8}\right)$
$\cos\left(\frac{7\pi}{8}\right)$	$-\cos\left(\frac{\pi}{8}\right)$
$\sqrt{\frac{1}{2}}\cos\left(\frac{2\pi}{8}\right)$	$\frac{1}{2}$
$\sqrt{\frac{1}{2}}\cos\left(\frac{6\pi}{8}\right)$	$-\frac{1}{2}$
$\cos\left(\frac{(8-k)\pi}{8}\right)$	$-\cos\left(\frac{k\pi}{8}\right)$
$\cos\left(\frac{(8+k)\pi}{8}\right)$	$-\cos\left(\frac{k\pi}{8}\right)$
$\cos\left(\frac{(16-k)\pi}{8}\right)$	$\cos\left(\frac{k\pi}{8}\right)$
$\cos\left(\frac{(16+k)\pi}{8}\right)$	$\cos\left(\frac{k\pi}{8}\right)$
etc.	

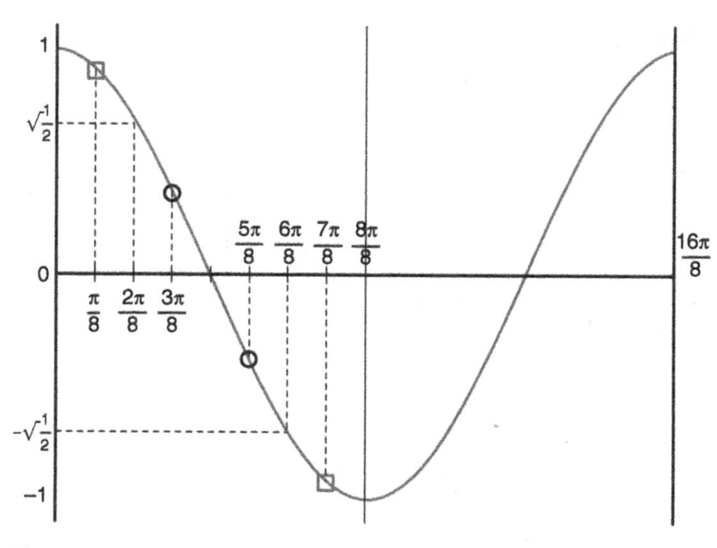

Figure 7.13 Useful cosine function relationships

Evaluating the cosines gives (approximately):

$$A = \begin{bmatrix} 0.5 & 0.5 & 0.5 & 0.5 \\ 0.653 & 0.271 & -0.271 & -0.653 \\ 0.5 & -0.5 & -0.5 & 0.5 \\ 0.271 & -0.653 & 0.653 & -0.271 \end{bmatrix} \quad (7.7)$$

Note that the weighting factors ± 0.271 and ± 0.653 are approximations to three decimal places of irrational numbers, i.e. $\sqrt{\frac{1}{2}}\cos\left(\frac{\pi}{8}\right) = 0.6532\ldots$ and $\sqrt{\frac{1}{2}}\cos\left(\frac{3\pi}{8}\right) = 0.2706\ldots$.

The 4×4 DCT, Eq. (7.1), can be expressed as the following matrix operation:

$$\overset{Y}{\begin{bmatrix} Y_{00} & Y_{01} & Y_{02} & Y_{03} \\ Y_{10} & Y_{11} & Y_{12} & Y_{13} \\ Y_{20} & Y_{21} & Y_{22} & Y_{23} \\ Y_{30} & Y_{31} & Y_{32} & Y_{33} \end{bmatrix}} = \overset{A}{\begin{bmatrix} 0.5 & 0.5 & 0.5 & 0.5 \\ 0.65 & 0.27 & -0.27 & -0.65 \\ 0.5 & -0.5 & -0.5 & 0.5 \\ 0.27 & -0.65 & 0.65 & -0.27 \end{bmatrix}}$$
$$\overset{X}{\begin{bmatrix} X_{00} & X_{01} & X_{02} & X_{03} \\ X_{10} & X_{11} & X_{12} & X_{13} \\ X_{20} & X_{21} & X_{22} & X_{23} \\ X_{30} & X_{31} & X_{32} & X_{33} \end{bmatrix}} = \overset{A}{\begin{bmatrix} 0.5 & 0.65 & 0.5 & 0.27 \\ 0.5 & 0.27 & -0.5 & -0.65 \\ 0.5 & -0.27 & -0.5 & 0.65 \\ 0.5 & -0.65 & 0.5 & -0.27 \end{bmatrix}} \quad (7.8)$$

What does this mean in practice? Equation (7.3) tells us that each coefficient of Y is generated by:

i) Multiplying every sample of X by a weighting factor, and
ii) Adding the results together.

The weighting factor of input sample x_{ij} to produce coefficient y_{xy} depends on the input sample position (i,j) and the coefficient position (x,y). Each weighting factor is calculated as:

$$C_x C_y \cos\frac{(2j+1)y\pi}{2N}\cos\frac{(2i+1)x\pi}{2N} \quad (7.9)$$

where (x,y) is the coefficient y position, (i,j) is the sample x position and C_x and C_y are calculated as before. The two cosine factors are the entries in the arrays A and A^T we saw earlier.

Example:

Consider the coefficient (0,2) of a 4×4 DCT. Substituting $x=2$ and $y=0$ into Eq. (7.9) gives the weighting factor array W_{02} for this position (Figure 7.14):

Coefficient Y_{02} is calculated as illustrated in Figure 7.15. Each position of W_{02} is multiplied with the corresponding sample X_{xy} using scalar multiplication shown as • in the figure. The 16 results are added together to generate Y_{02}:

$$Y_{02} = 0.25X_{00} - 0.25X_{01} - 0.25X_{02} + \ldots + 0.25X_{33}$$

---

1 This figure illustrates the process of calculating Y_{02} according to Eq. (7.3), i.e. evaluating the equation with $x=0$ and $y=2$.

0.25	−0.25	−0.25	0.25
0.25	−0.25	−0.25	0.25
0.25	−0.25	−0.25	0.25
0.25	−0.25	−0.25	0.25

Figure 7.14 Weighting factor array W_{02} for position (0,2), 4 × 4 DCT

Every coefficient Y_{xy} is calculated in a similar way, and each coefficient position (x,y) has its own weighting factor array W_{xy}. Once this process has been carried out for all 16 positions (0,0) to (3,3), we have generated a complete coefficient array **Y**. The complete set of 16 weighting factor arrays is shown in Figure 7.16. If we plot these arrays graphically, with larger numbers represented by brighter patches and smaller numbers represented by darker patches, we see a familiar set of basis patterns (see Figure 7.17).

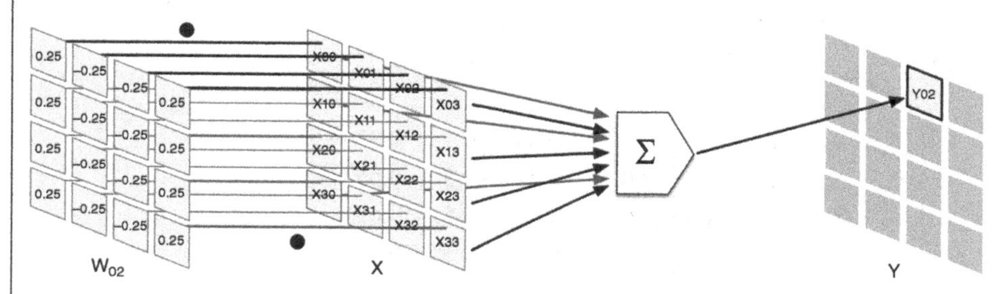

Figure 7.15 Calculating 4 × 4 DCT coefficient Y_{02} from weighting array W_{02} and input array X

0,0

0.25	0.25	0.25	0.25
0.25	0.25	0.25	0.25
0.25	0.25	0.25	0.25
0.25	0.25	0.25	0.25

0,1

0.33	0.14	−0.14	−0.33
0.33	0.14	−0.14	−0.33
0.33	0.14	−0.14	−0.33
0.33	0.14	−0.14	−0.33

0,2

0.25	−0.25	−0.25	0.25
0.25	−0.25	−0.25	0.25
0.25	−0.25	−0.25	0.25
0.25	−0.25	−0.25	0.25

0,3

0.14	−0.33	0.33	−0.14
0.14	−0.33	0.33	−0.14
0.14	−0.33	0.33	−0.14
0.14	−0.33	0.33	−0.14

1,0

0.33	0.33	0.33	0.33
0.14	0.14	0.14	0.14
−0.14	−0.14	−0.14	−0.14
−0.33	−0.33	−0.33	−0.33

1,1

0.43	0.18	−0.18	−0.43
0.18	0.07	−0.07	−0.18
−0.18	−0.07	0.07	0.18
−0.43	−0.18	0.18	0.43

1,2

0.33	−0.33	−0.33	0.33
0.14	−0.14	−0.14	0.14
−0.14	0.14	0.14	−0.14
−0.33	0.33	0.33	−0.33

1,3

0.18	−0.43	0.43	−0.18
0.07	−0.18	0.18	−0.07
−0.07	0.18	−0.18	0.07
−0.18	0.43	−0.43	0.18

Figure 7.16 4 × 4 DCT weighting factor arrays

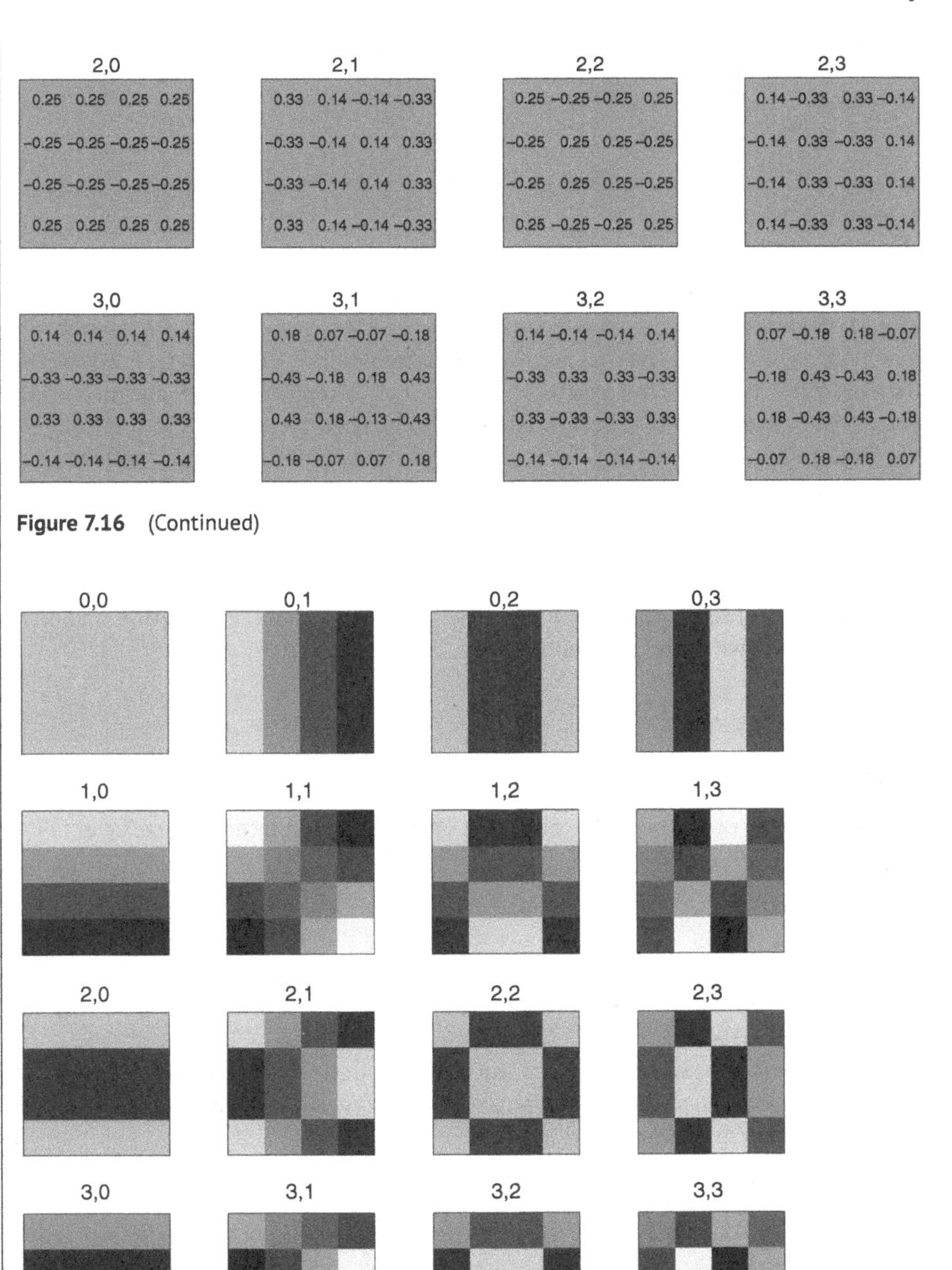

Figure 7.16 (Continued)

Figure 7.17 4 × 4 DCT weighting factor patterns

7.3.9 The Discrete Cosine Transform: What It Means

Let us consider the forward DCT process graphically.

For each coefficient position (x,y):

- Multiply each sample position (i,j) with the corresponding weight $W_{xy}(i,j)$ for the coefficient position (x,y)
- Add the results together.

This process has the effect of detecting each of the patterns shown in Figure 7.17. For example, if the block of samples **X** happens to look similar to pattern (0,2) with a dark vertical line in the centre, then all the elements of the summation Y_{02} will coincide and the DCT coefficient (0,2) will have a large positive value. If block **X** happens to look like the negative of pattern (0,2), with a light vertical line in the centre, then the DCT coefficient (0,2) will have a large negative value. If **X** is not at all similar to pattern (0,2), then coefficient (0,2) will have a very small magnitude.

This is another way of saying that the DCT calculates the sign and magnitude of spatial frequencies in the input block **X**. For the 4×4 DCT, there are 16 spatial frequencies that each correspond to one of the basis patterns in Figure 7.17.

If we increase the transform size to 8×8, we get a similar progression of weighting arrays and patterns, as shown earlier in Figure 7.11.

Examples

Figure 7.18 shows three examples of blocks (X) transformed into DCT coefficients (Y) using the 4×4 DCT.

Example (a)

The top-left coefficient, position (0,0), is calculated by multiplying every sample of block X by a constant and adding the results. This is the DC coefficient and represents the average of the block.

The coefficient at position (1,0) is calculated as follows. Each sample of block X is multiplied by the corresponding position in weight array W_{10}, and the results are added together, as shown in Figure 7.19.

The first line is:

$$(103 * 0.33) + (101 * 0.33) + (92 * 0.33) + (95 * 0.33) = 127.7$$

The second line is:

$$(133 * 0.14) + (149 * 0.14) + (143 * 0.14) + (145 * 0.14) = 77.1$$

The third line sums to −77.9, and the fourth sums to −127.0. Adding all the products gives a total of −0.16, the DCT coefficient at position (1,0). The very small coefficient magnitude indicates that there is almost no correlation between the image block X and the DCT weighting pattern (1,0).

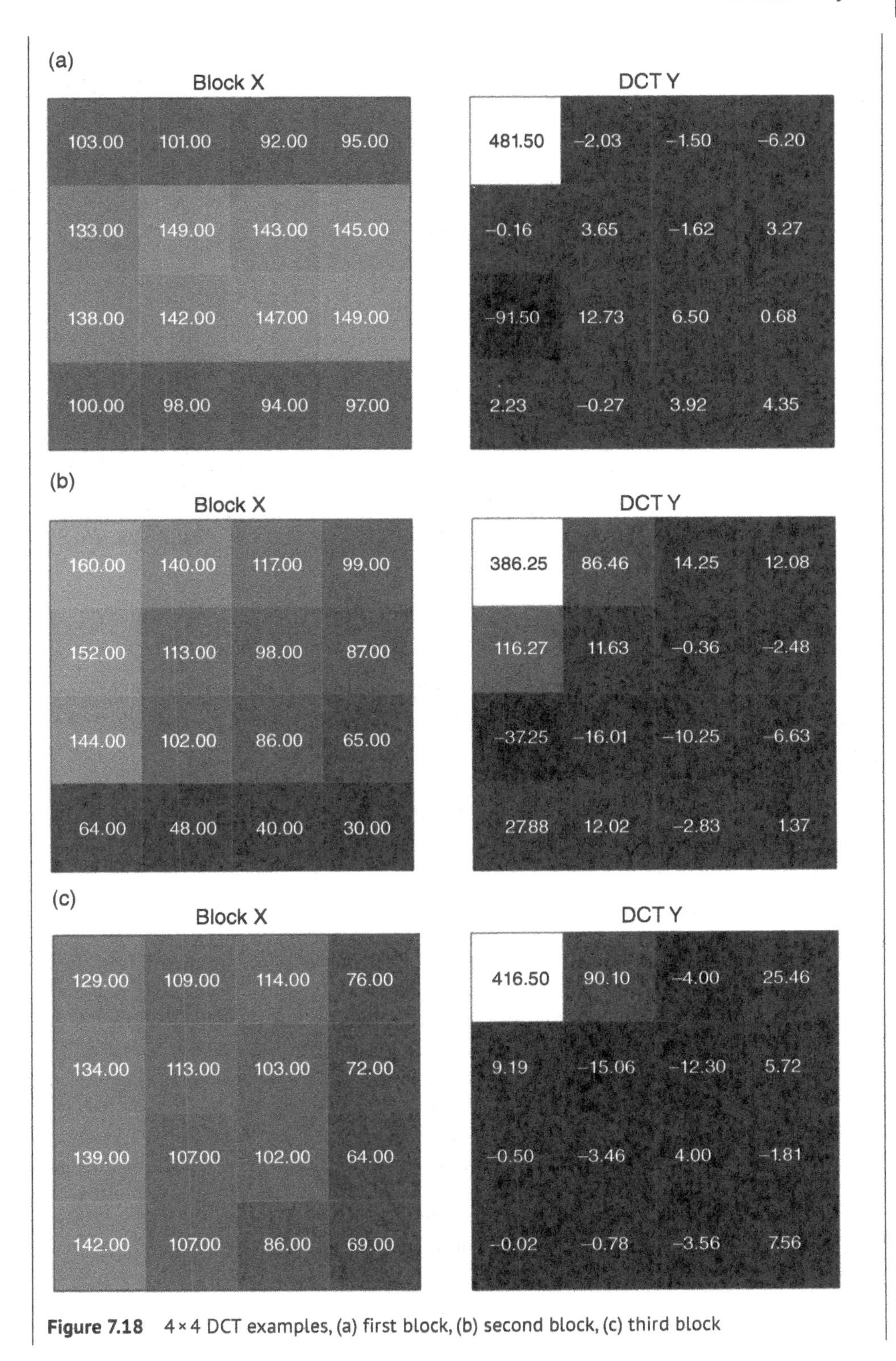

Figure 7.18 4 × 4 DCT examples, (a) first block, (b) second block, (c) third block

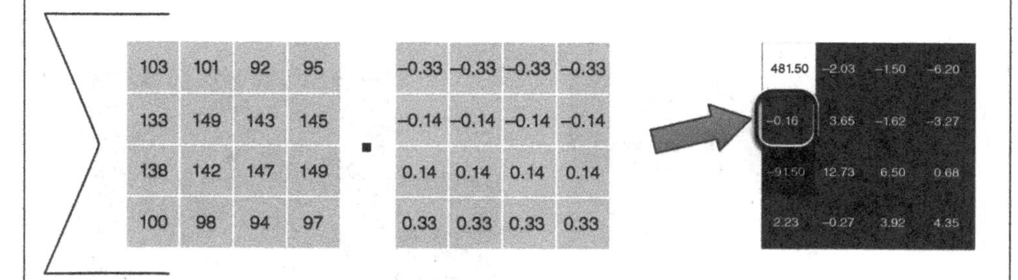

Figure 7.19 Example (a), calculating coefficient position (1,0)

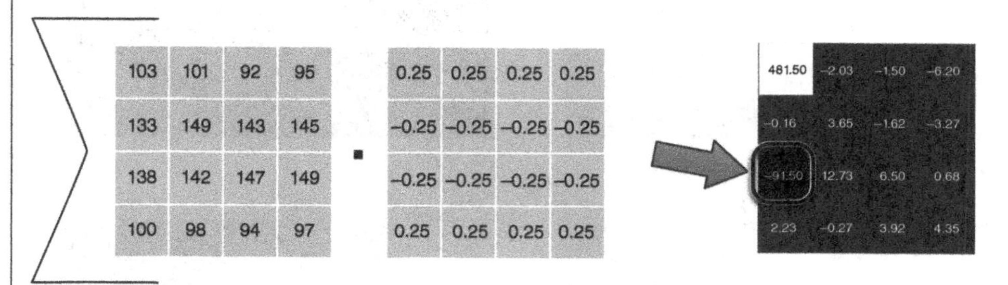

Figure 7.20 Example (a), calculating coefficient position (2,0)

Let us consider position (2,0). Coefficient (2,0), i.e. Y_{20}, is created by multiplying every sample of block X with the corresponding position in weight array W_{20} and adding the results, as shown in Figure 7.20.

The first line is:

$$(103 * 0.25)+(101 * 0.25)+(92 * 0.25)+(95 * 0.25)=97.75$$

The second line:

$$(133 * -0.25)+(149 * -0.25)+(143 * -0.25)+(145 * -0.25)=-142.5$$

The third line sums to −144, and the fourth line sums to 97.25. Adding all four lines gives a total of −91.5.

Visually, block X looks similar to an inverted version of the DCT basis pattern (2,0) (see Figure 7.21). Block X has a light horizontal strip across the centre, whereas the DCT pattern (2,0) has a dark horizontal strip across the centre. Because of this relationship, coefficient Y_{20} is a strong negative value (−91.5).

The remaining coefficients are calculated in the same way. Coefficient (1,2) has a value of +12.73, indicating a similarity between the image block and weighting pattern (1,2), and so on.

Similar behaviour is observed for the 8 × 8 DCT. Figure 7.22 shows two further example blocks (d) and (e), each transformed into an 8 × 8 block of DCT coefficients. The transformed blocks contain a few large-valued coefficients, mostly clustered about the top-left position, and a large number of small-valued coefficients.

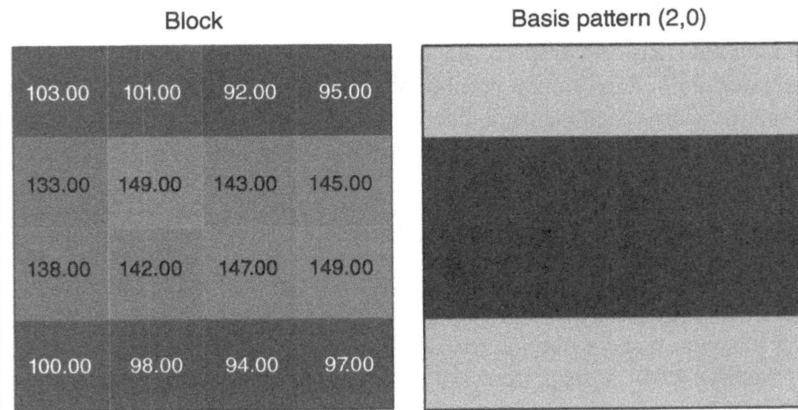

Figure 7.21 Block X and basis pattern (2,0)

(d) Block X

15	3	−7	−16	−23	−37	−22	−45
20	11	18	5	−5	−5	−12	−31
30	35	17	19	10	−9	−10	−20
33	35	17	15	9	0	2	−35
51	36	31	22	27	10	5	−21
76	82	63	72	50	39	53	26
105	95	91	81	81	51	64	41
86	97	78	73	73	63	54	39

DCT Y

226.4	139.3	−12.8	20.1	−10.6	14.3	−26.0	10.7
−251.0	−2.2	6.0	−0.7	−0.9	6.0	6.8	−0.1
36.9	−5.4	7.7	−7.4	0.7	−2.4	4.3	2.3
−5.4	0.3	8.4	−0.7	2.8	12.5	−2.8	5.7
−59.4	2.1	1.6	13.0	−4.9	−3.4	0.3	2.0
11.2	7.7	6.4	−2.2	3.1	2.9	−11.5	4.9
0.5	−2.6	5.1	−7.2	2.0	−4.5	−11.2	−11.3
−6.7	7.1	1.3	1.4	11.3	−5.6	4.0	9.0

(e) Block X

46	42	25	21	8	30	2	18
56	54	45	20	30	40	4	7
91	75	67	71	68	78	37	63
71	54	53	41	38	43	16	48
4	−3	−2	−17	−20	−12	−48	−22
5	4	−4	−7	−6	7	−30	−21
34	24	8	11	−5	11	−17	3
48	38	46	33	35	43	15	11

DCT Y

189.8	90.1	8.8	23.9	3.5	−24.4	44.0	−27.1
89.6	6.7	10.9	1.2	−0.8	−5.9	1.3	−2.4
25.4	1.1	−2.0	7.2	−11.6	5.3	−7.2	−0.2
−168.7	−1.5	4.5	−3.0	−5.5	−3.1	−3.9	−6.8
−16.0	−0.9	4.1	−9.6	−0.7	4.8	3.4	4.3
50.4	−5.8	18.1	−6.4	12.2	5.7	−2.5	−12.5
43.6	−14.5	−11.2	5.1	3.4	1.2	−0.5	−2.9
−14.7	−1.3	2.9	−3.6	9.0	−0.8	1.4	−6.9

Figure 7.22 8 × 8 DCT examples, (d) fourth block, (e) fifth block

7.3.10 The Discrete Cosine Transform: Significant Coefficients

As we saw in Section 7.3.1, a useful property of a block transform is that it can concentrate important visual information into a small number of coefficients. This means that instead of retaining all the transform coefficients, a video encoder can remove insignificant coefficients without losing too much image quality. We will look at an example.

Example: Significant coefficients of a 4 × 4 image block

Figure 7.23 shows a 4 × 4 block from a frame of the *Foreman* video clip. The original block at the lower left has luma pixel values ranging from 129, mid-grey, to 200, light grey. There is a diagonal feature running through the block. The DCT coefficients, i.e. the weights of each 4 × 4 DCT basis pattern, are shown at the lower right. Let us look at how each of the DCT basis patterns, shown in Figure 7.9, is scaled and combined to recreate the block. We will consider the first five coefficients from the most significant or largest magnitude to the fifth most significant.

Figure 7.24 shows the basis pattern for the DC coefficient Y_{00} (top line). This coefficient has a value of 668, which effectively scales the DC coefficient pattern to give a uniform luma sample value of 167. This is the average or DC level of the original block.

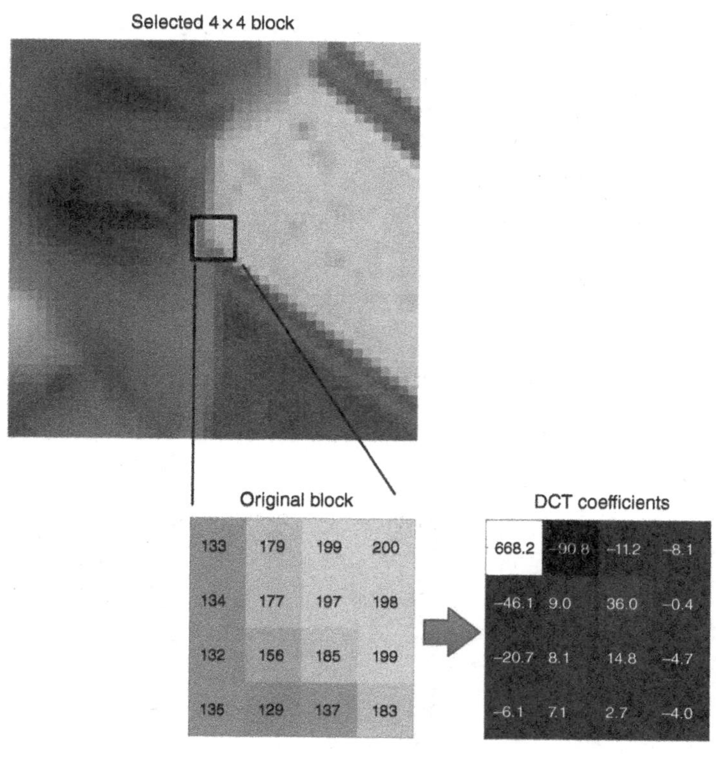

Figure 7.23 4 × 4 image block and DCT coefficients

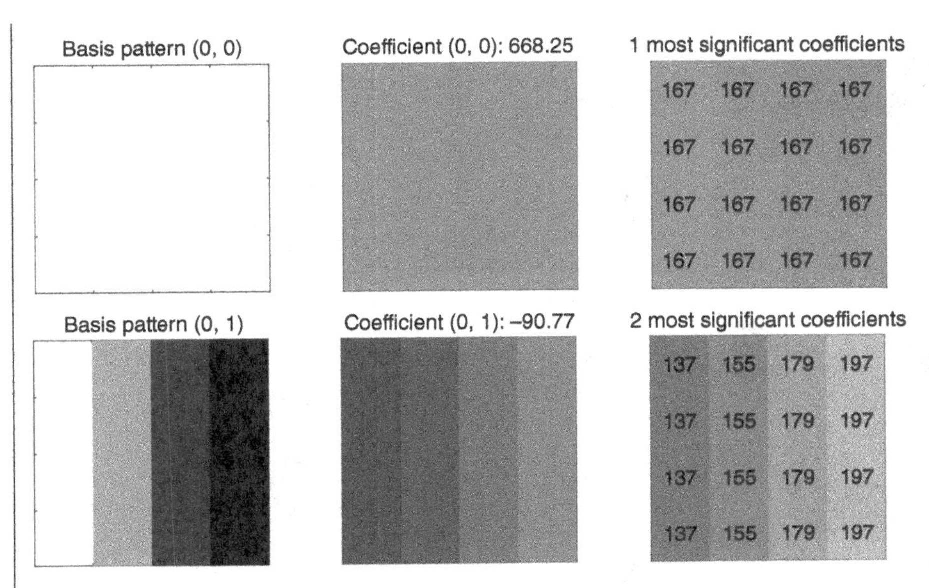

Figure 7.24 The two most significant coefficients and basis patterns

The next most significant basis pattern is (0,1), also shown in Figure 7.24 on the lower line. This coefficient has a value of −91, which scales and inverts the basis pattern. The original basis pattern is dark at the top and light at the bottom. Scaling the pattern gives a particular range of greys from light grey to dark grey. Inverting the pattern – applying a negative coefficient – flips the pattern so that we have light grey at the top and dark grey at the bottom. Combining the scaled (0,0) and (0,1) patterns gives us the block at the lower right of Figure 7.24, i.e. the block formed by scaling and adding the two most significant coefficients, (0,0) and (0,1). We can see that even with just two basis patterns, we get a block that looks somewhat like the original.

Continuing along the list of coefficients, the next most significant is coefficient (1,0) with a value of −46, shown at the top of Figure 7.25. Once again, the basis pattern is scaled and inverted. The value of the coefficient is significantly smaller than the previous coefficient, and so the pattern appears fainter. Adding together the first three coefficients, the scaled (0,0), (0,1) and (1,0) patterns, gives an output that already looks similar to the original block (see Figure 7.25, top-right).

This process continues with the fourth and fifth coefficients. With five coefficients, shown in Figure 7.25 at the bottom-right, we have an output block that looks almost identical to the original. However, the actual sample values are not quite the same, compared with Figure 7.23. This is an important point: with only five out of the 16 coefficients, we can recreate an almost identical copy of the original block.

Once all 16 basis patterns have been scaled and combined, we have a perfect copy of the original image block. However, this example shows that a visually similar block can be constructed by combining only a few of the most significant coefficients and basis patterns. Just how many coefficients we need depends on the image block.

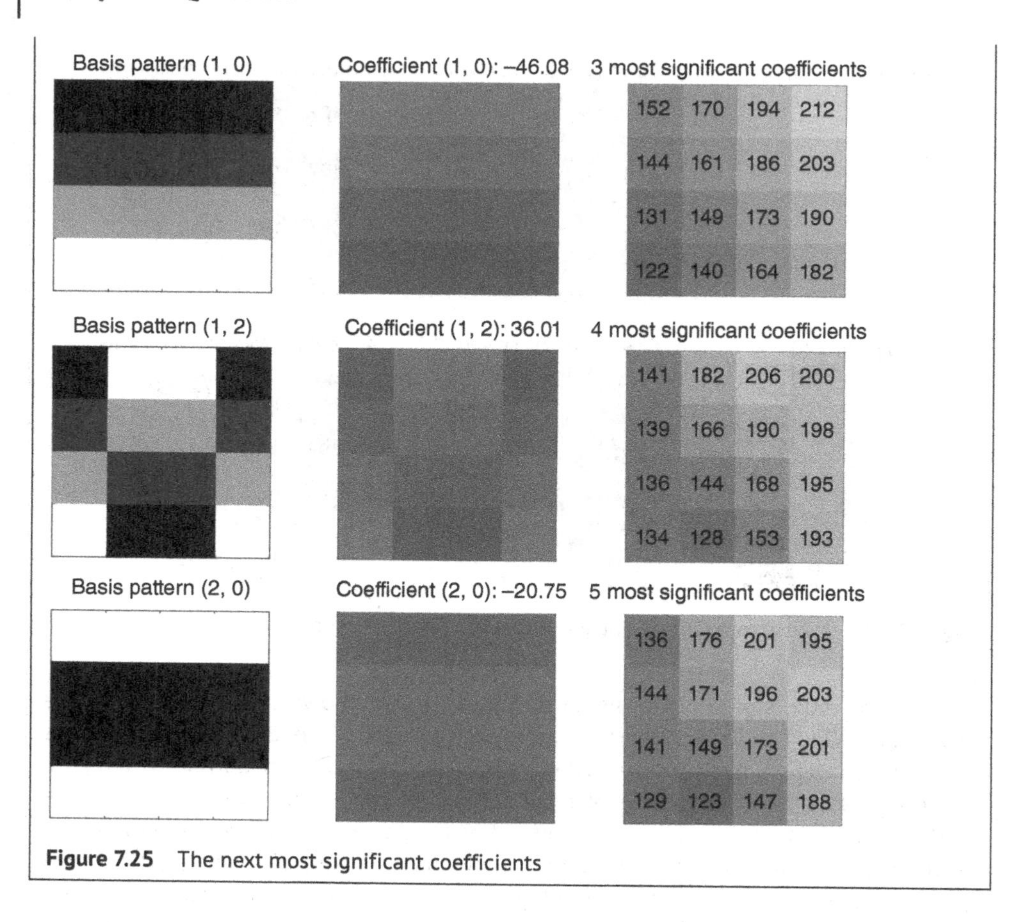

Figure 7.25 The next most significant coefficients

The next stage in the encoding process, quantisation, aims to remove the insignificant coefficients and retain only the significant coefficients in each block.

7.4 Quantisation

Once a block of samples has been transformed into a block of transform coefficients, the coefficients can be quantised. As the example in Figure 7.4 shows, quantisation reduces the precision of each coefficient, which means that large-valued coefficients lose precision and small-valued coefficients are set to zero.

Quantisation exploits properties of typical images and of the way we see images. For example, the image in Figure 7.26 shows an original frame from a video clip. If we discard all but the top-left four coefficients in each 8×8 DCT block, we get the image in Figure 7.27. This means that for every block of 64 coefficients, we discard all but the low-frequency coefficients (0,0), (1,0), (0,1) and (1,1), then decode and reconstruct the image. Much of the fine detail is lost, but the main features of the image are still easily recognisable, even

Figure 7.26 Original frame

Figure 7.27 Frame with only top-left four DCT coefficients of each 8 × 8 block, all other coefficients set to zero

though we have only retained four out of the 64 coefficients in each 8×8 image block. Much of the visually important information is carried in just a small number of transform coefficients.

Adjusting the quantisation process is an effective way of controlling both compression and image quality. Consider the example given in Figures 7.4 and 7.5. After quantisation, which in this case involved dividing by 16 and rounding towards zero, and rescaling by multiplying by 16, the coefficients can only have values that are multiples of 16, such as 0, ±16 and ±32. The gap between possible coefficient values, 16 in this case, is known as the quantiser step size. Increasing the step size has two effects. First, less information is retained in the quantised values, making them easier to compress. Second, the rescaled values are increasingly different from the original coefficients, which means that distortion is increased and the decoded image quality will be worse.

Examples

The original 776 × 440 pixel image of Figure 7.26 takes up 1.4 Mbytes of storage space in its uncompressed form. Compressing the image with JPEG compression involves transforming each 8 × 8 block of pixels using a two-dimensional DCT, followed by quantisation of each coefficient. JPEG's quantisation or quality parameter controls the amount of quantisation applied to each coefficient. It controls the quantiser step size or the gap between quantised and rescaled coefficient values. A medium quantiser step size retains much of the original information but results in a significantly compressed file size of 64 kbytes (see Figure 7.28). A large quantiser step size removes a lot of visual information (see Figure 7.29). The image is now clearly degraded, showing the characteristic block distortions of over-compressed DCT-based image compression. However, the file size has been reduced to 9 kbytes, and the image subject is still more or less distinguishable.

Figure 7.28 Image with medium compression/medium quantisation, 64 kbytes

Figure 7.29 Image with high compression/high quantisation, 9 kbytes

7.4.1 Combining Transform and Quantisation

Let us consider the transform and quantisation processes together and look at the way they interact with each other. Figure 7.30 shows an 8×8 block of image samples on the left. When this block is transformed using an 8×8 DCT, we get the block of coefficients shown in the middle. If we inverse transform the coefficients using an 8×8 inverse DCT, we get the output block shown on the right. There is no information loss in these steps, and so the output block is identical to the input block.

Let us look at what happens when we introduce the quantiser. In the forward process, the original block of samples or residuals is transformed by the DCT, and the resulting block of coefficients is quantised (see Figure 7.31). We apply a simple quantiser process, dividing each coefficient by a step size of 8 and rounding towards the nearest integer (see Figure 7.32). Notice how the magnitude of larger values is reduced, e.g. the top-left DC coefficient, originally +1040, is quantised to +130, and notice that values smaller than the step size of ±8 are set to zero.

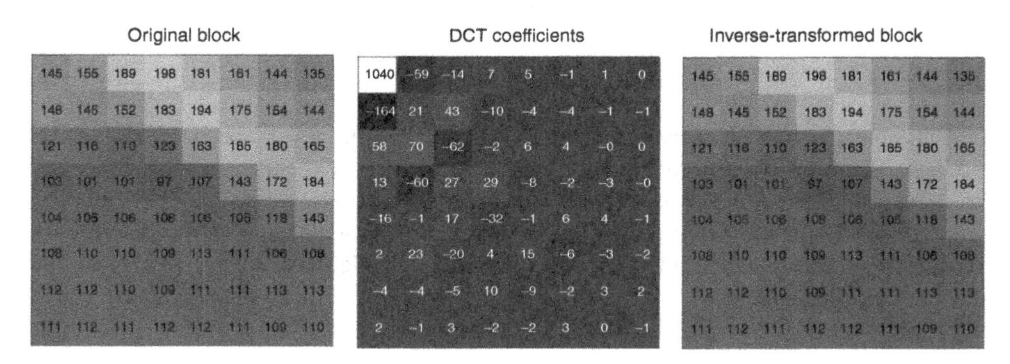

Original block

145	155	189	198	181	161	144	135
148	145	152	183	194	175	154	144
121	116	110	123	163	185	180	165
103	101	101	97	107	143	172	184
104	105	106	108	106	105	118	143
108	110	110	109	113	111	106	108
112	112	110	109	111	111	113	113
111	112	111	112	112	111	109	110

DCT coefficients

1040	-59	-14	7	5	-1	1	0
-164	21	43	-10	-4	-4	-1	-1
58	70	-62	-2	6	4	-0	0
13	-60	27	29	-8	-2	-3	-0
-16	-1	17	-32	-1	6	4	-1
2	23	-20	4	15	-6	-3	-2
-4	-4	-5	10	-9	-2	3	2
2	-1	3	-2	-2	3	0	-1

Inverse-transformed block

145	155	189	198	181	161	144	135
148	145	152	183	194	175	154	144
121	116	110	123	163	185	180	165
103	101	101	97	107	143	172	184
104	105	106	108	106	105	118	143
108	110	110	109	113	111	106	108
112	112	110	109	111	111	113	113
111	112	111	112	112	111	109	110

Figure 7.30 8 × 8 image block, forward and inverse transformed, no quantisation

Figure 7.31 Forward transform and quantise

Original block

145	155	189	198	181	161	144	135
148	145	152	183	194	175	154	144
121	116	110	123	163	185	180	165
103	101	101	97	107	143	172	184
104	105	106	108	106	105	118	143
108	110	110	109	113	111	106	108
112	112	110	109	111	111	113	113
111	112	111	112	112	111	109	110

DCT coefficients

1040	-59	-14	7	5	-1	1	0
-164	21	43	-10	-4	-4	-1	-1
58	70	-62	-2	6	4	-0	0
13	-60	27	29	-8	-2	-3	-0
-16	-1	17	-32	-1	6	4	-1
2	23	-20	4	15	-6	-3	-2
-4	-4	-5	10	-9	-2	3	2
2	-1	3	-2	-2	3	0	-1

Quantised coefficients, Q = 8

130	-7	-2	1	1	-0	0	0
-21	3	5	-1	-1	-1	-0	-0
7	9	-8	-0	1	0	-0	0
2	-8	3	4	-1	-0	-0	-0
-2	-0	2	-4	-0	1	0	-0
0	3	-2	0	2	-1	-0	-0
-1	-1	-1	1	-1	-0	0	0
0	-0	0	-0	-0	0	0	-0

Figure 7.32 Forward transform and quantise, Qstep = 8

Figure 7.33 Rescale and inverse transform

Quantised coefficients, $Q = 8$

130	−7	−2	1	1	−0	0	0
−21	3	5	−1	−1	−1	−0	−0
7	9	−8	−0	1	0	−0	0
2	−8	3	4	−1	−0	−0	−0
−2	−0	2	−4	−0	1	0	−0
0	3	−2	0	2	−1	−0	−0
−1	−1	−1	1	−1	−0	0	0
0	−0	0	−0	−0	0	0	−0

Rescaled coefficients, $Q = 8$

1040	−56	−16	8	8	−0	0	0
−168	24	40	−8	−8	−8	−0	−0
56	72	−64	−0	8	0	−0	0
16	−64	24	32	−8	−2	−0	−0
−16	−0	16	−32	−0	8	0	−0
0	24	−16	0	16	−8	−0	−0
−8	−8	−8	8	−8	−0	0	0
0	−0	0	−0	−0	0	0	−0

Inverse-transformed block, $Q = 8$

146	155	186	202	182	157	144	135
152	145	153	181	195	174	151	144
121	112	109	127	162	187	184	167
105	100	100	100	110	143	174	181
106	107	108	108	102	101	119	144
106	111	109	108	112	108	102	106
114	115	111	108	112	114	112	110
110	111	110	112	115	111	105	106

Figure 7.34 Rescale and inverse transform, Qstep = 8

In the inverse process shown in Figure 7.33, the block of quantised coefficients is rescaled, which in this case means multiplying each value by the same step size (8). The resulting block of coefficients is inverse-transformed to create an output block of samples, as shown in Figure 7.34. After rescaling, the rescaled coefficients have similar magnitudes to the original coefficients but are not identical to the originals. Consider the first few examples along the top row:

Coefficient	Quantised (/8, round towards nearest integer)	Rescaled (×8)
1040	130	1040
−59	−7	−56
−14	−2	−16
7	1	8
5	1	8
...	...	...

When the rescaled coefficients are inverse-transformed, the result is an output image block that looks similar to the original but is not identical. The output block in Figure 7.34 has a similar visual appearance to the original, with a lighter-shaded band in the top-right corner, but some of the pixel values are slightly different.

We will now consider three further quantisation step sizes: 16, 32 and 64. In each case, every coefficient is quantised by dividing it by the step size and rounding. The quantised coefficient is rescaled by multiplying it by the same step size. Figure 7.35 shows the results for a step size of 16. More coefficients are set to zero, and the difference between the original and output blocks is slightly greater. However, the output block is still visually similar to the original.

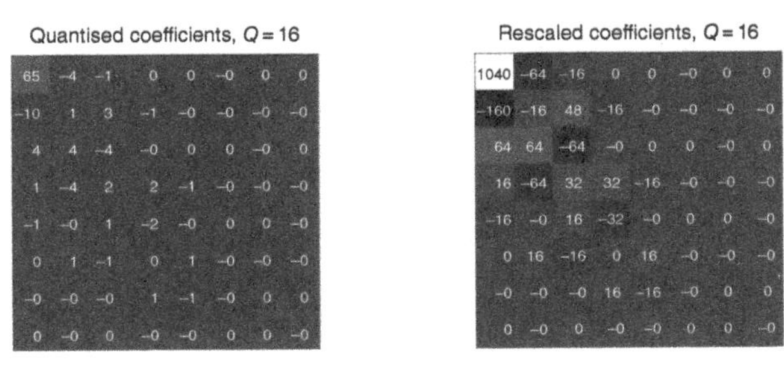

Figure 7.35 Rescale and inverse transform, Qstep = 16

Figure 7.36 Rescale and inverse transform, Qstep = 32

Figure 7.37 Rescale and inverse transform, Qstep = 64

As the step size increases to 32, shown in Figure 7.36, and further to 64, shown in Figure 7.37, the number of zero coefficients increases. With a step size of 64, only eight out of the original 64 coefficients are non-zero. The output block, shown in Figure 7.37, is now clearly different from the original image block shown in Figure 7.32.

This example illustrates the relationship between quantiser step size, compression and image quality. Increasing the quantiser step size reduces the amount of information that needs to be encoded but also reduces the visual quality of the decoded image block.

7.4.2 Designing the Quantiser

In the above-mentioned examples, we used a simple quantisation process, which we will call Quantiser (a), in which each coefficient is divided by a step size and rounded towards the nearest integer. This process of quantisation and rescaling is illustrated graphically in

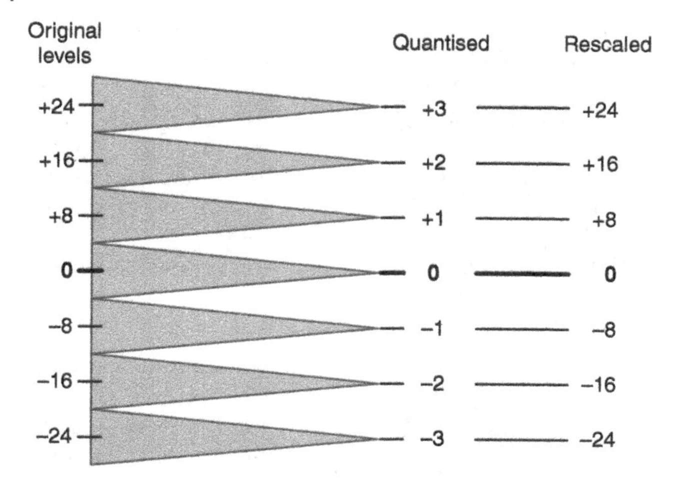

Figure 7.38 Quantiser (a), step size = 8

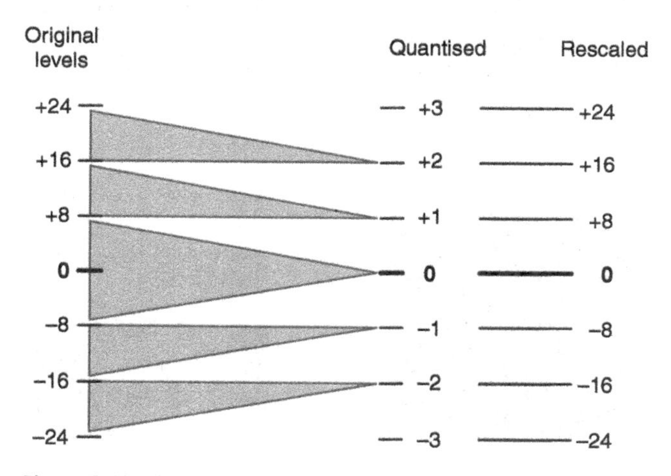

Figure 7.39 Quantiser (b), step size = 8

Figure 7.38. Coefficients in the range ±4 are quantised to zero and rescaled to zero; coefficients in the range +4 to +12 are quantised to +1 and rescaled to +8; and so on.

We can redesign this quantiser with the same step size but with a different mapping between original and rescaled coefficients (see Quantiser (b), shown in Figure 7.39). With Quantiser (b), coefficients in the range ±8 are quantised to zero and then rescaled to zero. Original coefficients less than +16 but ≥8 are quantised to +1, then rescaled to +8. Original coefficients in the range −8 to −16 are quantised to −1, then rescaled to −8, and so on. This alternative quantiser is more 'aggressive' than the quantiser of Figure 7.38 since more coefficients are quantised to zero. For example, a coefficient of −5 would be quantised to −1 and rescaled to −8 using Quantier (a), since it falls outside the band of values −7 to +7 that are mapped to zero. The same coefficient value of −5 would be quantised to 0 and rescaled to 0 using Quantiser (b).

Notice that even though the step size is the same, Quantiser (b) produces more zero-valued coefficients than Quantiser (a). We would therefore expect Quantiser (b) to compress video more and to introduce more distortion in the decoded video frames than Quantiser (a).

Instead of quantising every coefficient in the same way, it can be useful to weight the quantisation step size so that lower-frequency coefficients are quantised less and higher-frequency coefficients are quantised more. An example of frequency-weighted quantisation is shown in Figure 7.40. The quantiser step size increases towards the bottom right of an 8×8 block, where the coefficient frequencies are higher. Lower-frequency coefficients, towards the top left of the block, tend to be more important for visual quality, and a weighted quantiser can retain more information in the important lower frequencies whilst removing more information from the higher frequencies.

16	16	16	16	17	18	20	24
16	16	16	17	18	20	24	25
16	16	17	18	20	24	25	28
16	17	18	20	24	25	28	33
17	18	20	24	25	28	33	41
18	20	24	25	28	33	41	54
20	24	25	28	33	41	54	71
24	25	28	38	41	54	71	91

Figure 7.40 Frequency-weighted quantisation

7.5 Transform and Quantisation in Practice

The transform and quantisation stages in a video codec are closely connected. The transform does not provide compression by itself but is designed to work with the subsequent quantisation stage, which quantises each transform coefficient.

In a practical codec, the transform and quantiser should be designed with certain goals in mind, including:

- Maximising rate-distortion performance – maximum quality and minimum compressed bitrate.
- Minimising computational complexity.
- Ensuring compatibility between encoders and decoders.
- Providing flexibility, for example, in the selection of transform size and quantiser step size.

In recently developed video coding standards, these goals are satisfied by approaches such as the following:

- Specify a transform that can be implemented using a minimal number of basic arithmetic operations such as additions, subtractions and binary shifts, with as few multiplications as possible since these tend to be more computationally expensive than additions and shifts.
- Specify a transform that can be efficiently implemented in different ways to maximise efficiency on different software and hardware platforms.
- Specify a particular approximation to transform weighting factors such as the irrational numbers in the 4×4 DCT, as shown in Eq. (7.7).

- Avoid divisions and fractional arithmetic if possible.
- Limit the number of bits of precision required during computation. For example, if the input samples are 8 bits, what is the maximum dynamic range required during the computation of the transform and quantisation stages?
- Share aspects of the design across multiple transform sizes where possible.
- Specify the quantiser to give an effective level of control of the trade-off between quantiser step size, bitrate and quality.

We will examine the transform and quantiser design in the HEVC/H.265 standard and see how these goals are achieved in practice.

7.6 HEVC Transform and Quantisation

We saw in Chapter 4 how each HEVC coding unit (CU) is partitioned into one or more transform units (TUs). In a colour video sequence, each TU comprises a luma transform block (TB) and two chroma TBs. Luma TBs can range from 4×4 to 32×32 samples in size.

Each CU or TB can be processed in one of the following ways, as illustrated in Figure 7.41:

1) Transform the TB using a block transform and quantise the resulting coefficients. Code the coefficients using transform sub-block (TSB) coding.
2) Bypass the transform and quantise the TB samples directly. Code the coefficients, 4×4 at a time, using TSB coding, see Chapter 8.
3) Bypass both the transform and the quantisation stages and pass the TB samples directly for TSB coding.
4) In pulse code modulation (PCM) mode, directly code the CU samples in the bitstream.

The first option is probably the most common approach to coding a TB. We will discuss this in detail next.

7.6.1 Overview of HEVC Transform and Quantisation

The classic block-based transform and quantisation process is shown at the top of Figure 7.42. A block of samples **X** is transformed, for example, using a two-dimensional DCT, into a block of coefficients **Y**. The coefficients are quantised, for example, by dividing by a quantiser step and rounding, to produce a block of quantised levels **L**.

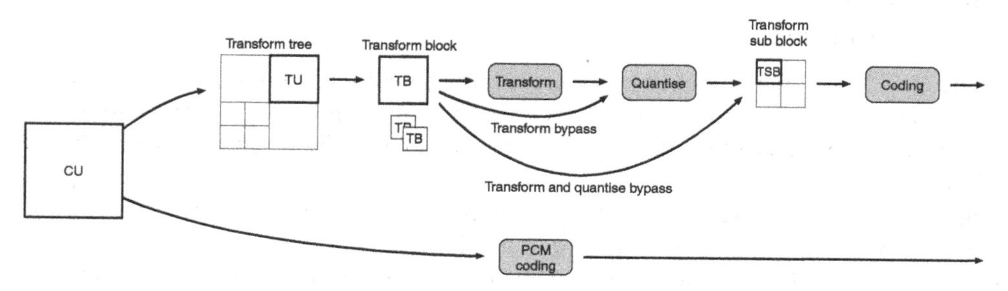

Figure 7.41 CU and TB processing in HEVC

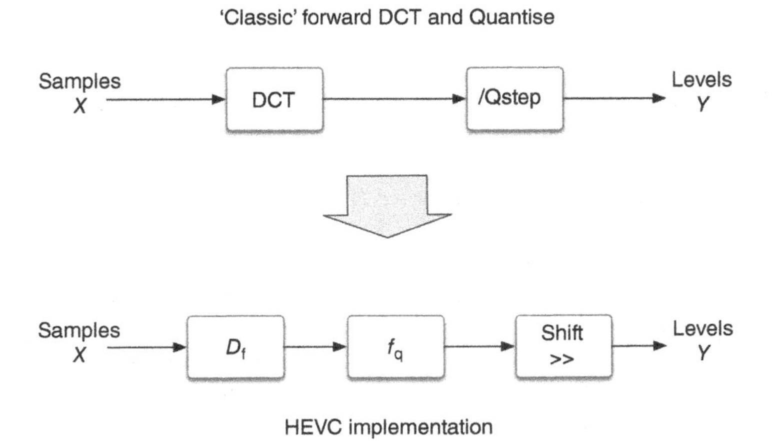

Figure 7.42 HEVC forward transform and quantise

The equivalent process in an HEVC encoder is shown at the bottom of Figure 7.42. The block of samples is transformed by D_f, a scaled approximate DCT. At the output of the transform, each coefficient is multiplied by a factor f_q and then bit-shifted down to the correct dynamic range to produce output level block **L**.

A classic inverse process is shown at the top of Figure 7.43. Each quantised level in block **L** is rescaled, for example, by multiplying with quantiser step Q_{step}, to produce a block of coefficients **Y′**. The coefficient block is inverse-transformed, using a two-dimensional inverse DCT, to produce an output block **X′**.

The equivalent process in an HEVC decoder, shown at the bottom of Figure 7.43, applies a rescaling factor g_q followed by a bit shift to rescale the coefficients. A scaled approximate inverse discrete cosine transform (IDCT) D_i processes an output block **X′**.

We will consider how the HEVC forward and inverse processes are developed from the 'classic' transform and quantise processes.

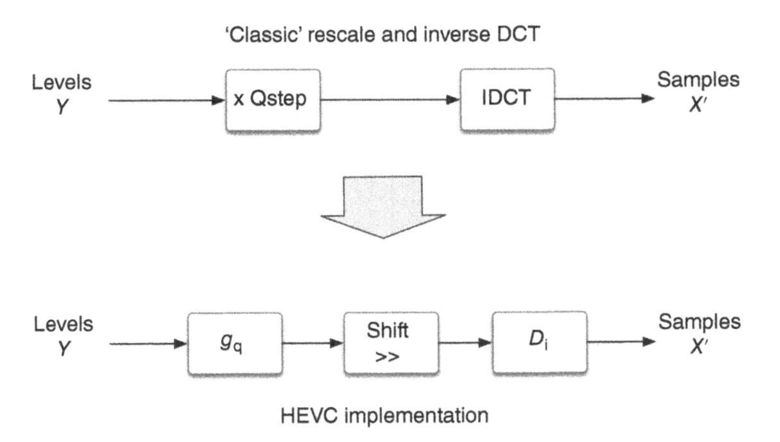

Figure 7.43 HEVC rescale and inverse transform

7.6.2 The HEVC Transforms

HEVC specifies square transforms with four different sizes for luma blocks: 4×4, 8×8, 16×16 and 32×32. These transforms are derived from the DCT and the DST using certain approximations. Luma TBs of size 4×4, which have been predicted using intra-prediction, are transformed using a 4×4 transform based on the DST, denoted D_{intra4}. All other TBs are transformed using a transform based on the DCT, denoted D_4, D_8, D_{16} or D_{32}.

7.6.2.1 DCT-Based Transforms

At the left of Figure 7.44 is the 4×4 DCT matrix **A** that we saw earlier in Section 7.3.5. Recall that some of the weighting factors are irrational numbers. We can develop a scaled integer version by multiplying each element of **A** by a scaling factor such as 128 and rounding the result to the nearest integer, i.e. Round(A*128), shown in the middle column of Figure 7.44. The transform matrix $\mathbf{D_4}$ used in an HEVC codec is shown at the right of Figure 7.44. This is almost, but not quite, a scaled 4×4 DCT [3]. The weighting factors shown in bold are slightly different from the scaled DCT matrix **A**, i.e. ±83 and ±36 instead of ±84 and ±35. We can describe the HEVC transform matrix $\mathbf{D_4}$ as a scaled approximation to the 4×4 DCT matrix **A**, i.e. $\mathbf{D_4} \approx \text{round}(\mathbf{A_4} * 128)$. Each transform matrix element can be represented using 8 bits, which means that each element of $\mathbf{D_4}$ fits within the range −128 to +127.

The 4×4 matrix $\mathbf{D_4}$ is embedded in the 8×8 HEVC matrix $\mathbf{D_8}$, shown in Figure 7.45. This means that the 16 elements of $\mathbf{D_4}$ make up the first half of each of the first, third, fifth and seventh rows of $\mathbf{D_8}$. Similarly, $\mathbf{D_8}$ is embedded in $\mathbf{D_{16}}$, as shown in Figure 7.46, and $\mathbf{D_{16}}$ is embedded in $\mathbf{D_{32}}$, as shown in Figure 7.47. This has certain advantages for efficient implementation, making it possible for a common hardware engine to process all four transform sizes [3]. All four HEVC transform matrices $\mathbf{D_N}$ are scaled approximations to the DCT matrices **A** of the same dimensions, as illustrated in Table 7.3.

The set of HEVC matrices $\mathbf{D_{4,8,16,32}}$ was developed according to certain design goals [4]:

- Each matrix should approximate but not necessarily be identical to a scaled DCT/IDCT.
- Each matrix should be approximately orthogonal[2].
- The forward and inverse matrices should be symmetrical.
- The matrices should be embedded, i.e. each smaller matrix $\mathbf{D_N}$ is a subset of a larger matrix.

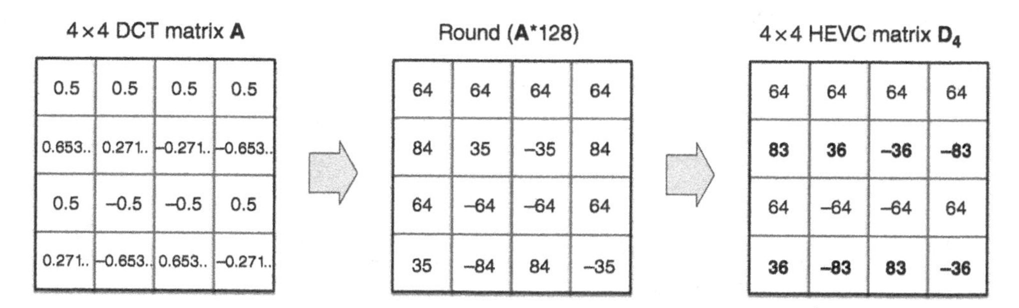

4×4 DCT matrix **A**

0.5	0.5	0.5	0.5
0.653..	0.271..	−0.271..	−0.653..
0.5	−0.5	−0.5	0.5
0.271..	−0.653..	0.653..	−0.271..

Round (**A***128)

64	64	64	64
84	35	−35	84
64	−64	−64	64
35	−84	84	−35

4×4 HEVC matrix $\mathbf{D_4}$

64	64	64	64
83	**36**	**−36**	**−83**
64	−64	−64	64
36	**−83**	**83**	**−36**

Figure 7.44 DCT and D_4 matrices

2 Orthogonal: the product of a matrix with its transpose results in a diagonal matrix.

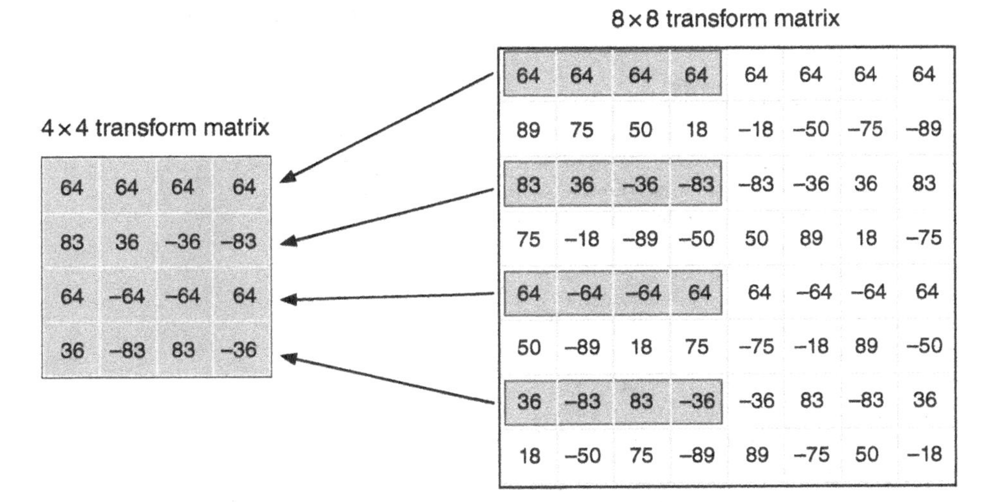

Figure 7.45 HEVC transforms: D_4 and D_8

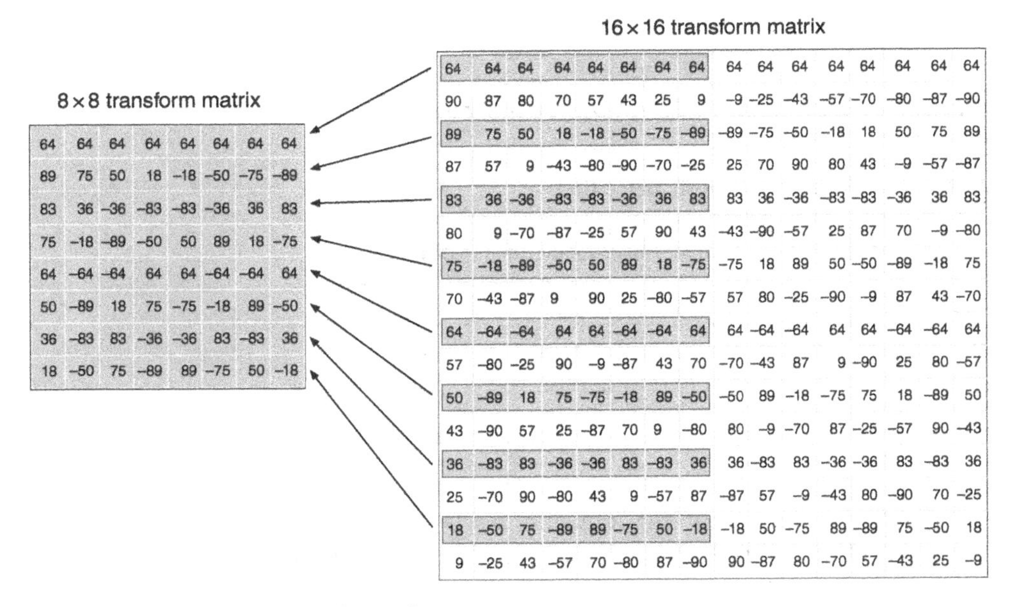

Figure 7.46 HEVC transforms: D_8 and D_{16}

- Each matrix should be separable. Row and column transforms can be applied separately to achieve the complete two-dimensional transform.
- Each transform should be capable of being implemented using limited precision arithmetic, using at most 16-bit multipliers and 32-bit accumulators in a hardware implementation.

The transforms specified in the HEVC standard were developed as a compromise between these design goals.

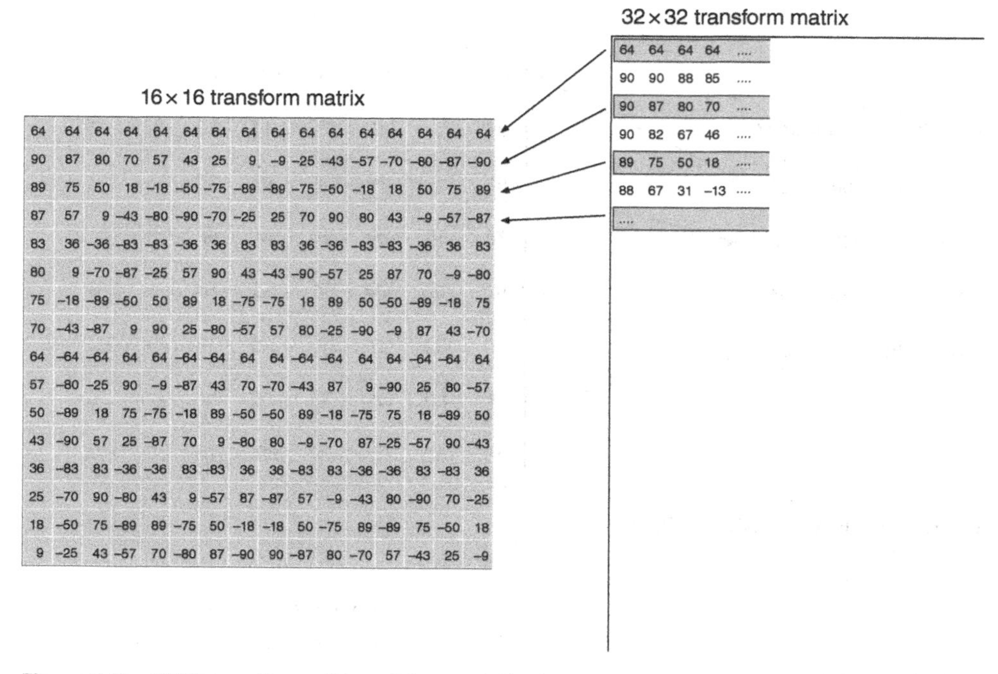

Figure 7.47 HEVC transforms: D_{16} and D_{32}, partially shown

Table 7.3 HEVC and DCT transform matrices

HEVC transform matrix	Approximate relationship to DCT matrix
$\mathbf{D_4}$	≈ round($\mathbf{A_4}$ * 128)
$\mathbf{D_8}$	≈ round($\mathbf{A_8}$ * 181)
$\mathbf{D_{16}}$	≈ round($\mathbf{A_{16}}$ * 256)
$\mathbf{D_{32}}$	≈ round($\mathbf{A_{32}}$ * 362)

Figure 7.48 shows the four transform matrices, $\mathbf{D_4}$, $\mathbf{D_8}$, $\mathbf{D_{16}}$ and $\mathbf{D_{32}}$, with each weighting factor plotted as a shaded square. In each case, the weighting factors are in the range of −128 to +127.

7.6.2.2 DST-Based Transform

An alternative 4×4 transform is used to process 4×4 luma TBs in intra-predicted CUs, based on the 4×4 DST [5][3]. This transform can be more effective for blocks that have been predicted using directional intra prediction modes (see Chapter 5).

3 Yeo et al. originally proposed a simplification of the KLT, but the matrix D_{intra4} adopted in the H.265 standard has a sinusoidal basis.

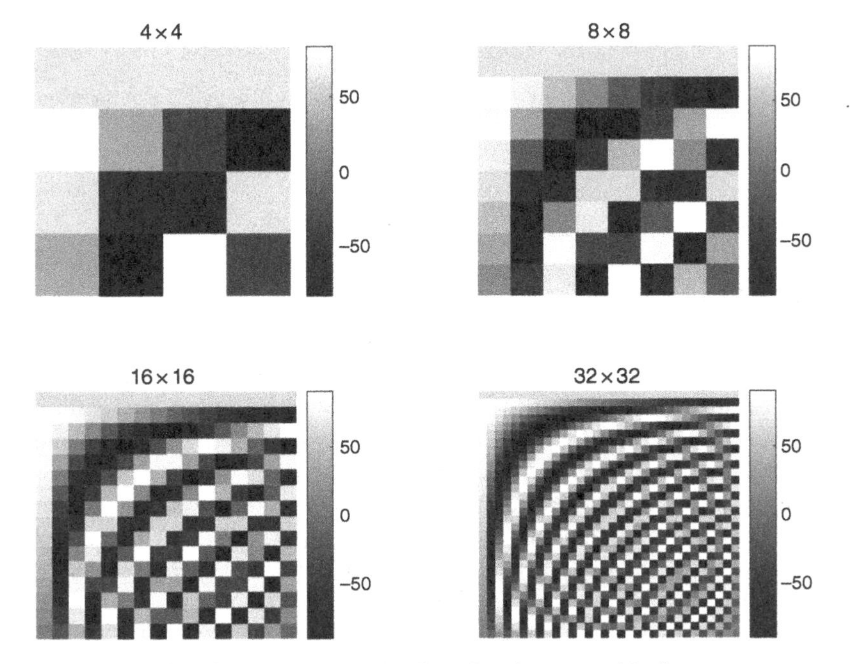

Figure 7.48 HEVC transform matrices D_4 to D_{32} shown graphically

7.6.2.3 The HEVC Forward and Inverse Transform and Quantisation Processes

The HEVC/H.265 forward transform process is:

1) Apply a one-dimensional column transform **D**
2) Bit shift to reduce the dynamic range
3) Apply a one-dimensional row transform $\mathbf{D}^T$
4) Bit shift to reduce the dynamic range.

This is shown together with the quantiser f_q and final bit shift, discussed in Section 7.6.3 in Figure 7.49.

The corresponding inverse transform process, defined in the H.265 standard, is:

1) Apply a one-dimensional column transform $\mathbf{D}^T$
2) Bit shift to reduce the dynamic range
3) Apply a one-dimensional row transform **D**
4) Bit shift to reduce the dynamic range.

This is shown following the rescale g_q and post-rescale shift in Figure 7.50.

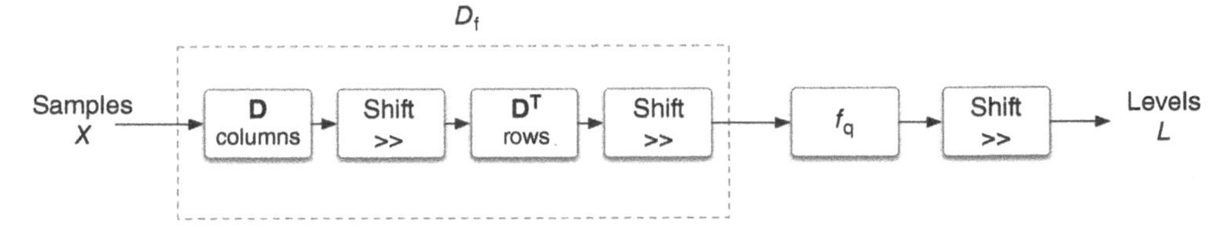

Figure 7.49 HEVC forward transform and quantise

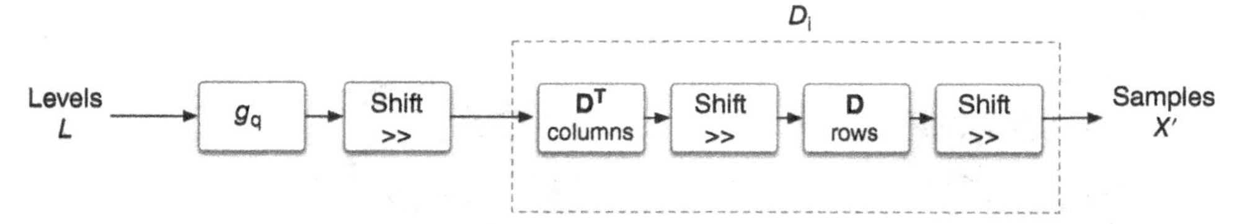

Figure 7.50 HEVC rescale and inverse transform

7.6.3 HEVC Quantisation and Rescaling

In an HEVC encoder, each block of coefficients **Y** is quantised to produce a block of levels **L**. The forward quantisation process is not specified in the H.265 standard. The HEVC decoder's process of rescaling levels **L** to produce a coefficient block **Y′**, a block that is similar to but not identical to the original block **Y**, is specified exactly in the H.265 standard and has certain implications for the forward quantisation process.

Both the quantisation process at the encoder and the rescaling process at the decoder utilise a quantisation step size Qstep. Quantisation maps each input coefficient Y to a level L and rescaling maps each L to an output coefficient Y' (see Figure 7.51). The input coefficient Y has a continuous range from a minimum Min(Y) to a maximum Max(Y). The output coefficient Y' has approximately the same range as Y but only has a certain number of discrete values, spaced by the step size Qstep. Increasing Qstep increases the spacing between discrete output values and reduces the number of possible values of Y' and vice versa. The step size Qstep influences the compressed bitrate of the coded HEVC sequence and the quality of the decoded video frames. A larger value of Qstep tends to produce higher compression and lower-quality decoded video, whereas a smaller Qstep gives less compression and higher-quality decoded video.

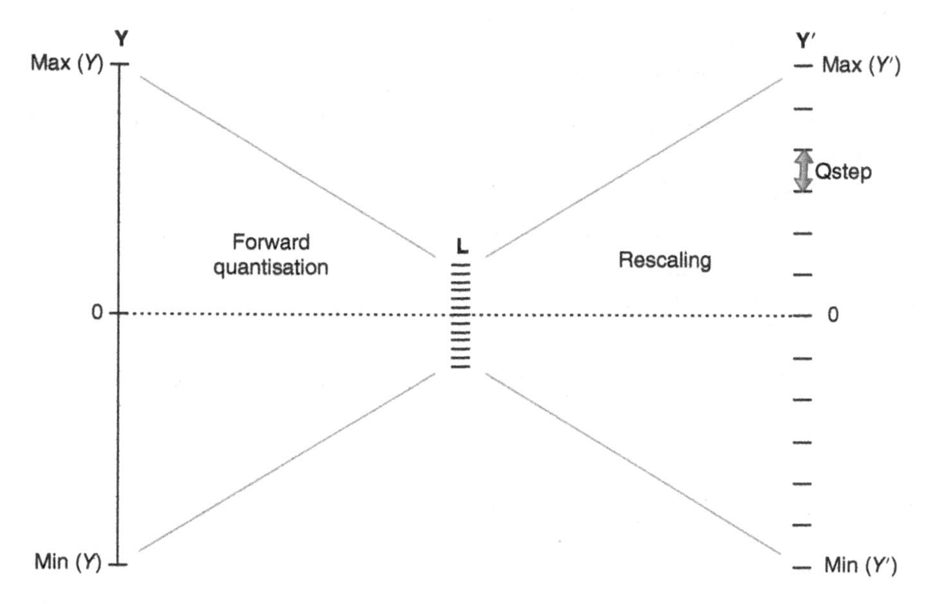

Figure 7.51 HEVC quantisation and rescaling: mapping input to output coefficients

7.6.3.1 HEVC QP and Qstep

Qstep is controlled by a quantisation parameter (QP) that can be chosen by an HEVC encoder. QP is signalled in the HEVC bitstream, which means that whatever value is chosen by the encoder is received and applied by the decoder. Hence, the encoder and decoder use the same QP and the same value of Qstep to quantise or rescale each block of coefficients. A change in QP is signalled at most once per CP by sending a delta QP parameter.

QP can take values between 0 and 51 for 8-bit video HEVC data. Each QP maps to a step size Qstep, as shown in Table 7.4. Each increment of QP corresponds to an increase in Qstep by a factor of approximately $2^{1/6}$. This means that Qstep doubles with every six increments of QP. A QP of six maps to a step size of 1.26 which will give a very modest amount of compression. Adding 6 gives us QP = 12, which has a step size of around 2.5, i.e. a doubling in step size. Adding 6 again, QP = 18, doubles the step size to around 5. The highest QP, 51, corresponds to a step size of around 228.

Table 7.4 HEVC QP, Qstep, f_q and g_q

QP	Qstep	$f_q \approx 2^{14}$/Qstep	$g_q \approx 2^6$ * Qstep
0	0.63	26,214	40
1	0.71	23,302	45
2	0.79	20,560	51
3	0.89	18,396	57
4	1.0	16,384	64
5	1.12	14,564	72
6	1.26	13,107	80
7	1.41	11,651	90
...	...	...	...
12	2.52	6554	160
...	...	...	...
18	5.03	3277	320
...	...	...	...
24	10.07	1638	640
...	...	...	...
30	20.16	819	1280
...	...	...	...
36	40.31	410	2560
...	...	...	...
42	80.63	205	5120
...	...	...	...
48	161.27	102	10,240
49	181.02	91	11,520
50	203.19	80	13,056
51	228.01	72	14,592

Instead of dividing by Qstep during forward quantisation or multiplying by Qstep during inverse quantisation, an HEVC codec uses the factors f_q and q_q, which we will discuss in Section 7.6.3.2.

Figure 7.52 shows the relationship between QP and Qstep for the range QP = 12 to QP = 18. As QP increases by one from 14 to 15, Qstep changes from 3.17 to 3.56, equivalent to multiplication by approximately $2^{1/6}$. The complete range of QP from 0 to 51 (for 8-bit video image data) is plotted in Figure 7.53.

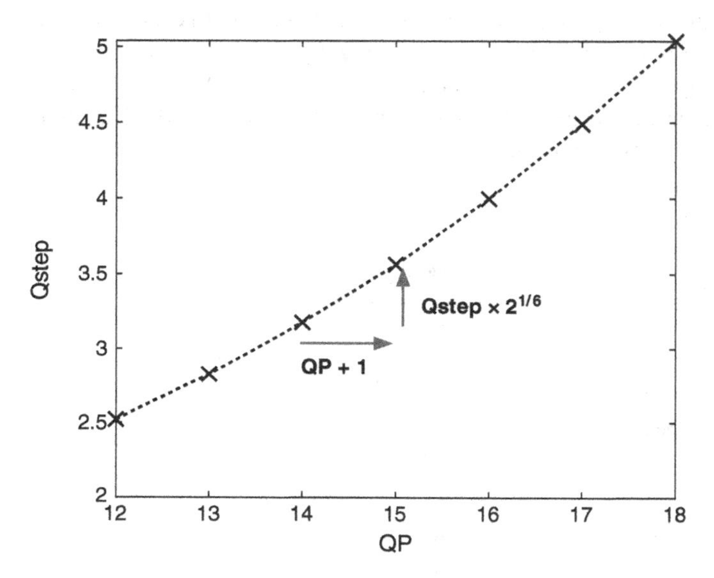

Figure 7.52 HEVC QP and Qstep, partial range

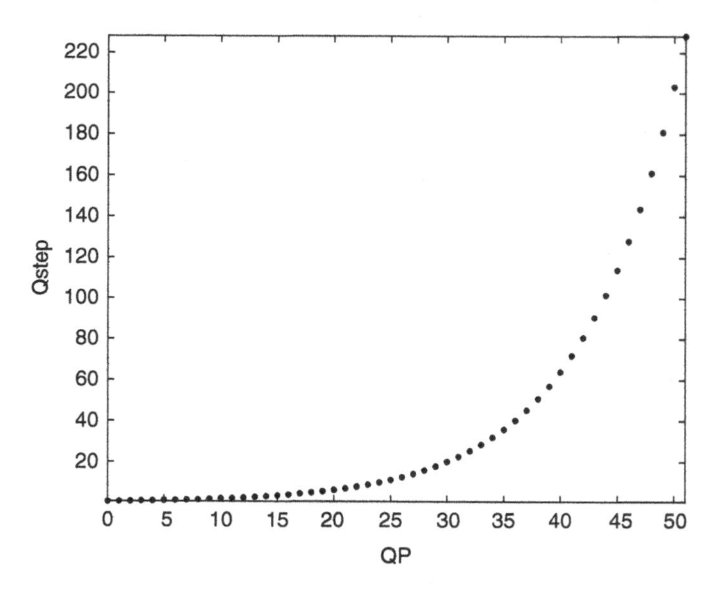

Figure 7.53 HEVC QP and Qstep, full range (8-bit image data)

The relationship between QP and Qstep is a compromise between controlling quantisation and optimising compression performance. An encoder can change QP in positive or negative increments and achieve an approximately consistent effect on compressed bitrate, at different bitrate ranges and for different video sources.

Example:

Let us look at the relationship between QP, quality and file size for a 1080p, 100-frame video clip, *Pedestrian Area*. The original, uncompressed file size is 311 Mbytes. Encoding the clip with the ×265 codec (see Chapter 11) using different QPs gives the file sizes shown in Table 7.5. At a QP of 20, the file is compressed by around 50× and the visual quality is good, as illustrated by the close-up of one frame on the left side of Figure 7.54. At a QP of 38, the file is compressed by nearly 700×. There is some distortion, as shown on the right side of Figure 7.54, but the subject is still recognisable.

In the example of Table 7.5, every time QP increases by 6, the quantisation step size doubles and the compression ratio increases by a factor of the order of 2×. The actual relationship between QP and compression ratio depends very much on the actual content of the video clip. This example demonstrates that we can use QP as a control mechanism to increase or decrease compressed file size and quality.

Table 7.5 Encoding 'Pedestrian Area' with different HEVC QP settings

QP	Compressed file size	Compression ratio
20	6.3 Mbytes	49×
26	1.9 Mbytes	163×
32	906 kbytes	343×
38	457 kbytes	680×

(a)

(b)

Figure 7.54 Close-up of frame from 'Pedestrian Area', QP = 20 (a), QP = 38 (b)

7.6.3.2 HEVC Forward Quantiser

We could quantise a coefficient Y by dividing by Qstep and rounding, e.g.:

$$L = \text{round}\left(\frac{Y}{\text{Qstep}}\right)$$

Division can be a computationally expensive process. Instead of dividing, an HEVC encoder can *multiply* by a factor $f_q \approx 2^{14}/\text{Qstep}$, then right shift by 14 binary places, which is the equivalent of dividing by Qstep with rounding, i.e.:

$$L = \left(Y \times f_q\right) \gg 14$$

Which is the same as:

$$L = \left(Y \times \frac{2^{14}}{\text{Qstep}}\right) \gg 14$$

The factor $f_q \approx 2^{14}/\text{Qstep}$ is shown in Table 7.4, for each value of QP and Qstep.

We can introduce a weighting factor w to adjust the quantisation of specific coefficient positions, see Section 7.6.3.4, where:

$w = 16$ indicates no weighting
$w > 16$ indicates an increase in quantiser step size, i.e. more quantisation
$w < 16$ indicates a reduction in quantiser step size, i.e. less quantisation.

The forward quantisation process becomes:

$$L = \left(Y \times \frac{16}{w} \times \frac{2^{14}}{\text{Qstep}}\right) \gg 14 \tag{7.10}$$

The forward quantisation process is not defined in the HEVC standard. In practice, the rescaling process, which is defined in the standard and introduced below, implies a forward quantisation process similar to Eq. (7.10).

We could adjust the forward quantisation process to optimise rate-distortion performance, for example, by adjusting the rounding process, as long as the rescaling process still works.

7.6.3.3 HEVC Rescaling

The rescaling process, which is defined in the H.265 standard, can be stated as:

$$Y' = L \times \text{Qstep}$$

As we saw in Table 7.4, small values of Qstep have fractional values. Replace Qstep with $g_q \approx 2^6 \times \text{Qstep}$, followed by a right shift of 6 binary places to compensate for the multiplication by 2^6:

$$Y' = \left(L \times \left(2^6 \times \text{Qstep}\right)\right) \gg 6$$

Introduce the weighting factor w, where $w = 16$ indicates no weighting:

$$Y' = \left(L \times w \times \left(2^6 \times \text{Qstep}\right)\right) \gg 10 \tag{7.11}$$

In Eq. (7.11), we have increased the right shift to 10 binary places to compensate for the factor of 16 ($=2^4$) introduced as *w*.

The factor $g_q \approx 2^6 \times$ Qstep is specified in the HEVC standard and is shown in Table 7.4 for each value of QP and Qstep.

The rescaling process is defined in the HEVC standard as a precise series of integer arithmetic operations:

```
d[x][y] = Clip3(coeffMin, coeffMax, ((TransCoeffLevel[xTbY]
  [yTbY][cIdx][x][y] * m[x][y] * levelScale[qP%6]<<(qP/6)) +
  (1 << (bdShift-1))) >> bdShift)
                                         [6], Equation 8-311
```

This is broadly equivalent to Eq. (7.11), with the following clarifications:

- d[x][y] is the rescaled coefficient Y′ at position (x,y) in the current block
- m[x][y] is the weighting factor w for the current position (x,y)
- levelScale[qP%6]<<(qp/6) is the factor $g_q \approx 2^6 \times$ Qstep
- bdShift incorporates the 10-bit shift from Eq. (7.11)
- the result is clipped to the minimum and maximum coefficient range (coeffMin, coeffMax).

7.6.3.4 HEVC Quantisation Matrices

The quantiser step size for each coefficient position (*x,y*) is weighted by a factor *w*. The default weighting matrices are defined in the HEVC standard and shown in Table 7.6. Notice that the array for 4×4 TBs is flat, which means that every coefficient position is weighted evenly. The arrays for 8×8 TBs are weighted such that higher-frequency coefficients, at the lower right, have a larger weighting, i.e. a larger Qstep and hence more quantisation. This takes advantage of a property of the human visual system: human viewers are less sensitive to high-frequency spatial information, and hence these higher-frequency coefficients can be quantised more than lower-frequency coefficients.

Recall that each matrix entry *w* is pre-scaled by a factor of 16. This means that the actual Qstep modification *W* for each coefficient position is as shown in Table 7.7, where a weight of 1.0 means no change to the quantiser step size and a weight >1.0 means a corresponding increase in quantiser step size.

Modifying the Default Matrices An encoder may choose to modify the default quantisation matrices, replacing the default frequency-dependent weighting shown above. Replacements for the default matrices can be sent in the Sequence Parameter Set (SPS) or Picture Parameter Set (PPS), i.e. the matrices can be changed at the sequence or picture level.

Quantisation Scaling Matrices for Larger Transform Sizes The transform size in HEVC can range from 4×4 to 32×32. For TBs larger than 8×8, the 8×8 default matrices for intra or inter predicted blocks are scaled up by copying or replicating each *w* value four or 16 times. Figure 7.55 shows an example for a group of four weighting values *w* within the default 8×8 inter array. Each value is replicated four times to create part of the 16×16 inter array and replicated 16 times to create part of the 32×32 inter array. An HEVC decoder generates each of the weighting arrays, i.e. 4×4, 8×8, 16×16 and 32×32, and selects the appropriate array depending on the transform size and prediction type of the current block.

Table 7.6 HEVC default quantisation weighting arrays *w*

w (all 4×4)							
16	16	16	16				
16	16	16	16				
16	16	16	16				
16	16	16	16				
w (8×8 intra)							
16	16	16	16	17	18	21	24
16	16	16	16	17	19	22	25
16	16	17	18	20	22	25	29
16	16	18	21	24	27	31	36
17	17	20	24	30	35	41	47
18	19	22	27	35	44	54	65
21	22	25	31	41	54	70	88
24	25	29	36	47	65	88	115
w (8×8 inter)							
16	16	16	16	17	18	20	24
16	16	16	17	18	20	24	25
16	16	17	18	20	24	25	28
16	17	18	20	24	25	28	33
17	18	20	24	25	28	33	41
18	20	24	25	28	33	41	54
20	24	25	28	33	41	54	71
24	25	28	33	41	54	71	91

Table 7.7 HEVC Qstep weights *W*

W (all 4×4)							
1	1	1	1				
1	1	1	1				
1	1	1	1				
1	1	1	1				
W (8×8 intra)							
1.0000	1.0000	1.0000	1.0000	1.0625	1.1250	1.3125	1.5000
1.0000	1.0000	1.0000	1.0000	1.0625	1.1875	1.3750	1.5625
1.0000	1.0000	1.0625	1.1250	1.2500	1.3750	1.5625	1.8125
1.0000	1.0000	1.1250	1.3125	1.5000	1.6875	1.9375	2.2500
1.0625	1.0625	1.2500	1.5000	1.8750	2.1875	2.5625	2.9375
1.1250	1.1875	1.3750	1.6875	2.1875	2.7500	3.3750	4.0625
1.3125	1.3750	1.5625	1.9375	2.5625	3.3750	4.3750	5.5000
1.5000	1.5625	1.8125	2.2500	2.9375	4.0625	5.5000	7.1875

Table 7.7 (Continued)

W (8×8 inter)							
1.0000	1.0000	1.0000	1.0000	1.0625	1.1250	1.2500	1.5000
1.0000	1.0000	1.0000	1.0625	1.1250	1.2500	1.5000	1.5625
1.0000	1.0000	1.0625	1.1250	1.2500	1.5000	1.5625	1.7500
1.0000	1.0625	1.1250	1.2500	1.5000	1.5625	1.7500	2.0625
1.0625	1.1250	1.2500	1.5000	1.5625	1.7500	2.0625	2.5625
1.1250	1.2500	1.5000	1.5625	1.7500	2.0625	2.5625	3.3750
1.2500	1.5000	1.5625	1.7500	2.0625	2.5625	3.3750	4.4375
1.5000	1.5625	1.7500	2.0625	2.5625	3.3750	4.4375	5.6875

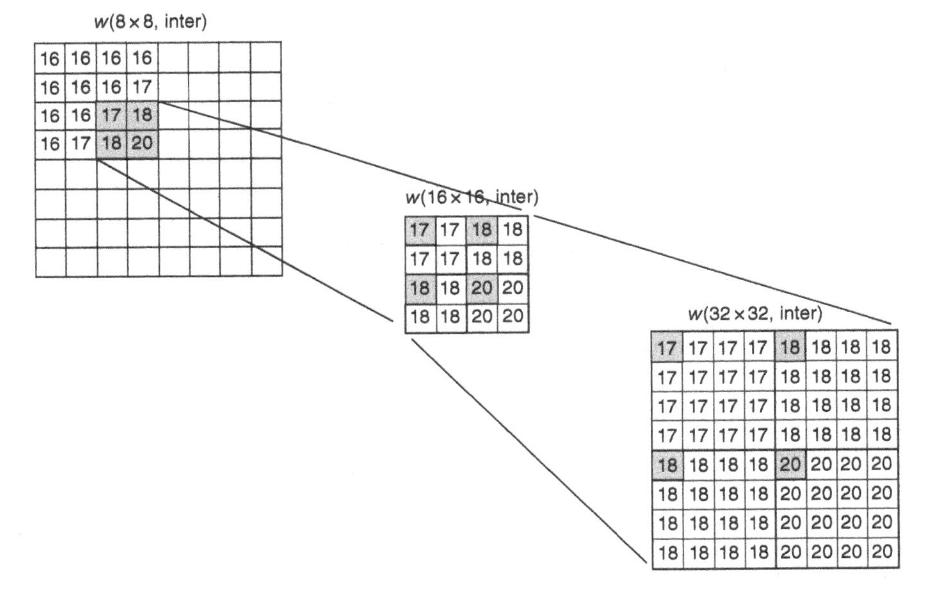

Figure 7.55 Replicating weighting factors for larger transform sizes

7.6.4 Example: The Complete HEVC Transform and Quantise Processes

Consider a block of inter-predicted residual samples **X**:

25	29	10	0
24	28	7	−3
10	12	3	8
15	16	0	9

We will look at what happens to the block during the HEVC forward and inverse transform and quantise processes[4].

4 Note: this example was developed using MATLAB approximations to the HEVC processes, so the numbers may not be identical to those produced by HEVC HM Reference Software.

7.6.4.1 Forward Transform and Quantise

Let us follow through the forward transform and quantise process shown in Figure 7.49.

1) Apply the HEVC core transform $\mathbf{D_4}$ to each column of $\mathbf{X}$:

4736	5440	1280	896
1334	1655	974	−1143
384	320	0	256
−802	−860	28	589

2) Apply the first bit shift, right shift 1 binary place (i.e. divide by 2):

2368	2720	640	448
667	827	487	−572
192	160	0	128
−401	−430	14	294

3) Apply the transposed transform $\mathbf{D_4}^T$ to each row:

395,264	234,240	−34,816	−103,520
90,176	115,077	−78,016	16,384
30,720	11,072	10,240	−10,976
−33,472	−73,669	19,776	11,832

4) Second bit shift, right shift 8 binary places:

1544	915	−136	−405
352	449	−305	64
120	43	40	−43
−131	−288	77	46

This is our forward transform output $\mathbf{Y}$.

5) Apply the quantiser f_q. Choose a quantiser parameter QP = 26 and a flat quantiser matrix for a 4×4 block. This gives a quantiser step size Qstep ≈ 12.64. We apply a scaled version of this, i.e. we multiply every value by 2^{14}/Qstep ≈ 1285:

1,984,040	1,175,775	−174,761	−520,426
452,320	576,965	−391,926	82,240
154,200	55,255	51,400	−55,256
−168,336	−370,081	98,945	59,110

6) Final post-quantiser bit shift, right shift 17 binary places:

15	8	−2	−4
3	4	−3	0
1	0	0	−1
−2	−3	0	0

This is our block of quantised coefficient levels **L** that can now be encoded and transmitted as part of a compressed HEVC bitstream. The top-left value is the DC coefficient and has the largest magnitude. As we move down and to the right, higher-frequency AC coefficients tend to have smaller magnitudes.

7.6.4.2 Rescale and Inverse Transform

Let us apply the rescale and inverse transform process of Figure 7.50 to the block of quantised coefficient levels **L**.

1) Rescale g_q. We use the same quantiser parameter QP = 26 and hence Qstep ≈ 12.64. We rescale by multiplying every value by $2^6 \times$ Qstep ≈ 816 and by weighting factor $w = 16$:

195,842	104,450	−26,110	−52,222
39,170	52,226	−39,166	2
13,058	2	2	−13,054
−26,110	−39,166	2	2

This is a scaled-up version of **L**, with an offset of 2 added so that the next bit shift rounds correctly.

2) First post-rescale bit shift, right shift 7 binary places:

1530	816	−204	−408
306	408	−306	0
102	0	0	−102
−204	−306	0	0

This is our decoded coefficient block **Y′**, ready for the inverse transform. Compare it with the original coefficient block **Y**, and you will notice that there are differences in the magnitude of the non-zero coefficients and that the zero coefficients remain zero after rescaling:

Y′ (decoded coefficients)				**Y (original coefficients)**			
1530	816	−204	−408	1544	915	−136	−405
306	408	−306	0	352	449	−305	64
102	0	0	−102	120	43	40	−43
−204	−306	0	0	−131	−288	77	46

1) Apply the transposed transform $\mathbf{D_4}^T$ to each column:

122,502	75,072	−38,454	−32,640
119,340	92,310	−24,072	−19,584
63,444	12,138	−2040	−19,584
86,394	29,376	12,342	−32,640

2) Second bit shift, right shift 7 binary places:

957	586	−301	−255
932	721	−189	−153
495	94	−16	−153
674	229	96	−255

3) Apply the transform $\mathbf{D_4}$ to each row:

81,442	122,773	38,251	2526
101,887	110,399	33,089	−6783
32,950	48,787	16,621	28,362
59,107	66,401	7583	39,453

4) Final bit shift, right shift 12 binary places:

19	29	9	0
24	26	8	−2
8	11	4	6
14	16	1	9

This is the decoded residual of our block **X**. Because the quantisation step f_q removes information, **X**′ is not identical to the original block **X**:

X′ (decoded samples)				X (original samples)				X−X′			
19	29	9	0	25	29	10	0	6	0	1	0
24	26	8	−2	24	28	7	−3	0	2	−1	−1
8	11	4	6	10	12	3	8	2	1	−1	2
14	16	1	9	15	16	0	9	1	0	−1	0

We can repeat the entire example by choosing different quantisers QP and see how the choice of QP affects the amount of non-zero data in the level block Y and the quality of the final decoded residual block **X**. When QP = 4, the quantiser step size is 1 and there is minimal change between **X** and **X′**, just ±1. As QP increases, the amount of data in the quantised block **L** reduces and the amount of difference or distortion between the original and decoded blocks **X** increases.

Starting from the original array **X** as mentioned above, applying a range of QP values gives the following results. Note how increasing the QP produces a greater difference between **X** and **X′**, i.e. more distortion due to compression, but also less information in the quantised block **L**, i.e. a more compressible signal.

QP	L				X				X−X′			
4	96	57	−9	−26	24	29	10	−1	1	0	0	1
	22	28	−20	4	24	28	7	−4	0	0	0	1
	7	2	2	−3	9	12	2	8	1	0	1	0
	−9	−19	4	2	14	15	−1	9	1	1	1	0
13	27	16	−3	−8	21	30	10	0	4	−1	0	0
	6	7	−6	1	25	28	7	−4	−1	0	0	1
	2	0	0	−1	9	12	1	8	1	0	2	0
	−3	−6	1	0	15	16	−1	8	0	0	1	1
26	15	8	−2	−4	19	29	9	0	6	0	1	0
	3	4	−3	0	24	26	8	−2	0	2	−1	−1
	1	0	0	−1	8	11	4	6	2	1	−1	2
	−2	−3	0	0	14	16	1	9	1	0	−1	0
31	8	5	−1	−3	15	29	4	−2	10	0	6	2
	1	2	−2	0	25	27	6	−4	−1	1	1	1
	0	0	0	−1	9	13	3	8	1	−1	0	0
	−1	−2	0	0	15	18	−3	9	0	−2	3	0
41	1	0	−1	−1	−34	45	4	−1	59	−16	6	1
	0	0	−1	0	21	41	28	−9	3	−13	−21	6
	0	0	0	−1	−22	−6	7	8	32	18	−4	0
	−1	−1	0	0	13	37	−15	20	2	−21	15	−11

7.6.5 Bypassing Transform and/or Quantisation in HEVC

As we discussed earlier with reference to Figure 7.41, the transform and/or quantisation processes can be optionally bypassed during HEVC encoding and decoding.

An encoder may choose to apply one of the following modes for individual TBs or CUs. For example, if the QP is very low and therefore there is very little compression being applied, transquant_bypass or PCM mode may actually be more efficient than applying the transform, quantise and coding processes.

7.6.5.1 Transform Bypass

If transform bypass is enabled, a flag transform_skip_flag is sent for certain sizes of TBs. If the flag is 1, no inverse transform is applied by the decoder. The transform can be selectively bypassed for each individual TB. Instead of applying the inverse transform, rescaled coefficient blocks $\mathbf{Y}'$ are simply further scaled to bring the samples up to the correct dynamic range.

7.6.5.2 Transform and Quantisation Bypass

If this option is enabled, syntax element cu_transquant_bypass_flag is sent for each CU indicating whether or not the entire transform and quantisation process is bypassed for this

CU. At the encoder, if the processes are bypassed, prediction is carried out as normal, then residual blocks **X** are copied directly to coefficient-level blocks **L**, which are then coded as 4×4 TSBs using the process described in Section 8.9.1. At the decoder, if the cu_trans-quant_bypass_flag flag is 1, decoded coefficient blocks **L** are passed directly to the reconstruction stage, with no inverse quantisation or inverse transform.

7.6.5.3 PCM Mode

If this option is enabled, a flag pcm_flag is sent for each Intra CU. If the flag is 1, the CU is coded in intra-PCM mode. Unlike the transformation and quantisation bypass mode, PCM mode codes the sample values of the CU directly into the bitstream, converting them directly to binary numbers. No prediction, transform or quantisation is applied.

7.7 Transform and Quantise in H.266 Versatile Video Coding

7.7.1 Transforms in H.266

The transform design in Versatile Video Coding (VVC) has similarities to the HEVC transform design, including transforms specified using integer arithmetic that can be processed using 16-bit operations, scaling of transform basis functions by approximately 64*($\sqrt{N}$), where N is the transform size, and embedding of smaller-sized transform matrices within larger-sized transform matrices [7].

VVC specifies multiple integer transform types, including different DCT forms and a form of the DST. Different transform types can be selected and combined when processing a two-dimensional block, e.g. different transform types can be applied to the block in the horizontal and vertical dimensions. The choice of transforms can be signalled explicitly or, for intra-blocks, derived implicitly, based on the dimensions of the block that is to be transformed.

A second stage transform, the low-frequency non-separable transform (LFNST), may be applied to the low-frequency transform coefficients of intra-predicted blocks, in an analogous way to H.264's DC coefficient transform.

The transform may be applied to a block that is a subset of the current CU, with the rest of the residual samples in the current CU set to zero. Four examples of CUs are shown in Figure 7.56, with the transform applied to a sub-block that is half or a quarter of the entire CU and the remaining samples zeroed. This can be efficient for residual blocks where most of the energy is concentrated into just part of the block.

7.7.2 Quantisation in H.266

The basic quantisation design in VVC [8] is similar to HEVC, with a quantiser selected by a QP, such that the quantiser step size doubles every time the QP increases by 6. Modifications to the quantisation process include:

Trellis-coded quantisation (TCQ) specifies two quantiser functions Q_0 and Q_1. The first coefficients in each block are decoded using quantised Q_0, and the decoder switches to quantiser Q_1 depending on the value of the previously decoded coefficient. This can improve the quality of decoded and reconstructed samples at the expense of increased processing complexity.

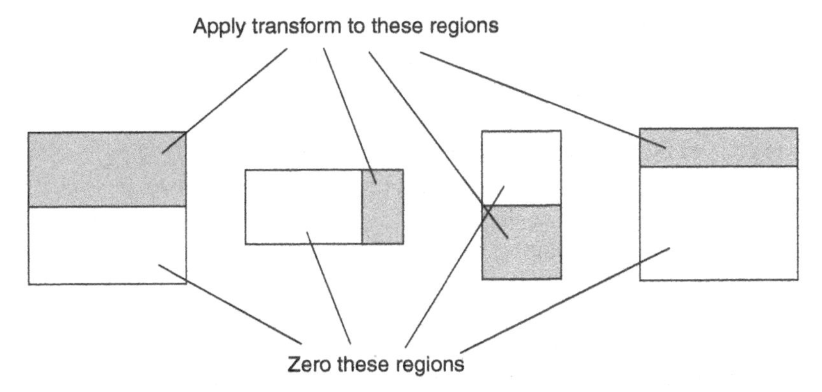

Figure 7.56 Sub-block transform examples in VVC

Quantisation weighting matrices can be defined using one of 28 Scaling Lists, which themselves can be signalled using the new Adaptation Parameter Set, discussed in Section 4.6. The QP range is extended to a maximum of 63, compared to HEVC's maximum of 51, which enables lower bitrate encoding than the previous standard.

7.8 Conclusions

This chapter has introduced the linked concepts of block transformation and quantisation in video coding, in which a block of pixels or residual samples is transformed into a spatial frequency domain and then quantised. This has the overall effect of converting a typical block in a video sequence into a sparse array of quantised coefficients, usually containing at most a few non-zero coefficients. With careful choice of transform and quantiser, this process retains important visual information whilst significantly reducing the amount of information in each block. This helps in the process of compressing the video data, at the expense of discarding some of the original visual information. Quantisation is one of the main levers that control the video compression process, providing a trade-off between lower and higher compression and corresponding higher and lower quality.

After transform and quantisation, we are left with a few non-zero coefficients in each block that must be encoded and sent as part of the compressed bitstream, together with all the parameters and other information needed by the decoder to decode and reconstruct the video sequence. Chapter 8 will look at entropy coding, which is the process of converting all of these values into a compressed video bitstream that can be stored or transmitted.

References

1 Lui, Y. and Ostermann, J. (2018). Scene-based KLT for intra coding in HEVC. *Proceedings of the Picture Coding Symposium*, San Francisco, CA, USA (24–27 June 2018). pp. 6–10.

2 Ahmed, N., Natarajan, T., and Rao, K.R. (1974). Discrete cosine transform. *IEEE Transactions on Computers*. 2 (1): 90–93. https://doi.org/10.1109/T-C.1974.223784.

3 Budagavi, M., Fuldseth, A., Bjøntegaard, G. et al. (2013). Core transform design in the high efficiency video coding (HEVC) Standard. *IEEE Journal of Selected Topics in Signal Processing.* 7 (6): 1029–1041. https://doi.org/10.1109/JSTSP.2013.2270429.

4 Fuldseth, A., Bjøntegaard, G., and Budagavi, M. (2011). CE10: Core transform design for HEVC, JCTVC-G495. *Joint Collaborative Team on Video Coding (JCT-VC) of ITU-T SG16 WP3 and ISO/IEC JTC1/SC29/WG11*, Geneva, CH (21–30 November 2011).

5 Yeo, C., Tan, Y.H., Li, Z. et al. (2010). Mode-dependent fast separable KLT for block-based intra coding. *IEEE International Symposium of Circuits and Systems (ISCAS).* pp. 621–624. https://doi.org/10.1109/ISCAS.2011.5937642.

6 ITU-T Recommendation H.265, Section 8.6.3, Equation 8-311 (2019).

7 Zhao, X., Kim, S.-H., Zhao, Y. et al. (2021). Transform coding in the VVC standard. *IEEE Transactions on Circuits and Systems for Video Technology.* 31 (10): 3878–3890. https://doi.org/10.1109/TCSVT.2021.3087706.

8 Schwarz, H., Coban, M., Karczewicz, M. et al. (2021). Quantization and entropy coding in the versatile video coding (VVC) standard. *IEEE Transactions on Circuits and Systems for Video Technology.* 31 (10): 3891–3906. https://doi.org/10.1109/TCSVT.2021.3072202.

8

Entropy Coding

8.1 Introduction

A video encoder produces many different symbols or syntax elements, containing information such as transform coefficients, prediction choices, block sizes and many other parameters. Some of the syntax elements occur many thousands of times in even a short video clip, while others occur occasionally or just once per video. Certain syntax elements such as transform coefficients can have a range of values, while others are binary. The actual value of a syntax element may be statistically related to other syntax element values and other information. We can take all of these factors into account when we are trying to find an efficient encoding for all the syntax elements in a coded video clip.

In this chapter, we will look at:

- Where entropy coding sits in a video codec and what it achieves.
- Pre-processing symbols before entropy coding.
- The main types of entropy coding used in video codecs.
- Arithmetic coding in detail.
- Probabilities, context models and adaptation.
- How entropy coding works in H.265/high-efficiency video coding (HEVC) and H.266/versatile video coding (VVC).

An entropy encoder converts a sequence of symbols or syntax elements, S1, S2, S3, etc., into a bitstream, and an entropy decoder converts the bitstream back into a sequence of symbols (see Figure 8.1). The entropy encoder and decoder can make use of certain information during encoding and decoding, such as:

- The expectation or probability that the current symbol S has value V, $P(S = V)$.
- Local statistics or context, e.g. what values have already been coded in the current region of the video frame or video sequence.
- The current entropy coder state (see Section 8.6).

Coding Video: A Practical Guide to HEVC and Beyond, First Edition. Iain E. Richardson.
© 2024 John Wiley & Sons Ltd. Published 2024 by John Wiley & Sons Ltd.
Companion website: www.wiley.com/go/richardson/codingvideo1

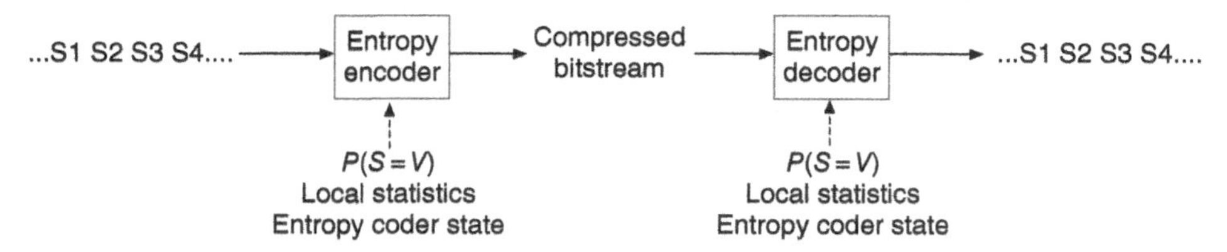

Figure 8.1 Entropy encoder and decoder

A video codec can increase the efficiency of entropy coding by:

- Pre-processing values before entropy coding, e.g.:
  - Scanning or processing quantised coefficients in a certain order, to capitalise on the expected statistical relationship of coefficients in a block.
  - Predicting a parameter value such as a motion vector or quantisation parameter (QP) from one or more previously coded values, and entropy coding the difference between the actual and predicted values.
  - Inheriting a parameter or value, such as a motion compensation mode and vector, from a previously coded block or region.
  - Creating a list of likely values and selecting from the list.
  - A combination of some of the above-mentioned points.
- Estimating the probability of a syntax element having a particular value, e.g. 0 or 1, based on how often it took that value in the past.
- Estimating the probability of a syntax element having a particular value, based on recent and/or surrounding statistics or context.
- Choosing an entropy coding scheme that matches the number of bits per symbol with the probability of that symbol value occurring.

We will look at two extracts from the HEVC/H.265 syntax and how they are entropy-coded using fixed-length codes, variable-length codes or binary arithmetic coding (BAC), all of which we will examine in detail later.

Example 1: Slice Segment Header

HEVC's slice segment header (see Table 8.1) occurs once per coded slice segment [1]. The header is made up of a series of syntax elements, each of which is a parameter describing some aspect of the coded bitstream. The parameter first_slice_segment_in_pic_flag is set to 1 when this is the first slice segment in a coded picture and set to 0 otherwise. Some of the parameters are conditional, so they may or may not appear in the bitstream depending on other parameters. The conditional decisions are not shown in Table 8.1.

The slice segment header is encoded using fixed-length and variable-length binary codes, shown as u(integer) and ue(v)/u(v), respectively. Flags such as pic_output_flag are coded as a u(1) value, which means a single bit, taking the value 0 or 1. Syntax element slice_segment_address is coded using a variable-length code u(v), and colour_plane_id is always coded using two bits, u(2). The syntax elements slice_pic_parameter_set_id and slice_type are coded using variable-length exp-Golomb codes, ue(v), a type of structured variable-length code that we will look at in Section 8.5.2.

Table 8.1 Extract from HEVC slice segment header syntax

Syntax element	Descriptor	Note
first_slice_segment_in_pic_flag	u(1)	Single-bit flag
...		
no_output_of_prior_pics_flag	u(1)	Single-bit flag
slice_pic_parameter_set_id	ue(v)	Variable-length code, exp-Golomb coded
...		
dependent_slice_segment_flag	u(1)	Single-bit flag
slice_segment_address	u(v)	Variable-length code, integer represented with variable number of bits
...		
slice_reserved_flag[i]	u(1)	Single-bit flag
slice_type	ue(v)	Variable-length code, exp-Golomb coded
...		
pic_output_flag	u(1)	Single-bit flag
...		
colour_plane_id	u(2)	Two-bit code
...		

... indicates conditional logic that is not shown.

Example 2: Motion Vector Difference

HEVC's motion vector difference (MVD) syntax (see Table 8.2) can be sent once for every motion vector that is coded in the bitstream, which means that it may be sent many thousands of times in a few seconds of video. A motion vector has an *x* component, here denoted [0], and a *y* component, here denoted [1]. Each component is a positive or negative integer, represented in the bitstream as a series of flags and values, indicating:

- abs_mvd_greater0_flag, indicating whether the component is non-zero.
- abs_mvd_greater1_flag, indicating whether the component is greater than +1 or less than −1.
- abs_mvd_minus2, indicating the absolute value minus 2.
- mvd_sign_flag, indicating the sign, + or −.

The first two flags are single-bit binary decisions. Syntax element abs_mvd_minus2 can take a range of integer values from 0 to the maximum vector magnitude minus 2. All the syntax elements are coded using BAC, represented by the descriptor ae(v). Because abs_mvd_minus2 has a range of possible values, it is converted into a binary codeword – binarised – before coding each bit of the codeword using BAC. We'll consider binarisation and BAC in Section 8.8.

Table 8.2 Extract from HEVC motion vector difference syntax

Syntax element	Descriptor	Note
abs_mvd_greater0_flag[0]	ae(v)	Single-bit flag, x component
abs_mvd_greater0_flag[1]	ae(v)	Single-bit flag, y component
...		
abs_mvd_greater1_flag[0]	ae(v)	Single-bit flag, x component
...		
abs_mvd_greater1_flag[1]	ae(v)	Single-bit flag, y component
...		
abs_mvd_minus2[0]	ae(v)	Positive integer value, x component
mvd_sign_flag[0]	ae(v)	Single-bit sign, x component
...		
abs_mvd_minus2[1]	ae(v)	Positive integer value, y component
mvd_sign_flag[1]	ae(v)	Single-bit sign, y component

... indicates conditional logic that is not shown.

8.2 Entropy Coding for Video Compression

The challenge of entropy coding is to efficiently encode every syntax element in a video sequence whilst satisfying several aims as far as possible, including:

- Maximising compression efficiency by using the smallest number of bits to represent the complete coded sequence.
- Minimising computational complexity, by coding and decoding within the capabilities of practical processors and devices. This might include, for example, ensuring that parallel processors can be used efficiently to handle encoding or decoding.
- Providing an efficient mapping to methods of storage or transmission, e.g. mapping suitable-sized chunks of coded data to network packets.

8.2.1 Entropy Coding to Maximise Compression Efficiency

Let's assume that we want to encode a symbol S that has one of a number of possible values, V. For example, symbol mvd_sign_flag[0] can have two values, $V=0$ or $V=1$. Symbol abs_mvd_minus2 can have multiple positive integer values of V.

The information content I of a symbol S with value V is given by [2]:

$$I = \log_2\left(\frac{1}{P(S=V)}\right)$$

$P(S = V)$ is the probability that symbol S has value V and the information content I is measured in bits. The smaller the value of $P(S = V)$, the less likely it is that the current symbol takes value V, the greater the information content of $S = V$, and hence, the greater the number of bits that may be required to code $S = V$.

Example:

Consider three symbols (a), (b) and (c) (see Table 8.3).

Symbol (a) is binary, and each value V is equally probable. The probability of each value is 0.5, i.e. $P(S = 0) = P(S = 1) = 0.5$. The information content of each possible value is 1 bit.

Symbol (b) is binary, this time with a 70% probability that $S = 1$ and a 30% probability that $S = 0$. In general, values that are less likely to occur have a higher information content. We see that $P(S = 0)$ has an information content of approximately 1.74 bits, whereas $P(S = 1)$ has an information content of approximately 0.51 bits.

Symbol (c) has four possible values, $V = 0, 1, 2,$ or 3, with varying probabilities. The information content of each option, $S = V$, changes depending on the probability of each value occurring.

We can consider how we might want to code symbol (c). If the value of (c) is 1, the most probable value, then we want to use 1 bit to represent (c). If the value of (c) is 3, the least probable value, then we want to use approximately 3.32 bits to represent (c).

The goal of an entropy coder is to represent S as efficiently as possible using a binary code. Ideally, the number of bits used to encode S should be as close as possible to the information content of S. For example (a) above, this would be satisfied by simply coding the value itself using a single bit. However, many symbols S have varying probabilities, such as examples (b) and (c). This leads to at least two challenges for an entropy coder:

i) Can we represent $S = V$ using a number of bits that are close to the information content I? For example, we might use an integral number of bits to represent symbol (b) above, e.g. 1 bit for $S = 1$, 2 bits for $S = 0$. Alternatively, certain entropy coding schemes such as arithmetic coding may be able to represent a single symbol S with a fractional number of bits, which may get us closer to the ideal information content of 0.51 or 1.74 bits.

Table 8.3 Example: symbols with multiple values

Symbols S	Value V	$P(S = V)$	$I(S = V)$, bits
(a) Binary symbol	0	0.5	1
	1	0.5	1
(b) Binary symbol	0	0.3	1.74
	1	0.7	0.51
(c) Non-binary symbol	0	0.2	2.32
	1	0.5	1
	2	0.2	2.32
	3	0.1	3.32

ii) Can we estimate $P(S = V)$ accurately? In a practical encoding scenario, we do not usually know the exact probability of each value V. We can use information such as past statistics and neighbouring values to estimate $P(S = V)$. The more accurate the estimate, the more efficient our entropy coding will become.

Representing each symbol with a code that is close to $I(S = V)$ bits in length and estimating or modelling the probability of each value $P(S = V)$ are two desirable properties of a practical entropy coder. In addition, the encoder and decoder should be computationally efficient and capable of starting and stopping/resynchronising at appropriate intervals.

8.2.2 Types of Entropy Coding

Let us consider a number of entropy coding schemes used in practical video codecs (see Figure 8.2). Note that these are not necessarily mutually exclusive. For example, HEVC uses a combination of fixed-length coding, variable-length coding and BAC for different sections of the coded syntax.

8.2.2.1 Fixed-Length Coding

Each symbol S is mapped to a fixed length codeword, i.e. the number of bits used to represent S does not change. In HEVC's Sequence Parameter Set, symbol sps_seq_parameter_set_id is an identifier parameter that can be in the range 0–15 and is always encoded with a 4-bit binary code. Fixed-length coding does not compress the data. It may be suitable for values such as sps_seq_parameter_set_id that only occur infrequently – in this case, once per coded sequence – or values for which the probability of each possible value is the same.

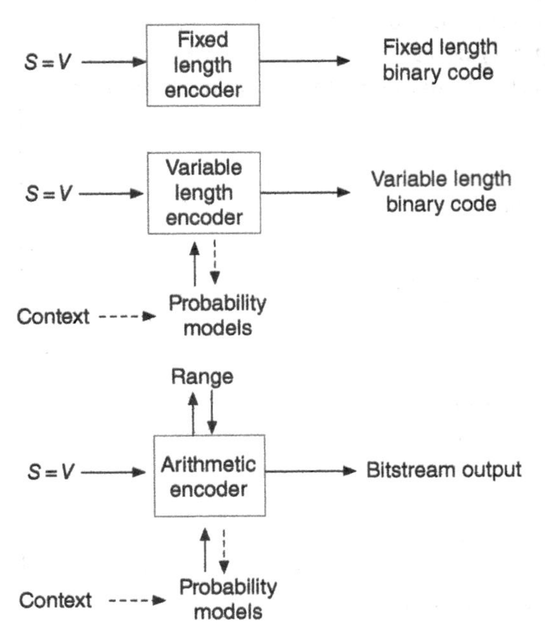

Figure 8.2 Fixed length, variable-length and arithmetic coding

8.2.2.2 Variable-Length Coding

Each symbol S is mapped to a variable-length codeword (VLC), which means that the number of bits used to represent S changes depending on the value V. For example, an exp-Golomb codeword has a variable number of bits depending on the value to be coded (see Section 8.5.2). The VLCs can be constructed based on a probability model, i.e. based on an estimate of the probability of each value V. This estimate may be fixed, as in the pre-defined tables of variable-length codes in earlier standards such as MPEG-2 Video, or it may change depending on the actual frequency of occurrence of each value V. The choice of probability model for encoding S may depend on context, i.e. on statistics within the local area of the coded sequence.

8.2.2.3 Arithmetic Coding

When an arithmetic coder encodes a symbol S, it causes a change to internal variables, namely a range and a position (see Figure 8.2). The position variable is a pointer that indicates a position within the range. As each symbol is coded, the actual value of the symbol V and the probability that the symbol is V cause a change in these internal variables. The position variable is communicated to the decoder as the output bitstream. As we will see later, it is possible for an arithmetic coder to approach the ideal fractional number of bits, which is the information content of each symbol. As with variable-length coding, the probability estimate or probability model may be dynamically updated based on previously coded statistics, and the choice of probability model may depend on context.

8.2.2.4 Binary Arithmetic Coding

A binary arithmetic coder is an arithmetic coder that only handles binary input symbols, i.e. the input symbol can only have the values 0 or 1 (Figure 8.3). This means that a non-binary symbol S must be converted to a string of binary values or bins, using a process known as binarisation. The binary arithmetic coder works in a similar way to the arithmetic coder described above. Each bit or bin is coded using a probability model that may be dynamically updated and based on context. In addition, the way in which a non-binary symbol is binarised may also depend on context.

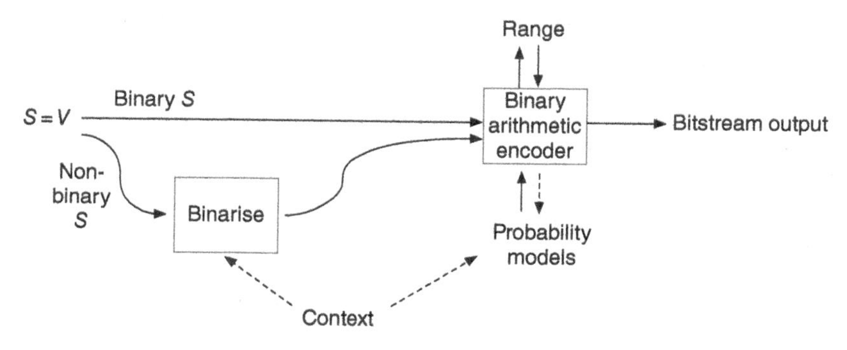

Figure 8.3 Binary arithmetic coding

8.2.3 Designing an Entropy Coder

An entropy coder for video compression should provide efficient compression. At the same time, it is important that the encoder and decoder do not take too much processing power. It is also important to handle practical issues such as synchronisation when a client starts decoding a video clip. A practical entropy coder for video coding, such as the entropy coders specified in standards H.264 and H.265, tends to involve trade-offs between some or all of the following issues.

8.2.3.1 Compression Efficiency

Variable-length coders and arithmetic coders make use of a probability $P(S = V)$, i.e. the probability that symbol S has value V. Accurately estimating $P(S = V)$ can improve compression performance. The probability estimate or probability model can be updated based on previously coded values of S. For example, if the actual coded symbol S is a 1, increase the probability $P(S = 1)$ so that the next time S is coded, the estimated probability of 1 is slightly higher.

Increasing the number of probability models, such as modelling $P(S = V)$ under different local conditions or contexts, may help to accurately model the probability of each symbol value. However, increasing the number of probability models may make it harder to build up accurate statistics of previously coded values of S, since each condition or context will occur less often.

Matching the coded bitrate to the information content of the coded symbols can improve compression performance. This is one of the reasons that arithmetic coding can potentially outperform variable-length coding since arithmetic coding has the capability of outputting a fractional number of bits per input symbol.

8.2.3.2 Computational Complexity

The total number of processing operations may be particularly influenced by frequently occurring syntax elements. If most of the actual syntax elements in a coded video sequence are quantised coefficients, then it is important to minimise the complexity of coding these values.

Most modern-day processors have the capability to pipeline and/or parallelise computational processes. In order to exploit this capability, it may be necessary to address certain dependencies in the coding process. For example, if operation B depends on operation A, then B and A cannot be processed in parallel.

8.2.3.3 Synchronisation

A decoder must be able to start decoding at certain defined points, e.g. at the start of a video sequence or at a key frame or intra-coded frame. It may be advantageous to make certain syntax elements such as higher-level syntax elements easy to decode quickly.

For example, the higher-level syntax in the HEVC standard is coded using fixed-length or variable-length codes, which makes it relatively easy for a decoder to quickly find a suitable starting point for decoding. In contrast, HEVC's lower-level syntax is coded using arithmetic coding, which can give better compression but is harder to jump into, in the middle of a coded sequence.

Both encoder and decoder must use the same probability estimates or probability models, which implies that the probability models must be reset at a synchronisation point, such as a point at which a decoder starts decoding a video frame. However, resetting the probability estimates too often may reduce compression efficiency.

8.3 Pre-processing

We have seen throughout the book that video codecs can make use of the relationships between nearby pixels, blocks and frames to compress video data. Entropy coding efficiency can be maximised by organising certain values produced by a video encoder, before processing them with an entropy encoder. Values such as transform coefficients and prediction parameters may be pre-processed prior to entropy coding, for example by:

- Scanning coefficients in a block to group together non-zero or similar values.
- Differentially predicting motion vectors before coding the difference.
- Grouping and coding certain features in multiple passes, such as sign bits.

Pre-processing can help to minimise the information to be encoded, to maximise the efficiency of context-adaptive entropy coding and to improve computational performance, for example making it easier to pipeline or parallelise entropy coding.

8.3.1 Pre-processing Examples

Example 1: Re-ordering

Figure 8.4 shows a block of transform coefficients with the coefficients arranged in a raster scan order, from the top left to the lower right. Re-ordering or scanning the block in a different order, e.g. reverse-diagonal as shown here, may help to group together coefficients in a way that can improve entropy coding efficiency.

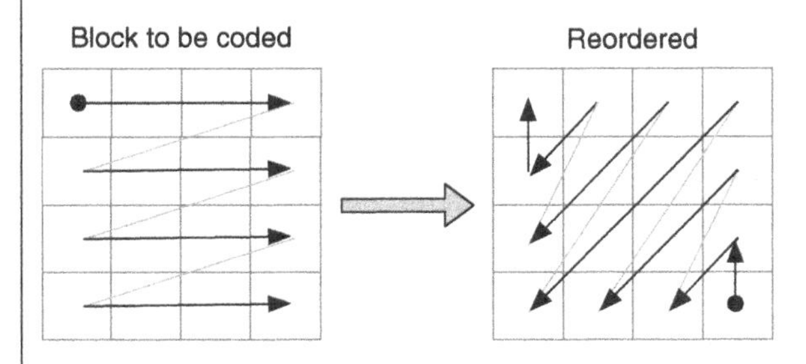

Figure 8.4 Pre-processing: re-ordering

Example 2: Coding in Passes

The block on the left of Figure 8.5 contains a mixture of zeros, positive and negative values, with several ±1 values. This is typical of a transformed and quantised image or video block. Coding the block in passes or stages can help to isolate and group certain types of information. For example, the first pass here identifies the zero and non-zero values with a 1 or a 0, producing a map of non-zero values, or a significance map. The second pass identifies the sign of each non-zero value. Note that by coding the significance map first, only 11 sign values need to be coded, one per non-zero coefficient. The third pass identifies which of the 11 non-zero values has a magnitude greater than 1. The fourth pass indicates the magnitude of the 6 values that remain.

A particular type of information is coded in each pass, e.g. non-zero or zero and sign. All the information in a pass may have similar statistics, which can improve the performance of entropy coding. Further, the amount of information and the range of information in each pass are limited. For example, the fourth pass only contains positive values that are greater than 1. This can also help the performance of entropy coding.

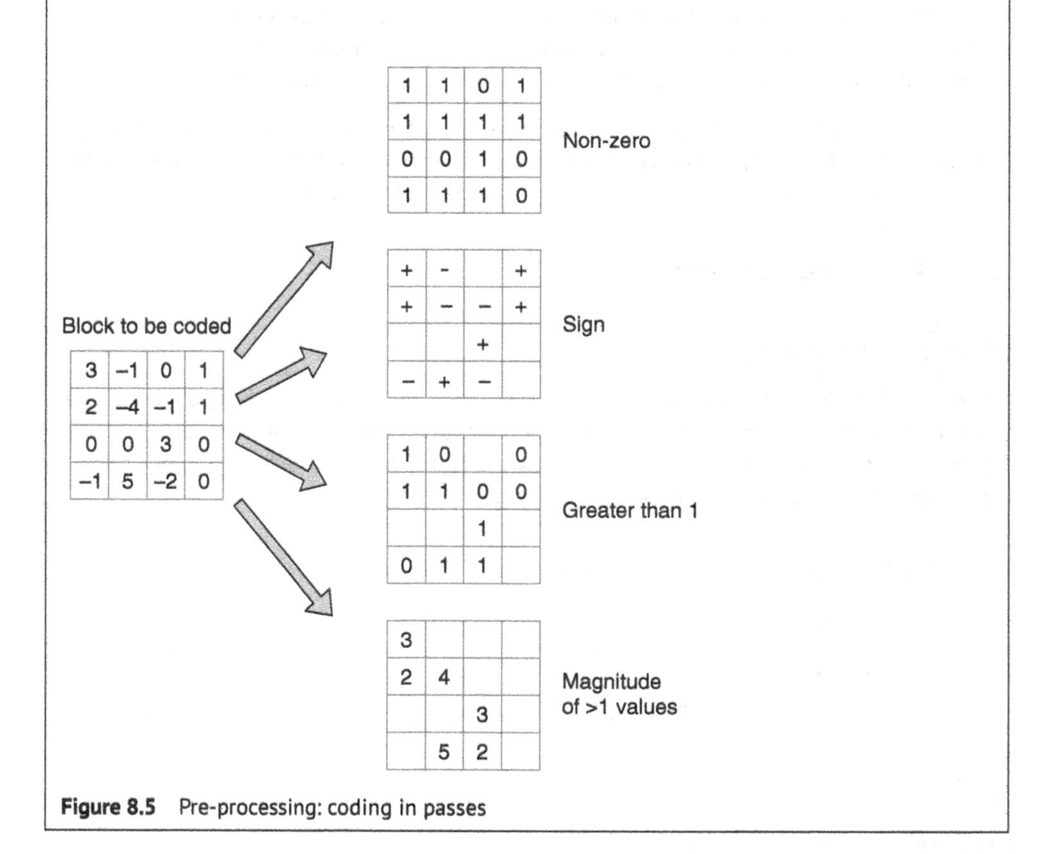

Figure 8.5 Pre-processing: coding in passes

Example 3: Predicting from Context

In Figure 8.6, a block value to be coded is surrounded by previously coded blocks, labelled 'context'. The value for the current block, e.g. a motion vector, coefficient and QP, is likely to be correlated with neighbouring block values. It is often possible to reduce the information in the current block value by predicting it from the context and coding just the difference. This approach has been used in some form since the early video coding standards.

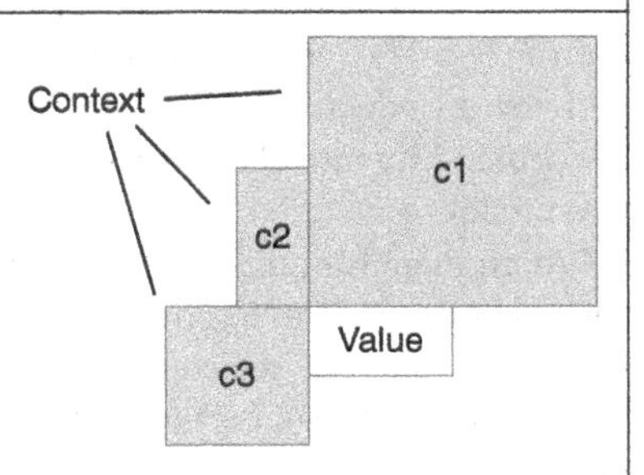

Figure 8.6 Pre-processing: predicting from context

Chapters 5 and 6 discuss the pre-processing of intra and inter prediction parameters in the HEVC standard, and we will look at the pre-processing of HEVC transform coefficients in Section 8.9.1.

8.4 Probability Models and Context Adaptation

An entropy encoder can estimate or model the probability that symbol, S, has a particular value, V. For example, if S is binary, a probability model estimates the probability that S is 0 or 1, i.e. $P(S = 0)$, $P(S = 1)$.

Figure 8.7 shows the general idea. A probability model or context model, shaded in the figure, maintains an estimate of $P(S = 0)$. Of course, $P(S = 0) + P(S = 1) = 1$. The entropy coder uses this estimate during the encoding of the actual value of S.

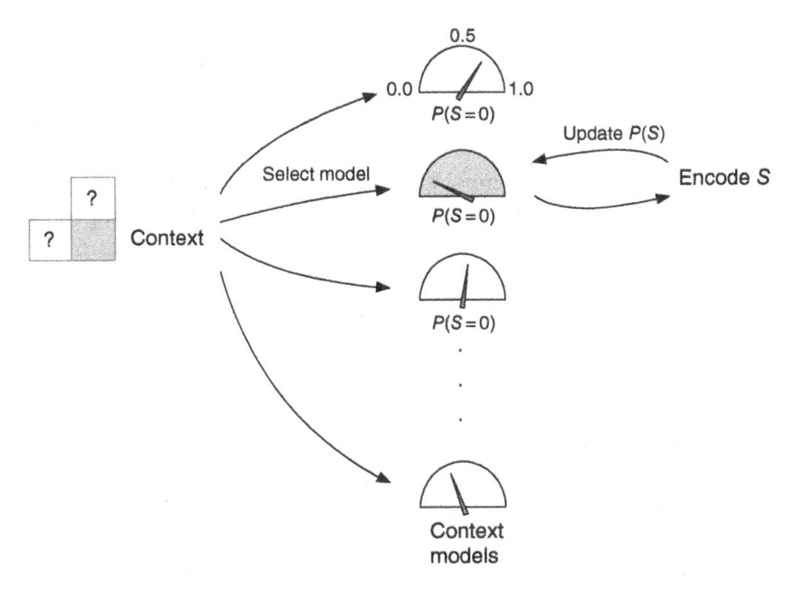

Figure 8.7 Context-adaptive probability models (binary symbol S)

The actual value of S is 0 or 1. If the actual value of S is 0, then the probability $P(S = 0)$ increases slightly, since we have coded one more value of $S = 0$. If the actual value of S is 1, then $P(S = 0)$ decreases slightly. We can update the probability model after encoding S, increasing or decreasing $P(S = 0)$ depending on whether a 0 or a 1 was encoded. In this way, the probability model adjusts $P(S = 0)$ based on the actual encoding statistics. It is an adaptive model.

Which model should we choose for encoding S? We could simply maintain one model for each syntax element, e.g. one model for abs_mv_greater0_flag and another model for mvd_sign. Alternatively, we could maintain multiple models for certain syntax elements and select a model based on the particular syntax element (e.g. mvd_sign) and/or what values have been coded in the local area of the video frame (the context). If all the recently coded blocks have a positive MVD (mvd_sign = +), then this may increase the probability that the current mvd_sign will also be positive.

Figure 8.7 illustrates how context-adaptive probability modelling of a binary symbol S works. The encoder uses recently coded statistics, the context, to select a probability model or context model, from among a number of possible models. The selected model provides an estimate $P(S = 0)$. The encoder uses this estimate to encode the current symbol, S. Depending on the actual value of S (0 or 1), the encoder updates the chosen context model. If the coded symbol was 0, then $P(S = 0)$ increases slightly and vice versa.

How many probability models should we maintain? This is a trade-off between several factors. If we have more probability models, we can, in theory, track the probability of a particular symbol value more accurately. On the other hand, more probability models mean that each individual model adapts more slowly since the actual encoding of a symbol using that model occurs less often. Each probability model requires storing and updating variables, so more models may increase storage and computational complexity.

In Section 8.8, we will look at probability modelling and context adaptation in context-adaptive binary arithmetic coding (CABAC).

8.5 Variable-Length Coding

A variable-length coding scheme maps each symbol value to a binary codeword with a variable number of bits. Preferably, symbol values that are more likely to occur are mapped to shorter codewords, and symbol values that are less common are mapped to longer codewords. The mapping for a particular symbol value to a particular VLC should be unique, and the codeword should be uniquely decodable. Consider the examples in Table 8.4. Each symbol value in the first column is mapped to a VLC, e.g. the codewords in Table (a) in the second column.

Table (a) fits the criterion that smaller symbol values are more probable. Smaller symbol values are mapped to shorter VLCs. Each VLC is unique and uniquely decodable. This means that when a decoder starts to decode each VLC, it can unambiguously work out which symbol value to decode. For example, if the decoder encounters 0, it decodes value 0 and stops processing this codeword. If it receives 1, 1, 0, 0, then it decodes value 3 and knows it is finished. Notice that Table (a) grows rapidly but also has a very short codeword for symbol value 0. This table might be suitable for a scenario where value 0 is highly probable.

Table 8.4 Variable-length code table examples

Symbol value	Table (a)	Table (b)	Table (c) (*Not valid!*)
0	0	00	0
1	100	010	*00*
2	101	011	10
3	1100	100	11
4	1101	110	*110*
5	11100	101	...
6	11101	1110	...
7	111100	11110	...
...	...	...	...

Table (b) also fits the criterion that smaller symbol values are more probable – every VLC is unique and can be unambiguously decoded. Notice that this table grows more slowly in size but has a longer codeword for the most likely value 0. This table might be more appropriate for a scenario where value 0 is slightly more probable than value 1, etc.

Table (c) would not work as a conventional VLC table because some symbol values are not unambiguously decodable. For example, a decoder encountering (0, 0) cannot work out whether this is the sequence of values 0, 0 or the single value 1. A decoder encountering (1, 1, 0) does not know whether to decode two values (3, 0) or to decode a single value 4. The problem occurs because some of the codewords (0, 11) are prefixes of other codewords (00, 110, shown in italics). Hence, a conventional VLC table has no codewords that are prefixes of other codewords in the same table.

We will look at a few practical VLC schemes that are used in video coding.

8.5.1 Unary Coding

A unary variable-length code consists of a string of 1s terminated with a 0, or a string of 0s terminated with a 1, such as the examples given in Table 8.5. Smaller symbol values are mapped to shorter VLCs, so this coding method may be suitable for a symbol where smaller values are more probable than larger values. Each codeword can be uniquely decoded. Encoding and decoding can be carried out with a simple algorithm. In the mapping of Table 8.5, the encoder converts a symbol value N into a string of N 1s followed by a 0. The decoder reads 1s until it encounters a 0 and decodes a value equal to the number of 1s. A potential disadvantage of unary VLCs is that the codeword length increases in direct proportion to the value, leading to large codewords for higher values. For example, a symbol value of 100 maps to a VLC consisting of 100 1s followed by a 0, i.e. a codeword that is 101 bits long.

8.5.2 Exponential Golomb Coding

Exponential-Golomb or Exp-Golomb coding is another coding scheme that is constructed according to a regular pattern. Table 8.6 shows an example of a mapping of symbol values to Exp-Golomb codewords. In this table, symbol value 0 is mapped to binary 0. For all

Table 8.5 Unary VLCs

Symbol value	Unary VLC
0	0
1	10
2	110
3	1110
4	11110
5	111110
6	1111110
7	11111110
...	...

Table 8.6 Exp-Golomb codewords: example

Symbol value	Exp-Golomb VLC
0	0
1	10**0**
2	10**1**
3	110**00**
4	110**01**
5	110**10**
6	110**11**
7	1110**000**
...	...

values >1, the Exp-Golomb VLC is constructed as a prefix and a suffix. The prefix here is a unary code, a series of M 1s followed by a 0[1]. The suffix is an M-bit binary number, shown here in bold type.

Once again, smaller symbol values are mapped to shorter VLCs. Each codeword is uniquely decodable. Encoding and decoding are slightly more complex than for unary codewords, but they can still be carried out algorithmically. An Exp-Golomb VLC can be decoded by reading consecutive 1s until a 0 is detected. The number of ones gives us M. The decoder then reads M further bits and constructs the symbol value.

Compared to unary codewords, Exp-Golomb codewords increase in size more gradually, increasing approximately with $\log_2$ of the symbol value. For example, using the scheme of Table 8.6 a value of 100 maps to the Exp-Golomb codeword 1111110100101, with a length of 13 bits.

1 Alternatively, the prefix may be a series of M 0s followed by a 1.

Table 8.7 Motion vector probability example

Motion vector value V	Probability $P(V)$	$\text{Log}_2(1/P(V))$
−1.5	0.014	6.16 bits
−1	0.024	5.38 bits
−0.5	0.117	3.10 bits
0	0.646	0.63 bits
0.5	0.101	3.31 bits
1	0.027	5.21 bits
1.5	0.016	5.97 bits

8.5.3 Huffman Coding

Huffman coding is a method of constructing a variable-length code table for a symbol, based on the probability of occurrence of each possible value. For example, Table 8.7 lists seven possible values V of a motion vector component. Value 0 is the most probable, and value −1.5 is the least probable. The information content of each value is shown in the right-hand column. Value 0 is very likely and therefore has the lowest information content. If we know the probabilities of occurrence of each value, we can apply the Huffman algorithm to construct a set of binary VLCs. The result is shown in Table 8.8. The most-probable value 0 is assigned the shortest codeword, a binary 1. The next most probable value, −0.5, is assigned a longer codeword, binary 00, and so on. A few important points about the resulting set of VLCs are shown in Table 8.8:

- Each codeword is uniquely decodable. It is not a prefix of any other codeword.
- The size of each codeword increases approximately with the information content.
- Huffman codewords do not usually exactly match the information content of each value, since Huffman VLCs, in common with other VLCs, are always an integer number of bits.
- Constructing the VLC table requires knowledge of the probability of occurrence of each value. If these probabilities $P(V)$ change, then the resulting VLC table changes.

Table 8.8 Huffman coded values

Value V	Huffman code	Bits (actual)	Bits (ideal)
0	1	1	0.63
−0.5	00	2	3.1
0.5	011	3	3.31
−1.5	01000	5	6.16
1.5	01001	5	5.97
−1	01010	5	5.38
1	01011	5	5.21

8.5.4 Precalculated VLC Tables

In practice, the probability of a symbol having a particular value, such as a motion vector having the values shown in Table 8.7, depends very much on the content being encoded. In a video frame containing a lot of movement, we might expect the probability of zero vectors, i.e. $MV = 0$, to be lower than in a video frame without much movement. One solution to this problem is to train the entropy coder with expected probabilities, based on a large set of representative video clips, and then to specify the codewords in a standard, by defining pre-calculated VLCs. This approach was taken in early video coding standards such as MPEG-1 video, MPEG-2 video and H.263. In these standards, each symbol is encoded with a pre-calculated, specified set of VLCs or fixed-length codes. Each VLC table is constructed based on a pre-determined estimate of probabilities. Table 8.9 shows the first few VLCs specified in the H.263 standard for the syntax element MVD.

8.5.5 Context-Adaptive VLC

Fixed tables of VLCs, such as Table 8.9, are not optimal for every video sequence, since the actual probability of each value will vary depending on scene content. One solution to this problem is context-adaptive variable-length coding (CAVLC), in which the mapping of symbols to VLCs depends at least partly on recent coding statistics.

Figure 8.8 is an overview of CAVLC encoding in the H.264 AVC (advanced video coding) standard. For certain syntax elements, the encoder chooses from among a set of pre-defined VLC tables, based on local statistics. For example, syntax element coeff_token is coded using one of four VLC tables, each containing a different set of VLCs, depending on how many non-zero coefficients there were in recently coded blocks. If neighbouring, previously coded blocks contained a small number of non-zero coefficients, the assumption is that the current block is likely to contain a small number of non-zero coefficients, and the encoder chooses a VLC table biased towards efficient coding of small numbers of non-zero coefficients in the current block. If neighbouring blocks contain a large number of non-zero coefficients, the encoder chooses a VLC table that is more efficient for coding larger numbers of non-zero coefficients.

Table 8.9 H.263 MVD variable-length codes (extract)

MVD	Variable-length code
0	1
0.5	010
−0.5	011
1	0010
−1	0011
1.5	00010
−1.5	00011
…	…

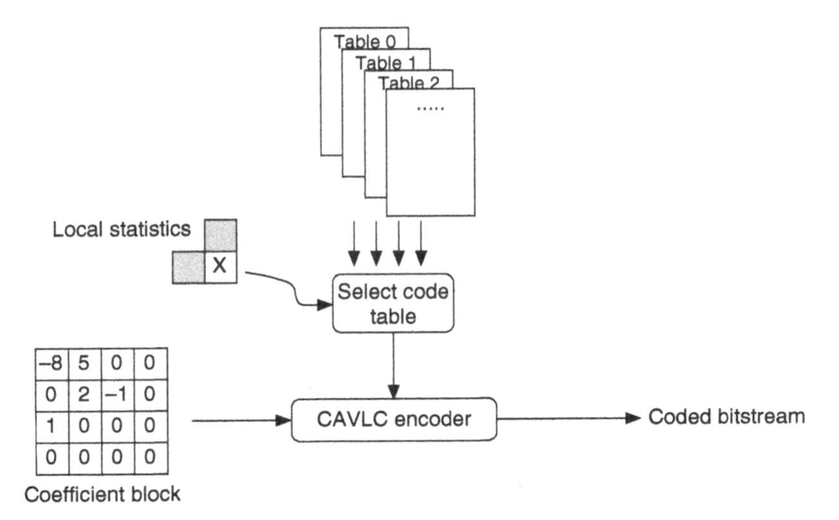

Figure 8.8 H.264 context-adaptive variable-length encoding

8.6 Arithmetic Coding

The variable-length coding schemes described in Section 8.5 share a disadvantage, namely that VLCs with integral lengths are likely to be sub-optimal since information content can be a fractional number of bits. The compression efficiency of variable-length codes can be particularly poor for symbols with probabilities greater than 0.5, as the best that can be achieved is to represent these symbols with a single-bit code. See the motion vector value 0 in Table 8.8, which has an information content of 0.63 bits but is mapped to a VLC with a length of 1 bit.

Arithmetic coding is an alternative to Huffman coding that can more closely approach theoretical maximum compression ratios [3]. An arithmetic encoder converts a sequence of data symbols into a single fractional number and can approach the optimal fractional number of bits required to represent each symbol.

We will now consider an example of arithmetic coding.

Table 8.10 lists five motion vector values (−2, −1, 0, 1 and 2) and their probabilities[2]. Each vector is assigned a sub-range within the range of 0.0–1.0, depending on its probability of occurrence. In this example, value (−2) has a probability of 0.1 and is assigned the sub-range 0–0.1, i.e. the first 10% of the total range 0–1.0. Value (−1) has a probability of 0.2 and is assigned the next 20% of the total range, i.e. the sub-range 0.1–0.3. After assigning a sub-range to each vector, the total range 0–1.0 has been divided amongst the data symbols – the vectors – according to their probabilities.

The procedure for encoding a sequence of *MV* values is as follows:

1) Set an initial range.
2) Based on the first syntax element, pick a sub-range corresponding to the value of the syntax element.

2 In this example, we assume that these are the only possible motion vector values.

Table 8.10 Arithmetic coding of motion vectors: probabilities and sub-ranges

Vector	Probability	$\log_2(1/P)$	Sub-range
−2	0.1	3.32	0–0.1
−1	0.2	2.32	0.1–0.3
0	0.4	1.32	0.3–0.7
1	0.2	2.32	0.7–0.9
2	0.1	3.32	0.9–1.0

3) Starting with this sub-range, pick a smaller sub-range corresponding to the value of the second syntax element.
4) Repeat for all the syntax elements.
5) The result is a narrow sub-range that can only be reached with this exact sequence of values. We can encode the result by sending a binary number that sits anywhere within the final sub-range.

Example: Encoding Procedure for Motion Vector Sequence (0, −1, 0, 2)

Encoding step	Range (L → H)	Symbol value	Sub-range (L → H)	Notes
1) Set the initial range	0 → 1.0			
2) For the first syntax element, find the corresponding sub-range (low to high)		(0)	0.3 → 0.7	
3) Set the new range (1) to this sub-range	0.3 → 0.7			
4) For the next data symbol, find the sub-range L to H		(−1)	0.1 → 0.3	This is the sub-range within the interval 0–1
5) Set the new range (2) to this sub-range within the previous range	0.34 → 0.42			0.34 is 10% of the range; 0.42 is 30% of the range
6) Find the next sub-range		(0)	0.3 → 0.7	
7) Set the new range (3) within the previous range	0.364 → 0.396			0.364 is 30% of the range; 0.396 is 70% of the range
8) Find the next sub-range		(2)	0.9 → 1.0	
9) Set the new range (4) within the previous range	0.3928 → 0.396			0.3928 is 90% of the range; 0.396 is 100% of the range

Each time a symbol is encoded, the range (L to H) becomes progressively smaller. At the end of the encoding process, we are left with a final range (L to H). The entire sequence of data symbols, four in this case, can be represented by transmitting any fractional number that lies within this final range. In the example mentioned above, we could send any number in the range of 0.3928–0.396, e.g. 0.394.

The initial range (0, 1) is progressively partitioned into smaller ranges as each data symbol is processed. After encoding the first symbol, value 0, the new range is (0.3, 0.7). The next symbol, value −1, determines the sub-range (0.34, 0.42) that becomes the new range, and so on. The final symbol, value +2, determines the sub-range (0.3928, 0.396), and the number 0.394 falling within this range is transmitted. The value 0.394 can be represented as a fixed-point fractional number using 9 bits, so our data sequence (0, −1, 0, 2) is compressed to a 9-bit quantity.

The decoder receives a pointer that lies within the final range; in this example, the binary fraction that represents 0.394. We can decode each of the four syntax elements by identifying the value corresponding to each of the relevant sub-ranges. For example, the pointer 0.394 is in the range 0.3–0.7, so the first value has to be (0).

Decoding procedure

Decoding procedure	Range	Sub-range	Decoded symbol value
1) Set the initial range	0 → 1		
2) Find the sub-range in which the received number falls. This indicates the first data symbol value.		0.3 → 0.7	(0)
3) Set the new range (1) to this sub-range	0.3 → 0.7		
4) Find the sub-range of the new range in which the received number falls. This indicates the second data symbol.		0.34 → 0.42	(−1)
5) Set the new range (2) to this sub-range within the previous range	0.34 → 0.42		
6) Find the sub-range in which the received number falls and decode the third data symbol		0.364 → 0.396	(0)
7) Set the new range (3) to this sub-range within the previous range	0.364 → 0.396		
8) Find the sub-range in which the received number falls and decode the fourth data symbol		0.3928 → 0.396	(2)

A key advantage of arithmetic coding is that the transmitted number, 0.394 in this case, which may be represented as a fixed-point number with sufficient accuracy using 9 bits, is not constrained to an integral number of bits for each transmitted data symbol. To achieve optimal compression, the sequence of data symbols should be represented with:

$$\log_2(1/P_0)+\log_2(1/P_{-1})+\log_2(1/P_0)+\log_2(1/P_2)\text{bits}=8.28\,\text{bits}$$

In this example, arithmetic coding lets us represent the sequence using 9 bits, which is close to optimum. A scheme using an integral number of bits for each data symbol such as Huffman coding is unlikely to come so close to the optimum number of bits, and so in theory, arithmetic coding can outperform VLC schemes such as Huffman coding.

In practice, arithmetic coding can be computationally intensive. Converting all the input values to binary numbers and coding each bit using arithmetic coding, so-called BAC, has certain computational advantages.

8.7 Binary Arithmetic Coding

A BAC encoder represents a series of bins, i.e. binary values to be encoded, as a binary fraction that lies somewhere within a Range. Each successive bin value makes the Range progressively narrower. The decoder can decode the bin values by identifying the Range pointed to by the binary fraction.

Each bin is associated with a probability model or context model that estimates the probability that the next bin value is a 0 or a 1, $P(0)$ and $P(1)$. Coding a bin involves sub-dividing a Range, in proportion to the probability estimates $P(0)$ and $P(1)$.

In this section, we will look at the workings of a BAC coder in detail. Note that certain BAC implementations such as the HEVC BAC do not function exactly this way, but the general concepts will be helpful for understanding specific implementations.

We will consider:

- Encoding and decoding bins
- Representing fractions as binary numbers
- Sending and receiving bits and renormalising

We will use the following example to illustrate each of these points:

Example:

A BAC encoder codes three bins, b0, b1 and b2. Each bin has a context model or probability model that estimates the probability of the bin being 0 or 1, as shown in Table 8.11. The BAC engine codes each bin and updates a range variable after each bin. A fractional value, pointing to a position within the final range, represents the complete sequence of coded bins, Figure 8.9.

Table 8.11 Example: bins and probability estimates

Bin	*P*(0)	*P*(1)
b0	0.3	0.7
b1	0.1	0.9
b2	0.5	0.5

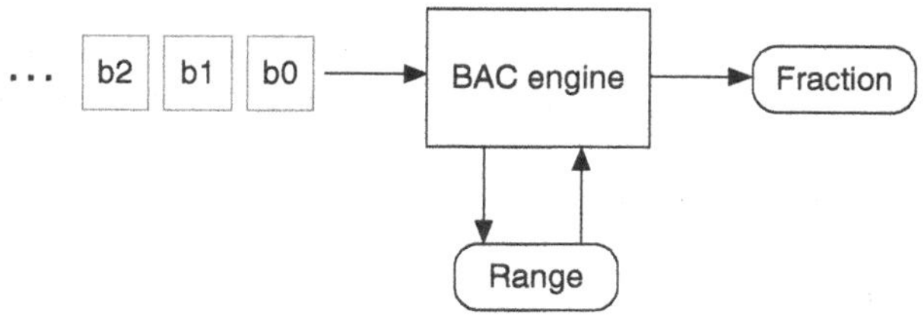

Figure 8.9 BAC engine coding a series of bins

8.7.1 Binarisation

A non-binary parameter to be encoded, such as a motion vector value, is first binarised, i.e. converted into a binary representation, before coding with the BAC engine. Each bit of the binary representation is a bin. In Table 8.12, a parameter with multiple integer values is mapped to a unary binary codeword (see Section 8.5.1). The first bit of this codeword is bin 1, the second, if present, is bin 2, and so on.

Each of the bins for this value, bins 1–9 shown here, is coded using the binary arithmetic coder. Each bin has an associated context model or probability model that is either unique

Table 8.12 Unary binarisation example

Parameter value	Binarisation								
0	0								
1	1	0							
2	1	1	0						
3	1	1	1	0					
4	1	1	1	1	0				
5	1	1	1	1	1	0			
6	1	1	1	1	1	1	0		
7	1	1	1	1	1	1	1	0	
8	1	1	1	1	1	1	1	1	0
Bin number	**1**	**2**	**3**	**4**	**5**	**6**	**7**	**8**	**9**

to that bin or shared with other bins. For example, bin 0 might have its own context model, and bins 2–9 might share a single context model.

In the example of Table 8.12, bin 1 is 0 if the parameter value is 0, 1 for all other parameter values. Bin 2 is 0 if the parameter value is 1, 1 for all parameter values greater than 1, and so on.

8.7.2 Encoding a Bin

Coding each bin b consists of the following steps (see Figure 8.10), which are essentially the same as the general arithmetic coding process discussed earlier:

1) Start with an initial range, i.e. a variable range with a minimum and a maximum value.
2) Determine two sub-divisions, each proportional to the probability that the value of the bin is 0 or 1.
3) Choose the sub-division that matches the actual bin value, e.g. the lower sub-division for $b = 0$, and the upper sub-division for $b = 1$.
4) Set this sub-division as the new range.
5) Repeat for the next bin.

The result is that the range gets narrower and narrower as each bin is coded. The final range can *only* be reached with the exact sequence of bins that were coded, i.e. the final range completely represents the sequence of bins.

In general:

- A less probable bin or sequence of bins results in a narrower range.
- A more probable bin or sequence of bins results in a wider range.
- It takes more bits to code a pointer to a narrower range than a wider range (see Section 8.7.5).
- A narrower range has more information content than a wider range.

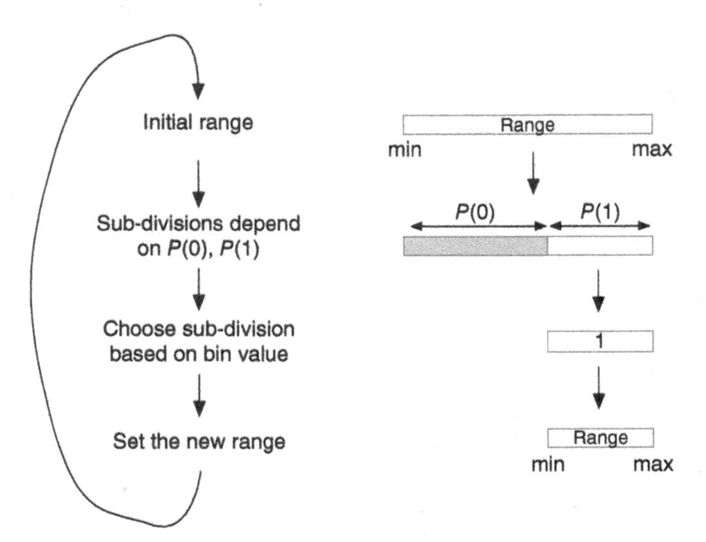

Figure 8.10 Coding a single bin

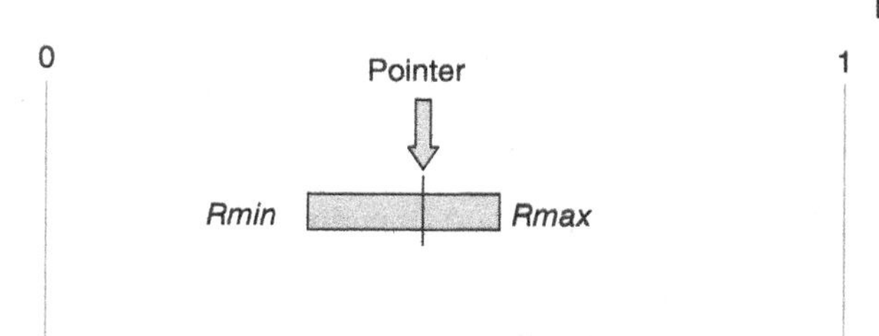

Figure 8.11 BAC: pointer to range

The encoder identifies a fractional number that points somewhere in the final range. It does not matter what the actual fraction is, as long as it is within the final range, and so the encoder will choose the binary fraction with the smallest number of bits that point within the final range. In Figure 8.11, the final range is Rmin to Rmax. The encoder chooses a binary fraction that points within this range and sends it to the decoder. The decoder uses the fraction to identify the range and decode the bins.

We will look later at how the encoder sends the binary fraction to the decoder. As we will see, the encoder can start to send the most significant bits (MSBs) of the fraction to the decoder after it has processed a certain number of bins, without waiting for the entire bin sequence.

8.7.3 Decoding a Bin

Conceptually, a BAC decoder decodes each bin b as follows (see Figure 8.12):

1) Start with an initial range and the fraction or pointer.
2) Determine the two sub-divisions, each proportional to the probability that the value of this bin is 0 or 1.
3) Choose the sub-division that contains the pointer – in this case, the upper sub-division.
4) Set this sub-division as the new range and output the bin value, 1 in this case.
5) Repeat for the next bin.

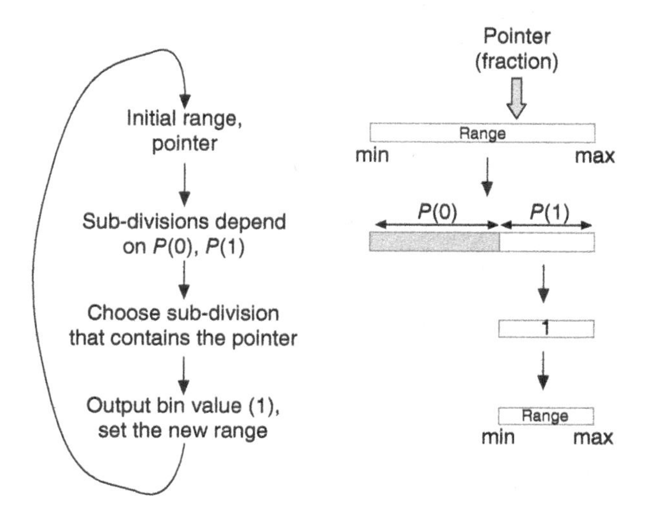

Figure 8.12 Decoding a single bin

In practice, as we will see later, the decoder can receive the binary fraction as a long series of bits and can start decoding and outputting bins before it receives the entire fraction.

8.7.4 BAC Encoding and Decoding: Example

Let us consider our three bins, b0, b1 and b2, with probabilities as before. When the BAC engine codes each bin, it calculates two sub-divisions of the range based on the probabilities $P(0)$ and $P(1)$. As Figure 8.13 shows, the sub-divisions are in proportion to $P(0)$ and $P(1)$. For example, the first bin b0 has $P(0) = 0.3$ and the range is sub-divided into the first 30% and the remaining 70%. If we code a zero in bin b0, we choose the first sub-division, 0–0.3, and if we code a one, we choose the second sub-division, 0.3–1.0. Conceptually, we adjust the range by a lesser amount if we code the most probable symbol (MPS) – in this case, coding a one means that we reduce the range slightly, to 70% of its original span. We adjust the range by a greater amount if we code the least probable symbol (LPS).

We will look at what happens if we code the message: b0 = 1, b1 = 0, b2 = 1 (see Figure 8.14). We start with a Range of (0, 1).

Encode b0:
The initial range (0, 1) is sub-divided into a lower sub-division (0, 0.3) and an upper sub-division (0.3, 1), corresponding to b0 being 0 and 1, respectively. The actual value of b0 is 1, which is the most probable value, so we choose the upper sub-division (0.3, 1). This is the new range.

Encode b1:
Sub-divide the range (0.3, 1) into lower and upper sub-divisions. This time, the lower sub-division is 10% of the range (0.3, 0.37) and the upper sub-division is 90% of the range, corresponding to the probabilities of b1 being 0 and 1, respectively. The actual value of b1 is 0, which is the least probable value, there is only a 10% chance of this. We choose the lower sub-division, and the new range becomes (0.3, 0.37).

Encode b2:
Subdivide (0.3, 0.37) once more, this time into two equal-size sub-divisions, since $P(0) = P(1) = 0.5$. Choose the upper sub-division for the coded value b2 = 1. The final range becomes (0.335, 0.37).

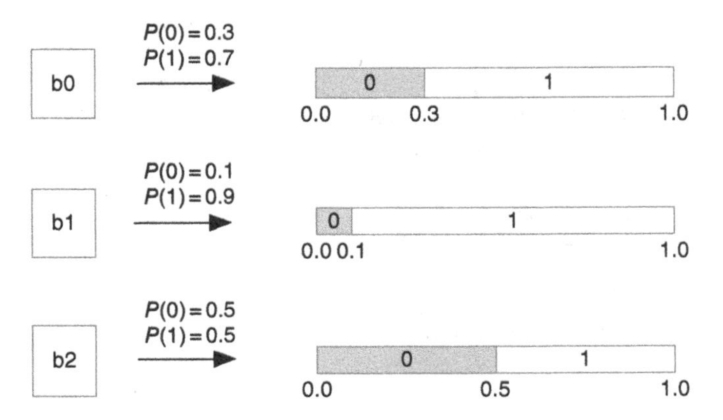

Figure 8.13 Range sub-divisions depending on bin probability

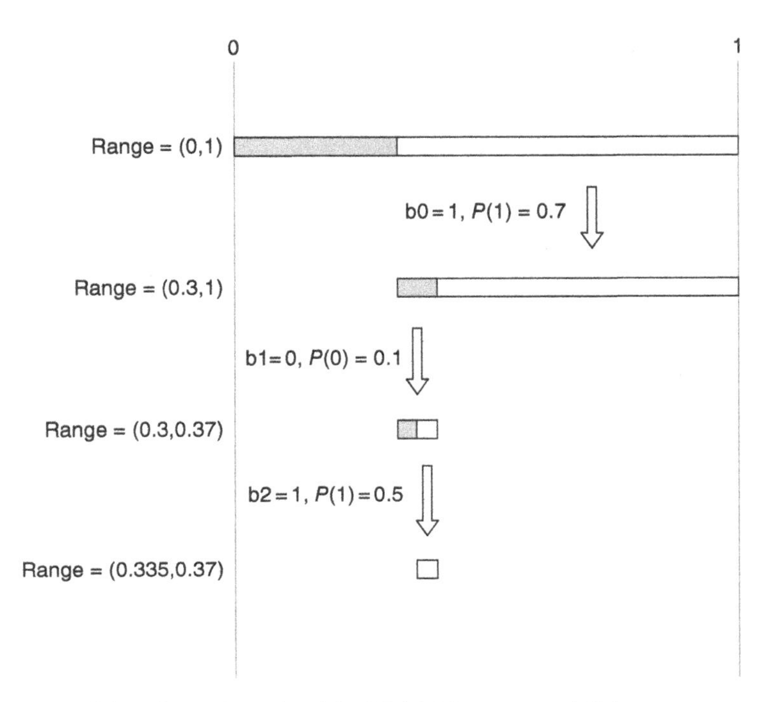

Figure 8.14 Example: coding bins b0, b1, b2: sequence 1, 0, 1

The encoder sends to the decoder a fractional number that points somewhere in this range (0.335, 0.37). The binary fraction 0.01011b, which is decimal 0.34375, is the shortest-length binary fraction that falls within the final range. In the following discussion, a binary fraction will be shown as a series of 0s and 1s followed by b, e.g. '0.10011100b'.

Bin	Value	Initial range	Lower sub-division (0)	Upper sub-division (1)	Final range
b0	1	(0, 1)	(0, 0.3)	(0.3, 1)	(0.3, 1)
b1	0	(0.3, 1)	(0.3, 0.37)	(0.37, 1)	(0.3. 0.37)
b2	1	(0.3, 0.37)	(0.3, 0.335)	(0.335, 0.37)	(0.335, 0.37)

The decoder receives the binary fraction, 0.01011b, which is decimal 0.34375. Armed with a knowledge of the probabilities $P(0)/P(1)$ for b0, b1 and b2 and this fraction pointing to the final sub-range, the decoder can successfully identify the final sub-range and therefore identify the sequence b0, b1 and b2 (see Figure 8.15). As with the general arithmetic coding example earlier, the decoder identifies the sub-range for each successive bin. b0 has two sub-ranges between 0 and 1, and our fraction falls within the upper sub-range, so we decode b0 = 1. b1 has two sub-ranges between 0.3 and 1, and the fraction points to the lower sub-range, so we decode b1 = 0. b2 has two sub-ranges. This time we are in the upper sub-range, so we decode b2 = 1.

8.7.4.1 All the Possible Coding Outcomes

What would happen if we coded a different sequence of bin values? Figure 8.14 demonstrated that coding the sequence 1, 0, 1 resulted in a range from 0.335 to 0.37. If we repeat this

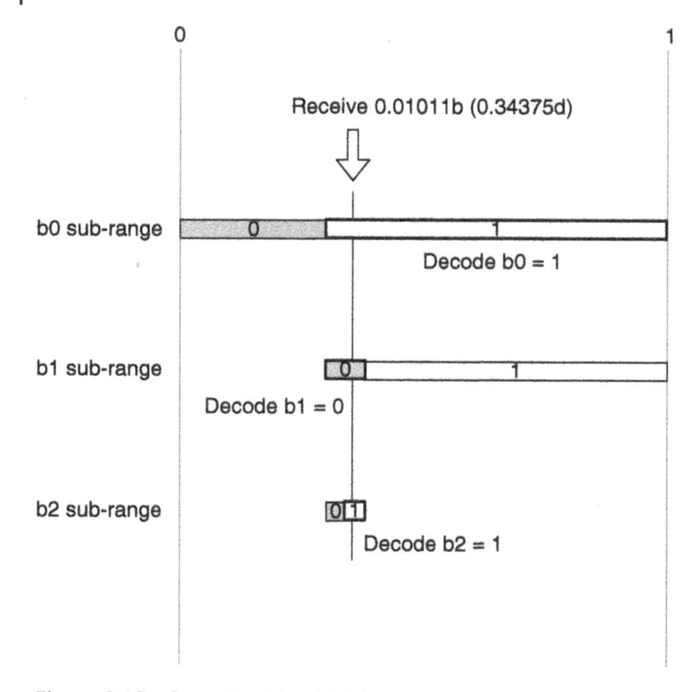

Figure 8.15 Decoding bins b0, b1, b2: sequence 1, 0, 1

process for all the possible sequences b0, b1 and b2, then we end up with eight different sub-ranges as listed in Table 8.13 and shown in Figure 8.16. The first bin b0 divides the initial range into either (0,0.3) or (0.3,1.0), in proportion to the probabilities of b0 being 0 and 1. The second bin b1 divides the sub-range further, in proportion to the probabilities $P(\text{b1} = 0) = 0.1$ and $P(\text{b1} = 1) = 0.9$. There are four possible combinations b0, b1 and therefore there are four possible sub-ranges after the second bin has been coded. The third bin b2 divides the sub-range further, in proportion to the probabilities of b2 being 0 or 1, both 0.5. Each combination of bins (b0, b1 and b2) leads us to a different and unique sub-range. Figure 8.17 shows the lower and upper boundaries of each of these sub-ranges for our example.

Table 8.13 Sub-ranges and binary fractions for all the sequences b0, b1, b2

b0,b1,b2	Lower range	Upper range	Shortest-length binary fraction
000	0	0.015	0.0b = 0.0d
001	0.015	0.03	0.000001b = 0.015625d
010	0.03	0.165	0.001b = 0.125d
011	0.165	0.3	0.01b = 0.25d
100	0.3	0.335	0.0101b = 0.3125d
101	**0.335**	**0.37**	**0.01011b = 0.34375d**
110	0.37	0.685	0.1b = 0.5d
111	0.685	1.0	0.11b = 0.75d

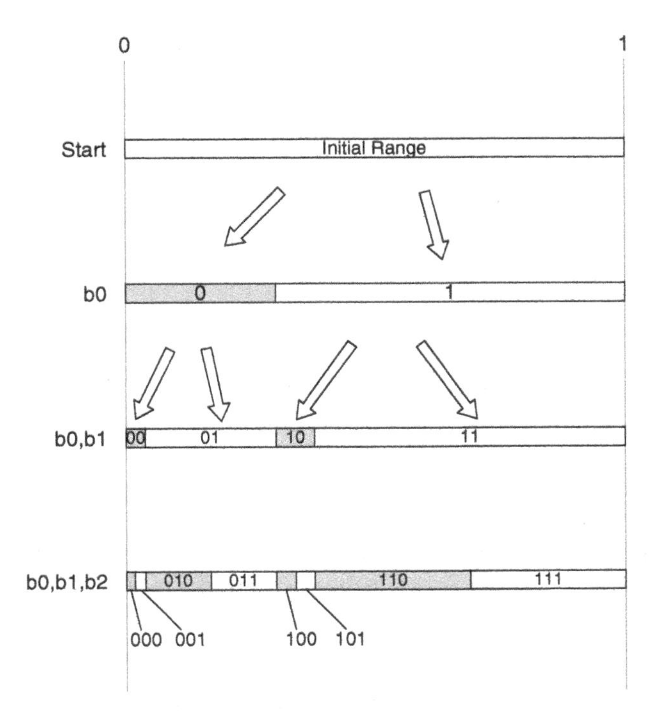

Figure 8.16 Bins b0, b1, b2: all the possible coding choices

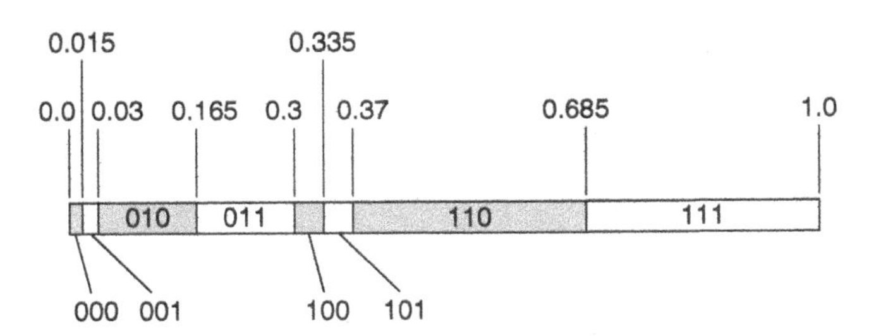

Figure 8.17 Bins b0, b1, b2: lower and upper ranges of all the coding choices

Table 8.13 lists the shortest-length binary fractions for each possible sequence, b0, b1 and b2, and Figure 8.18 illustrates these. Depending on the final range, the shortest-length binary fraction contains a varying number of bits. The sequence 1,1,0 gives the range 0.37–0.685. Conveniently, the fraction 0.5 (decimal) fits into this range, so we can simply send 0.1 (binary). Sequences 0,0,1 and 1,0,1 require much longer binary fractions, 0.000001b and 0.01011b, respectively. Note that the length of the binary fraction depends on the width of the final range and, therefore, on the probability of the specific sequence occurring. The sequence 1,1,0 is much more likely than the sequence 0,0,1, and as a result, the binary fraction, i.e. the number of bits required to represent the sequence, is a lot shorter. For more details on binary fractions, see Section 8.7.5.

We will look at what the BAC encoding process has actually achieved. The encoder wants to encode n bins, each with an estimated probability $P(0)/P(1)$. The encoder partitions the initial range into 2^n sub-ranges, each proportional in size to the probability of occurrence

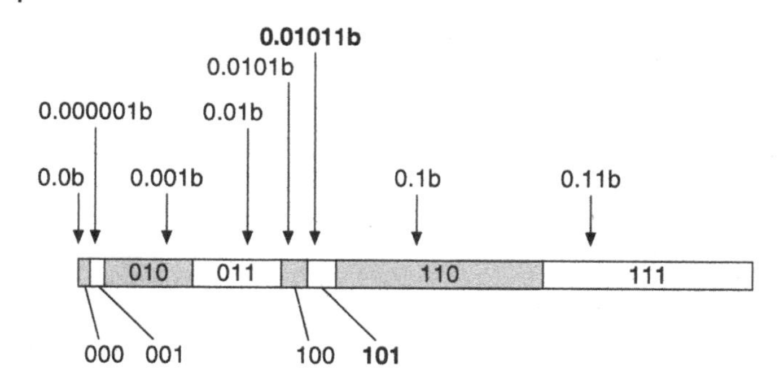

Figure 8.18 Binary fractions identifying each sub-range b0b1b2

of a particular bin sequence b0b1...bn. In this example, we encoded a 3-bit sequence that produced a partitioning into $2^3 = 8$ sub-ranges, each proportional in size to the probability of a particular bin sequence b0b1b2. For example, the probability that b0b1b2 = 101 is $(0.7 \times 0.1 \times 0.5) = 0.035$, which is exactly the final partition size (0.335, 0.37).

Each sub-partition is associated with a unique bin sequence, b0b1...bn. By identifying a position within a specific sub-partition, e.g. by sending the binary fraction 0.01011b = 0.34375d, the encoder gives the decoder enough information to decode the exact bin sequence.

8.7.5 Binary Fractions

We have seen so far that BAC involves identifying a binary fraction and communicating this from the encoder to the decoder. Table 8.14 lists the binary fractions with up to four

Table 8.14 Binary fractions with four or less significant bits

Binary	Decimal
0.0b	0.0
0.0001b	0.0625
0.001b	0.125
0.0011b	0.1875
0.01b	0.25
0.0101b	**0.3125**
0.011b	**0.375**
0.0111b	0.4375
0.1b	0.5
0.1001b	0.5625
0.101b	0.625
0.1011b	0.6875
0.11b	0.75
0.1101b	0.8125
0.111b	0.875
0.1111b	0.9375
(1.0b)	(1.0)

significant bits. 0.0b, 0.1b and 1.0b are 0.0, 0.5 and 1.0 decimal. We can see from this table that we need more than four significant bits to identify the range (0.335, 0.37). 0.0101b is too small since it is less than 0.335d. 0.011b is too large since it is greater than 0.37d. In our example mentioned above, the shortest-length binary fraction the encoder can choose is 0.01011b = 0.34375d. In a similar way, the encoder chooses the shortest-length binary fraction to represent any of the possible sequences b0b1b2. Table 8.13 lists these binary fractions, and Figure 8.18 shows them graphically.

Both encoder and decoder assume that the integer part is always 0, i.e. we will not send a fractional pointer of 1.0b or greater. Hence, the encoder sends the 5-bit binary sequence 01011 to represent the fraction 0.01011b. Recall that our original bin sequence b0b1b2 = 101 has a probability of 0.035, given the probabilities of each bin being 0 or 1. The information content of such a sequence is $I = \log_2(1/P) = \log_2(1/0.035) = 4.84$ bits. Our binary arithmetic coder has succeeded in representing a sequence with an information content of 4.84 bits, using exactly 5 bits.

8.7.6 Sending Bits: BAC Encoder

How do the encoder and decoder send and receive binary fractions? The encoder needs to send a bit for each bit position in the fractional part. In our example mentioned above, we send 01011, i.e. 5 bits. The decoder receives this and constructs the fraction 0.01011b, which points to the range corresponding to the bin sequence 101.

The BAC encoder and decoder face the following two challenges:

1) If we have a long sequence of bins to code, we do not want to wait until the end of the sequence before we send anything from the encoder to the decoder.
2) As the range gets narrower, we need increasingly high-precision variables in the encoder and decoder to keep track of the range and the pointer. We saw in the example mentioned above that some narrow ranges required much longer binary fractions than other, wider ranges.

We can address both issues by scaling the range or renormalising once the MSB of the fractional pointer becomes unambiguous, i.e. once it is unambiguous that the MSB is 1 or 0, then outputting the corresponding MSB (0 or 1) from the encoder to the decoder.

In this way, the encoder sends MSBs as and when they become unambiguous and avoids its range becoming too narrow. The decoder receives MSBs, scales its own range and decodes bins when enough fractional MSBs have been received.

Example: Choosing a Binary Fraction

Figure 8.19 shows how the binary fractions develop as the sub-range progressively narrows. The first bin value, b0 = 1, gives the range (0.3,1). The binary fraction 0.1b fits within this range, but the MSB is ambiguous because other binary fractions such as 0.0101b with a different MSB also fit within the range.

The second bin value, b1 = 0, gives a new range of (0.3, 0.37). The two MSBs are now unambiguous because any binary fraction within this range has to start with 0.01b. The encoder can output a 0 and renormalise the fraction, then output a 1 and renormalise once more.

The third bin value, b2 = 1, narrows the range to (0.335, 0.37).

The entire bin sequence 101 can now be represented by 0.01011b as before. The left-hand two bits of the fractional part [01] may be output during encoding rather than waiting until the end of the message. The encoder outputs the five bits: 0, 1, 0, 1, 1.

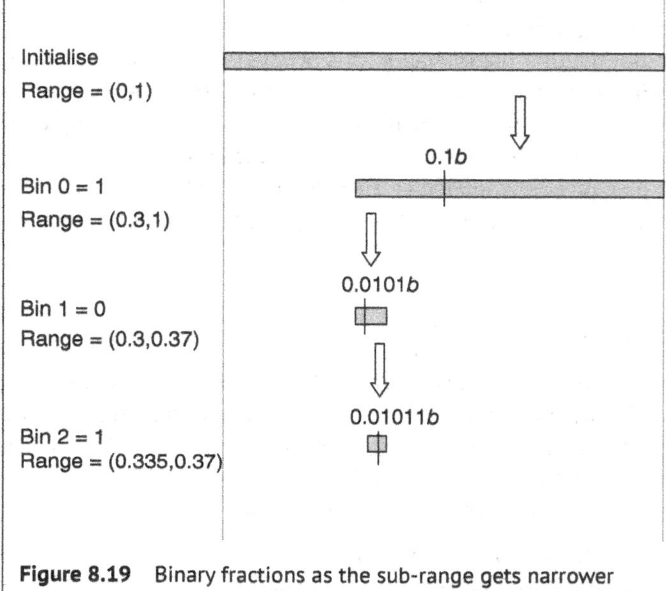

Figure 8.19 Binary fractions as the sub-range gets narrower

8.7.6.1 Renormalisation and Outputting Bits at the Encoder

For our binary arithmetic encoder, the rules for renormalisation depend on the minimum and maximum values of the range (see Figure 8.20). Assuming the starting range is from 0 to 1.0:

a) If the minimum range is <0.5 and the maximum range is >0.5, then the MSB is still ambiguous. It could be 0 or 1 depending on future bins, so we cannot renormalise.
b) If the maximum range is <0.5, then the MSB has to be 0. Renormalise by multiplying minimum and maximum range values by 2, output a 0.
c) If the minimum range is >0.5, then the MSB has to be a 1. Renormalise by multiplying minimum and maximum range values by 2 and subtracting 1, output a 1.

Figure 8.21 shows the encoder normalisation decisions as a flowchart. If the maximum range is less than 0.5 or the minimum range is greater than 0.5, renormalisation occurs and

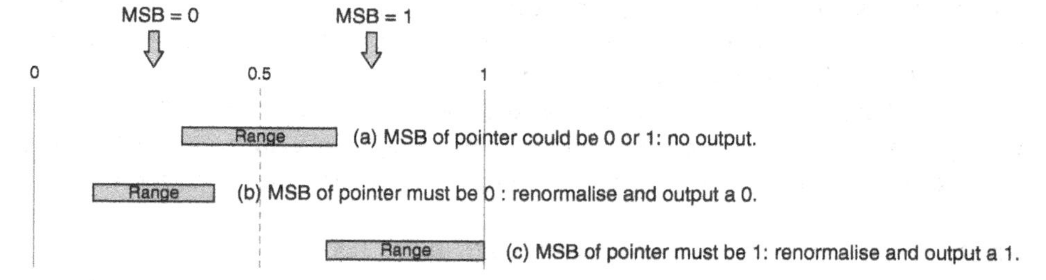

Figure 8.20 Renormalising and outputting a bit at the encoder

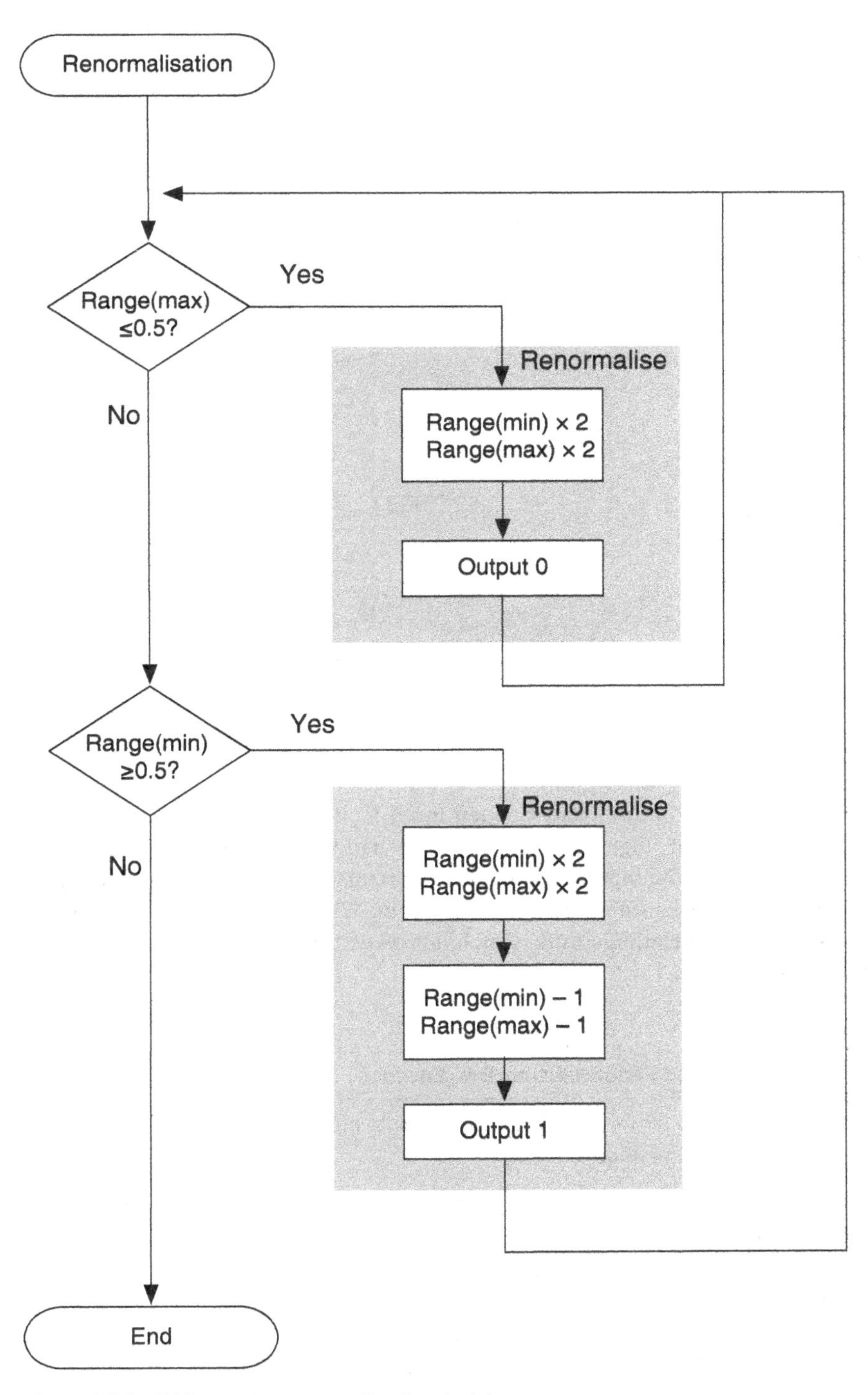

Figure 8.21 BAC encoder renormalisation decisions

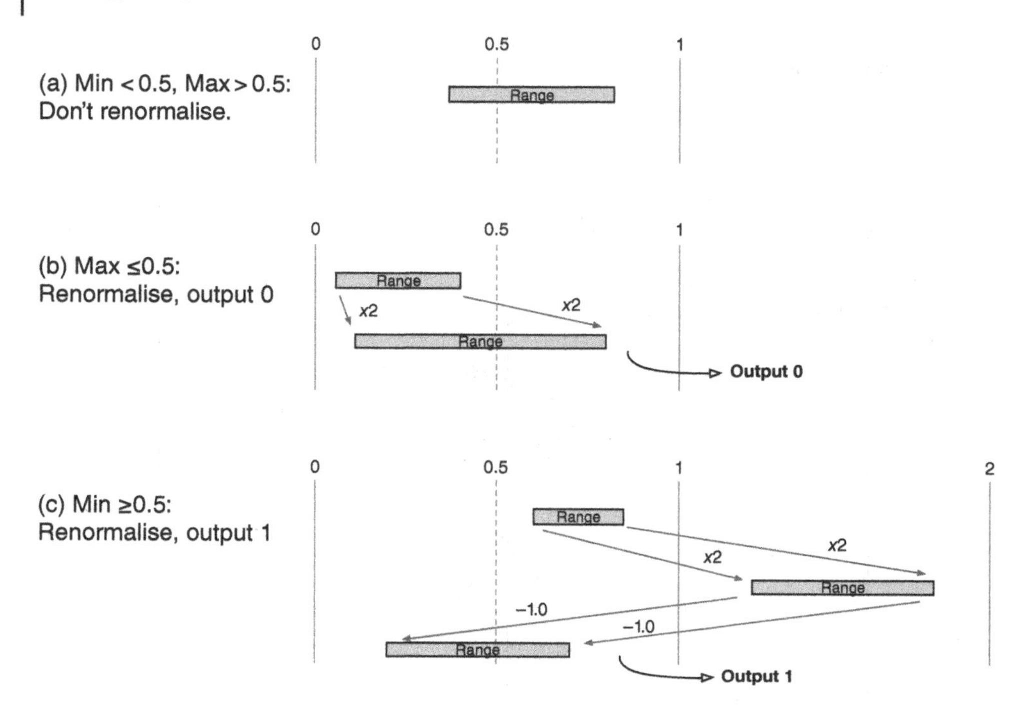

Figure 8.22 BAC encoder renormalisation examples, (a) no renormalisation, (b) renormalise and output 0, (c) renormalise and output 1

a 0 or a 1 is output. Note that this may happen multiple times until we get to the point where the range overlaps 0.5. Figure 8.22 illustrates each of the three possible outcomes. If the range overlaps 0.5, i.e. the minimum range is less than 0.5 and the maximum range is greater than 0.5, option (a), normalisation doesn't occur. When the encoder renormalises, options (b) and (c), normalisation occurs, which means that the range doubles in size and the encoder outputs a 0 or a 1.

Example: Sending Bits and Renormalising, BAC Encoder
Look at how our BAC encoder renormalises and outputs bits during encoding of our bin sequence b0b1b2 = 101 (see Figure 8.23): 1) **Encode b0:** The new range is (0.3,1). The range overlaps 0.5, so there is no renormalisation. 2) **Encode b1:** The new range is (0.3, 0.37). The encoder renormalises three times before moving on: a) Range = (0.3, 0.37). Rmax ≤ 0.5. New range = (0.6, 0.74), output a 0. b) Range = (0.6, 0.74). Rmax ≥ 0.5. New range = (0.2, 0.48), output a 1. c) Range = (0.2, 0.48). Rmax ≤ 0.5. New range = (0.4, 0.96), output a 0. The range now overlaps 0.5, so we don't renormalise.

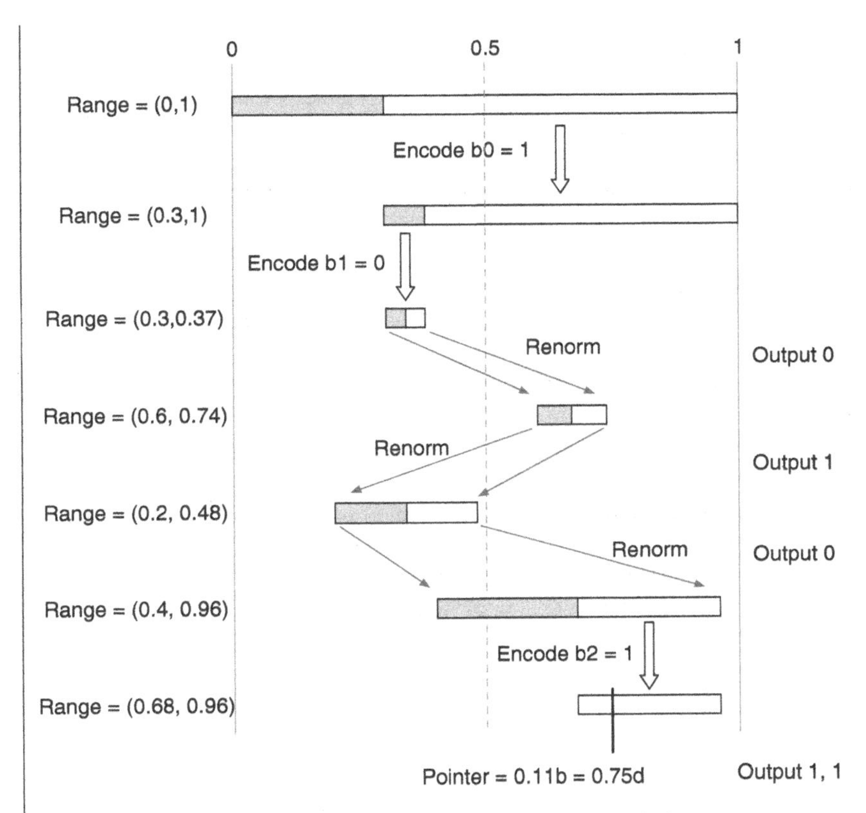

Figure 8.23 Renormalising and sending fractional bits (encoder)

3) **Encode b2:** Divide the range (0.4, 0.96) into lower and upper sub-ranges (0.4, 0.68) and (0.68, 0.96). To encode a 1, choose the upper range (0.68, 0.96). The smallest binary fraction that fits in this range is 0.11b (=0.75), so output the fractional part (1, 1) to finish.

During the complete process, the encoder outputs fractional bits 0, 1, 0, 1, 1. The difference from the earlier example is that the encoder starts outputting bits before encoding is complete, every time it renormalises the Range. The Range is always scaled up before it gets too small, and so the number of bits required to represent the Range is minimised.

8.7.7 Receiving Bits at the BAC Decoder

A BAC decoder receives bits, constructs a binary fraction and decodes a bin string.

From the decoding perspective, renormalisation occurs when the decoder has received enough of the fractional pointer to cause the MSB of the bin string to be unambiguous. The decoder repeats the following process:

1) Receive a fractional bit (0 or 1).
2) Is the next bin of our bin string unambiguous, i.e. can it only be 0 or only 1? If so, decode the bin (0 or 1), scale the range and repeat until the next bin is ambiguous.

Example: BAC Decoding with Renormalisation and Output of Decoded Bins

How does the decoder handle partial decoding of our bin sequence b0b1b2? As each fractional bit is received, the possible range of the pointer reduces. On the left side of Figure 8.24, we see the decoder progressively receiving more MSBs of the fractional pointer until the entire pointer is received (0.01011). The decoder carries out the following steps:

1) **Receive fractional bit 0:** Our pointer must lie within the range (0.0b, 0.1b), or (0.0, 0.5) in decimal. From Figure 8.17, bin b0 is still ambiguous, because it can be 0 or 1 within the pointer's range (0, 0.5).
2) **Receive fractional bit 1:** Our pointer must lie within the range (0.01b, 0.1b), which is (0.25d, 0.5d). Bin b0 is still ambiguous.
3) **Receive fractional bit 0:** Our pointer must lie within the range (0.010b, 0.011b), which is (0.25d, 0.375d). Bin b0 is still ambiguous.
4) **Receive fractional bit 1:** The pointer is within the range (0.0101b, 0.011b), which is (0.3125d, 0.375d). Bin b0 must be 1, since b0 = 1 for every pointer above 0.3. We can decode bin b0 = 1. Bins b1 and b2 are still ambiguous.
5) **Receive the final fractional bit 1:** Our final pointer is 0.01011b = 0.34375d, which exactly identifies bins b1 and b2 as being 0, 1.

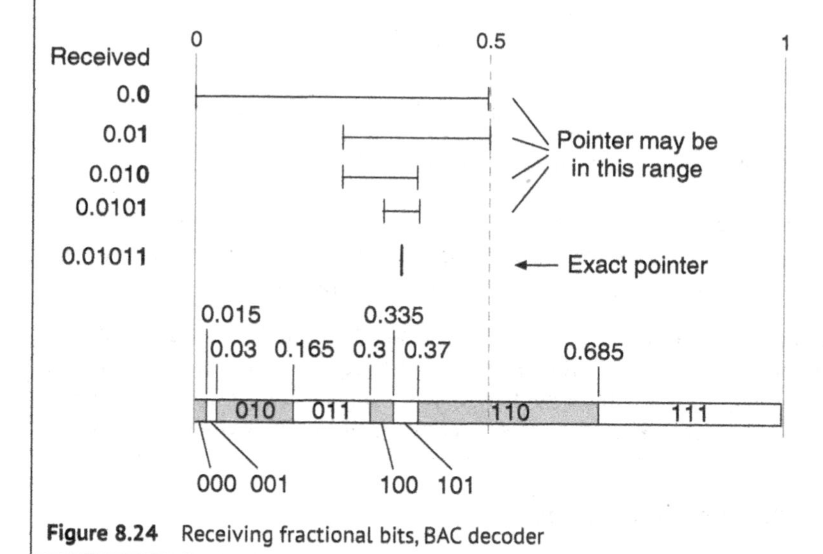

Figure 8.24 Receiving fractional bits, BAC decoder

8.7.8 Binary Arithmetic Coding: Putting It All Together

The encoder can output the MSBs of the fractional pointer as it continues encoding bins, and the decoder can decode bins as and when it receives enough bits of the fractional pointer. For a long sequence of bins such as a coded video frame, this means that the encoder is continuously processing bins and outputting MSBs, whilst the decoder is continuously receiving MSBs and decoding bins.

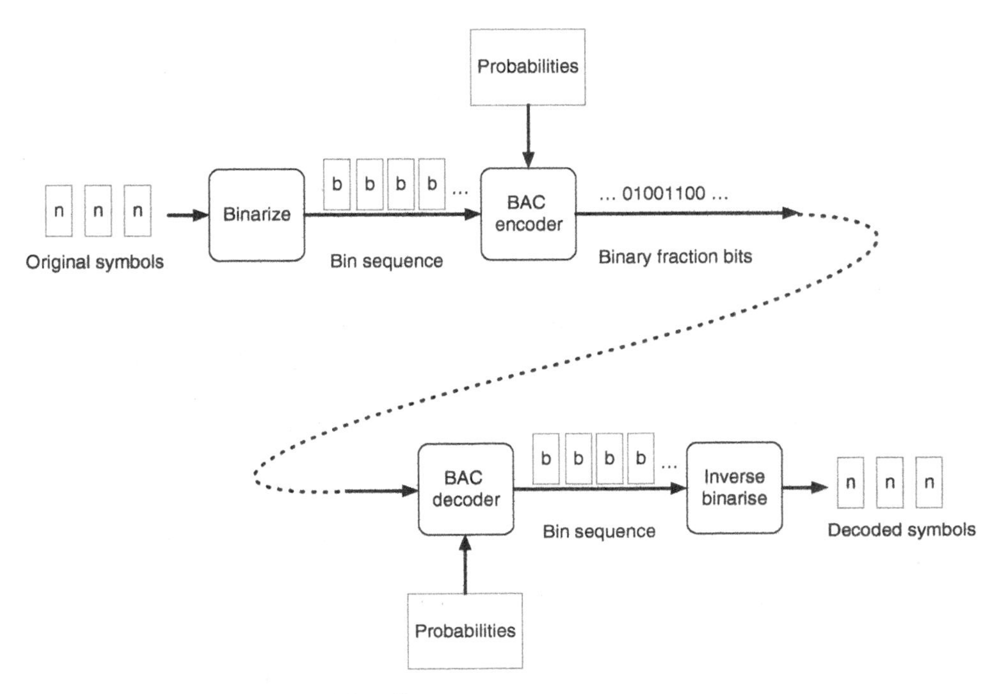

Figure 8.25 BAC encoding and decoding

Figure 8.25 shows the general idea. A series of symbols n are binarised to form a bin sequence b, unless they are already binary. Each bin is encoded by the BAC encoder, according to a table of probabilities that indicates the probability of the bin being a 0 or a 1. The result is a growing binary fraction or range pointer. Each time the MSB of the fraction becomes unambiguous, it is output as a coded bit. The BAC decoder receives the sequence of coded bits and reconstructs the binary fraction. Each time the next bin is unambiguous, because it has to be a 0 or a 1, the BAC decoder outputs that decoded bin. Decoded bins are grouped and inverse binarised to recreate the series of symbols n.

We will look next at how the probability models or context models work, giving us the complete CABAC system.

8.8 Context-Adaptive Binary Arithmetic Coding (CABAC)

CABAC is an approach to arithmetic coding that was introduced in the H.264/AVC standard and further developed in the H.265/HEVC and H.266/VVC standards. CABAC has the following features:

- A BAC engine. BAC is a special case of arithmetic coding, in which each symbol to be coded can only take the values 0 or 1.
- Non-binary symbols are mapped to strings of binary values known as bins in a process known as binarisation.

- Each bin is coded by the BAC engine using a probability model or context model. The context model estimates the probability of the bin being 0 or 1. After coding the bin, the context model is updated based on the actual value of the bin.
- The context model for a bin may be selected based on statistics of recently coded video data, known as the context.

8.8.1 The CABAC Encoder

The input to the encoder (Figure 8.26) is a series of symbols, for example, coefficients, motion prediction information and other syntax elements. Some of these are binary, having just two possible values, whilst others have multiple possible values. Symbols with non-binary values, i.e. with more than two possible values, are binarised to create a bin string that is passed to a binary arithmetic coder. Symbols with binary values are passed directly to the BAC.

As we saw in Section 8.7.2, processing each bin consists of:

- Updating a Range interval, expressed as a binary fraction, based on the probability of the bin being 0 or 1 and the actual value of the bin (0 or 1).
- Optionally, renormalising the Range interval and outputting the MSB of the binary fraction as a coded bit.

The binary arithmetic coder converts a series of bins into a coded bitstream, based on the probability of each bin being 0 or 1 and the actual value of the bin, 0 or 1. Each bin is coded either in bypass mode or context-adaptive mode. In bypass mode, it is assumed that it is always equally probable that the bin is 0 or 1, i.e. $P(0) = P(1) = 0.5$. In context-adaptive mode, the probability of the bin being 0 or 1 is determined using a context model. The probability depends on the coding decisions for previously coded neighbouring syntax elements – the context – and on the previous coding decisions for the same context model. In other words, the encoder uses the context to select a particular context model and then

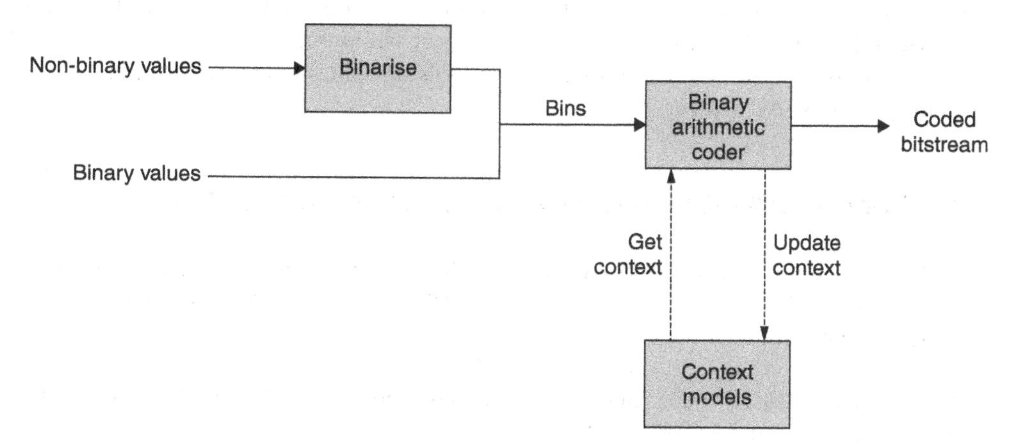

Figure 8.26 CABAC encoder

uses the probability stored in this model to code the bin. After coding a bin in context-adaptive mode, the selected context model is updated based on the actual bin value.

8.8.2 The CABAC Decoder

The input to the decoder is a coded bitstream. The decoder (Figure 8.27) processes the coded bitstream to decode a series of bins. Each bin is decoded either in bypass mode or context-adaptive mode. In bypass mode, the decoder assumes that the probability of the bin being 0 or 1 is always exactly 0.5. In context-adaptive mode, the binary arithmetic decoder determines the probability of the bin being 0 or 1 based on a context model. Mirroring the operation of the encoder, recently coded statistics are used to select a particular context model and the decoder uses the probability information stored in this model to decode the bin. The selected context model is updated once the actual bin value has been decoded. The decoded bins are converted to non-binary values – known as inverse binarisation – or output directly if they started out as binary values.

8.8.3 CABAC Context Models

Section 8.4 introduced the concept of a probability model, known as a context model or just a context. A context model estimates the probability that a symbol S has a value V, i.e. $P(S = V)$. The arithmetic coder uses the estimate $P(S = V)$ to code a symbol S with an actual value of V.

Information about recently coded values – the context – is used to select a probability model from among a number of possible models. The selected model provides an estimate $P(S = 0)$. The encoder uses this estimate to encode the current binary symbol S. Depending on the actual value of S (0 or 1), the encoder updates the chosen context model. If the coded symbol was 0, then $P(S = 0)$ increases slightly and vice versa (see Figure 8.7, Section 8.4).

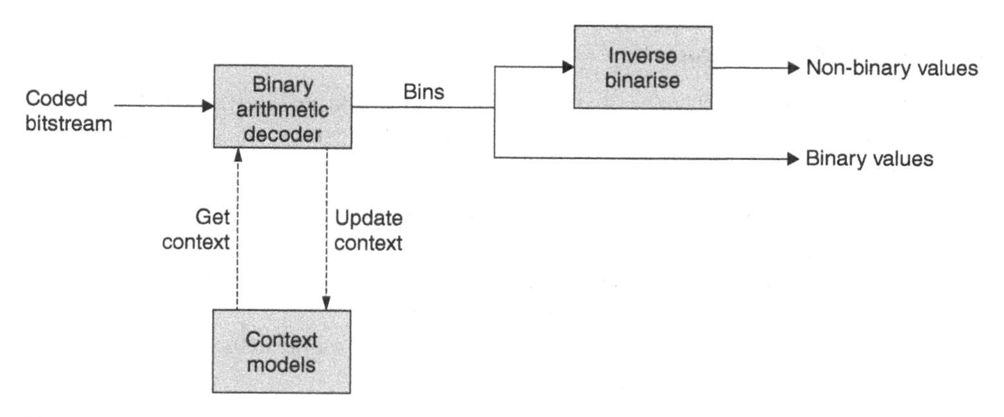

Figure 8.27 CABAC decoder

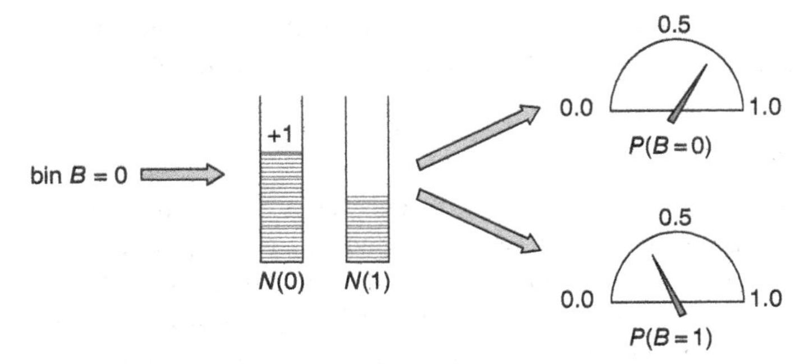

Figure 8.28 Probability model: tracking $P(B=0)$, $P(B=1)$

We will look at a simple binary probability model or context model in more detail (see Figure 8.28). The encoder maintains a count of the number of zeros and ones coded so far, using this context model. If the current bin is a 0, then (in this example) the number of zeros $N(0)$ is incremented by one. The encoder can estimate the probability of a bin being zero or one as follows:

$$P(0)=\frac{N(0)}{N(0)+N(1)}$$

and

$$P(1)=\frac{N(1)}{N(0)+N(1)}$$

Alternatively, the encoder can calculate one of these, e.g. $P(B=0)$, and derive the other: $P(B=1)=1-P(B=0)$.

8.8.4 CABAC Context Adaptation

The CABAC encoder and decoder use the context to select a probability model. Each probability model has a unique identifier, the context index. The process of context coding and adaptation is:

1) Select a context index based on a defined set of rules based on the values of recently coded syntax elements.
2) Retrieve $P(\text{bin}=0)$ and $P(\text{bin}=1)$ from the context model identified by the context index.
3) Code the current bin using the retrieved probabilities.
4) Update the probabilities $P(\text{bin}=0)$ and $P(\text{bin}=1)$ based on the actual bin value 0 or 1, for the selected context model.

We will look at a specific example from the HEVC standard.

Example: Context Adaptation for HEVC Parameter coded_sub_block_flag

Each transform block (TB) in an HEVC picture is scanned and coded as a series of 4×4 transform sub-blocks (TSBs) (see Section 8.9.1). The encoder sends a flag for each sub-block, coded_sub_block_flag, indicating whether the sub-block contains non-zero coefficients, flag is 1, or contains all zero coefficients, flag is 0. To encode this flag using BAC, we need to know (1) the current value of the flag, 1 or 0, and (2) the probability that the flag is 1 or 0. We can estimate the probability of the flag depending on recently coded values, i.e. the local context. In this case, the context depends on two neighbouring sub-blocks (see Figure 8.29). Because we scan the TB from the lower right, to the upper left, we will usually already have encoded the sub-blocks to the right and below the current 4×4 sub-block. How many of these sub-blocks contain non-zero coefficients? The answer can be 0, indicating that neither sub-block has coefficients, 1, indicating that just one of the neighbours has coefficients, or 2, indicating that both neighbours have coefficients.

The number of neighbouring sub-blocks containing non-zero coefficients is either 0, 1 or 2. This number is used to select one of three contexts, which means identifying one of three context indices. Each of the three contexts for this sub-block has a different probability model. In this example, for context 0, $P(0)$ is large and $P(1)$ is small. This means that the probability that the current coded_sub_block_flag is 0, given that both neighbouring blocks contain zero coefficients, is relatively high. $P(0)$ is smaller for context 1, which is $P(0)$ given that one of the neighbours contains non-zero coefficients, and smallest for context 2, which is $P(0)$ given that both neighbours contain non-zero coefficients. This all makes intuitive sense: if both the neighbours have no non-zero coefficients, it is likely that this block has no non-zero coefficients.

The encoder carries out the following steps for the current block:

1) Determine the context, labelled 0, 1 or 2 in Figure 8.29.
2) Derive the probabilities $P(0)$, $P(1)$ for the selected context[3].

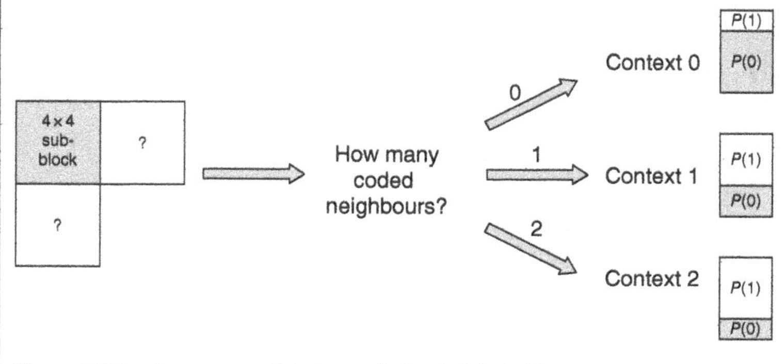

Figure 8.29 Context models for coded_sub_block_flag

3 Since $P(0) + P(1) = 1.0$, it is only necessary to store one probability. An H.264 or HEVC codec stores P(MPS), the probability of the most probable symbol, and val(MPS), the value of the most probable symbol (0 or 1), see Section 8.9.3.1 later.

3) Encode coded_sub_block flag using the probabilities $P(0)$ and $P(1)$ and the actual flag value, which is 0 or 1.
4) Update the probabilities $P(0)$ and $P(1)$ for the selected context[3]. If the actual flag value was 0, then $P(0)$ increases slightly. If the actual flag value was 1, then $P(1)$ increases slightly.

We have now looked at all of the basic components of CABAC: binarisation, the BAC encoder and decoder, selecting probability models and updating probability models. We will consider the specifics of entropy coding in HEVC.

8.9 Entropy Coding in HEVC

The syntax of an HEVC-coded video sequence is specified in Section 7.3 of the standard, Syntax in Tabular Form [1]. As we saw earlier in Section 8.1, the standard specifies syntax elements that are coded using fixed-length codes, variable-length codes and CABAC. For example, as we saw in Table 8.1, syntax element pic_output_flag is coded directly in the bitstream as a single bit and slice_type is coded as a variable-length code using an Exponential Golomb mapping. In Table 8.2, we saw that the syntax element abs_mvd_minus2 is coded using CABAC. This syntax element, abs_mvd_minus2, is an example of an integer-valued syntax element that is binarised before coding with BAC.

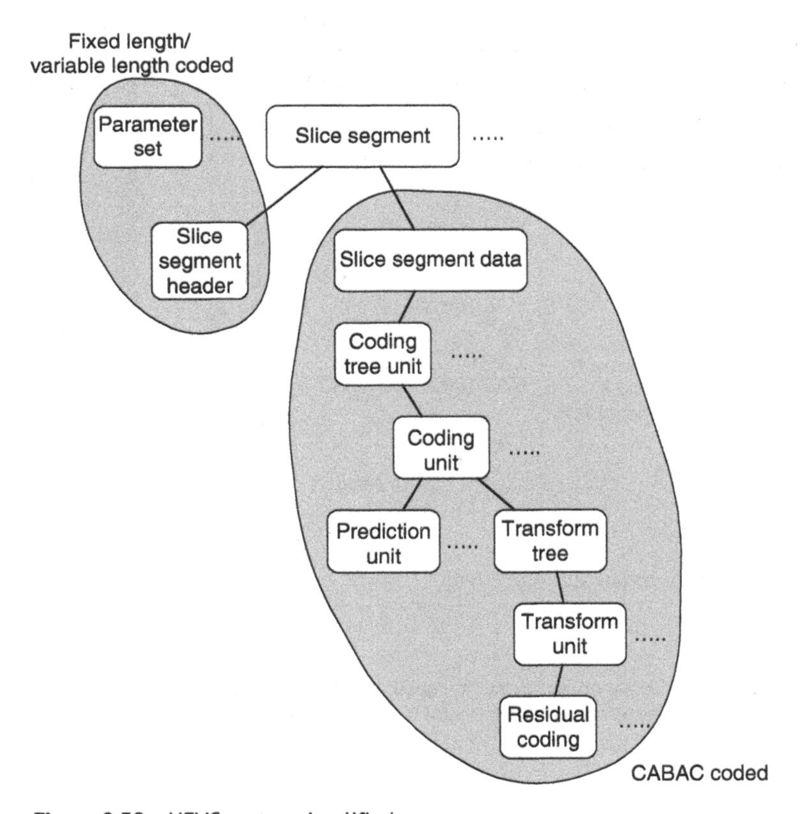

Figure 8.30 HEVC syntax, simplified

Figure 8.30 shows a simplified illustration of the HEVC syntax. At the top, at the raw byte sequence payload (RBSP) level, are structures such as Parameter Sets and Slice Segments. Each of these units is packaged in its own network adaptation layer unit (NALU). A Slice Segment consists of a Slice Segment Header and Slice Segment Data. The Slice Segment Data structure and everything it contains, including coding tree units (CTUs), coding units (CUs) and so on, is coded using CABAC. The Slice Segment Header and higher-level structures such as Parameter Sets are coded using fixed and variable-length codes.

Each Slice Segment Data structure is coded using CABAC (see Figure 8.31). At the start of coding each Slice Segment Data structure, the CABAC process is initialised (see Section 8.9.5), setting each of the context models to a pre-defined probability. The CTUs that make up the slice segment are coded, with the CABAC engine tracking and updating probabilities as each bin is encoded. The CABAC process is terminated at the end of the Slice Segment, which means that the arithmetic coding process is flushed so that the final bits of the fractional pointer, i.e. the range, are encoded or decoded. The process begins again at the start of the next Slice Segment.

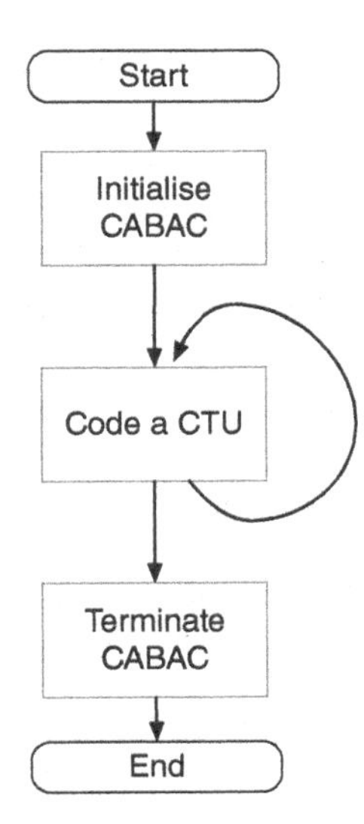

Figure 8.31 Coding an HEVC Slice Segment Data structure

8.9.1 Pre-processing in HEVC

As discussed in Section 8.3, certain values and parameters may be pre-processed to minimise the information to be encoded, for example, by rearranging and/or predicting values.

8.9.1.1 HEVC Transform Blocks and Sub-blocks

An HEVC CU is encoded into one or more TBs, each containing quantised transform coefficients. An encoder processes these blocks to maximise entropy coding efficiency, splitting larger TBs into 4×4 TSBs, scanning coefficients and sub-blocks, and processing each sub-block in a number of passes.

A TB is a block of quantised transform coefficients, with a size of 4×4, 8×8, 16×16 or 32×32. In a typical TB, many or most of the coefficients are zero. Non-zero coefficients tend to be clustered around the top-left, known as the DC coefficient. For example, the 8×8 TB shown in Figure 8.32 has a DC coefficient value of +34 and 10 further non-zero coefficients that are mostly placed near the top-left of the block, at lower spatial frequencies. Higher frequency coefficients, towards the lower-right, are mostly zero or have small values such as ±1.

34	−2	0	−1	0	0	0	0
8	0	−6	0	0	0	0	0
−18	0	0	5	0	0	0	0
0	2	0	0	0	0	1	0
0	0	0	0	0	0	0	0
0	0	0	−1	0	0	0	0
1	0	0	0	0	0	0	0
0	0	0	0	0	0	0	0

Figure 8.32 HEVC 8×8 transform block

HEVC TBs larger than size 4×4 are coded as a series of TSBs of size 4×4. The 8×8 TB of Figure 8.32 is coded as four 4×4 TSBs. The TSBs within a TB are coded in a scan order that starts from TSB towards the lower-right of the TB, which contains the highest-frequency coefficients, and finishes with the upper-left TSB, which contains the lowest frequency AC coefficients (i.e. the coefficients other than the DC coefficient), and the DC coefficient[4]. There are three possible scan orders: diagonal, horizontal and vertical (see Figure 8.33). An 8×8 TB contains four 4×4 TSBs, as shown in the top row. Note that the diagonal and vertical scans are identical for this block size, so there are just two distinct TSB scan orders within an 8×8 TB. A 16×16 TB contains sixteen 4×4 TSBs, as shown in the middle row with all three possible scans shown. A 32×32 TB contains 64 4×4 TSBs, as shown in the bottom row with only the diagonal scan shown.

The choice of scan pattern, whether diagonal, horizontal or vertical, depends on the prediction type of the prediction unit corresponding to the TB. Inter-predicted blocks always use the diagonal scan. Intra-predicted blocks use one of the three scan types, depending on the intra-prediction mode. The same scan pattern is used for scanning the TSBs (Figure 8.33) and for scanning the coefficients within each TSB.

Coding a TB consists of the following main steps:

1) Code the position of the Last Significant Coefficient, i.e. the highest-frequency non-zero quantised coefficient in the TB. This is the first coefficient encountered when decoding

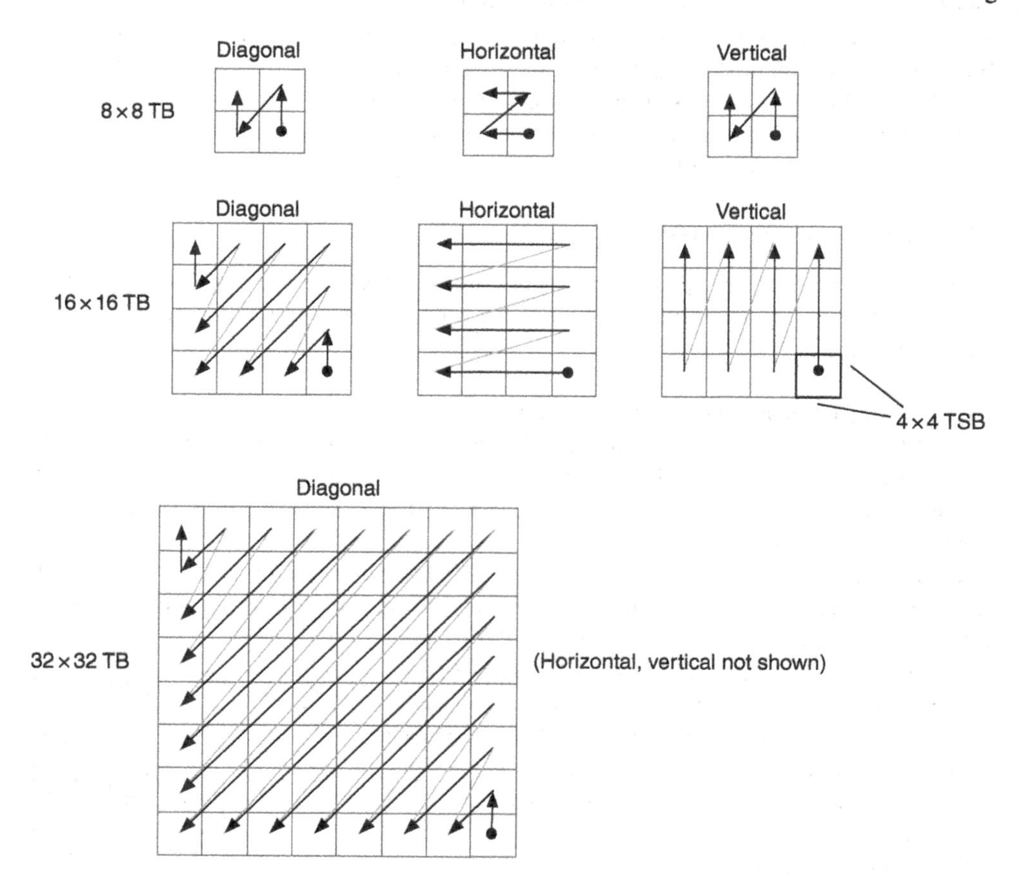

Figure 8.33 Scanning HEVC TSBs within a TB

4 The likelihood of a TSB containing non-zero coefficients increases as the scan proceeds towards lower frequencies. The context models for TSB and coefficient coding reflect this.

the TSBs, since the scan pattern starts from the highest frequency and ends at the lowest frequency. Code the (*x*,*y*) coordinates of this coefficient position, relative to the top-left of the TB, using BAC.

2) Starting with the TSB that is immediately after the TSB containing the Last Significant Coefficient in the scan order, code a flag for each TSB position, coded_sub_block_flag, indicating whether the TB contains any non-zero coefficients. No coded_sub_block_flag is coded for TSB (0,0), the TSB at the top-left position containing the DC coefficient.
3) For each TSB that contains non-zero coefficients, code the TSB coefficients (Section 8.9.1.2).

Examples:

Example 1. 32×32 TB, Last Significant Coefficient at position (9, 11), horizontal scan (see Figure 8.34a). TSBs marked 0 contain all zeros, whereas shaded TSBs contain non-zero coefficients. The coded_sub_block_flags, in scan order, are as follows:

1, 0, 1, 1, 1, 0, 1, 1, 0, 1, 1, 0, 1, 1, 1, 0, 1

excluding the first and last TSBs, i.e. the TSB containing the Last Significant Coefficient and the top-left TSB.

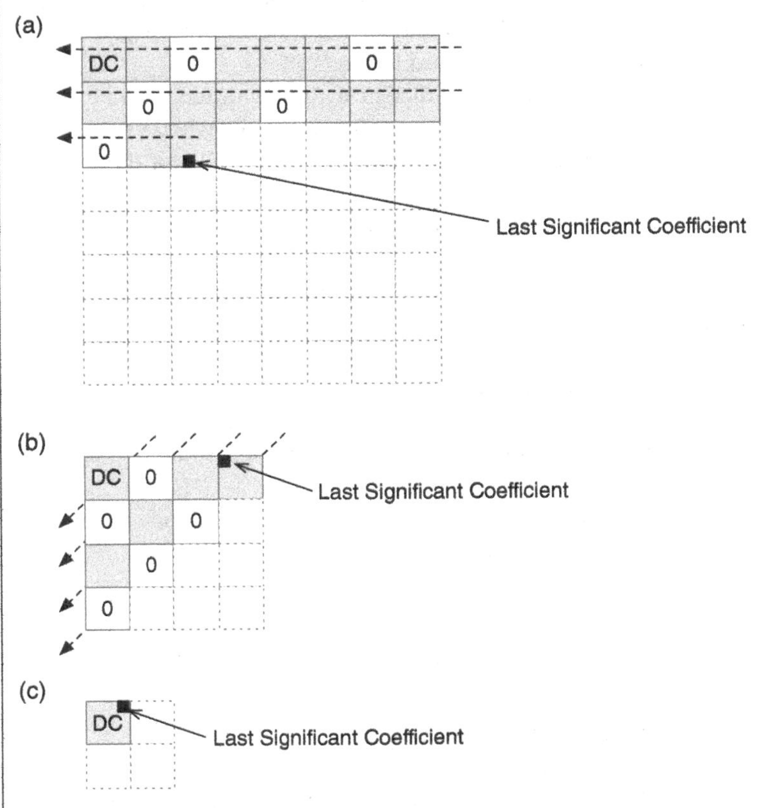

Figure 8.34 HEVC TSB examples, (a) 64 TSBs, (b) 16 TSBs, (c) 4 TSBs

Example 2. 16×16 TB, Last Significant Coefficient at position (12, 0), diagonal scan (see Figure 8.34b). The coded_sub_block_flags, in scan order, are as follows:

0, 0, 0, 1, 1, 1, 0, 0

Example 3. 8×8 TB, Last Significant Coefficient at position (3, 0), vertical scan (see Figure 8.34c). Only the top-left DC TSB contains non-zero coefficients so no coded_sub_block_flag syntax elements are sent.

8.9.1.2 Coding Transform Sub-blocks

Each HEVC TSB containing non-zero coefficients is coded using up to five passes. Where applicable, each of the following five processes is applied to the TSB in turn, using the scan order, diagonal, horizontal or vertical, that was used for the TSBs within the TB.

1) **Significant coefficients:** Starting at either coefficient 15, lower-right, or the coefficient after the Last Significant Coefficient, if this exists in the current TSB, code sig_coeff_flag for each coefficient position. A 1 indicates a non-zero coefficient, and a 0 indicates a zero coefficient. This flag is not coded for the final 'DC' coefficient in the final TSB within the TB.
2) **Coefficients with magnitude >1:** Starting at the first non-zero (significant) coefficient in scan order, code coeff_abs_level_greater_1_flag for each significant coefficient. This flag indicates either a coefficient magnitude greater than one (1) or equal to one (0). A maximum of eight of these flags are coded for the TSB.
3) **Coefficients with magnitude >2:** For a coefficient with magnitude >1, as indicated by the previous process, coeff_abs_level_greater_2_flag indicates whether this coefficient has a magnitude greater than 2. This flag is coded once per TSB, for the first coefficient with magnitude greater than one.
4) **Coefficient sign:** For each non-zero coefficient, coeff_sign_flag indicates the sign of the coefficient, either negative, indicated by 1, or positive, indicated by 0.
5) **Remaining absolute level:** For each non-zero coefficient value not processed above, coeff_abs_level_remaining codes the remaining level value. For example:
   - If coeff_abs_level_greater_2_flag is 1 for this coefficient: subtract 2 from the absolute value of the coefficient, this is the remaining absolute level.
   - If coeff_abs_level_greater_1_flag is 1, and the coefficient position is after the coeff_abs_level_greater_2_flag position: subtract 1 from the absolute value of this coefficient, this is the remaining absolute level.
   - If coeff_abs_level_greater_1_flag was not coded for this coefficient: code the absolute value of this coefficient as the remaining absolute level.

Examples:

Three TSBs are coded (Figure 8.35) using diagonal scan (a), horizontal scan (b) and vertical scan (c).

a) TSB example with diagonal scan (see Figure 8.36).
   Significant coefficients: Last Significant Coefficient is not present in this block, so this flag is coded for all 16 coefficients. There are a total of 8 non-zero coefficients.
   Greater than 1: This flag is coded for all 8 non-zero coefficients, which is the maximum number of allowed occurrences.

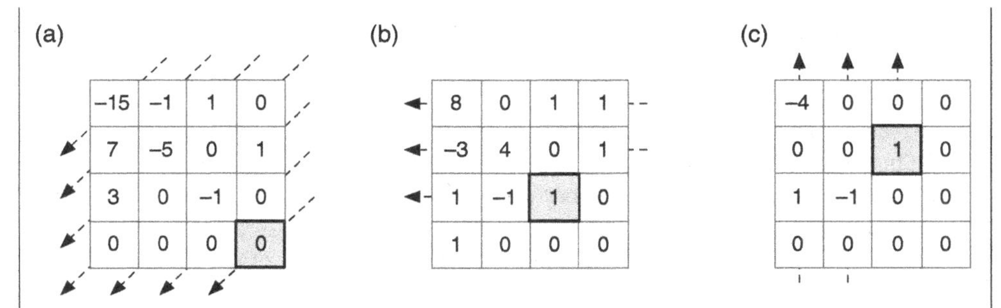

Figure 8.35 HEVC TSB coding examples, (a) diagonal scan, (b) horizontal scan, (c) vertical scan

Coefficient position	15	14	13	12	11	10	9	8	7	6	5	4	3	2	1	0
Value	0	0	0	1	–1	0	0	0	0	0	1	–5	3	–1	7	–15
Significant coefficients	0	0	0	1	1	0	0	0	0	0	1	1	1	1	1	1
Greater than 1 (max 8)				0	0						0	1	1	0	1	1
Greater than 2												1				
Sign				0	1						0	1	0	1	0	1
Remaining abs level												3	2		6	14

Figure 8.36 TSB example (a)

Greater than 2: The first coefficient greater than 1, coefficient position 4, is also greater than 2.
Sign: Coded for all non-zero coefficients.
Remaining absolute level: Coefficient position 4 is known to be greater than 2, so the encoder subtracts 2 and codes the absolute value of 3. Coefficient positions 3, 1 and 0 are known to be greater than 1, so the encoder subtracts 1 from each of these and codes the remaining absolute level.

b) TSB example with horizontal scan (see Figure 8.37)
Significant coefficients: Last Significant Coefficient is present (highlighted in Figure 8.35), and so this flag is coded for the 9 coefficients after the Last Significant Coefficient which was at position 10.
Greater than 1: This flag is coded for the first 8 non-zero coefficients and omitted for any further non-zero coefficients. In this case, it is omitted for coefficient position 0.

Coefficient position	15	14	13	12	11	10	9	8	7	6	5	4	3	2	1	0
Value						1	–1	1	1	0	2	–3	1	1	0	8
Significant coefficients							1	1	1	0	1	1	1	1	0	1
Greater than 1 (max 8)						0	0	0	0		1	1	0	0		
Greater than 2											0					
Sign						0	1	0	0		0	1	0	0		0
Remaining abs level												2				8

Figure 8.37 TSB example (b)

Greater than 2: The first coefficient greater than 1, coefficient position 5, is exactly 2 so is not greater than 2, hence the 'greater than 2' flag is 0.

Sign: Coded for all non-zero coefficients.

Remaining absolute level: Coefficient position 5 is known to have a magnitude of 2 (its greater than one but not greater than two), so there's no need to encode a remaining absolute level. Coefficient position 4 is greater than one, so the encoder subtracts 1 and codes the absolute value of 2. Coefficient position 0 is non-zero but no information has been coded indicating its magnitude, so the full magnitude of 8 is coded.

c) TSB example with vertical scan (see Figure 8.38).

Significant coefficients: Last Significant Coefficient is present (highlighted in Figure 8.35), and so this flag is coded for the 8 coefficients after the Last Significant Coefficient, which is at position 9.

Greater than 1: This flag is coded for the four non-zero coefficients.

Greater than 2: Coefficient position 0 is greater than 2.

Sign: Coded for all non-zero coefficients.

Remaining absolute level: Coefficient position 0 is known to be greater than 2, so a value of (4 – 2) = 2 is coded.

Each pass through the TSB can be thought of as capturing and encoding successive waves of information. Consider example (a) given above. Once the Significant Coefficients are identified, the zero-valued coefficients have been captured, i.e. fully represented (see Figure 8.39). Once the coefficients greater than 1 are identified, we can subtract 1 from the remaining coefficients. After identifying the first coefficient greater than 2, i.e. position 4, we can subtract 1 again from this coefficient. After coding the sign of each coefficient, the remaining absolute values (3, 2, 6, 14) can be coded.

To summarise, an HEVC encoder processes TBs to achieve efficient entropy coding. Larger TBs are scanned as a series of 4×4 TSBs. Each TSB is scanned in multiple passes. The result is that statistically related information tends to be grouped together and multiple processing steps are capable of being carried out in a parallel or pipelined way.

Coefficient position	15	14	13	12	11	10	9	8	7	6	5	4	3	2	1	0
Value							1	0	0	–1	0	0	0	1	0	4
Significant coefficients								0	0	1	0	0	0	1	0	1
Greater than 1 (max 8)							0			0				0		1
Greater than 2																1
Sign							0			1				0		0
Remaining abs level																2

Figure 8.38 TSB example (c)

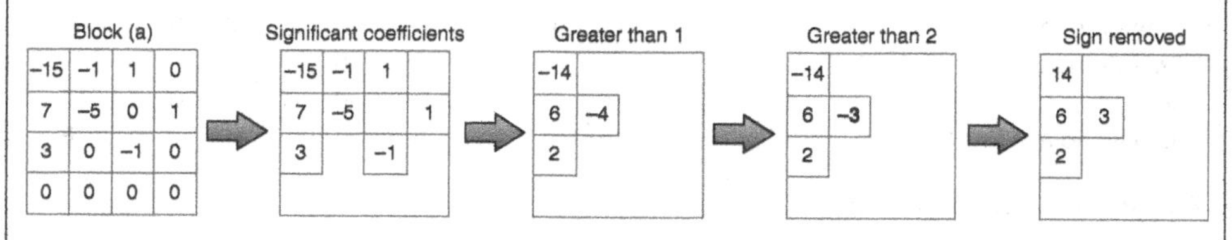

Figure 8.39 TSB example (a), information captured in each pass

8.9.1.3 Pre-processing HEVC Prediction Parameters

In Chapters 6 and 7, we saw how intra and inter prediction mode choices in HEVC are processed prior to entropy coding. An HEVC encoder uses the statistical relationships between blocks to try and reduce the amount of information to be encoded. The general approach is to generate a candidate list, which is a set of parameters from previously coded blocks that are likely to be used for the current block. We can choose from the candidate list and/or send actual parameters for the current block.

Intra prediction parameters:
An intra-predicted block has an intra prediction mode index that indicates one of up to 35 possible angular or planar prediction modes. Intra modes tend to be similar to neighbouring block modes. The encoder generates a list of three candidate modes, which are the most probable modes based on neighbouring, previously coded blocks. Either the index of one of these candidate modes is encoded or the actual intra mode is encoded if it is not in the candidate list.

Inter prediction parameters:
Figure 8.40 shows the general approach for processing inter prediction parameters prior to entropy coding. A current block has a choice of motion vector and reference picture index. An HEVC encoder generates a list of candidates, each of which is a motion vector and reference index, from a previously coded block. The encoder selects a candidate, e.g. a candidate that has motion parameters that are similar to the current block. In Merge mode, only the candidate index is encoded, index 1 in this example. In Advanced Motion Vector Prediction mode, a candidate index, an MVD, and a reference index are encoded. In both modes, the relationship between spatially and temporally adjacent blocks is used to reduce the amount of information that has to be sent.

8.9.1.4 Pre-processing HEVC Quantisation Parameters

The QP is applied by an HEVC decoder to rescale the decoded transform coefficients prior to the inverse transform. An encoder can signal a change in QP by sending a Delta QP in the bitstream.

Delta QP can be sent at most once per CU. The actual QP for an HEVC CU is calculated as follows:

1) Calculate a predicted luma QP based on the QP of previously coded CUs above and to the left of the current CU, and/or on the QP of the previous CU in decoding order.
2) Calculate the actual luma QP based on the predicted QP and the delta QP.
3) Calculate the chroma QP based on the luma QP.

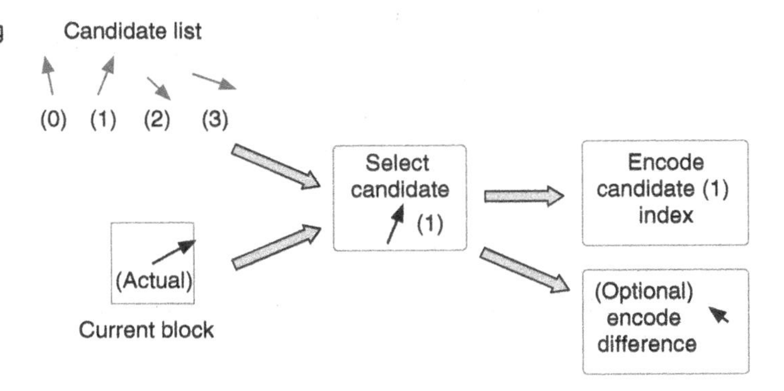

Figure 8.40 Encoding HEVC inter-prediction parameters

Multiple CUs may optionally be grouped into a Quantisation Group, determined by the parameter diff_cu_qp_delta_depth in the picture parameter set (PPS), in which case (1) delta QP is sent at most once per Quantisation Group and (2) all the CUs in the Quantisation Group share the same predicted luma QP.

8.9.2 HEVC Binarisation

There are over 50 different syntax elements that can be coded in a slice segment data structure. Each syntax element is binarised prior to CABAC encoding using binarisation methods specified in the H.265 standard. These methods include:

Fixed length: The syntax element value is converted directly into a binarisation with the specified length, e.g. a single bin for a flag variable.

Truncated rice: For smaller syntax element values, this is a unary binarisation (see Section 8.5.1), which may be followed by suffix bits. At a certain maximum value, the binarisation is truncated to a string of binary ones.

Exponential Golomb/Exp-Golomb: The syntax element value is mapped to an Exp-Golomb binarisation (see Section 8.5.2).

Special cases: Certain syntax elements have particular binarisations as specified in the H.265 standard. For example, the non-binary value coeff_abs_level_remaining signals the magnitude of coefficient levels greater than 2 (see Section 8.9.1.2). This syntax element occurs frequently in a coded bitstream and has a specific binarisation comprising a Truncated Rice binarisation for smaller indices and a composite codeword for larger indices.

8.9.3 The HEVC Binary Arithmetic Coder

The HEVC standard specifies a binary arithmetic coder with similar properties to the generic binary arithmetic coder described earlier in Section 8.7, but with certain specific features. Recall that the arithmetic coded sequence is represented by a binary fraction that points to somewhere within a Range. In the HEVC CABAC design:

- The lower and upper bounds of Range are represented by Low, identifying the lower bound of Range, and Range, identifying the width of the range, so that the upper bound of Range is given by (Low + Range).
- Low and Range can each take values between 0 and 1023, so they have 10-bit precision, with 0 representing decimal 0.0 and 512 representing decimal 1.0. Values greater than 512 are allowed during the normalisation process.
- Low is initialised to 0 and Range to 510, equivalent to decimal 0.0 and ~0.998.
- Instead of subdividing the Range depending on whether the bin is 0 or 1, we subdivide the Range depending on whether the bin is the LPS or the MPS. LPS is 0 if the probability of the bin being 0 is less than 0.5 (i.e. a 0 is less probable than a 1) and LPS is 1 if the probability of the bin being 0 is more than 0.5 (i.e. a 1 is less probable than a 0).
- The Range can be sub-divided using a table look-up based on the coded bin value.

8.9.3.1 HEVC BAC Encoding

The encoder keeps track of the state of the arithmetic coder using the variables Low and Range. The arithmetic coded interval falls somewhere between Low and Low + Range (see Figure 8.41).

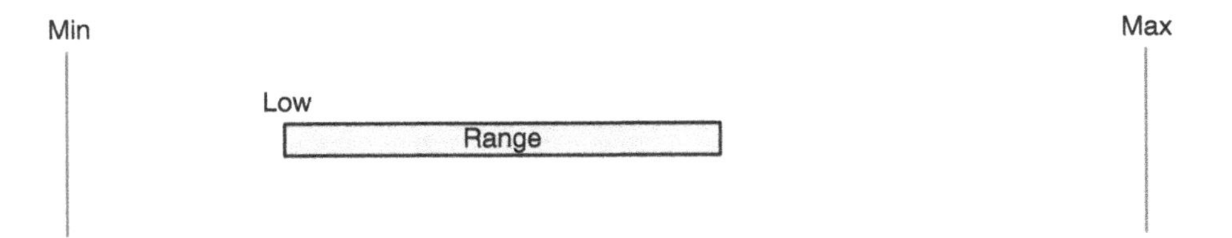

Figure 8.41 HEVC BAC: Low and Range

To encode a bin, the encoder splits the current Range depending on the bin value, 0 or 1, and the probability of that value occurring. If the probability is expected to be 0.5, the bin is encoded using Bypass coding, otherwise the encoder does the following:

1) Based on the context index, which is an index to a probability model, and the current value of Range, determine valMPS and LpsRange, which is the sub-division of range if the LPS is coded[5].
2) If the current bin value is the MPS (Figure 8.42):
   a) New Low′ = previous Low
   b) New Range′ = MpsRange = (Range − LpsRange)
3) If the current bin value is the LPS (Figure 8.43):
   a) New Low′ = (Low + Range) − LpsRange
   b) New Range′ = LpsRange
4) If the new Range′ is less than a threshold, renormalise it and output a 0 or a 1. Renormalising involves:
   a) Doubling Low
   b) Doubling Range
   c) Optionally shift Low
   d) Output a 0 or 1 (Figure 8.44)
5) Update the context model based on the actual bin value. If the MPS was coded, reduce the LPS probability. If the LPS was coded, increase the LPS probability. If the LPS probability reaches 0.5, swap LPS and MPS, see Section 8.9.4.

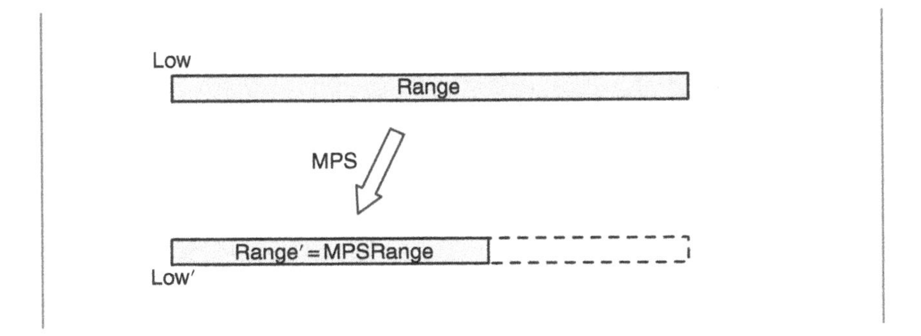

Figure 8.42 HEVC BAC: coding an MPS

5 Between them, valMPS (the value of the MPS, either 0 or 1) and LpsRange specify the probability of the bin being 0 or 1, i.e. P(0) and P(1).

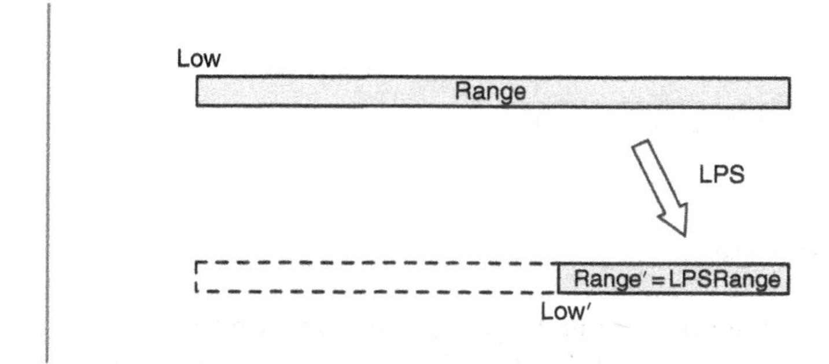

Figure 8.43 HEVC BAC: coding an LPS

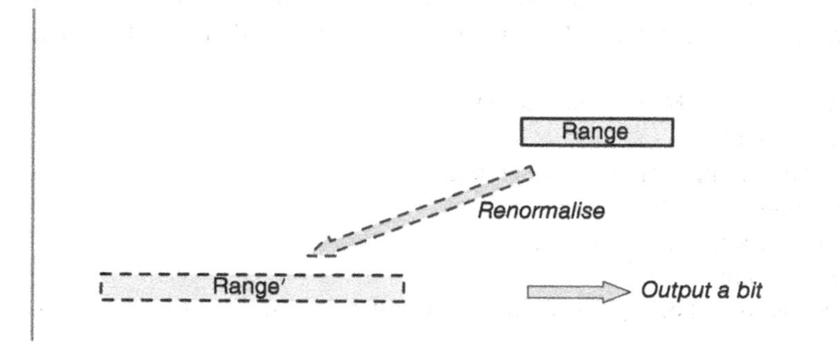

Figure 8.44 HEVC BAC: renormalising at the encoder

Bypass coding of a bin is equivalent to the following steps:

1) Divide the current Range into a lower or upper half, depending on whether the bin is 0 (see Figure 8.45), or 1 (see Figure 8.46).
2) Double Low and Range.
3) Optionally, shift Low and output a single bit, 0 or 1.

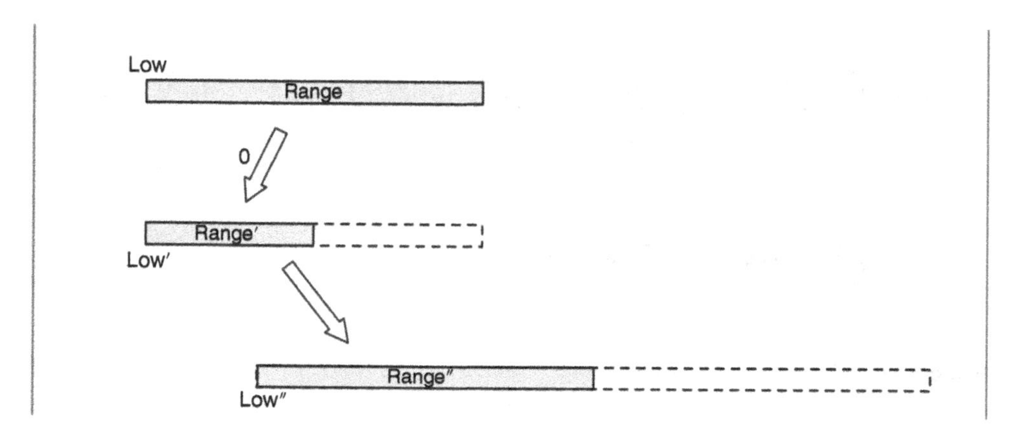

Figure 8.45 HEVC bypass coding '0'

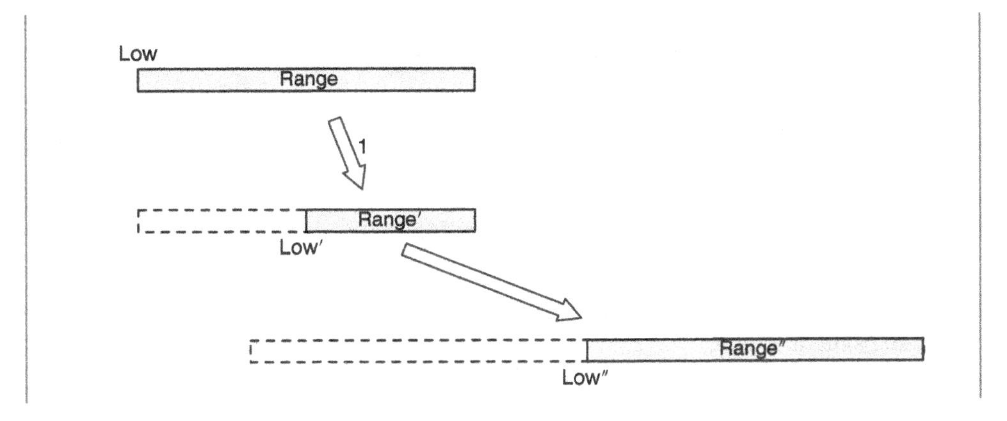

Figure 8.46 HEVC bypass coding '1'

The result of Bypass coding is that the Range does not change. An implementation of HEVC encoding can achieve the same result simply by manipulating the value of Low and leaving Range unaltered.

8.9.3.2 HEVC BAC Decoding

At the decoder, the input bitstream represents bits of a binary fraction, which is a pointer to a series of bin decisions. The input bitstream determines an Offset, i.e. a pointer into a Range. The decoder repeatedly does the following (see Figure 8.47):

- Sub-divides the range into a MPSRange and a LPSRange.
- Determines which division the Offset points to, corresponding to MPS or LPS.
- Decode a bin, 0 or 1, and set the new Range according to the division.

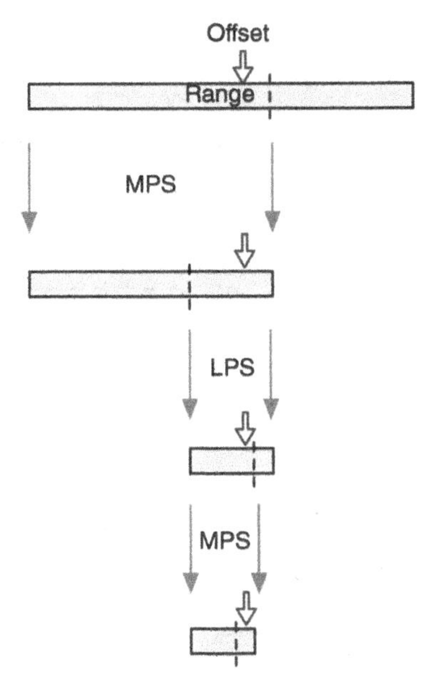

Figure 8.47 HEVC BAC: basic decoding process

- If the Range falls below a threshold, read another bit from the bitstream, renormalise the Range and Offset and refine the offset with the new bit.

In more detail, the decoding process is as follows:

The arithmetic decoder state is initialised to a Range of 510 and an Offset, which consists of the first 9 bits received from the bitstream. This gives us an Offset to 9 binary places, i.e. a 9-bit binary number in the range 0–512. This is equivalent to a binary fraction with 9 decimal places, pointing within the Range (0.0, 1.0).

Decoding a non-Bypass bin:

1) Based on the Context and the current Range, determine valMPS and LpsRange, i.e. the sub-division of Range.
2) If the Offset falls within MpsRange (see Figure 8.48):
   a) Decoded bin = MPS
   b) New Range′ = MpsRange
3) If the Offset falls within LpsRange (see Figure 8.49):
   a) Decoded bin = LPS
   b) New Range′ = LpsRange
   c) New Offset′ = Offset − MpsRange

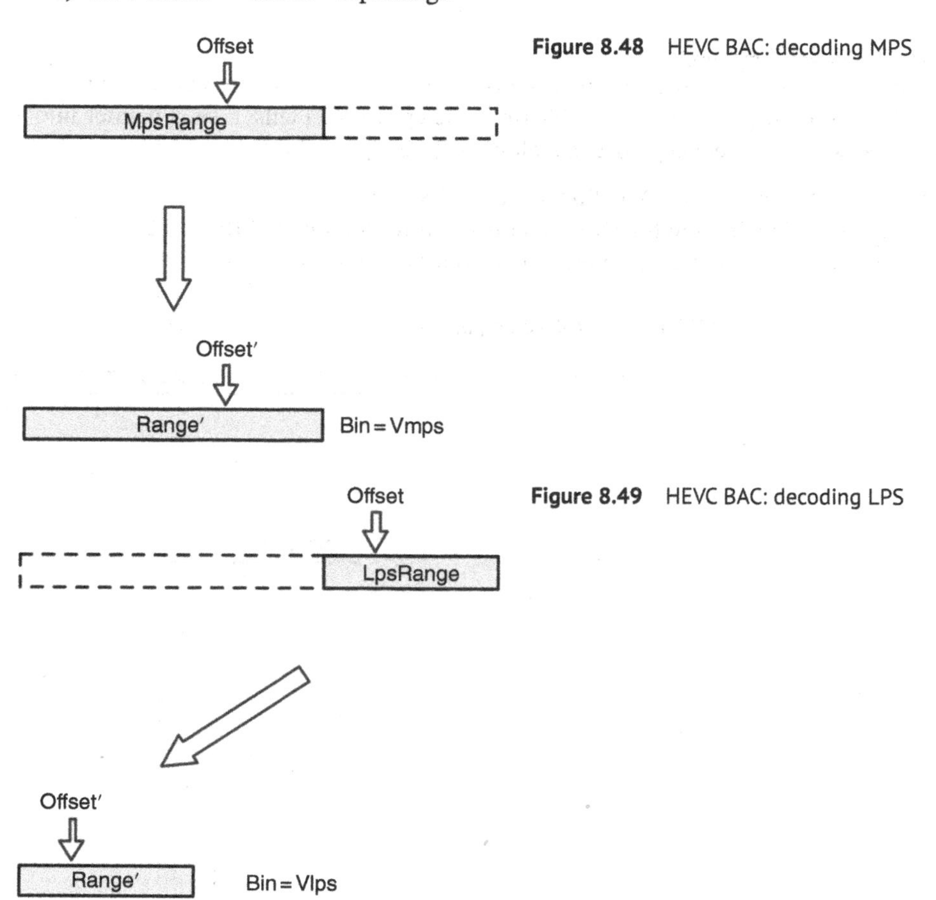

Figure 8.48 HEVC BAC: decoding MPS

Figure 8.49 HEVC BAC: decoding LPS

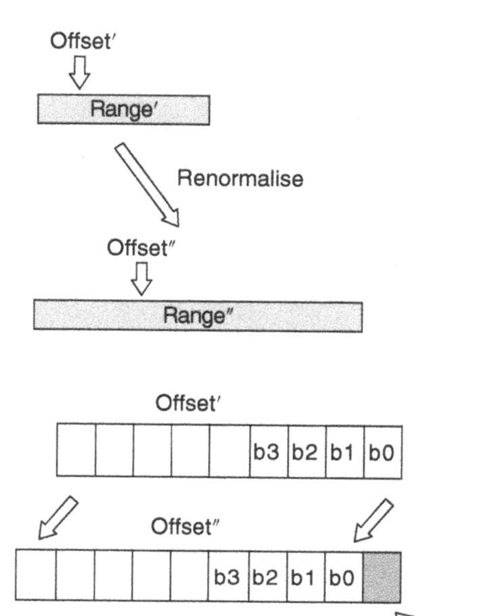

Figure 8.50 HEVC BAC: decoder renormalisation

4) Is the new Range′ < 256? If so, renormalise, repeating the following until Range′ ≥ 256 (see Figure 8.50):
   a) Double Range′ and Offset′
   b) Read a bit from the bitstream and set it as the LSB of Offset″
5) Update the context model based on the bin decoded, MPS or LPS.

8.9.3.3 HEVC BAC Coding Example

We will look at an example of HEVC BAC encoding.

We want to encode the following sequence of bin values:

0, 0, 1, 0

The BAC encoder has stored entries for each of the bins, indicating the value of the most probable symbol, ValMPS, and the Range for the least probable symbol, LpsRange (see Table 8.15).

Table 8.15 HEVC BAC coding example: values and ranges

Bin	Value	ValMPS	LpsRange
1	0	0	102
2	0	1	110
3	1	0	80
4	0	0	242

Figure 8.51 shows the process of encoding four bins.

1) Initialise (Low, Range) to (0, 510).
2) **Encode bin 1:** Look up the context model to find ValMPS and LpsRange. The bin value to be coded is 0, which is the same as the value of MPS. Low′ remains unchanged and Range′ becomes MpsRange (410)[6].
3) **Encode bin 2:** The bin value to be coded is 0, which is the opposite of valMPS, so this is an LPS. Low′ becomes 300 and Range′ becomes LpsRange (110).
4) Range is now less than 256 so the encoder renormalises until Range ≥ 256:
   a) Low is ≥256, so subtract 256 from Low → 44
   b) Double Low, Range → (88, 220)

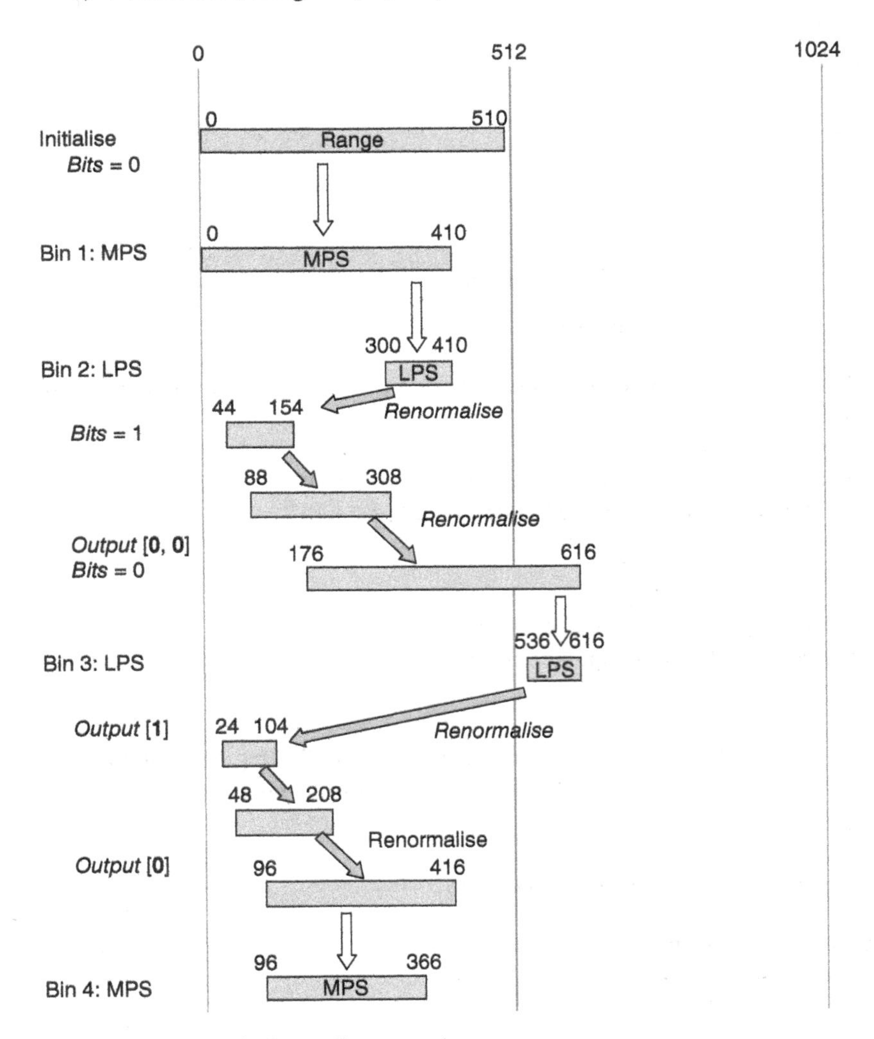

Figure 8.51 HEVC BAC encoding example

6 MpsRange = 512 − LpsRange, since MpsRange + LpsRange = 512.

   c) Range is still <256
   d) Double Low, Range → (176, 440)
   e) Output 0, 0
5) **Encode bin 3:** The bin value to be coded is LPS. Low′ becomes 536 and Range′ is LpsRange, which is 80.
6) Range is less than 256, renormalise until Range ≥ 256:
   a) Low is ≥512, so subtract 512 from Low → 24
   b) Double Low, Range → (48, 160)
   c) Range is still <256
   d) Double Low, Range → (96, 320)
   e) Output 1, 0
7) **Encode bin 4:** The bin value to be coded is MPS, Low′ remains at 96, Range′ is MpsRange (270).

We have processed four bins, with the result that the encoder has output MSBs 0, 0, 1, 0 of our binary Offset pointer and has updated Low and Range to the current values of (96, 270). The encoder can continue to process bins, updating Low and Range and outputting Offset MSBs whenever it renormalises the Range.

The equivalent BAC decoding process will read Offset bits and decode and output bins.

8.9.4 HEVC Context Modelling

There are two practical challenges with the context modelling example shown earlier in Figure 8.28. First, $P(B=0)$ and $P(B=1)$ can each take values from 0.0 to 1.0. However, one of them will always be ≤0.5 and the other will always be ≥0.5. It may be beneficial to estimate one of these ranges with greater accuracy than the other. Second, calculating a probability $P(0)$ or $P(1)$ involves a division operation, which is computationally intensive.

We can address the first problem and represent the probability estimate more efficiently by tracking only the smaller probability. In the HEVC standard, this is recorded as *P*(LPS), the probability of the LPS. P(LPS) is always ≤0.5. So instead of tracking $P(0)$ and $P(1)$, as we saw in Figure 8.28, the encoder tracks *P*(LPS) (see Figure 8.52), as a quantity LpsRange, which is the sub-division of Range if the LPS is coded. *P*(LPS) always stays between 0.0 and 0.5. Each context model (context) stores LpsRange and a second piece of information, val(MPS), which simply notes the ValMPS, 0 or 1. Table 8.16 lists a few examples.

The smaller of the two probabilities, whichever one is between 0 and 0.5, becomes *P*(LPS), and the higher probability value becomes val(MPS).

Let us look at the last row of Table 8.16. The MPS is 0. However, the probability of the LPS (1) is currently sitting at 0.49, very close to the threshold of 0.5. Assume that we receive a few more binary ones. Eventually, the number of ones will exceed the number of zeros and $P(B=1)$ will exceed 0.5. We can assume that $P(B=1)$ reaches 0.52. Val(MPS) changes from 0 to 1, since 1 is now the MPS and 0 is the LPS. $P(0) = 1.0 - P(1) = 0.48$. Hence, whenever *P*(LPS) increases to the point where it is greater than 0.5, val(MPS) flips to its opposite value and *P*(LPS) is reduced to less than 0.5.

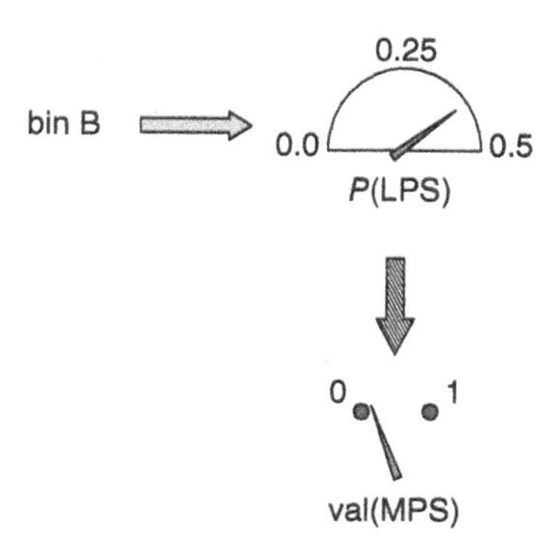

Figure 8.52 Probability model: adjusting *P*(LPS) and val(MPS)

Table 8.16 Examples: *P*(LPS), Val(MPS)

P(*B* = 0)	*P*(*B* = 1)	*P*(LPS)	Val(MPS)
0.2	0.8	0.2	1
0.6	0.4	0.4	0
0.51	0.49	0.49	0

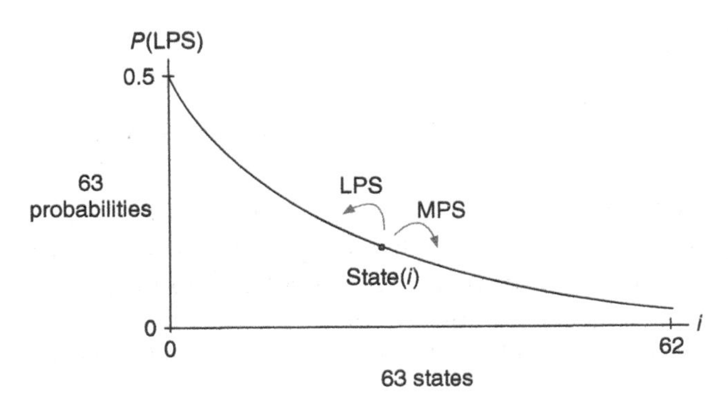

Figure 8.53 HEVC context model: state and *P*(LPS)

We can deal with the second problem, i.e. division being computationally expensive, by replacing the division operation with a state transition. In HEVC, *P*(LPS) is estimated using a state transition model. The current value of *P*(LPS) is represented by a state index *i*, which has a value from 0 to 62 (Figure 8.53). Each state *i* corresponds to a probability *P*(LPS) in the range 0.0–0.5. Smaller indexes correspond to larger probabilities *P*(LPS) and larger indexes correspond to smaller *P*(LPS). *P*(LPS) is updated each time a new bin is encoded.

When a new bin is encoded, its value will either be equal to LPS or MPS. If the value is LPS, then *P*(LPS) should *increase* slightly. This is achieved by decrementing the index *i*. Conversely, if the new value is MPS, then *P*(LPS) should *reduce* slightly, by incrementing the index *i*. The state transitions for coding an LPS or an MPS bin are listed in a table in HEVC (Table 9.53) in the 2018 version of the standard. A few examples are listed here in Table 8.17. Note the following:

- For smaller values of the current state *i*, the state is incremented (MPS) or decremented (LPS) by similar amounts.
- For larger values of the current state *i*, the state is incremented (MPS) by a small amount or decremented (LPS) by a large amount. This tends to push the index towards lower values/higher *P*(LPS).
- State 0 is a special case. If the model is currently in state 0 and receives an LPS, this flips Val(MPS), i.e. if Val(MPS) was a 0, it is now 1 and vice versa. The next state remains at 0, but now with a different Val(MPS).
- State 62 is also a special case. If the model is in state 62 and receives an LPS, the state jumps down to 38. However, if it receives an MPS, it stays in state 62 as the minimum *P*(LPS) has been reached.
- State 63, not shown, is reserved for terminating the arithmetic coding process.

Table 8.17 Selected state transitions from HEVC

Current state	Next state, receive LPS	Next state, receive MPS
0	0	1
1	0	2
2	1	3
....	....	....
16	13	17
17	13	18
18	15	19
....	....	....
48	33	49
49	33	50
50	34	51
....	....	....
61	38	62
62	38	62

In practice, the HEVC context model outputs a sub-division of the Range (see Section 8.9.3), rather than *P*(LPS). This means that instead of the two steps of calculating *P*(LPS) and then dividing the Range according to *P*(LPS), the context model simply calculates a Range sub-division based on the context State.

8.9.5 HEVC Context Models and Initialisation

An HEVC encoder or decoder maintains several hundred context models, each indexed by a context index ctxIdx and each containing variables pStateIdx, which indicates *P*(LPS), and valMps, which indicates the ValMPS, 0 or 1. The context models are organised into context tables (ctxTable). There are 56 different syntax elements that can be coded in a slice segment. Eleven are coded using only bypass coding, and the remaining 45 each correspond to a context table, such as the example shown in Figure 8.54. Each table contains between 2 and 132 context models.

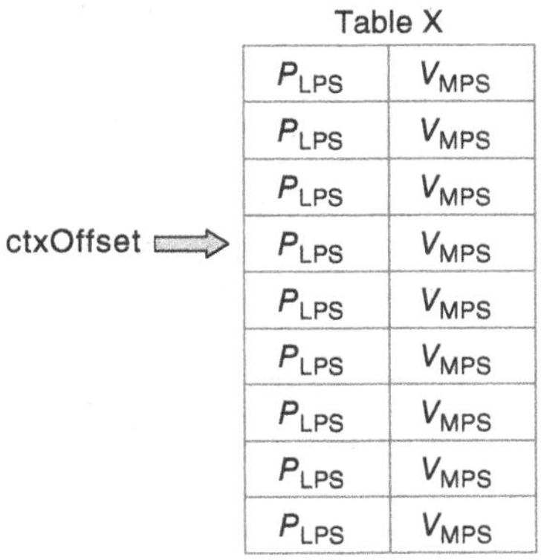

Each of the context models is initialised at the beginning of a coded slice segment. The values of pStateIdx and valMPS, and therefore P_{LPS} and Val_{MPS}, are initialised based on the slice-level QP. An offset parameter ctxOffset depends on the slice type, I, P or B.

During coding of the slice segment, each syntax element is binarised. For each bin, one context model

Table X

	P_{LPS}	V_{MPS}
	P_{LPS}	V_{MPS}
	P_{LPS}	V_{MPS}
ctxOffset ⇨	P_{LPS}	V_{MPS}
	P_{LPS}	V_{MPS}
	P_{LPS}	V_{MPS}
	P_{LPS}	V_{MPS}
	P_{LPS}	V_{MPS}
	P_{LPS}	V_{MPS}

Figure 8.54 Context table example

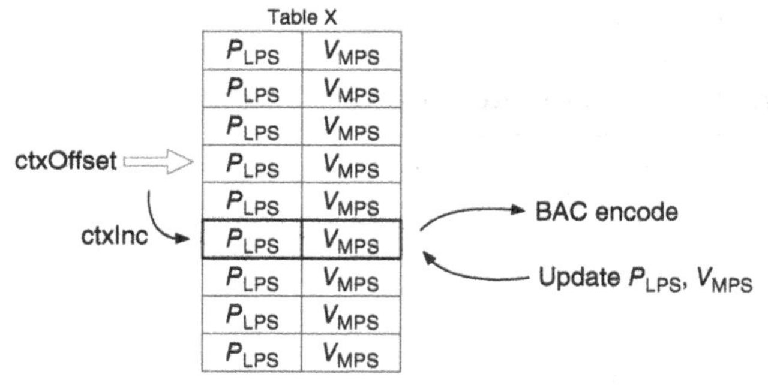

Figure 8.55 Context coding a bin

from one context table is selected and the bin is coded using that context model. The way in which the context model is selected is specified in the H.265 standard and can depend on a number of factors, including:

- The syntax element.
- The bin number in the binarisation (0, 1, 2, etc.).
- The slice type (I, P or B).
- The size of the current TB.
- The content of the current TSB.
- Local statistics such as the availability or content of neighbouring, previously coded blocks.

The process of context-based coding is as follows:

1) Select a context table for the current syntax element X.
2) Binarise the syntax element into 1 or more bins.
3) For each context-coded bin (see Figure 8.55):
   a) Calculate a context increment, ctxInc, either a fixed value or a value depending on local statistics.
   b) Calculate a context index ctxIdx = ctxOffset + ctxInc, which identifies a specific context model. Code the bin using the stored values pStateIdx and valMPS for this context model.
   c) Update the values of pStateIdx and valMPS for this context model, depending on the result of coding the bin.

Example: Selecting a Context for coded_sub_block_flag

This is a single-bit flag, so the binarisation consists of a single bin, bin 0.

The context table for this flag specifies a total of 12 contexts, numbered 0–11. An initial offset, ctxOffset, is selected during initialisation and is set to 0, 4 or 8 depending on the slice type, I, P or B.

When a new value of coded_sub_block_flag is coded, a context increment ctxInc is derived based on the sub-block location and the previously decoded values of coded_sub_block_flag. Based on this derivation, ctxInc is set to 0, 1, 2 or 3.

The chosen context index for coded_sub_block_flag is ctxIdx = ctxOffset + ctxInc. The value is entropy-coded using this context index, and the variables pStateIdx (which models P_{LPS}) and Val_{MPS} for this context index are updated (see Figure 8.56).

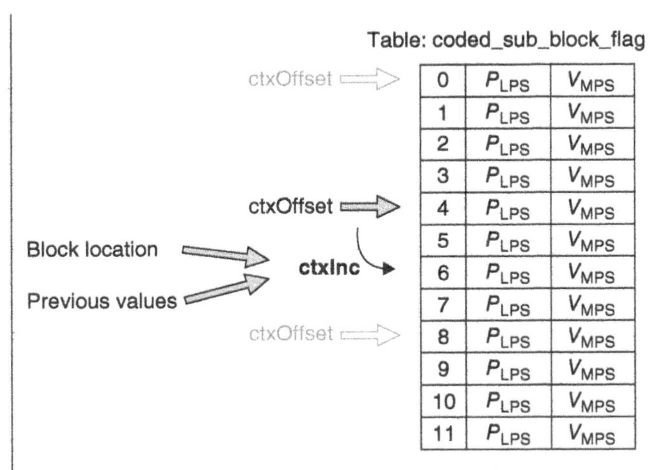

Figure 8.56 Context selection for coded_sub_block_flag

Example: Selecting a Context for cu_qp_delta_abs

This syntax element, representing the magnitude of the change in QP for the current CU, is a non-binary value that is binarised using a particular mapping that is defined in the H.265 standard. The binarisation, shown in Table 8.18, starts with a unary code for lower values of cu_qp_delta_abs and switches to an Exp-Golomb code with a unary prefix for higher values.

The context table for this variable specifies six contexts, numbered 0–5. An initial offset ctxOffset is set to 0, 2 or 4 depending on the slice type.

The first bin of the binarisation, bin 0, has a context increment ctxInc = 0. Bins 1, 2, 3 and 4 of the binarisation, if they exist in the binarised codeword, have a context increment ctxIdx = 1. This means that bins 1–4 share the same context model, i.e. the same P_{LPS} and Val_{MPS}.

Table 8.18 Binarisation for cu_qp_delta_abs

Index	Binarisation
0	0
1	10
2	110
3	1110
4	11110
5	111110
6	11111100
7	11111101
8	1111111000
....	....

Bins 0–4, if they exist, are each coded using a context index ctxIdx = ctxOffset + ctxInc. Bins 5 and higher, if they exist, are bypass coded (see Table 8.19).

Table 8.19 Context coding of cu_qp_delta_abs

Bin	ctxInc
0	0
1–4	1
5+	Bypass

8.10 Entropy Coding in H.266/VVC

The basic design of the VVC entropy coder is similar to HEVC [4]. Like HEVC, VVC codes higher-level syntax elements using fixed-length or variable-length codes and codes syntax elements at the slice layer and below using CABAC.

TBs are partitioned into coefficient groups (CGs). Like HEVC's TSBs, each CG has exactly 16 coefficients, but unlike HEVC, CGs can have sizes of 4×4, 8×2, 2×8, 16×1 or 1×16 depending on the height and width of the TB. The coefficient levels in each CG are encoded in a series of scan passes, including significance – which indicates 0 or non-zero, greater than one, parity – which relates to quantisation, greater than three and remaining value. The position of the last non-zero coefficient in the TB is signalled.

In terms of CABAC coding, non-binary syntax element values are binarised and the bins are BAC encoded in a similar way to HEVC, using context-based or bypass coding. Two probability estimates [5] are stored and updated for each context model, instead of just one, $P(\text{LPS})$, in HEVC. Each of the two probability estimates is updated at a different adaptation rate, based on the statistics of previously coded bins, and the probability index pState and MPS valMPS used to encode or decode a bin are derived from the two probability estimates.

8.11 Conclusion

A video encoder carries out processes including prediction, transform and quantisation to produce a series of syntax elements, which are values to be encoded. Entropy coding converts these syntax element values into a bitstream for transmission or storage. Important considerations include compression efficiency, computational efficiency and ease of access or recovery.

Recent standards such as HEVC and VVC have continued to develop the method of entropy coding that was introduced with the H.264 standard, CABAC.

In Chapter 9, we will look at one of the final pieces of the puzzle, the use of digital filtering to improve prediction in a video codec.

References

1 ITU-T Recommendation H.265. (2019). Section 7.3.6, Slice Header Syntax.

2 Borda, M. (2011). *Fundamentals in Information Theory and Coding*, 9. Heidelberg: Springer-Verlag.

3 Witten, I., Neal, R., and Cleary, J. (1987). Arithmetic coding for data compression. *Communications of the ACM* 3 (6): 520–540.

4 Schwarz, H., Coban, M., Karczewicz, M. et al. (2021). Quantization and entropy coding in the versatile video coding (VVC) standard. *IEEE Transactions on Circuits and Systems for Video Technology* 31 (10): 3891–3906. https://doi.org/10.1109/TCSVT.2021.3072202.

5 Alshin, A., Alshina, E., Park, J. et al. (2011). CE1: Multi-parameter probability up-date for CABAC. *JCT-VC document JCTVC-G764, 7th Meeting of the JCT-VC.*

9

Coded Video Filtering

9.1 Introduction

Video filtering involves applying a digital filter to pixels in a video image to modify or enhance the image in some way. Filtering can be applied for a number of purposes. We will concentrate in particular on filtering that is applied within a video encoder and a video decoder, sometimes described as loop filtering or in-loop filtering. The main purpose of an in-loop filter is to improve the performance of the video codec itself, by improving the prediction source and therefore improving the performance of prediction. As Figure 9.1 illustrates, reference frames in an encoder can be filtered before creating a prediction for a block to be encoded. The decoder has to create an identical prediction, so the same filter is applied to reference frames in the decoder (see Figure 9.2).

Two examples of in-loop filters used in video coding are deblocking filters and deringing filters. A deblocking filter attempts to reduce distortion at block boundaries. This is the characteristic blockiness, introduced by the lossy encoding process, that is familiar from over-compressed videos and images. A deringing filter attempts to reduce visible ripples near strong edges in a video image.

In this chapter, we will look at these and other types of in-loop filters, with a particular focus on how they are used in the H.265/HEVC standard.

In-loop filters affect the visual appearance of the filtered image. Visual quality is inherently subjective, which means that the extent to which such filters improve the displayed image is a matter of individual opinion. However, if an in-loop filter does an effective job of improving prediction and compression performance, then it also enables coding of higher-quality video at a given bitrate.

Lossy compression can provide significant amounts of compression but introduces distortion, sometimes described as quantisation noise, into the decoded video image. This distortion often appears in characteristic forms known as encoding artefacts. For example, block-shaped distortions or blocking artefacts have been a characteristic feature of lossy image and video coding, since early standards such as H.261 and JPEG were introduced. Figure 9.3 shows part of an original frame from the *Jockey* test clip. The same frame, after encoding and decoding using a relatively high quantisation parameter (QP), is shown in Figure 9.4. Blocking artefacts – visible square or rectangular blocks that appear superimposed on the image – are very obvious in Figure 9.4.

Coding Video: A Practical Guide to HEVC and Beyond, First Edition. Iain E. Richardson.
© 2024 John Wiley & Sons Ltd. Published 2024 by John Wiley & Sons Ltd.
Companion website: www.wiley.com/go/richardson/codingvideo1

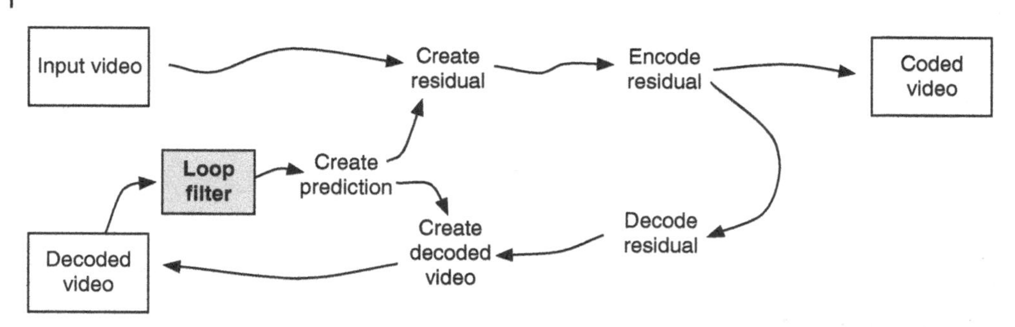

Figure 9.1 Video encoder with in-loop filter

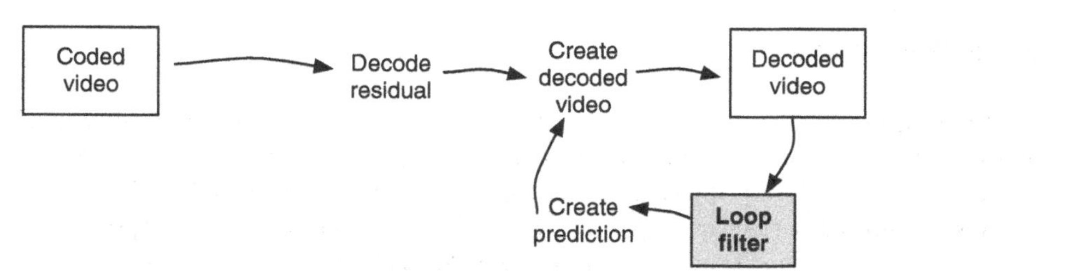

Figure 9.2 Video decoder with in-loop filter

Figure 9.3 Close-up of frame from *Jockey* clip, uncompressed

An in-loop filter uses knowledge about the video encoding process to first discriminate coding artefacts from genuine image features and then to correct or reduce these artefacts whilst attempting to retain genuine image features. For example, blocking artefacts occur next to coding block boundaries, and the magnitude of the artefact, or the amount by which sample values are distorted, is related to the quantiser parameter. A decoder can detect a blocking artefact based on its knowledge of the block boundaries and the current QP.

Figure 9.4 Decoded frame from *Jockey* showing compression artefacts

In-loop filters can be considered to be a way of reducing noise, such as coding artefacts in the decoded video signal, by exploiting knowledge of the characteristics of the noise, such as the way in which the video was coded.

9.2 Filtering and Video Coding

Filtering can be implemented in at least three places in a video codec: as a pre-filter before encoding video, as an in-loop filter within the encoding and decoding processes and/or as a post-filter after decoding and before display or further processing.

A pre-filter can be used to reduce camera noise or other unwanted variations in the video signal before encoding. Capturing video in low-light situations can lead to camera noise or graininess in the video image. Such noise may have high-frequency characteristics, which in turn introduce unwanted high-frequency transform coefficients that can increase the bitrate after compression. Camera shake during video capture can introduce unpredictable motion between frames, which may reduce the effectiveness of motion-compensated prediction, and increase the bitrate after compression. A pre-filter that reduces spatial noise, such as graininess, and/or temporal variation, such as camera shake, may result in more efficient video compression. A pre-filter that is implemented prior to encoding does not need to be defined in a standard.

A post-filter is any filtering that occurs after decoding and before subsequent processing or display. This could include, for example, filtering to further reduce compression artefacts such as blocking, ringing or magnitude shifts.

In this chapter, we will concentrate on in-loop filters. An in-loop filter attempts to improve the compression performance of the codec by reducing artefacts in decoded or reconstructed reference frames (see Figures 9.1 and 9.2). Because identical filtering must be carried out in the encoder and decoder, in-loop filters may be defined in video-coding standards. An in-loop filter is designed with the specific aim of improving compression performance.

Consider an area of a video frame containing a current block (see Figure 9.5). The encoder attempts to find a prediction for the current block in a previous frame. If the encoder made a prediction from an original, uncoded previous frame, it could find a very accurate prediction using the reference block shown in Figure 9.6, since this reference block is nearly identical to the block in the current frame. However, the decoder would not be able to create the same prediction, since the decoder only has access to decoded previous frames. This means that the ideal prediction reference block in Figure 9.6 is not available in a practical video codec. Instead, the encoder and decoder have to create predictions from decoded and reconstructed previously coded frames. If the decoded frame is distorted due to quantisation (see Figure 9.7), then the reference block is no longer identical to the current block, and it may not be possible to create an accurate prediction. Even if the encoder compensates perfectly for motion, the residual – that is the difference between the current and reference blocks – will contain a lot of information due to the extra distortion in the reference block.

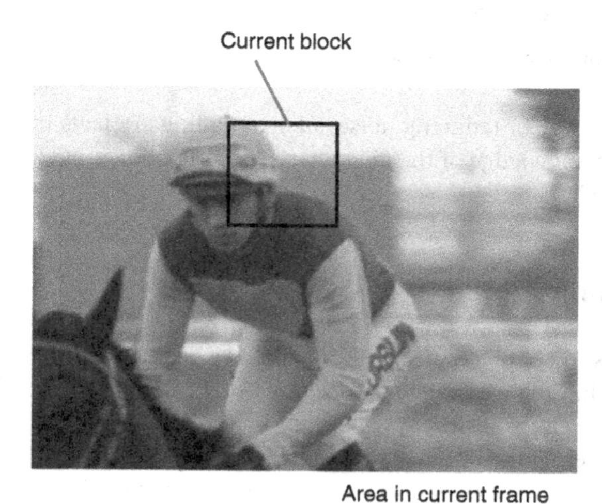

Figure 9.5 Current block, to be predicted

Figure 9.6 Finding a reference block in an original, uncompressed previous frame

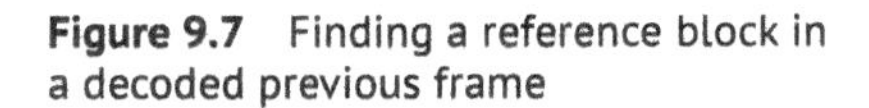

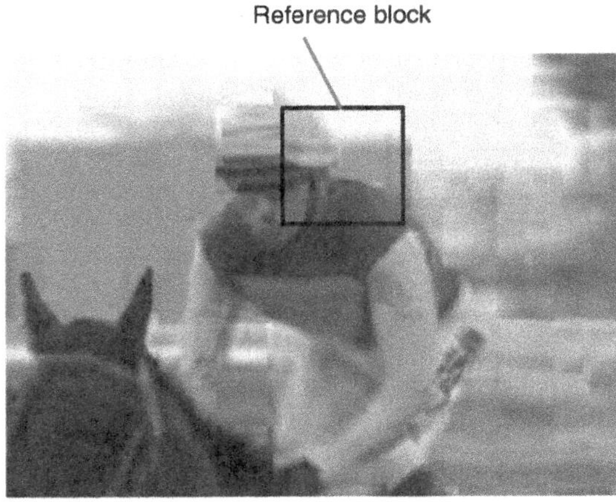

Figure 9.7 Finding a reference block in a decoded previous frame

The reference block of Figure 9.7 is distorted because some information has been removed through quantisation. This cannot be fully restored, but the prediction process might be improved if some of the distortions of Figure 9.7 can be reduced through filtering. This is the job of an in-loop filter, to improve the prediction process by reducing distortion in a reconstructed reference frame.

Some types of distortion are predictable and are a function of the way the video-coding process works. If the codec can identify a distortion or artefact that was introduced by encoding, then it can attempt to filter it to reduce distortion. In the next section, we will consider typical video-coding artefacts and how they arise.

9.3 Detecting and Correcting Video Coding Artefacts

9.3.1 Blocking Artefacts

9.3.1.1 Characteristics of Blocking Artefacts

Blocking artefacts are discontinuities across the boundaries between transform and/or prediction blocks in a coded video image. Blocking artefacts have a characteristic tiling appearance, such as in the middle image of Figure 9.8, as if a grid of tiled glass has been placed on the image. The visible edges of a blocking artefact align with the positions of transform blocks and/or prediction blocks when they first appear[1].

Consider two adjacent blocks with a boundary shown as a dotted line (see Figure 9.9). The top version has no artefact – there is a continuous graduation of tone from light on the left to dark on the right. The lower version illustrates a blocking artefact which appears as a discontinuity in the tone across the block boundary. I have drawn a line A–B across each block.

---

1 If another frame is predicted from a region containing a blocking artefact, using motion-compensated prediction, then the original blocking artefact edges may appear in the subsequent frame, shifted due to the motion vector offset.

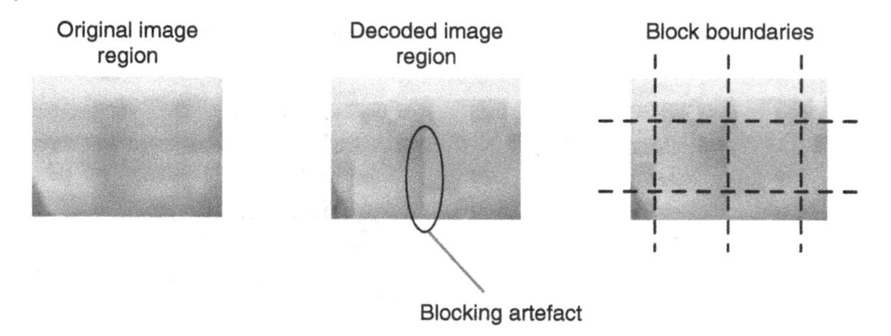

Figure 9.8 Original region, decoded region showing blocking artefact, block boundaries

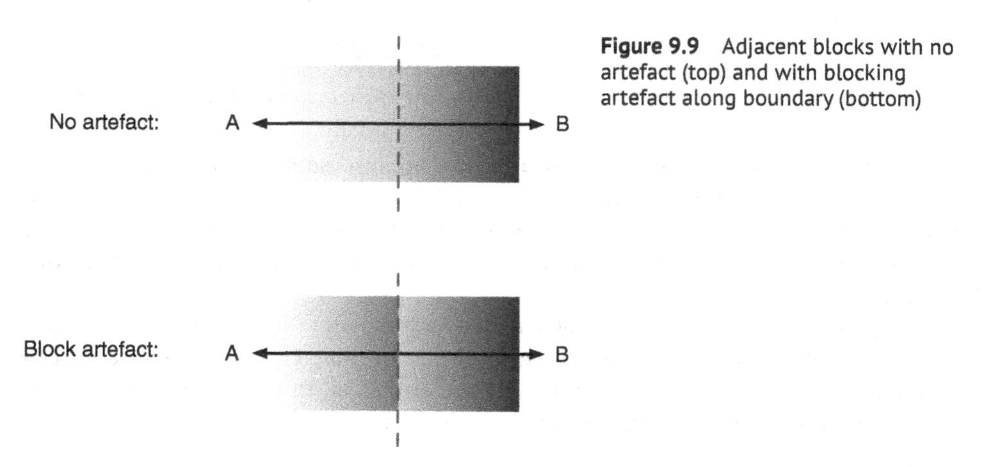

Figure 9.9 Adjacent blocks with no artefact (top) and with blocking artefact along boundary (bottom)

Consider the luminance amplitude along line A–B (see Figure 9.10a). In the distortion-free upper version, the amplitude varies smoothly and continuously from A to B. In the distorted lower version, there is a discontinuity in the amplitude. When this is depicted in the pixels of the blocks, we see the classic block boundary artefact.

In a digital image, pixel values are sampled at discrete intervals, so that the actual sample values behave like the examples in Figure 9.10b. With no distortion, the samples change smoothly across the block boundary. With a blocking artefact, there is a discontinuity as the samples cross the boundary. To compensate for such a blocking artefact, we want to adjust the pixel values nearest the boundary to make the progression more continuous.

9.3.1.2 Causes of Block Artefacts

What causes this type of artefact? A block transform converts a block of samples or pixels into a set of transform coefficients, each of which can be considered as a scaling factor that is applied to a basis pattern. Consider the low-frequency basis pattern shown in Figure 9.11. When there is no artefact, the basis pattern is scaled by coefficient values x and y to produce the decoded blocks as shown[2], with a continuous graduation in tone from left to right.

2 To create these particular blocks, I scaled the DC basis pattern too.

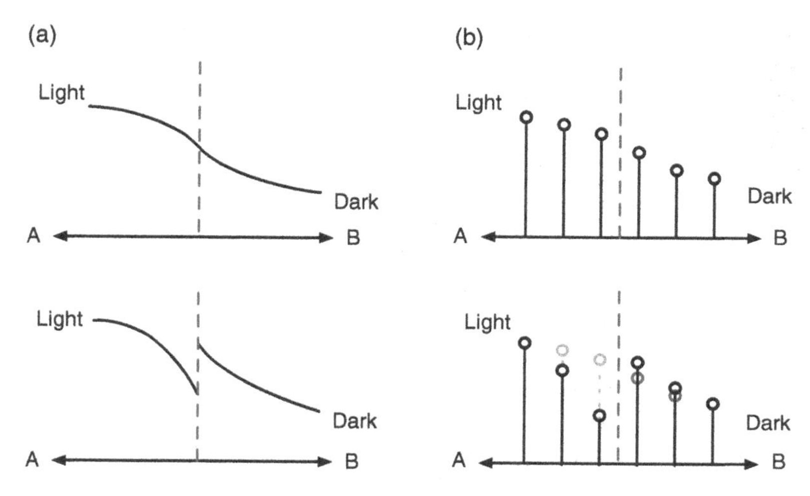

Figure 9.10 (a) Amplitude along line A–B, (b) Samples along line A-B

However, quantisation changes the decoded values of x and y, giving us x' and y' at the decoder. When these modified coefficient values are applied to scale the basis pattern(s), the result can be a discontinuity across the boundary (see Figure 9.12). When discontinuities like this are significant enough to be visible, we get the characteristic blocking effect in the decoded image (see Figure 9.4).

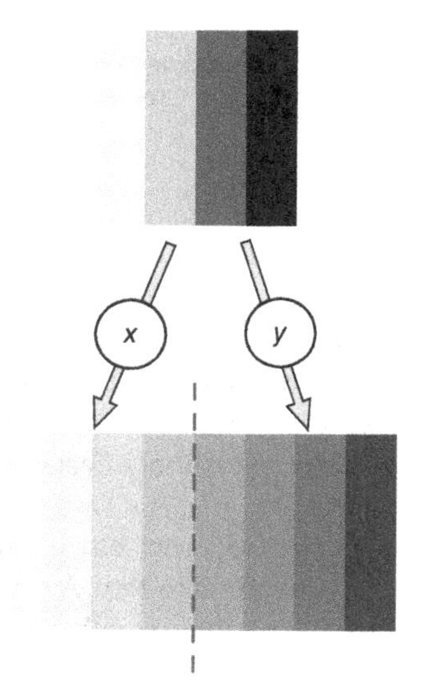

Figure 9.11 Transform basis pattern, scaled to produce left and right blocks, no artefact

9.3.1.3 Detecting and Correcting Block Artefacts

The aim of a deblocking filter is to identify and reduce block boundary discontinuities that were introduced by the compression process, preferably without affecting features that are in the original image and that happen to occur next to a block boundary.

Unless the encoder sends extra information to the decoder, the decoder must identify and correct blocking artefacts, based on known information about the decoded video image. There are certain aspects of blocking artefacts that are helpful here. First, blocking artefacts occur next to or near block boundaries such as transform block or prediction block edges[3]. Thus the effect of a blocking artefact may be strongest in the samples immediately adjacent to the boundary and weaker in samples further away from the boundary. Second, the magnitude of a blocking artefact discontinuity is affected by certain coding parameters such as the quantiser parameter. Hence, a blocking effect may be identified if there is a discontinuity across a block boundary and if the

3 Motion-compensated prediction may cause blocking artefacts to travel away from block edges, e.g. when predicting a current block from a region containing a blocking artefact, with a non-zero motion vector.

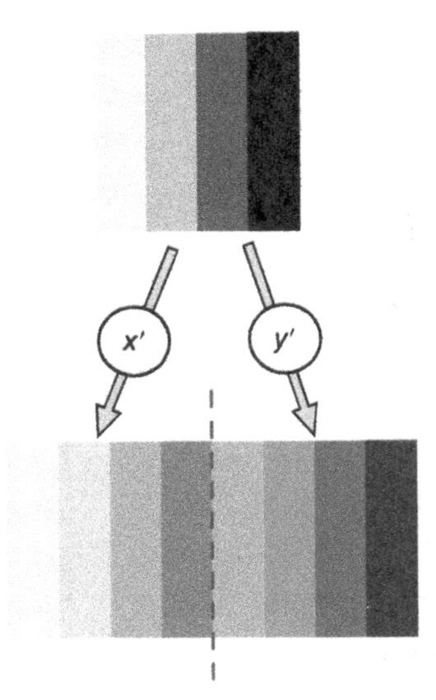

Figure 9.12 Transform basis pattern, quantised and rescaled coefficient values produce an artefact

magnitude of the discontinuity is below a certain threshold, which means that it is likely to have been caused by quantisation effects and not by an image feature. In Section 9.4.1, we will see how an H.265 codec uses these characteristics to detect and filter blocking distortions.

9.3.2 Ringing Artefacts

9.3.2.1 Characteristics and Causes of Ringing Artefacts

Ringing or ripple artefacts are often found adjacent to strong edges in the original video image and appear as dark or light ripples that did not exist in the original image. Figure 9.13 illustrates this. The left-hand image shows a strong edge in an area of an original image, in this case, an edge between a dark and light region of the picture. After encoding and decoding, ripple-like artefacts can appear near the edge, such as the ripples in the right-hand image.

We will consider a real example, comparing a region of an uncoded frame and a coded and decoded frame (see Figure 9.14). An extra shadow, a dark line, has appeared in the right-hand, decoded image – this has the appearance of a ringing, or ripple artefact, perhaps partly influenced by motion-compensated prediction. We can draw a line A–B and look at the sample amplitudes along the line. In simplified form, the uncoded samples along A–B look something like the left-hand diagram of Figure 9.15, a square wave. The decoded samples along A–B look something like the right-hand diagram of Figure 9.15. There is a ripple or overshoot effect, which is analogous to the Gibbs phenomenon found in digital filtering.

Higher-frequency block transform coefficients tend to have smaller magnitudes and are often quantised to zero. A square wave comprises a sum of increasingly higher frequencies, with increasingly smaller magnitudes. In a similar way, a strong edge in an image can be

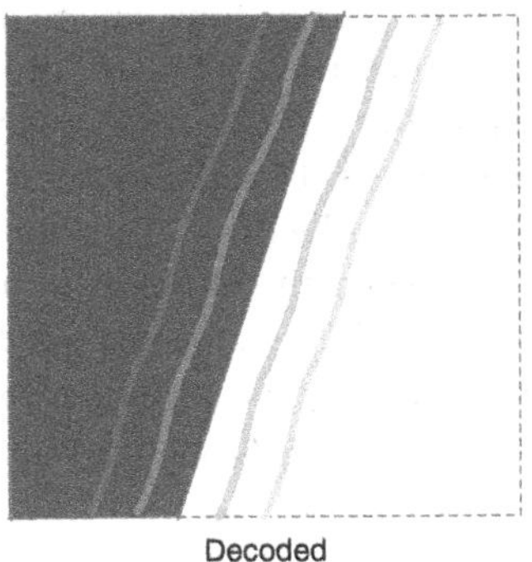

Figure 9.13 Area of image showing ringing artefacts (illustration)

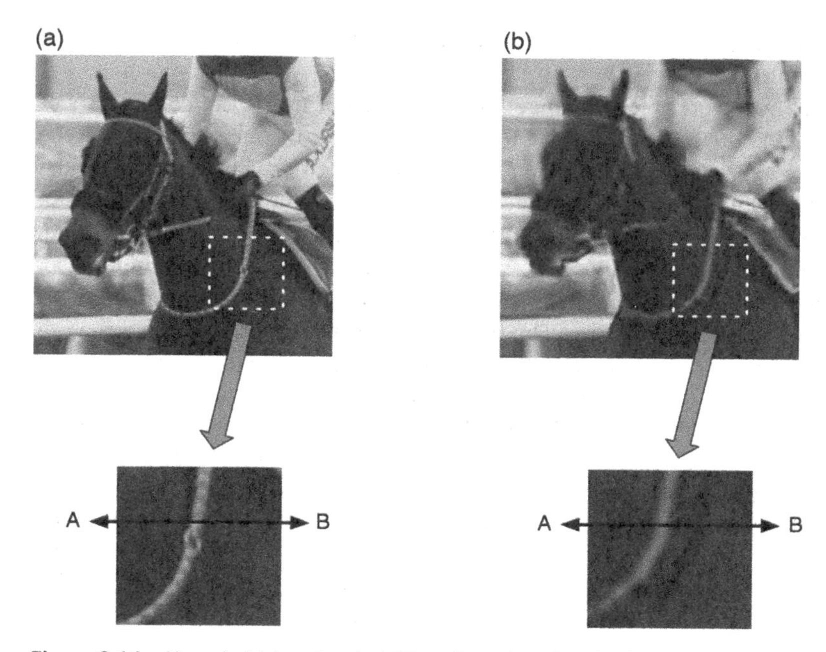

Figure 9.14 Uncoded (a) and coded (b) regions showing ringing artefact (b)

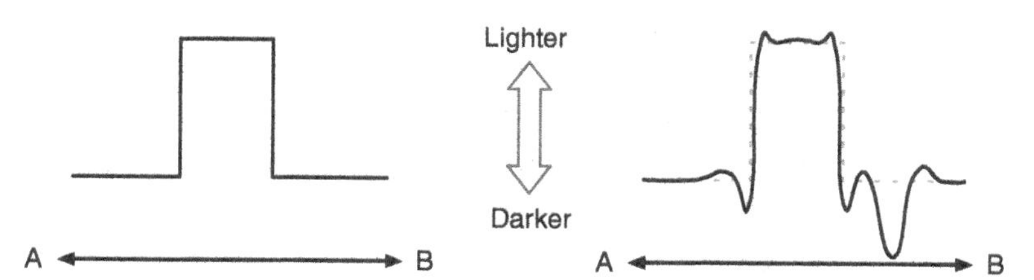

Figure 9.15 Uncoded (left) and coded (right) samples illustrating ringing artefact (right)

represented as a sum of increasingly smaller magnitudes and higher frequency transform coefficients. Removing the higher frequency coefficients by quantising them to zero means that the reconstructed block may exhibit ringing or ripple artefacts, as in this example.

9.3.2.2 Detecting and Correcting Ringing Artefacts

The ringing artefact example of Figure 9.15 has certain characteristic properties. Each of the distorted segments is a local peak or trough shape. By detecting these distinctive shapes in the reconstructed block, an encoder and a decoder can identify and correct the distortions. We will discuss how H.265/HEVC handles this in Section 9.4.2.

9.3.3 Magnitude Offset Artefacts

9.3.3.1 Characteristics and Causes of Magnitude Offset Artefacts

A magnitude offset can be defined as a change in the overall or average magnitude of samples in a video image block. For example, a block or region of samples may appear lighter or darker than the original samples due to a shift in the magnitude of luma samples. Alternatively, a region may appear to have shifted in colour due to a change in the magnitude of chroma samples.

Quantisation in a video codec can change the magnitude of transform coefficients or set them to zero. When the quantised transform coefficients are inverse transformed, the decoded block may experience a shift in magnitude that manifests as a change in the brightness or in the colour of the reconstructed image. Consider the shift in brightness in the example of Figure 9.16. At the top is an original 8×8 block of luma samples, with an average magnitude of 184.7. The block is transformed using an 8×8 DCT, quantised, rescaled and inverse transformed. The magnitude of the decoded samples changes depending on the quantisation step size. The figure shows three different examples, with quantisation step sizes set to 17, 28 and 40. When the quantisation step size is 17, the decoded block has an average magnitude of 182.3, so the average magnitude has been reduced by 2.4 compared to the original block. When the quantisation step size increases to 28, the difference between the original and decoded average magnitude is 3.2, and when it increases to 40, the difference is 5.2. In each example, the average brightness of the block has reduced due to quantisation, which may be visible as a darker patch in the decoded image.

Magnitude shifts can also occur due to motion compensation. For example, Figure 9.17 shows a region of the *Jockey* clip, encoded and decoded using HEVC with a high QP value

Original block

120	157	196	197	190	205	205	201
121	163	193	196	200	204	202	200
124	162	185	193	204	204	204	204
129	172	194	196	202	203	202	202
133	179	199	200	203	202	200	200
134	177	197	198	201	202	202	200
132	156	185	199	198	200	203	202
135	129	137	183	203	200	198	201

Mean = 184.7

Decoded block, Q = 17

123	164	194	191	187	198	204	199
127	156	183	190	192	199	202	199
134	152	175	191	197	200	200	199
140	159	181	194	199	200	199	198
143	169	192	198	198	200	200	197
142	168	192	198	197	200	201	197
138	154	176	192	199	200	199	198
134	139	159	187	202	201	197	199

Mean = 182.3

Decoded block, Q = 28

135	152	174	189	193	192	191	192
138	155	177	192	196	195	194	195
142	159	181	196	200	198	198	199
144	161	183	198	202	201	201	201
144	161	183	198	202	201	201	201
142	159	181	196	200	198	198	199
138	155	177	192	196	195	194	195
135	152	174	189	193	192	191	192

Mean = 181.5

Decoded block, Q = 40

133	152	177	191	195	194	197	201
133	152	177	191	195	194	197	201
133	152	177	191	195	194	197	201
133	152	177	191	195	194	197	201
133	152	177	191	195	194	197	201
133	152	177	191	195	194	197	201
133	152	177	191	195	194	197	201
133	152	177	191	195	194	197	201

Mean = 179.5

Figure 9.16 Original versus decoded block, varying Q step size

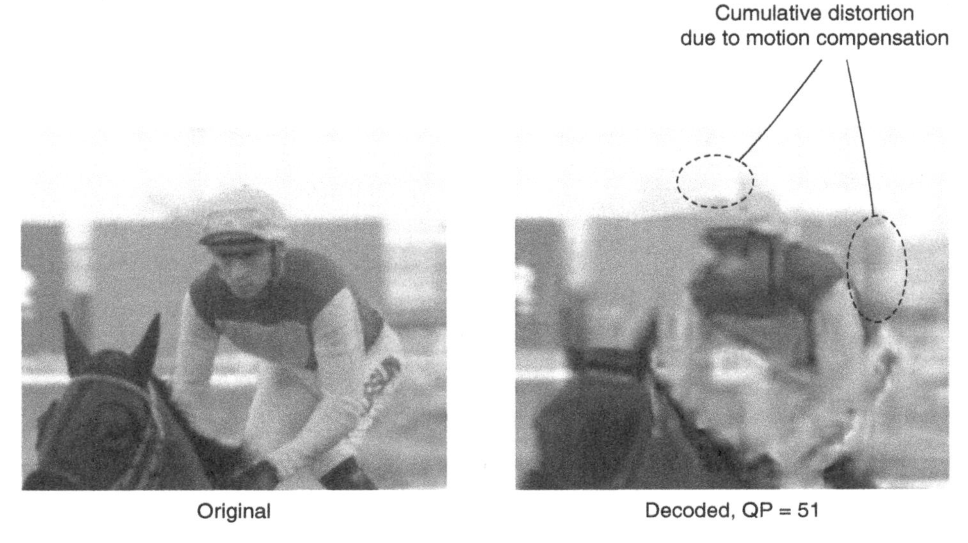

Figure 9.17 Original and decoded regions, showing cumulative distortion in decoded picture

of 51. This is a B-picture that has been predicted using motion compensation from reference frames, which themselves are highly quantised and therefore distorted. When the QP is high, the encoder may be forced to choose distorted regions to predict from, which can lead to cumulative distortions such as the areas highlighted here. The result can be a shift in apparent brightness, such as the lighter area in the jockey's hat in the decoded image, or a shift in colour representation (not shown).

9.3.3.2 Detecting and Correcting Magnitude Offset Artefacts

Magnitude offsets, such as the example of Figure 9.16, may be difficult to identify at the decoder. For instance, the relationship between quantisation step size and magnitude shift is not linear and depends on the image content of the original block. The distortions in the example of Figure 9.17 depend on the content of the current and reference pictures and on the prediction choices made by the encoder. As we will see in the next section, the H.265 filtering process can handle magnitude shifts by signalling band offsets in the encoded bitstream. The decoder uses these offsets to attempt to compensate for the magnitude change(s).

9.4 HEVC In-Loop Filtering

The H.265/HEVC standard specifies a deblocking filter and a sample adaptive offset (SAO) filter process, which includes edge and band offset filtering. Deblocking and SAO filtering are applied to reconstructed image samples before storing decoded pictures in the reference picture buffer (see Figure 9.18). Inter-coded blocks are predicted from filtered samples in the decoded picture buffer, whereas intra-coded blocks are predicted from unfiltered, reconstructed samples.

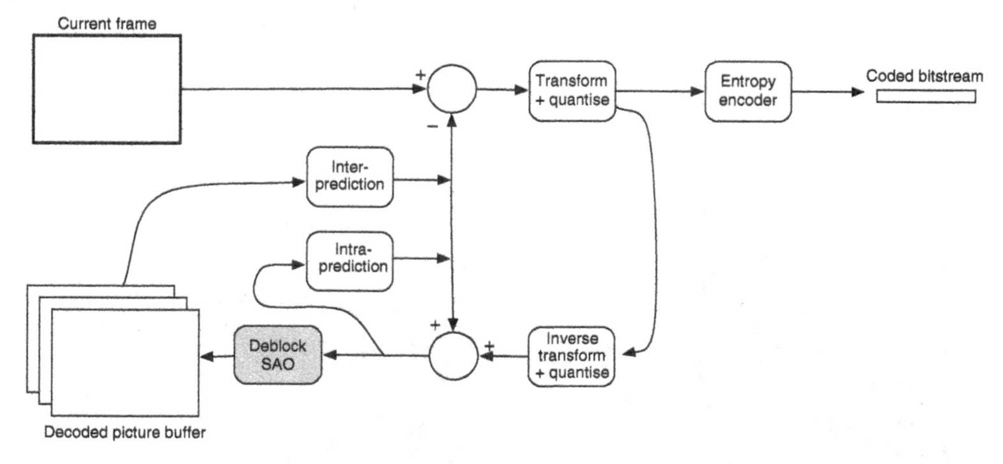

Figure 9.18 HEVC decoder showing deblocking and sample adaptive offset filters

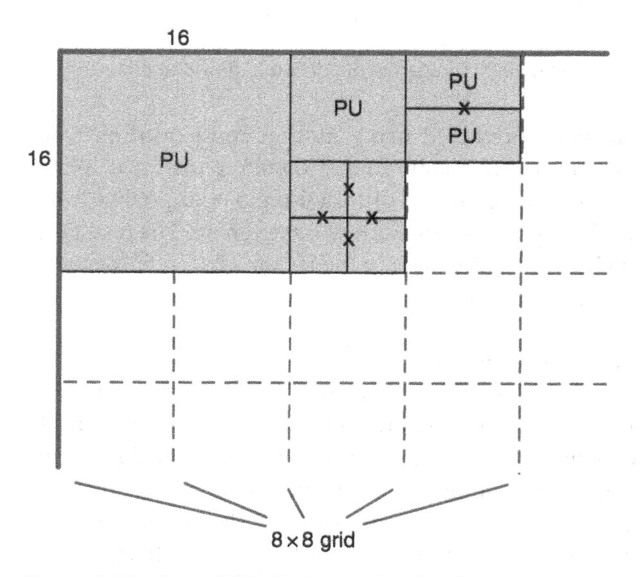

Figure 9.19 Part of H.265 picture showing 8x8 grid and filtered block boundaries

9.4.1 Deblocking Filter

HEVC's deblocking filter attempts to identify and reduce blocking artefacts adjacent to boundaries of transform units and/or prediction units in a reconstructed video frame [1]. When the deblocking filter is active, boundaries of prediction units (PUs) and/or transform units (TUs) that fall on an 8×8 grid may be filtered. Figure 9.19 shows PUs in an H.265 picture. PU or TU boundaries that fall on the dotted lines of the 8×8 grid may be filtered. PU or TU boundaries that do not align with the 8×8 grid – marked with x – are not filtered.

For each four-sample segment along the 8×8 grid, i.e. along the dotted lines in Figure 9.19, the encoder and the decoder make filtering decisions. Figure 9.20 shows a vertical

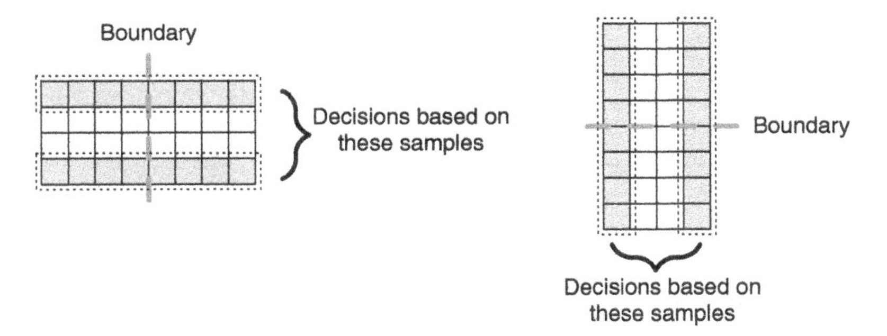

Figure 9.20 H.265 deblocking decision samples

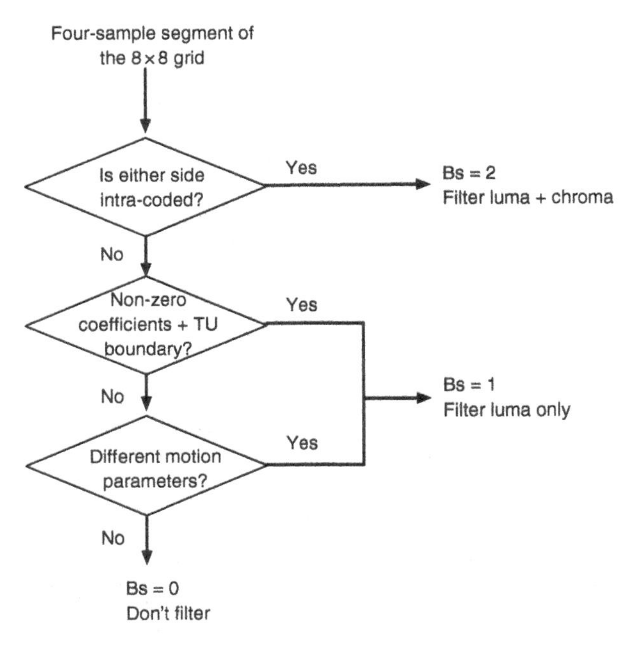

Figure 9.21 H.265 deblocking filter: Boundary Strength Bs

four-sample segment and a horizontal four-sample segment. Filtering decisions are made based on the values of 16 samples perpendicular to the four-sample boundary, as shown in Figure 9.20.

A boundary strength parameter Bs is calculated, ranging from 0 to 2 (see Figure 9.21). If either adjacent 4×4 region is intra-coded, Bs is set to 2, and both luma and chroma samples are filtered. Otherwise, if the two adjacent regions have different motion parameters, or if either region has non-zero coefficients and the boundary is a transform block boundary, then Bs is set to 1 and only luma samples are filtered. If none of these conditions are true, Bs is set to 0, and no filtering takes place.

A further decision process (see Figure 9.22) determines whether a strong filter, a weak filter or no filter is applied to the samples on either side of the grid boundary. If Bs is 0 or if the segment is not actually on a TU or PU boundary, no filtering is applied.

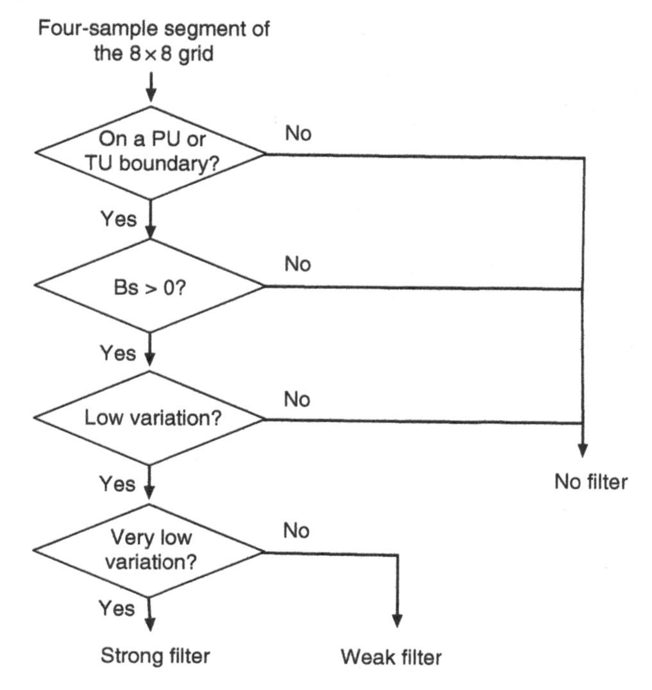

Figure 9.22 H.265 deblocking filter decision process

The next two tests depend on the variation in the samples on either side of the boundary. If there is a lot of variation on either side, we may be in an area where there is a lot of texture and the filter should not be applied. On the other hand, if the variation is low, then the areas on either side of the boundary are probably smooth, such as a flat region or a smooth gradient, and the filter may be applied. If the variation is very low, such as a particularly smooth region on either side of the boundary, and there is a relatively low change across the boundary itself, then a strong filter is applied; otherwise, a weak filter is applied. The strong filter smooths two samples on either side of the block boundary, and the weak filter smooths 0, 1 or 2 samples on either side (see Figure 9.23).

The test for whether neighbouring samples have a low or a very low variation depends on the QP. As we discussed earlier, the maximum magnitude of blocking artefacts is related to the QP. Therefore, if the variation in sample values is above a certain threshold dependent on QP, it is likely to be part of the original image and not due to blocking or other compression-related artefacts.

9.4.1.1 HEVC Deblocking Filter Example

The *Jockey* test clip (see Figure 9.3) was encoded using an H.265 encoder with a QP of 46. This is a relatively high QP, resulting in high compression but also significant distortion of the decoded image. Without any deblocking filter, the decoded pictures show obvious artefacts at block boundaries (see Figure 9.24a). By applying the H.265 deblocking filter, these blocking artefacts are reduced (see Figure 9.24b). Notice that the filter has done a reasonably good job of smoothing the block edges whilst retaining genuine edges in the image.

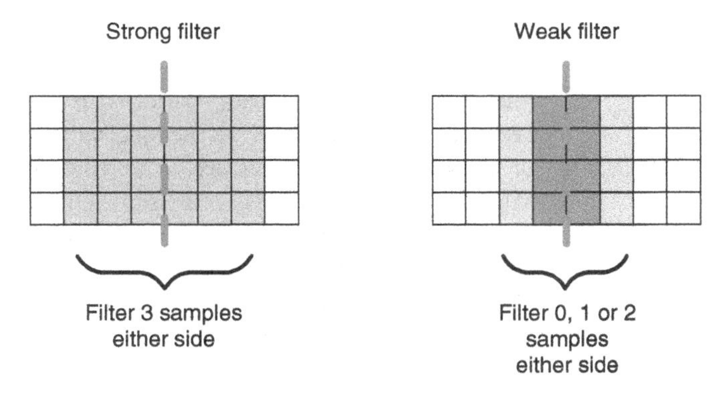

Figure 9.23 H.265 deblocking filter – affected samples

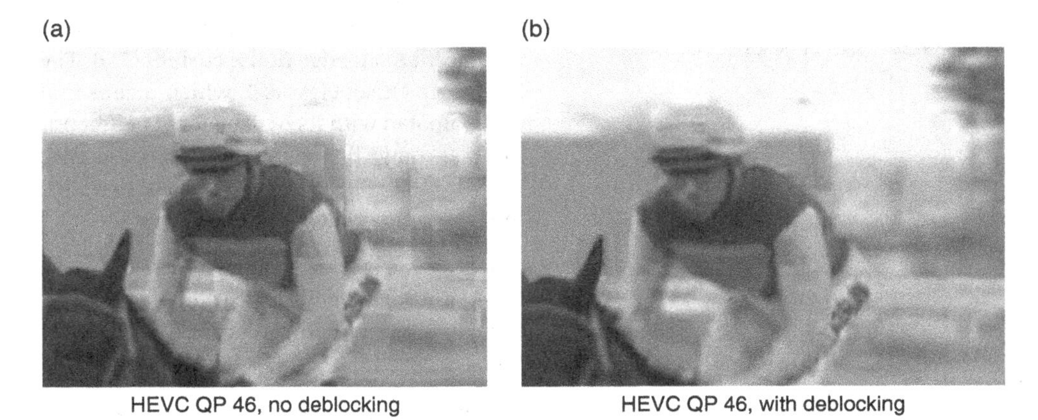

Figure 9.24 Decoded regions, without and with H.265 deblocking

When the deblocking filter is switched on, as shown in the right-hand image, prediction of future coded frames is made from the filtered version of the frame, resulting in an increase in prediction accuracy and therefore in compression performance.

9.4.2 Sample Adaptive Offset (SAO) Filter

The HEVC standard introduced the concept of SAO filtering [2]. HEVC SAO detects certain features in the reconstructed video image, namely strong edges in the image and magnitude offsets, and attempts to correct artefacts related to these features. Unlike HEVC deblocking, SAO is not limited to samples adjacent to block boundaries. Another difference from HEVC deblocking is that SAO parameters are sent in the bitstream for each coding tree unit (CTU), effectively directing the decoder how to apply SAO filtering. If it is in use, SAO is applied after the deblocking filter (see Figure 9.18).

If SAO is enabled for a CTU, the type of SAO is set to Edge Offset, Band Offset or None for the luma component and for the chroma components. For Edge Offset or Band Offset types, further information is sent in the bitstream, which specifies how to apply SAO to the current CTU.

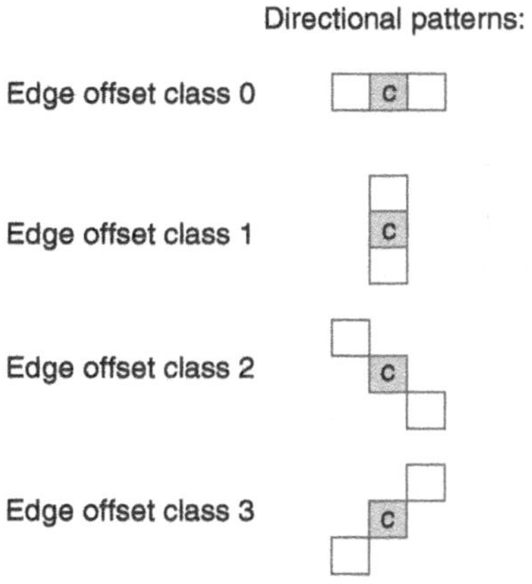

Directional patterns:

Edge offset class 0

Edge offset class 1

Edge offset class 2

Edge offset class 3

Figure 9.25 SAO Edge Offset classes and patterns

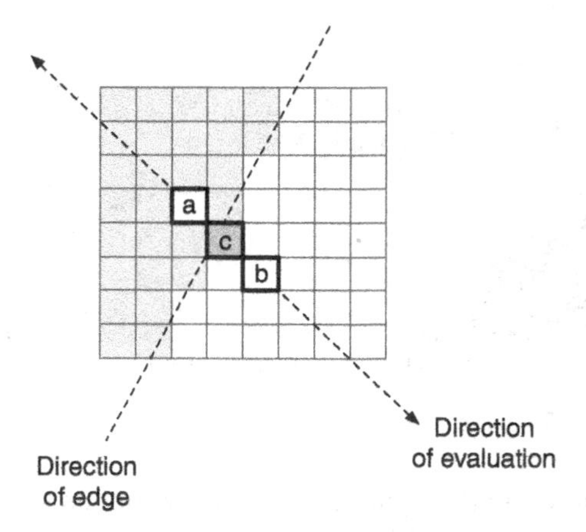

Figure 9.26 SAO: evaluating samples along an edge

9.4.2.1 SAO Edge Offset

The Edge Offset mode of the SAO filter attempts to identify certain patterns in the reconstructed block, such as the local valleys or peaks that are characteristic of ringing artefacts. For each CTB, one Edge Offset class can be signalled by the encoder. The EO class can have one of four values, each associated with a directional pattern (see Figure 9.25). Within the CTB, each sample position c is evaluated with two neighbouring samples, along the line of the selected directional pattern.

Figure 9.26 shows an example. A sample **c** is adjacent to an edge in the current CTB. The SAO Edge Offset class is 2, which means that c is evaluated with its two diagonal neighbour samples a and b. Sample c is compared with its neighbours to determine whether it fits into one of four categories (see Figure 9.27):

Category 1: c is a local valley or trough
Category 2: c is a concave corner
Category 3: c is a convex corner
Category 4: c is a local peak.

If sample c is classified as category 1 or 2, an edge offset is added. If sample c is classified as category 3 or 4, an edge offset is subtracted. The result is that local peaks or troughs, which may be due to ringing artefacts, tend to be smoothed out and reduced.

9.4.2.2 SAO Band Offset

The Band Offset mode of the SAO filter adds or subtracts an offset to all the samples within a certain band, specifically, a certain range of amplitudes. The range of possible sample amplitudes is divided into 32 equal bands. For 8-bit samples, each band from 0 to 31 contains eight possible values (see Figure 9.28). For all the samples in a CTU, a range of four consecutive bands is signalled, for instance, bands 14, 15, 16 and 17 or bands 22, 23, 24 and 25, together with a positive or negative offset within each band. The decoder shifts the samples within the designated four bands by the offset. This enables the Band Offset mode to correct certain shifts in magnitude within the CTU.

Example: Consider the original and decoded blocks from an earlier example shown in Figure 9.29. The decoded block samples have shifted in magnitude due to

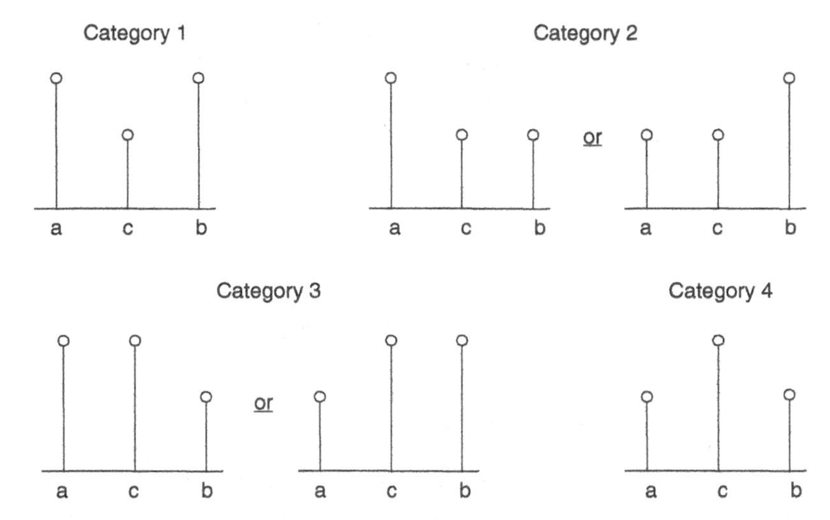

Figure 9.27 SAO: sample categories

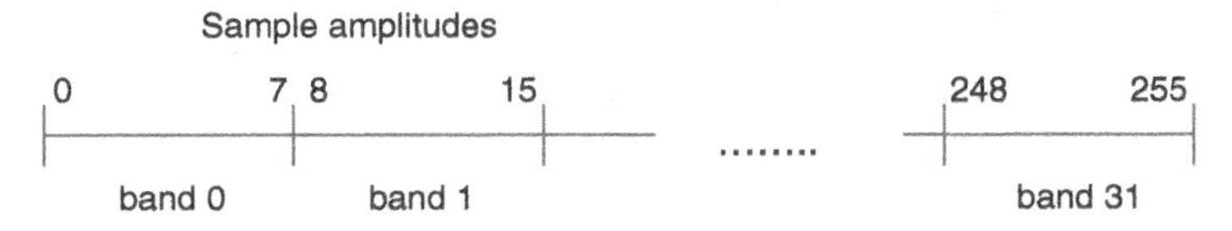

Figure 9.28 SAO bands (8-bit samples)

Original block

120	157	196	197	190	205	205	201
121	163	193	196	200	204	202	200
124	162	185	193	204	204	204	204
129	172	194	196	202	203	202	202
133	179	199	200	203	202	200	200
134	177	197	198	201	202	202	200
132	156	185	199	198	200	203	202
135	129	137	183	203	200	198	201

Decoded block Q = 40

133	152	177	191	195	194	197	201
133	152	177	191	195	194	197	201
133	152	177	191	195	194	197	201
133	152	177	191	195	194	197	201
133	152	177	191	195	194	197	201
133	152	177	191	195	194	197	201
133	152	177	191	195	194	197	201
133	152	177	191	195	194	197	201

Samples in range 176–207

Figure 9.29 SAO band offset example

quantisation. An HEVC encoder is limited to correcting the offset in just four consecutive bands. In this example, the encoder can choose bands 22, 23, 24 and 25, which cover the magnitude ranges 176–207, and send an offset for each of these bands. This enables the decoder to correct the magnitude shift of the samples shown in the dotted box in Figure 9.29.

9.5 VVC Filtering

H.266/versatile video coding (VVC) specifies three loop filtering stages, namely deblocking, SAO and adaptive loop filter (ALF), and a process known as Luma Mapping with Chroma Scaling (LMCS) (see Figure 9.30) [3].

The deblocking filter operates in a similar way to the HEVC deblocking filter described in Section 9.4.1, and the SAO filter is identical to that of HEVC, described in Section 9.4.2. VVC's ALF performs block-based filtering using a Wiener filter, with the filter coefficients derived by classifying each 4×4 block of samples based on the gradients between samples in the block.

LMCS is not, strictly speaking, a coding artefact filter. Instead, it maps sample values to a new range in order to better exploit the entire available dynamic range. A particular video source might only use a portion of the available dynamic range, for instance, a 10-bit luma sample can take values in the range of 0–1024, but a particular video signal may use only part of the range. Remapping the luma samples to the full dynamic range may result in better compression and decoded image fidelity. For inter-predicted blocks, luma samples from both the current block and the reference frame are mapped prior to motion-compensated prediction, and the luma mapping is reversed before applying the other loop filters and saving the decoded picture into the decoded picture buffer (see Figure 9.30).

The parameters that control LMCS and ALF can be signalled in an adaptation parameter set (APS). This is a Parameter Set signalled in a network adaptation layer unit (NALU) that specifies filter and other parameters that can be shared across multiple coded slices. In this

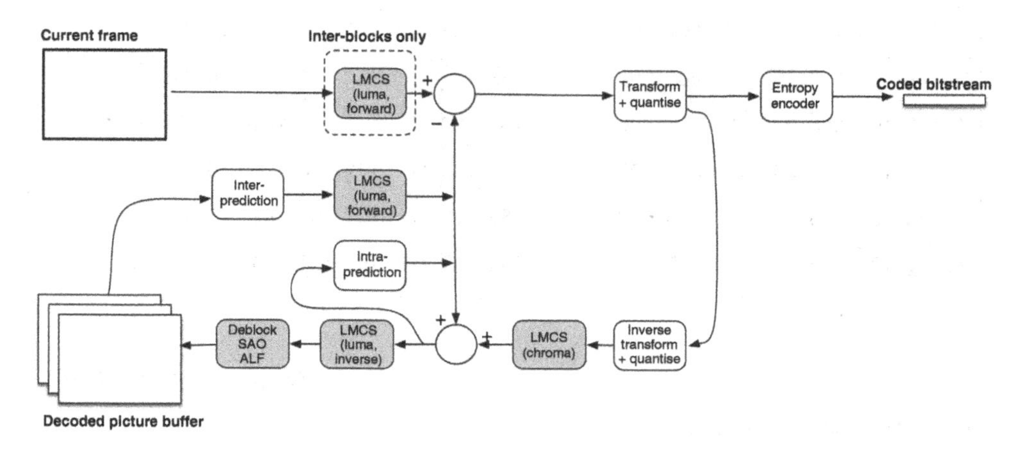

Figure 9.30 In-loop filters in a VVC encoder

way, one or more sets of filter parameters can be set up by sending APS NALUs. Coded slices can then refer back to one or other of the signalled APS, to efficiently signal changes in filtering parameters.

9.6 Conclusions

Loop filtering is used in standard-based video codecs in order to improve the performance of interframe prediction and, therefore, of the overall video codec. In-loop filtering has become increasingly sophisticated with each new video-coding standard. Its importance can be quantified by comparing the performance of a video codec with loop filtering switched on, to the same codec with the filtering switched off. In the VVC standard, loop filtering can improve compression performance by up to 10% depending on the source content [4].

We have now reached the end of our tour of the main processing steps in a video encoder or decoder. In the next chapter, we will look at how the video codec behaves in its natural habitat, specifically, how it enables efficient transmission and storage of digital video.

References

1 Norkin, A, Bjøntegaard, G., Fuldseth, A. et al. (2012). HEVC deblocking filter. *IEEE Transactions on Circuits and Systems for Video Technology* 22 (12): 1746–1754. doi: https://doi.org/10.1109/TCSVT.2012.2223053.

2 Fu, C.-M., Alshina, E., Alshina, A. et al. (2012). Sample adaptive offset in the HEVC standard. *IEEE Transactions on Circuits and Systems for Video Technology* 22 (12): 1755–1764. doi: https://doi.org/10.1109/TCSVT.2012.2221529.

3 Karczewicz, M., Hu, N., Taquet, J. et al. VVC in-loop filters. *IEEE Transactions on Circuits and Systems for Video Technology*. 31 (10): 3907–3925. doi: https://doi.org/10.1109/TCSVT.2021.3072297.

4 Bross, B., Chen, J., Ohm, J.-R. et al. (2021). Developments in international video coding standardization after AVC, with an overview of versatile video coding (VVC). *Proceedings of the IEEE* 109 (9): 1463–1493. doi: https://doi.org/10.1109/JPROC.2020.3043399.

10

Storing and Transporting Coded Video

10.1 Introduction

Video coding is a key technology for storing and transmitting video. The development of video coding standards since the early 1990s has enabled an increasingly diverse range of applications for digital video. These range from digital TV and digital versatile disc (DVD) video, to today's use of video for day-to-day communication and leisure activities, using a wide range of consumer devices. For example, the movement restrictions put in place during the COVID pandemic in the 2020–2022 period led to an unexpected growth in the use of video calling and video conferencing, as many people shifted to working from home [1]. Alongside the development of video coding, it has been necessary to develop file formats for storing digital video and methods for transmitting coded video and multimedia data across digital networks. The last decade has seen a continual increase in the use of streaming video. This has been made possible by developments in network capacity, streaming protocols, and video compression.

In this chapter, we will look at:

- How video codecs are used in practical scenarios in which a coded video stream is stored and/or transmitted over a network or a channel.
- How coded video can be stored in a container file such as an MP4 file.
- The differences between broadcast, user-to-user and streaming video.
- How adaptive bitrate streaming works.
- Ways of controlling the bitrate of coded video and dealing with transmission errors.

Figure 10.1 shows some typical examples. Video can be captured live from a camera, encoded and stored as a file, and/or transmitted over a network. A streaming server can deliver a pre-encoded video programme across a network. A client device can decode and display video that was received over a network and/or stored in a file.

Compression can be used to store video clips efficiently. A video clip takes up less space on disc or in memory if it is stored in a compressed form as a coded video file. In the consumer world, video and audio are often stored in a compressed, rather than an uncompressed form. When you record a video on a smartphone, the video is captured by the camera, processed, encoded with lossy compression and stored as a compressed file in the smartphone's memory (see Figure 10.3a). There may not be an option to store or retrieve

Coding Video: A Practical Guide to HEVC and Beyond, First Edition. Iain E. Richardson.
© 2024 John Wiley & Sons Ltd. Published 2024 by John Wiley & Sons Ltd.
Companion website: www.wiley.com/go/richardson/codingvideo1

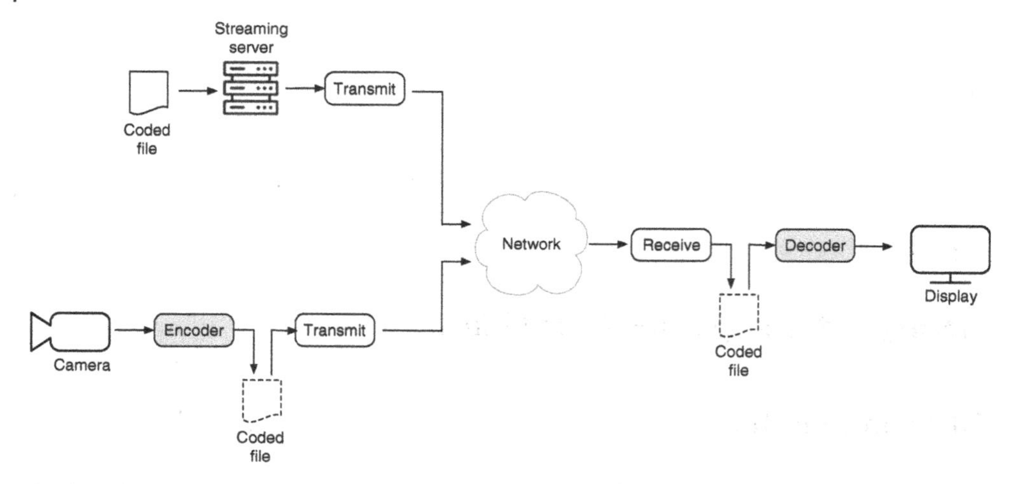

Figure 10.1 Transmission and storage scenarios

the original, uncompressed video, which means that the video file is 'born compressed' which means that it starts life as a compressed file. When lossy video compression is used, what is available to the user is always a reduced-quality version of what was captured by the camera. If the video encoder keeps the compressed visual quality relatively high, then the compressed version is good enough for many consumer applications such as sharing the video, editing, uploading and playback on a large screen.

In professional scenarios such as TV and film production, it is important to keep video quality as high as possible during shooting, storage, editing and post-production. Once the content is ready for distribution, it may then be lightly encoded, using a lossy codec but with very high quality, and then further encoded or transcoded into one or more versions for broadcast or transmission to consumers.

A video file containing a complete compressed sequence can be stored or transmitted as a single, complete file. It is then possible to share a video file by email or file transfer in the same way that any other file would be sent, using a protocol such as transmission control protocol (TCP) to packetise the file, send the packets and reassemble a complete file at the destination.

Face-to-face video calls over services such as FaceTime, Zoom, Teams and Skype require video to be delivered from a source, such as a computer's camera, to a destination, such as someone else's screen, in a short period of time, preferably a fraction of a second, so that conversation feels as natural as possible.

When broadcasting a TV channel, a coded version of a TV programme is created in a suitable format for transmission to the home. The video file is typically packetised and transmitted over a channel such as terrestrial, cable or satellite. The end-user device, e.g. a TV set, receives the transmitted packets, decodes them and displays the decoded video. The received signal may be stored in a compressed form on a digital video recorder (DVR) for later playback.

Streaming involves sending compressed video over a network to end users so that they can start watching the streamed content before an entire file or programme has been downloaded. Popular examples include services such as Amazon Prime, Netflix, Disney+ and Hulu, which offer TV-like content on demand, and platforms such as YouTube, Facebook, Instagram and TikTok, which stream user-generated video content. Unlike broadcast TV, which uses dedicated transmission channels such as the cable TV network, streaming services use the internet to send video to users. From the user's perspective, the streaming experience involves selecting a video to watch, clicking a button to start playback, and then after a few seconds delay, watching the video. To achieve this experience, a streaming server or network of servers delivers a video sequence such as a TV programme in smaller pieces or chunks, over the internet. As soon as the end-user client device has received enough of the file, it starts decoding and playing back the content. As long as the client continues to receive enough coded video data, it can keep playing back video continuously.

The size of the video file and/or the number of bits that must be sent per second can have a significant impact on these storage, delivery and playback scenarios. If the coded video file is too large, the available storage capacity may run out. If it takes too many bits to send each second of video, the capacity of the channel might be exceeded. The file size and bitrate may be reduced by compressing the file more, but then the quality of the decoded video may be unacceptable to the user. Balancing compression and quality and attempting to work within the constraints of the storage or transmission medium, whilst maintaining good quality for the user, is known as video rate control.

The internet is built upon the Internet Protocol (IP), an inherently unreliable packet transmission protocol, which means that packets are not guaranteed to reach the recipient. Data can be sent using a transport protocol such as the TCP or the User Datagram Protocol (UDP). TCP cannot guarantee how quickly the packets arrive, so the delivery delay can vary depending on network conditions. UDP may provide a more predictable delay but cannot guarantee successful delivery of packets. Video streaming and video calling applications have to be able to handle lost or delayed packets. As we saw in the earlier chapters, coded video is inherently interdependent. Blocks in a frame can be predicted from previously coded data in the same frame and in previously transmitted frames. If a packet of coded video is lost, it may be difficult or impossible for the decoder to continue successful decoding. Methods of handling transmission errors include providing regular resynchronisation points and limiting dependencies between coded video blocks.

In this chapter, we will look at file formats for coded video, transmission scenarios and protocols and transmission challenges such as rate control and error handling.

10.2 Storing and Delivering Coded Video

Consider the scenario shown in Figure 10.2. A video source such as a camera feed is encoded, transmitted across a network, decoded and displayed. Video is captured by the camera at a certain resolution and frame rate. Each video frame is encoded and placed in the encoder output buffer. As we have seen in the earlier chapters, coded video frames often vary in size. For example, an intra-coded frame is usually much larger than an inter-coded frame, due to the greater efficiency of inter-frame prediction. This means that the encoder output bitrate is likely to vary from frame to frame.

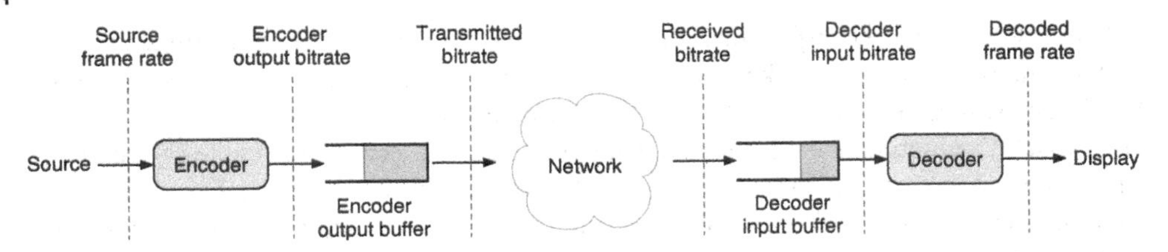

Figure 10.2 Delivering coded video: source to display

Coded frames are transmitted from the encoder output buffer across a network or a channel. The rate of transmission depends on the network connection or channel capacity and might vary during the communication session. The encoder output buffer decouples the variable encoded bitrate from the transmission bitrate somewhat, but it is common to use bitrate control feedback from the encoder buffer to the encoder to prevent the encoder output buffer from overflowing or underflowing (see Section 10.5).

At the receiver, video data are received from the network at a certain bitrate. Depending on the network, this may be the same as the transmitted bitrate, or it might vary depending on the network conditions. The received data are buffered in the decoder input buffer. The decoder retrieves data from this buffer, typically at a rate that mirrors the output bitrate from the encoder. Coded frames are decoded and presented for display, preferably at a constant frame rate so that the video appears to the viewer to play smoothly. The decoded frame rate may be the same as, or less than, the source frame rate. The encoder might choose to drop frames because it does not have enough computational capacity, or as part of a rate control strategy, so that the decoded frame rate will be lower than the source frame rate. Frames may also be dropped at the decoder, perhaps due to lost or delayed packets (see Section 10.6), or because the decoder is running out of computational capacity.

The scenario in Figure 10.2 could apply to applications such as video calling (see Section 10.4.2) or to live video broadcasting or live streaming, where video is encoded and transmitted as soon as it is generated. Of course, many applications use pre-recorded and pre-encoded video, as shown in Figure 10.3. In Figure 10.3a, the source video is encoded and stored in a coded video file. Most consumer devices have limited storage capacity, so when the user starts recording a video, the video content is usually routed through a video encoder so that the video is stored in a compressed form.

In Figure 10.3b, a coded video file is decoded and displayed locally. This is what happens when a consumer plays back compressed video stored on a device such as a mobile phone/cell phone or a computer. The video is stored in a coded form, so it needs to be decompressed in real time in order to play it back.

Figure 10.3c shows a coded file being transmitted across a network, decoded and displayed. A scenario that is not a live video call or live broadcast essentially works this way. A programme is edited and then encoded in a format, and at a bitrate that is suitable for broadcasting, then saved in one or more coded files. When it is time to broadcast the programme, the coded files are transmitted over the network. A client device such as a TV receives the coded file(s) as a series of packets, then decodes and displays the video data (see Section 10.4.1). This is also how streaming video works. A YouTube clip or Netflix episode is stored on a content server. When the user chooses to play the video, it is streamed across the Internet via a Content Delivery Network (CDN) as a series of packets that are

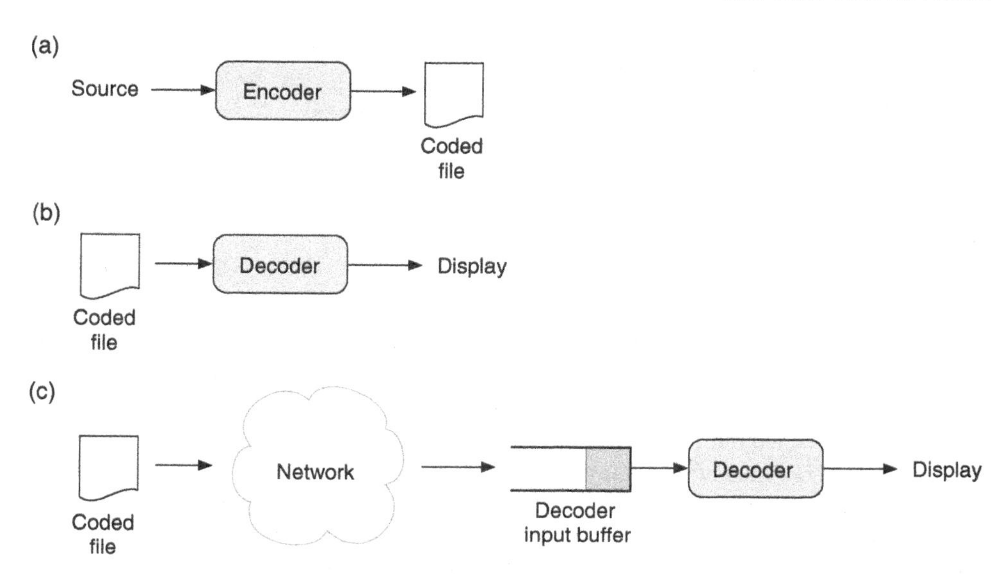

Figure 10.3 Encoding and decoding scenarios with coded video files, (a) encoding a file, (b) decoding a file, (c) downloading or streaming and decoding a file

decoded and displayed at the user's client device (see Section 10.4.3). Data received from the network are stored in the decoder input buffer. As we saw in Figure 10.2, the decoder processes the buffered data at a rate that should preferably ensure continuous real-time playback to the viewer.

These examples illustrate some of the important challenges of storing and communicating coded video, including:

- Storing video in a file format that is suitable for storage, playback and/or transmission.
- Controlling the output of the encoder so that the output buffer does not overflow or underflow and the encoded data are transmitted in a timely manner.
- Handling network issues such as variable transmission rate, variable network delay and packet losses.
- Controlling transmission so that the decoder input buffer does not overflow or underflow and the decoder can maintain continuous playback.
- Synchronising video with audio and any other necessary information.
- Providing as good a quality of experience as possible to the user by playing video at an acceptable quality, avoiding lengthy delays and maintaining continuous playback.

10.3 Coded Video File Formats

Coded video can be stored in a file to:

- Record a video captured on a consumer device, such as a mobile phone/cell phone.
- Store source, edited and processed video files for TV production.
- Store a video programme for broadcast or streaming transmission.
- Record a received broadcast or streamed video programme on a DVR.

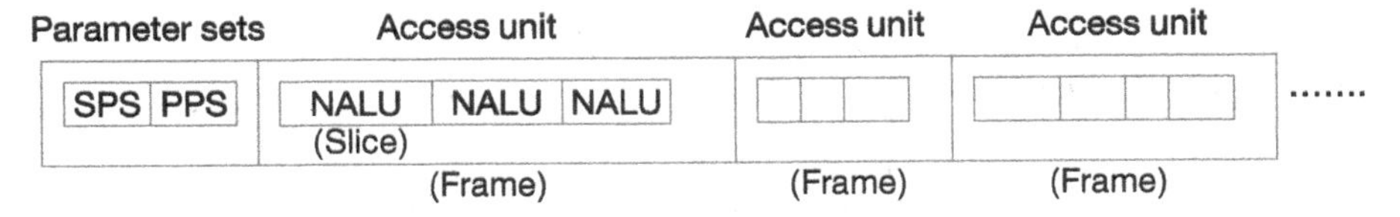

Figure 10.4 H.264/H.265/H.266 sequence structure

Figure 10.4 illustrates a typical structure for an H.264, H.265 or H.266 coded sequence, described further in Section 4.2. Information common to some or all of the frames is stored in parameter sets that may be sent or stored at the start of the sequence in Picture, Sequence and Video Parameter Sets. Each video frame is coded as an access unit (AU), which is a term for a single coded frame. An AU is coded and sent as one or more network abstraction layer units (NALU), each of which is a data structure defined in the standard that contains a coded slice, which is all or part of a coded frame. A decoder can process the coded sequence by receiving the information from start to finish, decoding the set of slices that make up each AU and outputting decoded frames for playback.

A compressed video sequence, such as a complete H.265 sequence, can be stored in a single file. For example, the HEVC model (HM) reference encoder, a software implementation of H.265, produces a file containing an encoded sequence, and the HM reference decoder processes and decodes this file. The file produced by the HM encoder contains the coded AUs stored sequentially (see Figure 10.5).

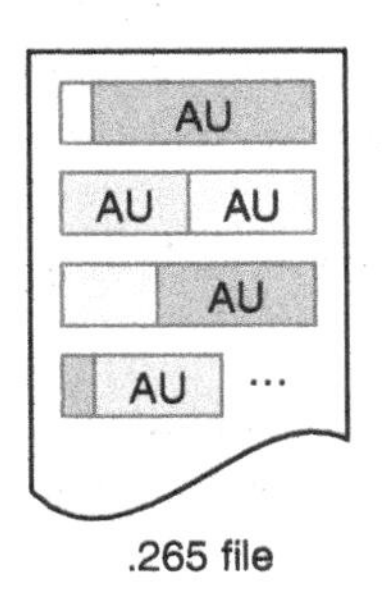

Figure 10.5 H.264/H.265 sequence stored in a file

Video is typically accompanied by audio – perhaps multiple channels of audio – and other information such as metadata and subtitle tracks. One way of storing all this information in a file is to simply concatenate the data (see Figure 10.6). Such a file contains all the information necessary to decode and play a video and audio presentation. However, a simple file structure like this may not be practical for all purposes. With this structure, a decoder needs to have most or all of the file available before it can start playback, since the audio is placed after the video at the end of the file. This means that the decoder has to receive most of the file data before decoding and presenting the video and accompanying audio. This is not ideal for progressive download or streaming, where it is usually preferable to start decoding and playback as soon as possible after the file starts to download.

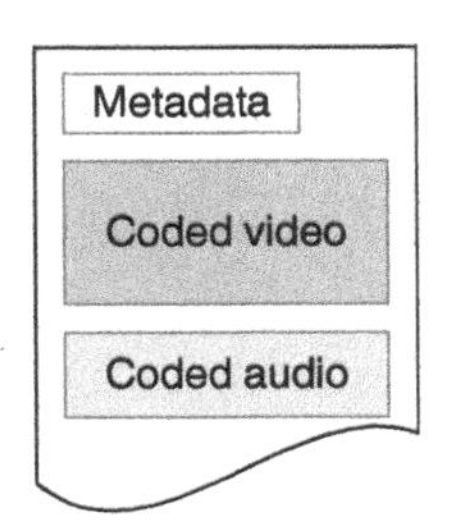

Figure 10.6 Coded video and audio concatenated in a file

Scenarios such as progressive download, broadcast and streaming may benefit from a more sophisticated file format, with interleaving of audio and video data and metadata to make it easier for a client to decode, playback and navigate within the file. We will look at one such file format in more detail: the Moving Picture Experts Group-4 (MPEG-4) file format. Other popular file formats include the MPEG-2 transport stream (TS) format (see Section 10.4.1) and the Matroska video (MKV) format.

10.3.1 The MPEG-4 File Format

The ISO base media file format (BMFF) is a standardised format for storing audiovisual data in media files [2, 3]. The popular MP4 file format, specified in MPEG-4 Part 14, is based on the BMFF [4]. Another document in the same file format standards family [5] specifies how coded H.264, H.265 or H.266 video is mapped to the ISO BMFF/MP4 file structure. Each coded AU becomes one Sample in an MP4 file, as we will see later.

Figure 10.7 illustrates how these file formats are related. The ISO BMFF evolved from the Quicktime format. The MP4 file format is an extension of the ISO BMFF and can optionally use the specification of the advanced video coding/high -efficiency video coding (AVC/HEVC) file format.

A complete movie or video programme can be stored in one or more ISO BMFF or MP4 files. Each of these files contains an audiovisual clip or sequence that is organised as a set of tracks. A file might contain a single video track, with video only and no audio, or a single audio track, with audio only and no video. The file could have a simple audiovisual file with one video track and one audio track, or it could have multiple video and/or audio tracks, e.g. a movie with multiple audio channels for surround sound and different languages.

The ISO BMFF or MP4 file contains a movie or *moov* metadata part, which describes the tracks and their relationships, and a media data or *mdat* part, which consists of a series of coded samples, such as coded audio samples or coded video NALU. Figure 10.8 shows an example of the metadata and media samples that can be stored in an MP4 file. The metadata acts like an index, identifying which tracks exist, how they are interrelated and where to find the coded media samples for each track.

Each track identifies a sequence of samples. These are units of coded media such as coded video or audio, with associated timing information. A player device can use the track information to identify, decode and present samples in the correct time order. To put it another way, the track information tells a player how to play back the video and audio correctly and points to the actual coded video and audio data.

The ISO BMFF, MP4, AVC and HEVC file format specifications leave some flexibility as to how the metadata and coded samples should be arranged in a file. As we will see, files intended for progressive download or streaming may benefit from being organised in certain ways.

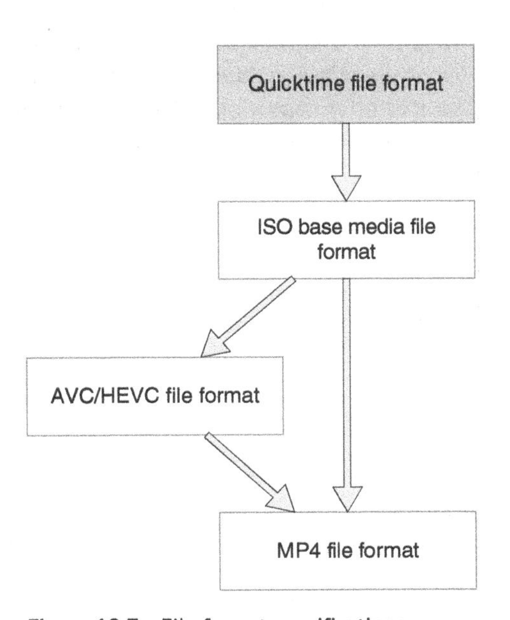

Figure 10.7 File format specifications

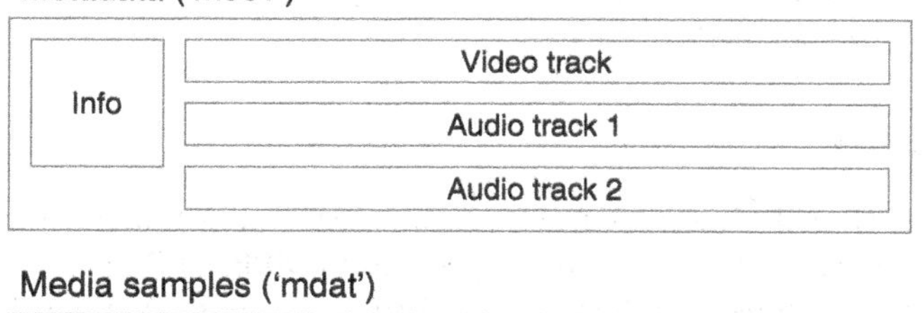

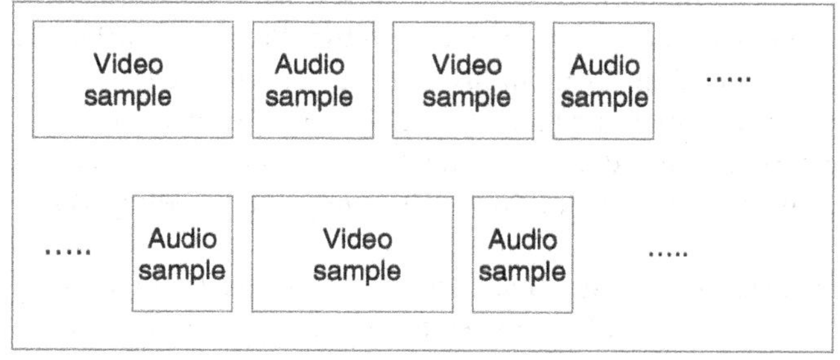

Figure 10.8 Structure of an ISO BMFF or MP4 file

Example: Mobile Video Capture

Consider the following example, Figure 10.9, top diagram. A 6-s video is captured on a mobile device such as an iOS device. The resolution is 1920×1080 pixels (1080p), and the video is compressed using H.265/HEVC. The file size is 6.5 Mbytes, corresponding to a coded bit rate of around 8.5 Mbps. After the file type and version fields (*ftyp*), the samples of captured audio and video are stored in the *mdat* part, followed finally by the *moov* metadata including the video and audio track information. This is a suitable file structure for capturing video, since the iOS device waits to write the *moov* metadata until all the video and audio have been encoded and saved. However, it is not ideal for

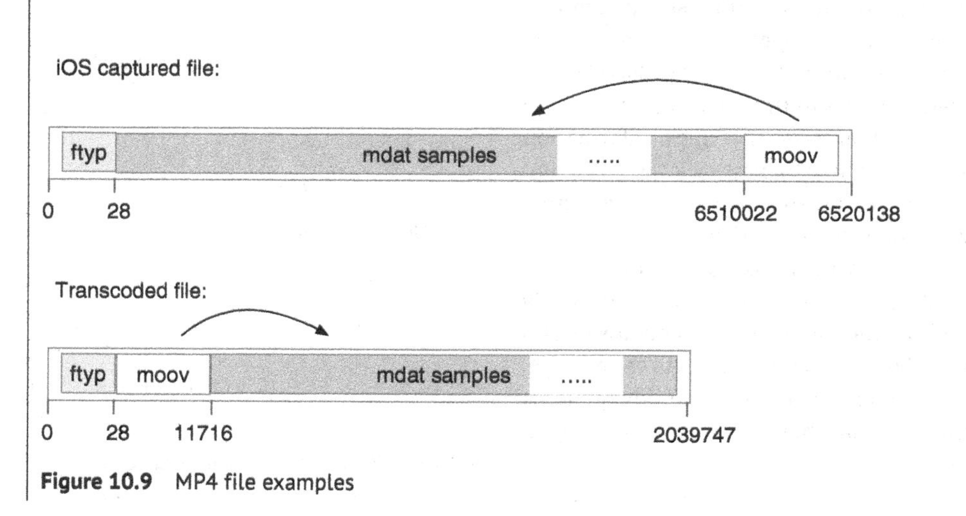

Figure 10.9 MP4 file examples

progressive download or streaming. A client has to download the entire file before decoding and playback can start, since the *moov* data are crucial for presenting the decoded video and audio data and are contained at the end of the file.

The lower diagram in Figure 10.9 shows the same video clip after converting it into a streaming-ready form. The clip has been transcoded – decoded and re-encoded – with higher compression so that the file size is reduced to around 2 Mbytes. It has been converted into a form suitable for progressive download or streaming by repositioning the *moov* element in front of the *mdat*-coded samples. Using progressive download or streaming, a receiver can start decoding and presenting the decoded audio and video data – the *mdat* samples – as soon as they are received, since the necessary *moov* track information is received first.

10.4 Transport of Coded Video

Coded video is sent from a source such as a server to a destination such as a client device according to a set of protocols. A set of protocols is a set of rules and procedures that enable the source and destination to communicate effectively over a network. Delivering real-time video and audio to a user's device requires certain quality of service (QoS) constraints to be met.

For face-to-face applications such as video calling using Zoom, FaceTime and Skype, end-to-end latency, which is the delay from capturing to displaying the video, should be as short as possible and preferably well under a second. For these types of applications, audio should be prioritised over video, and video image quality may be less important than low latency and reliability.

For broadcast or streaming applications such as digital TV, Netflix and YouTube, a longer latency is acceptable. The user is likely to tolerate a few seconds delay between selecting a programme and playback starting. High-quality video and audio are important, though users may tolerate lower quality at the very start of a programme. Stalling/buffering should be avoided if at all possible, since breaks in the programme can be more annoying to the user than variations in playback quality. A 2017 survey found that stalling during a programme was considerably more frustrating than low quality video [6].

10.4.1 Broadcast

Broadcast delivery involves sending coded video and audio across a network or channel to multiple destinations or clients. In a typical workflow (see Figure 10.10), a source or contribution file, such as a master file for a TV programme, is encoded or transcoded into a suitable form for broadcasting. A master file may be stored in uncompressed form or with light compression, so that the source is lossless or near lossless in appearance and takes up a large amount of storage space. Transcoding involves decoding and reencoding the source into a more highly compressed version, depending on the capacity of the transmission channel or network. The European Broadcasting Union (EBU) recommends that high-definition television (HDTV) video should be encoded at bitrates of at least 50 Mbit/s for studio production and around 9 Mbit/s for broadcast transmission [7].

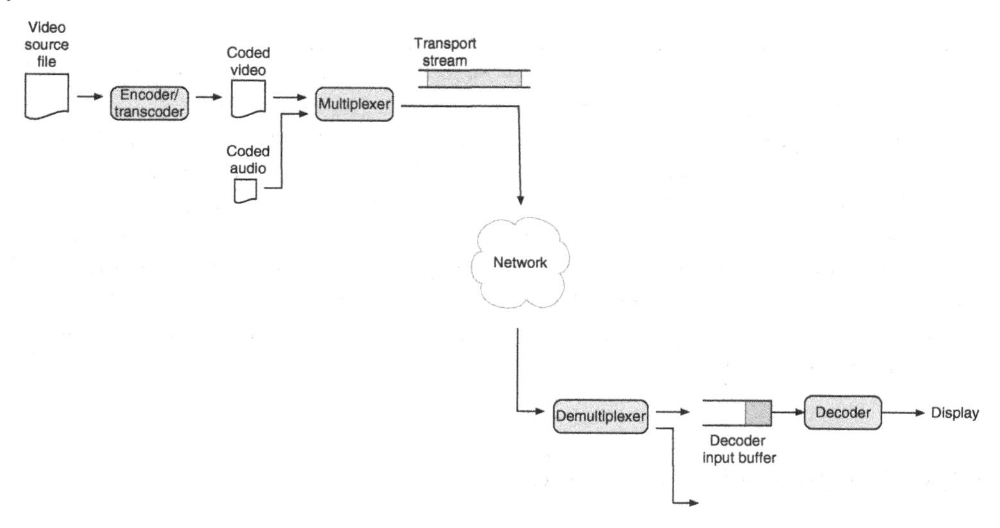

Figure 10.10 Broadcast encoding and multiplexing

Coded video and audio are multiplexed into a TS and then broadcast over a network or channel, such as Cable TV, Satellite or Terrestrial transmission, to multiple clients. Each client device, for instance, a TV set, receives the TS, extracts and buffers the coded video and then decodes and displays the video.

A transport protocol such as MPEG-2 Systems [8] specifies how video and audio are multiplexed for transmission and demultiplexed for decoding. MPEG-2 Systems defines methods of multiplexing coded video and audio into a TS. Coded video and audio are each converted into a sequence of packetised elementary stream (PES) packets. Video and audio PES packets are then interleaved to form a TS, together with timing, synchronisation and optional error protection information. The original version of MPEG-2 Systems was developed to carry MPEG-2 Video programmes, but later versions of the Systems standard added support for H.264 and H.265 video. MPEG-2 Systems remains a popular set of protocols nearly 30 years after its initial release.

10.4.2 Video Calling

Face-to-face communication applications like FaceTime, Zoom, Teams and Skype require video to be delivered from a source, say, a laptop camera, to a destination which could be someone else's screen. The source captures video and audio, encodes them into a compressed form and sends packets to the destination using a set of real-time protocols such as H.323. The destination receives the packets, decodes the compressed video and audio and plays them back to the user. With a careful choice of coding parameters and protocols and with a good network connection, the delay or latency between source and destination can be kept to a few hundred milliseconds or less, so that conversation appears as natural as possible.

The H.323 standard specifies a set of protocols for video calling (see Figure 10.11) [9]. H.323 defines protocols for initiating, controlling and carrying out a voice and/or video call over the Internet. A terminal, such as a personal computer (PC) or mobile device running

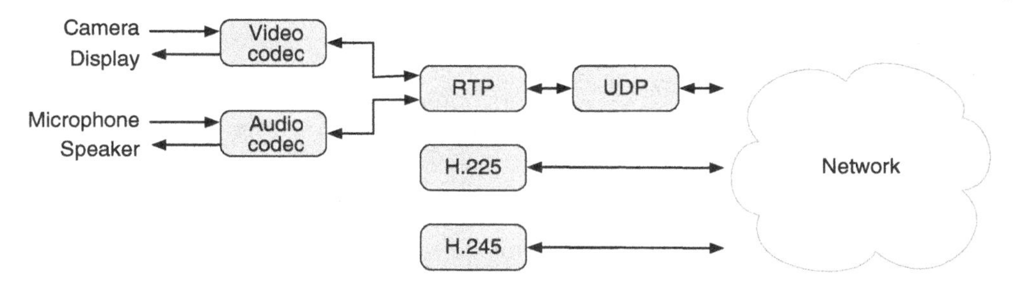

Figure 10.11 An H.323 terminal (simplified)

a video calling application, can communicate using the protocols shown in Figure 10.11. Video and audio are captured and encoded using a video codec and audio codec, then sent as a series of packets using the RTP and UDP. To receive video and audio, the terminal receives the packets, demultiplexes them into a video stream and an audio stream and decodes the video and audio. The H.225 protocol handles setting up and initiating a call. The devices use the H.245 protocol to send and receive messages, to control the opening and closing of video and audio streams, to agree call parameters and so on.

10.4.3 Video Streaming

Streaming involves sending compressed video and audio over a network to end users so that they can start watching the streamed content before an entire file or programme has been downloaded. An important concern for a streaming system is to be able to deliver continuous video and audio to a user once playback has started.

10.4.3.1 Progressive Download

Progressive download involves sending a multimedia file progressively across the network to a client device. Once enough of the file has been downloaded, the client starts decoding and playback. Figure 10.12 shows an example. A file that has been created for progressive download or streaming starts with track information such as an mp4 moov section, which is followed by interleaved video and audio samples, e.g. an mp4 'mdat' section. The client begins to download the file and starts to buffer the received bytes of data. Once the client has received the track information and any other metadata needed to decode the file and has downloaded enough coded and video audio data, it starts decoding and presenting, or playing back, the media. In order to continuously play back the media, the client buffer has to receive the rest of the samples before they are needed for decoding and playback.

Whether or not the user experiences continuous, successful video playback depends on a number of factors including the size of the video file, the transmission rate of the network and how the decoder buffer is managed.

After a certain amount of data has been received in the input buffer, the client starts decoding and playback. From this point onwards, unless the user pauses the programme, the decoder is continuously emptying the buffer in order to decode and display frames at the correct time intervals. If the decoder empties the buffer faster than the data are received

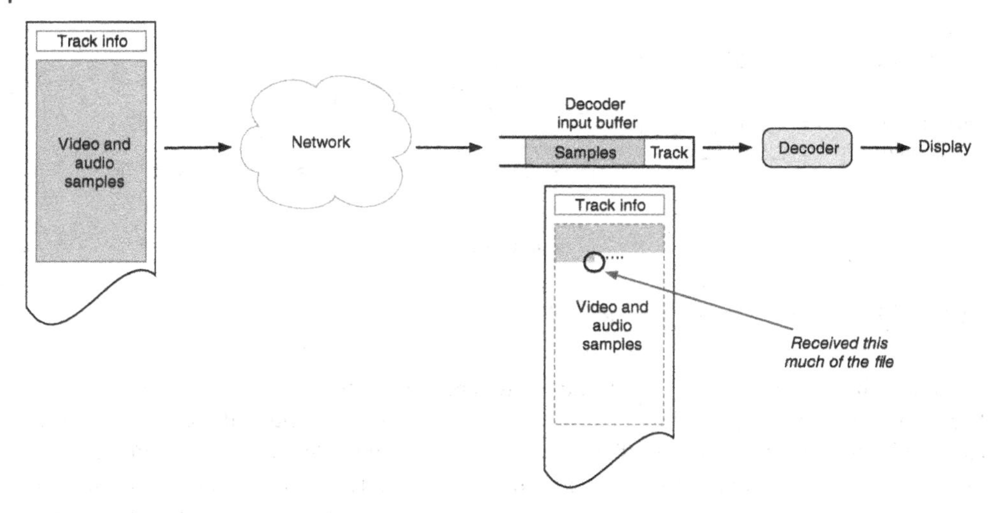

Figure 10.12 Progressive download

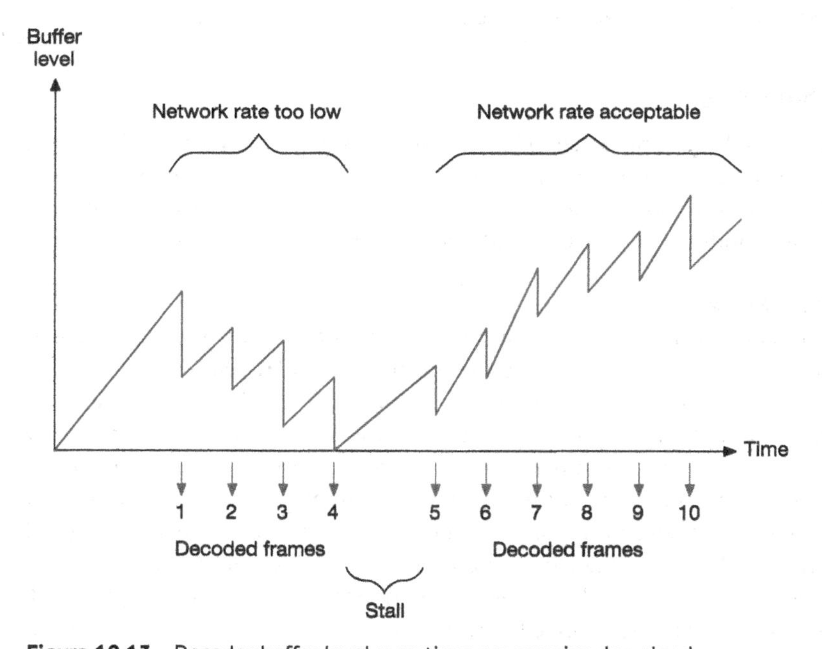

Figure 10.13 Decoder buffer level over time: progressive download

over the network, eventually a stall will occur. This means that the buffer runs out of data or underflows, the decoder cannot continue decoding and the user experiences a freeze or buffering of the video.

Figure 10.13 shows an example of how the decoder buffer level might change over time during a progressive download session. The decoder waits until the buffer fills to a certain level before decoding frames. As data are received from the network, the decoder buffer fills. This is shown as diagonal lines. The rate of fill – the slope of the diagonals – depends on the network rate. Each time a frame is decoded, a quantity of data is removed from the buffer, corresponding to one coded frame. This is shown as vertical lines on the graph.

From the start of the sequence to frame 4, the network rate is too low to fill the decoder buffer, i.e. it is lower than the video bitrate. As each frame is decoded and removed from the buffer, the buffer contents drop. After decoding frame 4, the decoder has emptied the buffer and must wait until enough data are received to decode frame 5, which means that the user experiences a short pause or stall in playback. After frame 5, the network data rate increases and is high enough to keep filling the buffer while decoding continues.

To ensure continuous playback, the rate at which data are received over the network should be higher than the compressed video bitrate, averaged over a period of time. For example, a network connection with an average bitrate of 100 Mbps should be able to deliver video that was encoded at bitrates of up to 100 Mbps. In practice, of course, internet bitrates are variable and hard to predict, especially for users on home or mobile internet connections. Choosing a very low video bitrate is one way to resolve this problem, but this means the user will experience low video quality. Waiting for a substantial portion of the video to download before playback starts reduces the likelihood of stalling, but this may lead to unacceptable delays for the user.

10.4.3.2 Streaming Servers

Progressive download involves a complete video file being downloaded to a client device, with the potential for the client to start decoding and playback once part of the file has been received. A streaming server can deliver media to a client in real time, with the client decoding and playing the media as it is received. Since the 1990s, a number of streaming server systems have been proposed and implemented.

Unlike progressive download, it may not be necessary for the client to store an entire media file. Instead, media data from the server fill a buffer at the client, and a video decoder on the client empties this buffer, decoding and playing back the video as it is received. Conceptually, this looks similar to Figure 10.12, except that the received video is not necessarily stored as a single file.

One example is Apple's Quicktime Streaming Server (QTSS) [10]. Multiple client devices connect to a server running the QTSS software. Using On-Demand streaming, each client connects to the server and requests a stream. In response, the server delivers coded media data over the network to each client using protocols such as the RTP and real-time streaming protocol (RTSP) (see Figure 10.14). In a live streaming scenario, the server typically begins streaming as the event starts, and clients may join the stream after the start.

Some streaming servers such as RealNetworks' SureStream server [11] are, or were, capable of adapting the bitrate of media delivered to each client. The server may store multiple copies of the video source that have been encoded at varying resolutions and bitrates. Each client may request a higher or lower bitrate version, depending on the estimated bandwidth between the server and that particular client.

A potential limitation of early streaming servers was the requirement for special purpose server software such as QTSS or RealServer and the use of streaming protocols such as RTSP/RTP and/or proprietary protocols. Since the early 2000s, networks of servers known as CDNs have been used to deliver content efficiently across the internet, but these were

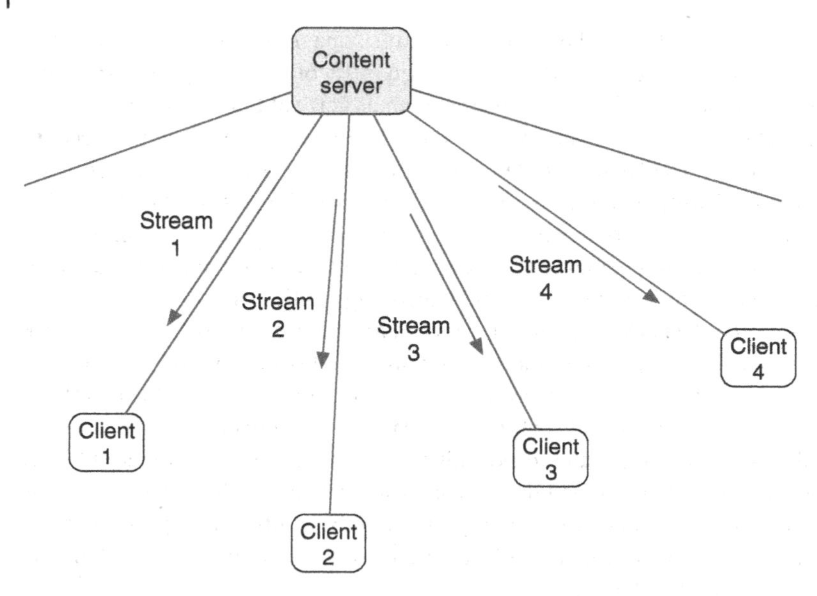

Figure 10.14 Streaming server and clients

initially designed to carry web content and support the HTTP protocol rather than support streaming protocols. Streaming protocols are often blocked by corporate firewalls.

A further limitation of some streaming server designs is that the load on the server increases with each client device that connects to the server. If the server has to interact with each individual client by establishing a connection, providing the requested stream and perhaps monitoring the QoS, then each new client adds a processing load to the server. This may make it challenging to scale up such a system to cope with large-scale deployments such as today's worldwide streaming services.

10.4.3.3 DASH and HLS: HTTP Adaptive Bitrate Streaming

In contrast to streaming server systems such as QTSS, adaptive bitrate streaming systems such as dynamic adaptive streaming over HTTP (DASH) and HTTP live streaming (HLS) can operate using conventional web servers. In a DASH or HLS system, a server stores available media streams as sets of coded media files and provides a menu or manifest to the client. A client initiates streaming by requesting the manifest and then requesting a series of relatively small media files that correspond to the stream that the viewer wants to watch.

The basic operations of DASH or HLS – the client first requesting a manifest file and then requesting a series of media files – can be carried out using well-known protocols such as HTTP. The server does not need to establish and monitor a dedicated connection with each client. It simply responds to HTTP GET requests for files and delivers them on demand. The responsibility for starting and then managing the connection is shifted from the server to the client. This approach can scale up effectively by simply adding more servers as network demand increases. Storing and sending coded media as a series of small files means that each small media file can be treated as a normal file and can be delivered using HTTP rather than by special-purpose streaming protocols such as RTP. By making streams available at multiple bitrates, each client can choose the bitrate that is appropriate at any given time, effectively managing variable transmission bitrates without intervention from the server.

The general operation of a DASH or HLS system is as follows:

Server-side:
A piece of media content such as a TV show or a movie is encoded and stored as multiple versions or representations, each of which is encoded at a different bitrate. Each version is split into smaller files or chunks, and each of these typically represents a few seconds of coded video and audio. Each chunk may be stored as a self-contained media file, as a segment of a larger media file or it may be contained within a larger file and accessed using an HTTP byte-range request.

Figure 10.15 illustrates an example of this. At each time period – 4 s in this case – four different representations are available, each stored as a separate file. Rate 0 is the lowest bitrate representation, and rate 3 is the highest bitrate representation. Naturally, higher-bitrate representations are larger files. The chunks are stored on one or more servers, together with a manifest or index file. The manifest file specifies the available chunks – four for each time period in this example – and the uniform resource locator (URL) of each chunk. Hence, the 16 chunks shown in Figure 10.15 are each stored on a server and are each accessible via an individual URL.

Client-side:
To stream a piece of content, the client first retrieves the manifest for that content from the server. The client selects a starting bitrate – for example, a lower bitrate version – and requests the first chunk of the content at the selected bitrate. Each requested chunk is delivered by the server to the client. The client can decode and play back the media in the chunk whilst at the same time requesting chunks for future time periods in the stream.

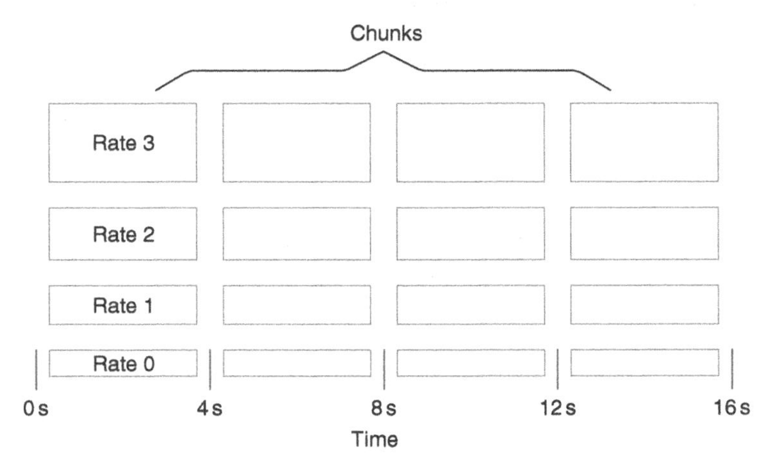

Figure 10.15 HTTP adaptive streaming: chunks and representations

Bitrate adaptation:
If multiple bitrate representations are available, a client can choose to change the requested bitrate. If the client determines that it has a relatively high bandwidth connection, it may request a higher-bitrate representation. If the bandwidth drops, the client may request a lower-bitrate representation. Thus, the client has the ability to dynamically adapt to changing bandwidth by changing the bitrate it requests from the server. With careful choices, a client can maintain constant video playback even if the network bandwidth is changing.

Example:

A client device streams the video content that was illustrated in Figure 10.15 in response to the user selecting a particular TV show for playback. The client retrieves the manifest that lists the four representations available at each time period. At the start of the stream – time 0 – the client chooses a representation, which in this example is the lowest-bitrate representation. The client requests the rate 0 chunk by making an HTTP GET request to the URL, which is specified in the manifest. The server responds by delivering the requested chunk to the client (see Figure 10.16).

For the next time period, which is 4–8 s, the client chooses a higher-rate chunk from the rate 1 representation. For the third time period, which is 8–12 s, the client chooses the rate 2 representation, once again by requesting the chunk at the URL specified in the manifest. For the 12–16 s time period, the client decides to stay at the rate 2′ level and requests the appropriate chunk.

It is up to the client which representation to choose at a particular time period. A client might estimate the available bandwidth and use this to determine which rate to select for the next chunk. With suitable choices of representation, a client can attempt to download the highest-rate chunk that the current bandwidth allows. If the available bandwidth increases, a client can respond to this by stepping up to a higher bitrate representation, which typically results in a higher-quality and/or higher-resolution display to the user. If the bandwidth drops, the client can step down to a lower-bitrate representation. This will typically appear as a lower-quality image to the viewer, but it is usually more important to maintain constant video playback, even at a lower quality, than to run out of buffered video, which will lead to a stall.

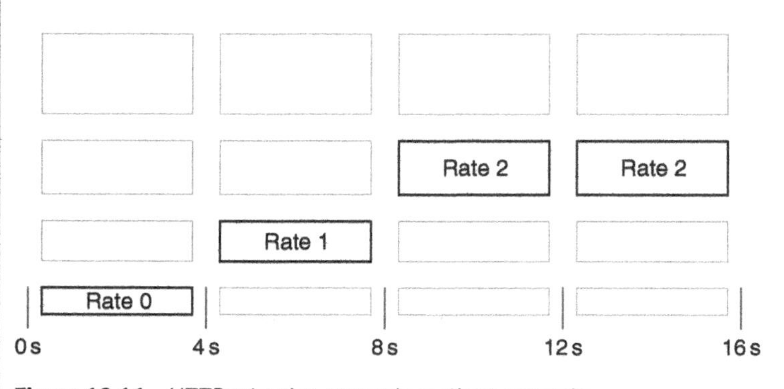

Figure 10.16 HTTP adaptive streaming: client example

10.4.3.4 Content Delivery Networks

Since 2010, HTTP adaptive bitrate streaming has become increasingly widespread, partly because it can be delivered at scale to handle hundreds of thousands, or millions, of clients through CDNs. The first CDNs emerged in the late 1990s/early 2000s and were developed to efficiently distribute HTTP web content. Since 2010, CDNs have been used increasingly to deliver adaptive bitrate content such as HLS or DASH streams.

Figure 10.17 shows a CDN architecture that may be used to distribute DASH/HLS content. An origin server hosts original coded versions of each piece of media content, each of which is stored as a set of coded chunks with an associated manifest. The origin server connects to the CDN. Each server in the CDN is capable of storing or caching chunks of coded media. Each individual client connects to an edge server rather than directly to the origin server. A client requests a chunk of media by making an HTTP GET request for the URL provided in the manifest. The request is initially handled by the edge server. If the requested chunk is already stored on the edge server, it is delivered to the client. If not, the request is relayed to other servers in the CDN, until a server is found that has the requested chunk. Each server can cache chunks that have been requested and served. The result is

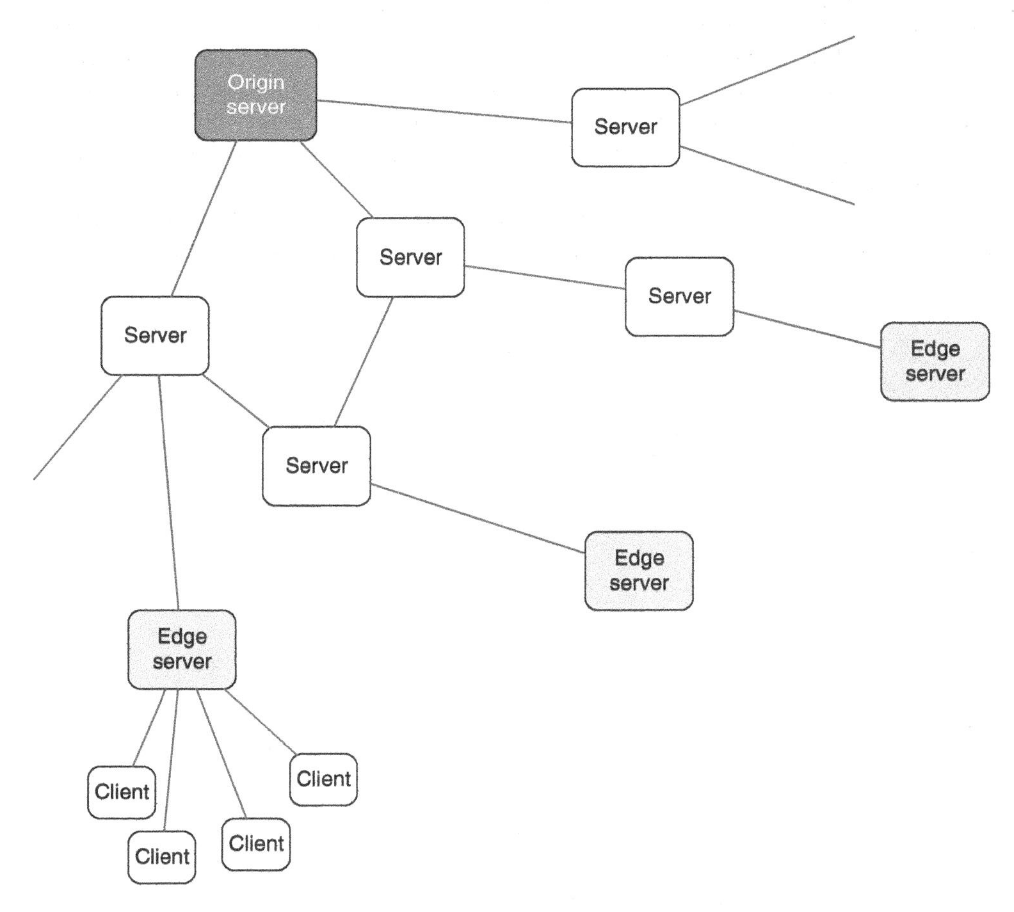

Figure 10.17 Content delivery network

that popular data, such as the coded media chunks corresponding to a popular TV show, tend to propagate through the CDN so that they are replicated on multiple servers. Less-popular media may need to be sourced from the origin server.

This type of architecture can scale up efficiently. More clients can be handled by adding more servers to the CDN and/or by increasing the capacity of each server. The individual servers in the CDN each act more or less as conventional internet servers by simply caching and serving small files on request. The CDN architecture, together with HTTP adaptive streaming, is used by providers including Netflix, Amazon, YouTube and many others to deliver media to hundreds of millions of client devices.

10.5 Video Rate Control

Coded video typically consists of frames that have been encoded using a mix of prediction types, for example, I-pictures encoded using only intra prediction, P-pictures encoded using prediction from one previously coded frame and B-pictures encoded using prediction from two or more previously coded frames. Inter prediction tends to be more efficient than intra prediction, and biprediction tends to be more efficient than single reference prediction. This means that coded I, P and B pictures can often vary in size. Figure 10.18 illustrates the relative coded sizes of pictures from a typical video sequence. The first picture I_0 is coded using intra prediction only. Because intra prediction is relatively inefficient, the size of the coded picture is relatively large. Pictures P_4 and P_8 are each predicted from a single reference picture per block, using inter prediction. Their coded sizes are smaller than the intra-coded picture I_0. The remaining pictures are coded using biprediction using up to two reference pictures per block. This is more efficient still, producing smaller coded pictures. Note also that pictures B_1, B_3, B_5 and B_7 each have a close temporal neighbour that is a prediction source. This means,

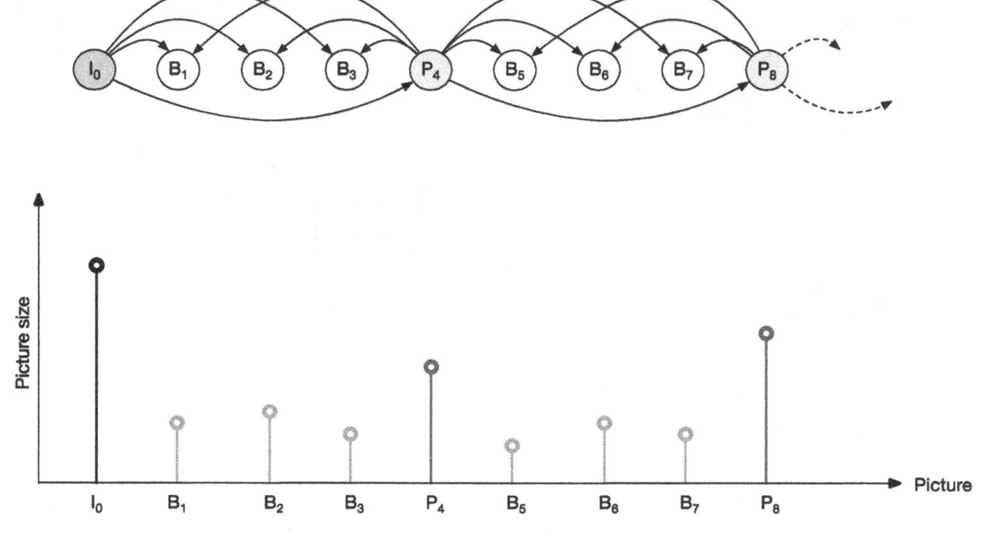

Figure 10.18 Picture types and relative coded sizes

for example, that B_1 might be expected to have high compression efficiency since it is predicted from the immediately adjacent reference frame I_0 and can use biprediction, forward prediction and intra prediction to find the best prediction option for each block.

Coded video frames will often vary in size due to the content of the scene, as well as due to the choice of prediction types. A structurally simple scene may be easier to predict than a complex scene. When the prediction is easier and more effective, the compressed frame tends to be smaller. In an inter-coded frame, scenes that have limited or predictable motion from one frame to the next are easier to predict than scenes with random motion. All of these factors mean that the bitrate of a coded video clip can vary considerably from frame to frame and over time.

Figure 10.19 shows the coded frame sizes for a 100-frame video clip, with the quantisation parameter (QP) set to a fixed value. From this figure, we can see:

- A large first frame, coded using intra prediction only.
- Three distinct sets of inter-coded frames, corresponding to P-pictures that are the largest inter-frames, B-pictures with no adjacent neighbours and B-pictures with one adjacent neighbour.
- A variation in coded frame size during the sequence. Frames in the first half of the clip tend to be larger, and frames in the second half tend to be smaller, which corresponds to a change in scene complexity.

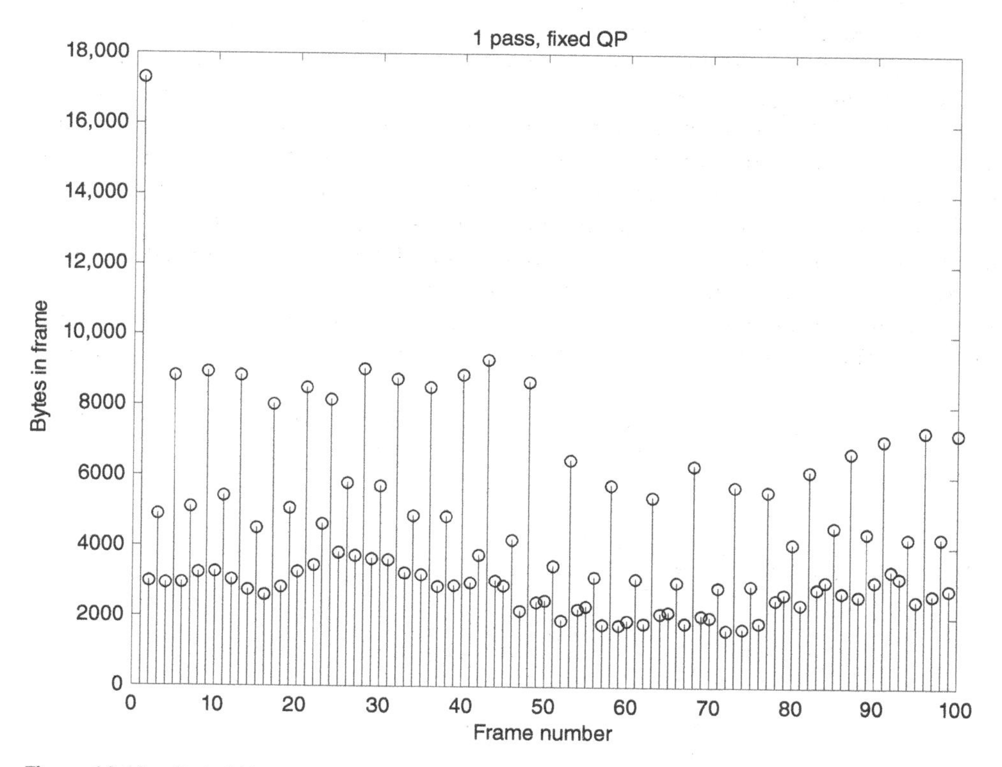

Figure 10.19 Coded bitrate example: fixed quantisation parameter (QP)

This variation in bitrate can pose problems for storage and transmission. If storage capacity is limited, it may be desirable to control the video bitrate so that the coded file is less than a certain target size. When transmitting coded video, it is important to control the video bitrate so that it is less than, or equal to, the available channel capacity.

Video bitrate control attempts to control the encoding process to meet constraints such as a target file size or a target transmission bitrate. A simple but effective way of controlling bitrate is to adjust the QP. Increasing the quantisation step size by increasing QP leads to higher compression but lower image quality, whereas reducing QP has the opposite effect. The output of a video encoder can be sent to a buffer, which has the effect of smoothing out some of the short-term size variations between coded video frames. The coded video bitrate can be controlled by measuring the contents of the buffer (Figure 10.20), increasing QP when the buffer fills up and reducing QP when the buffer empties. The result will be a more consistent output bitrate, at the expense of a variation in QP and therefore in video frame quality.

The same video sequence of Figure 10.19 was encoded with a target average bitrate of 1 Mbps using the ffmpeg x265 encoder with a single rate control pass (see Figure 10.21). This means that the encoder attempts to adjust the rate based on the current frame statistics and the current output buffer level. By using one pass rate control (Figure 10.21), the encoder attempts to keep the frame sizes reasonably consistent over time. The first I-picture is still significantly larger than the other coded pictures. The output rate is 896 kbps, which is around 10% below the target of 1 Mbps.

Figure 10.22 shows the same clip encoded using x265 with two-pass rate control. As the name suggests, the encoder processes the clip twice: the first time to gather statistics and the second time to apply those measured statistics during encoding. In the first pass, each frame is encoded using an estimate for the most suitable QP per frame, and the encoder collects statistics such as the actual coded size per frame. During the first pass, the encoded output is not stored or transmitted. In the second pass, the encoder uses the target bitrate together with the statistics from the first pass and adjusts the QP to more accurately meet the target output bitrate. The difference in size between the first I-picture and the remaining coded frames is reduced. The output rate is 968 kbps, which is closer to the target of 1 Mbps. Two-pass rate control can provide more accurate control of coded bitrate and a bitrate that is closer to the target bitrate, at the expense of increased processing time and increased computation.

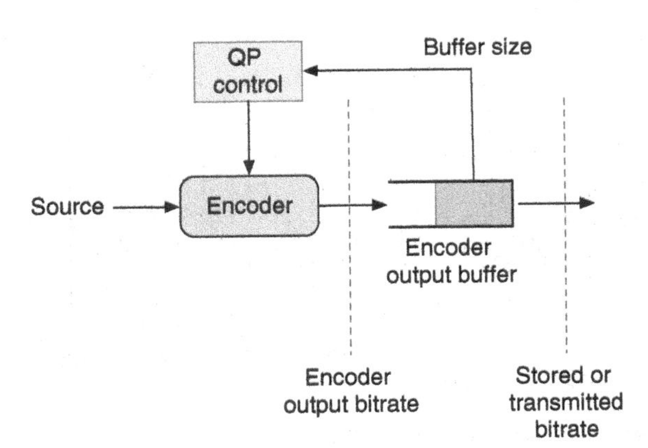

Figure 10.20 Controlling QP based on buffer contents

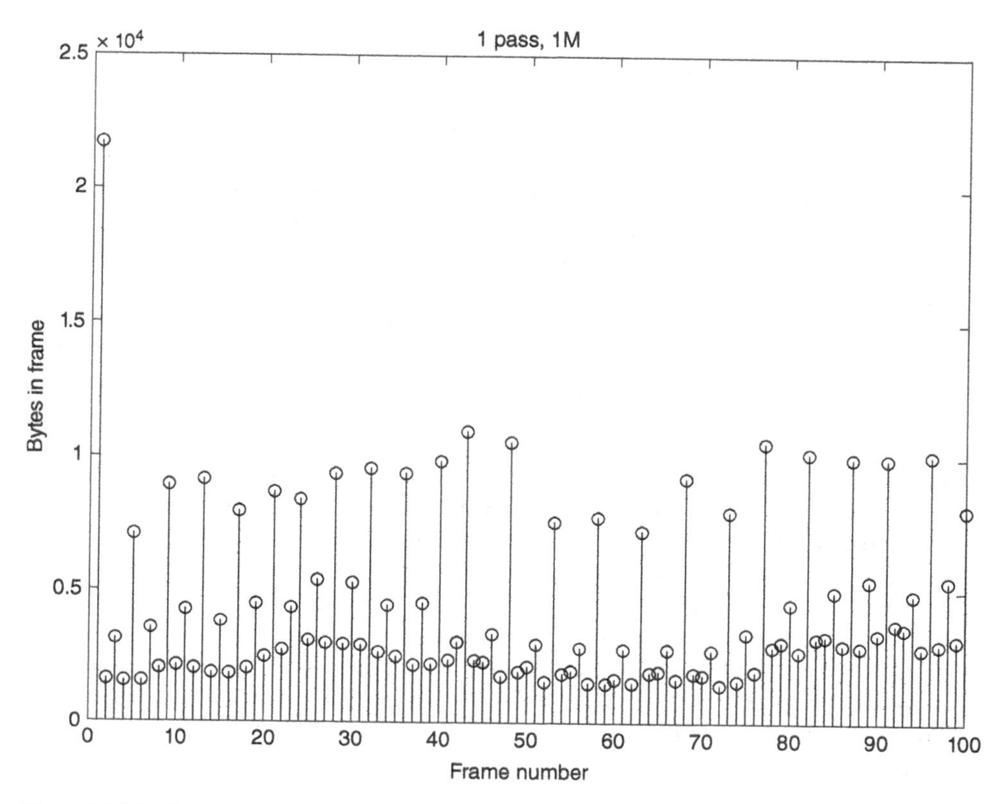

Figure 10.21 Rate control, one pass

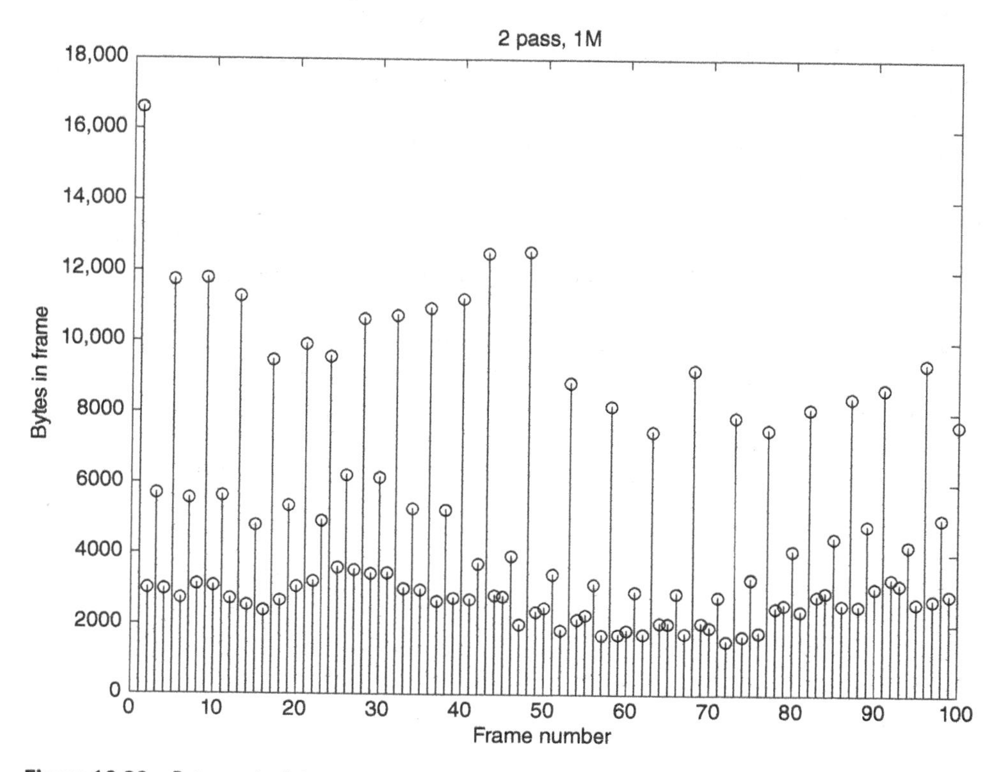

Figure 10.22 Rate control, two pass

Video rate control algorithms have developed over many years of research. Controlling the rate, maximising video quality and minimising quality differences between coded frames is a difficult problem, and many solutions have been proposed in the literature. A comprehensive survey of developments in video rate control can be found in ref. [12].

The rate control problem is further complicated by the many different storage and transmission scenarios. To give an example, live video calling with limited processing capacity, a low delay requirement and no opportunity for multi-pass rate control has very different challenges to on-demand streaming, where the streaming content provider can apply significant levels of processing to maximise bitrate and quality performance. Some possible scenarios with a variety of rate control requirements include:

- **Broadcast TV**: Each satellite, cable or terrestrial channel has an allocated bitrate. The average bitrate of the coded video should fit within the channel bitrate.
- **Video streaming**: If adaptive bitrate streaming is employed using DASH, HLS or similar (see Section 10.4.3.3), then multiple versions of the video are produced, each of which is coded at a different bitrate.
- **Video calling**: The coded video bitrate should fit within the available transmission bitrate. It is important to keep the end-to-end delay low, which may mean minimising bitrate variation, since only a fraction of a second of video is buffered before decoding.
- **Storage**: The total size of the encoded video file should fit within the target amount of storage.

Coded video streams may be described as constant bit rate (CBR) or variable bit rate (VBR). In practice, the frame-to-frame bitrate will vary as illustrated in the examples mentioned above. This short-term variation is typically smoothed out by the encoder output buffer and decoder input buffer. The question is then whether the bitrate sent over the channel is CBR or VBR.

Finally, video coding standards such as H.264, H.265 and H.266 specify limits that affect the coded bitstream. These limits are defined in terms of the operation of a hypothetical decoder, known as the hypothetical reference decoder (HRD). For example, Annex C of H.265/HEVC specifies an HRD model that can be used to check conformance to the standard. A simplified diagram of the HRD is shown in Figure 10.23. A coded picture buffer (CPB), which is the decoder input buffer, receives the coded bitstream. An H.265 decoder processes and decodes the stream to produce a set of decoded pictures, stored in a decoded picture buffer (DPB). The HRD can be used to check whether a coded bitstream conforms to the standard. As an example of this process, for a particular video Level limit, the CPB has a defined maximum size and the decoder processes coded video at a particular rate, which is a number of coding tree units (CTUs) per second. While decoding the bitstream, the CPB should not run out of data or

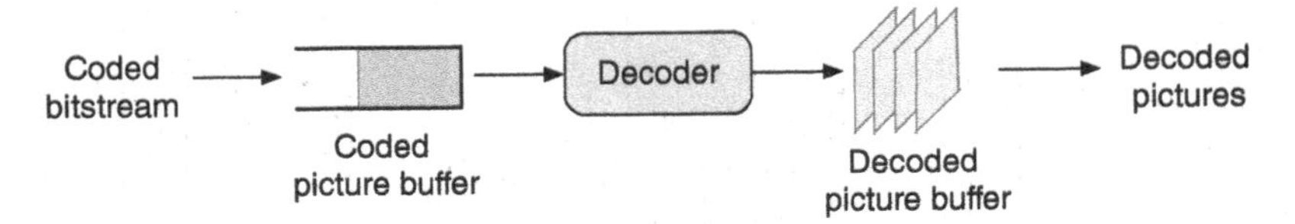

Figure 10.23 Hypothetical reference decoder (HRD), simplified

underflow, nor exceed its capacity, that is, overflow. The HRD places yet another constraint on an encoder rate control algorithm, since any bitstream produced by a conforming encoder should not break the restrictions of the HRD.

10.6 Error Handling

A video coding algorithm attempts to remove redundancy in visual data. It removes statistical redundancy by predicting data from previously coded data and through entropy encoding, and it removes subjective redundancy by removing perceptually unimportant data such as high-frequency transform coefficients. One result is that the data that remain in a coded video bitstream are susceptible to errors. Transmission errors or file errors may result in one or more bits of data in the coded video file being lost or corrupted. If the coded video syntax is no longer decodable, then part of the decoded video image is lost or damaged. Many video communication systems use inter prediction, which means that video frames are coded using prediction from previously coded video frames. This means that an error in the data corresponding to one frame may cause damage or corruption in further coded frames that are predicted from the original errored frame. Figure 10.24 shows an example. A transmission error occurs in the coded data representing frame 1. When frame 1 is decoded, a macroblock or CTU cannot be successfully decoded due to the error. Prediction within the frame, such as intra prediction and/or differential coding of parameters, means that blocks following the damaged block are incorrectly decoded, which results in a distorted or missing area of the frame. Frame 2 is predicted from frame 1, which means that the damaged area propagates to frame 2. Frame 3 is predicted from frame 2, and so the damaged area continues to propagate. Because the blocks of frame 2 are likely to be predicted using motion-compensated prediction, the damaged area may appear to move around in frame 2, rather than staying in exactly the same place.

Errors in a coded video stream can occur because of channel transmission errors and network problems, such as congestion or delay. A bit error is one or more bits that have the

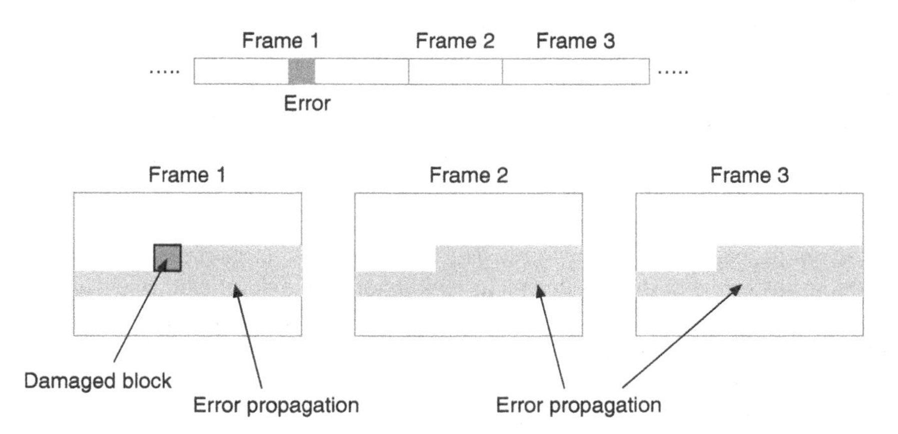

Figure 10.24 Error propagation in decoded video frames

incorrect polarity, i.e. a one has been changed to a zero or vice versa. Bit errors may occur in a network link or transmission channel, due to interference in a wireless link. Bit errors in a packet may be detected by a server or a receiver, perhaps because a checksum is incorrect or an error is detected in a cyclic redundancy code (CRC). A packet containing errors may be discarded by an intermediate server between the sender and receiver of coded video data. Internet servers may also discard packets due to congestion in the network if a server cannot handle the rate of packets it is currently receiving.

Video transmission applications such as streaming, broadcast and video calling require video to be delivered within a certain time frame – as quickly as possible in the case of video calling or within a few seconds for one-way applications such as streaming and broadcast. Once the end-user has started playback of a stream, it is important to keep the series of packets continuously arriving at the client to minimise the chance of stalling or buffering. Given the likelihood of channel or network problems and therefore lost or damaged packets, errors are inevitable in many practical scenarios. It is therefore necessary to handle transmission errors in coded video. Ideally, the system should continue to decode and display video if at all possible and should be designed to minimise the visual effect of errors. Strategies for handling errors include resynchronisation, restricting dependencies and error concealment.

Resynchronisation: An error may make it impossible for a decoder to continue decoding video if the decoder has lost its place in the video bitstream syntax. In this case, a decoder can abandon decoding and restart at a synchronisation point. In an application where errors are likely to occur, such as most video transmission scenarios, independently decodable video frames may be inserted at regular intervals. An instantaneous decoder refresh (IDR) picture in an H.264, H.265 or H.266 bitstream is coded without any dependency on previously coded pictures, using only intra prediction. When a decoder encounters a catastrophic error, it can wait for the next IDR picture and restart successful decoding at this point. Because they must use intra prediction, IDR pictures are less efficiently coded than other types of pictures. Hence, the frequency of IDR pictures is a trade-off between error resilience and compression efficiency. Frequent IDR pictures result in a higher bitrate bitstream, which is less efficient, but they make it possible for a decoder to rapidly recover from an error. Less frequent IDR pictures result in a more efficiently compressed bitstream but in a longer gap between resynchronisation points.

Restricting dependencies: Consider the example picture structure in Figure 10.18. P_4 is predicted from I_0 and P_8 is predicted from P_4, which means that P_4 depends on I_0 and P_8 depends on both P_4 and I_0. An error in I_0 could propagate to P_4, P_8 and all of the B-pictures including B_1, B_2, etc. An error in any picture that is used as a reference, such as I_0, P_4 and P_8 in this example, can propagate to other pictures. On the other hand, in this example, none of B_1, B_2, B_3, B_5, B_6 or B_7 are used to predict any other pictures. An error in one of the B-pictures will not propagate to other pictures. The B-pictures in this example are less sensitive to errors, so any errors in these coded pictures will not have a significant visual effect. From a user's point of view, an error in a picture that is not used for reference will only appear on the screen for one frame period, whereas an error in a picture that is used for reference may appear for many frame periods and be much more obvious to the viewer. Designing a repeating picture structure, such as the example shown in Figure 10.18, has an impact on temporal dependencies and temporal error propagation.

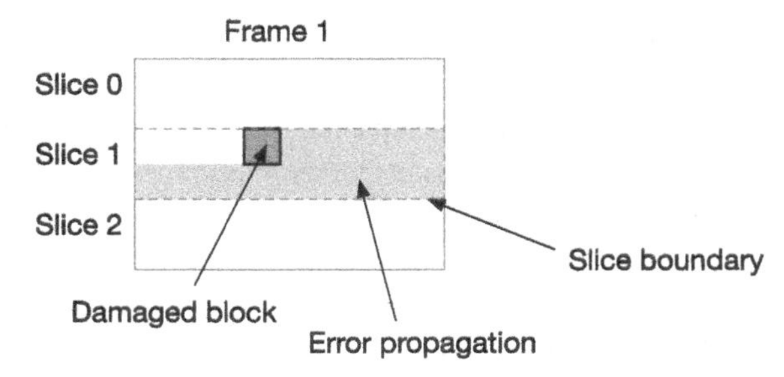

Figure 10.25 Error propagation and slice boundaries

Spatial dependencies can be restricted through the use of structures such as slices. A slice in H.265/HEVC contains an independent slice segment followed optionally by a number of dependent slice segments. Predictions such as intra prediction, motion vector prediction or parameter prediction cannot cross slice boundaries. This means that an error that occurs in one slice should not propagate to subsequent slices so that the following slices can be decoded without any reference to data in the slice containing the error. Figure 10.25 shows the example of Figure 10.24 with slice boundaries marked. The picture is divided into three slices. The error occurs in a block within slice 1 and propagates to all the subsequent blocks in the same slice. Slice 2 is coded without any dependencies upon slice 1, and so the error does not propagate into slice 2. Increasing the number of slices in a picture can reduce the effect of errors. However, more slices in a picture reduces compression efficiency since each slice requires header information, and the lack of inter-slice prediction reduces the overall prediction efficiency. Once again, the use of slice structures is a trade-off between compression efficiency and error resilience.

Error concealment: An error in the decoded bitstream may be signalled externally by a transport protocol message identifying a packet error, or it may be detected by the video decoder when the expected series of syntax elements are not received. A decoder can attempt to reduce the visual impact of an error through the use of error concealment and may use neighbouring, error-free regions of the frame or temporally adjacent frames to attempt to predict and fill in the damaged region. When error concealment is successful, the result is a less obviously distorted frame [13].

10.7 Conclusions

Video compression coding has evolved alongside practical applications for storing and transmitting video. Each video coding standard was developed to take account of practical requirements, such as storing video on consumer devices, sending and receiving video in real time during a video call or streaming video to consumers. These applications require formats and methods for storing and transmitting compressed video along with compressed audio so that video and audio can be delivered efficiently to end users.

In Chapter 11, we will look at how video codecs are implemented and integrated into practical applications and devices. We will also consider how to start experimenting with and developing video codecs.

References

1 Tudor, C. (2022). The impact of the COVID-19 pandemic on the global web and video conferencing SaaS market. *Electronics* 11 (16): 2633. https://doi.org/10.3390/electronics11162633.

2 ISO/IEC 14496-12 (2022). *Information Technology — Coding of Audio-Visual Objects — Part 12: ISO Base Media File Format*. Geneva: International Organization for Standardization.

3 Singer, D. and Stockhammer, T. (2018). Overview of the ISO base media file format. A Powerpoint presentation from *ISO/IEC JTC1/SC29/WG11 N18093*. White paper, Macau, China (October 2018).

4 ISO/IEC 14496-14 (2022). *Information Technology — Coding of Audio-Visual Objects — Part 14: MP4 File Format*. Geneva: International Organization for Standardization.

5 ISO/IEC 14496-15 (2022). *Information Technology — Coding of Audio-Visual Objects — Part 15: Carriage of Network Abstraction Layer (NAL) Unit Structured Video in the ISO Base Media File Format*. Geneva: International Organization for Standardization.

6 MUX: San Fransisco and London. (2017). Video Streaming Perceptions Report. https://web.archive.org/web/20220119130916/https://static.mux.com/downloads/2017-Video-Streaming-Perceptions-Report.pdf (accessed 6 December 2023).

7 EBU Recommendation R132 (2011). *Signal Quality in HDTV Production and Broadcast Services*. Geneva: European Broadcasting Union.

8 ISO/IEC 13818-1 (2019). *Information Technology — Generic Coding of Moving Pictures and Associated Audio Information — Part 1: Systems (MPEG-2 Systems)*. Geneva: International Organization for Standardization.

9 ITU. (2009). Packet-based multimedia communication systems. ITU-T Recommendation H.323.

10 Apple Quicktime Streaming. https://web.archive.org/web/20081218075016/https://images.apple.com/server/docs/QuickTime_Streaming_TB_v10.4.pdf, Apple Computer, Inc, originally published 2005 (accessed 21 December 2023).

11 Conklin, G.J., Greenbaum, G.S., Lillevold, K.O. et al. (2001). Video coding for streaming media delivery on the internet. *IEEE Transactions on Circuits and Systems for Video Technology* 11 (3): 269–281. https://doi.org/10.1109/76.911155.

12 Chen, L., Yang, M., Hao, L., and Rawat, D. (2017). H.265/HEVC rate control in real-time transmission over 5G mobile networks. *Proceedings of the 10th International Conference on Mobile Multimedia Communications*, Chongqing, China (July, 2017). Brussels, Belgium: ICST. pp. 192–198.

13 Usman, M., He, X., Xu, M., and Lam, K.M. (2015). Survey of error concealment techniques: research directions and open issues. *2015 Picture Coding Symposium (PCS), Proceedings*. Cairns, Australia (31 May–3 June 2015). New York: IEEE. pp. 233–238. https://doi.org/10.1109/PCS.2015.7170081.

11

Implementation and Performance

11.1 Introduction

This book has examined in detail how video codecs work. An important consideration for practical video codecs is the trade-off between compression, video quality and computation. A video coding standard, such as H.264 or H.265, provides a set of available coding tools, which are specified in the profiles that are defined in each standard. The actual performance of a video codec, even when it conforms to a standard, depends very much on how it is implemented in software or hardware, as well as on the decisions taken by the video encoder. Different applications can and often do require different encoding and/or implementation strategies. Compare the following for some examples where these might be required:

- Two-way video calling or video conferencing, such as FaceTime, Zoom and Teams, requires low-delay, reliable video transmission, with encoding and decoding taking place simultaneously and in real time at both ends of the call. Minimising both end-to-end delay and computational cost may be more important than maximising video quality. Video encoding and decoding have to be carried out within the constraints of consumer devices, such as processing power and memory limitations.
- When streaming a popular TV show to an international audience, the same coded video clip is going to be transmitted and decoded perhaps many millions of times. During encoding, it may be more important to maximise video quality and minimise coded bitrate, even if this takes time and computing power to achieve. Encoding is likely to take place on dedicated, rack-mounted hardware, and decoding is carried out on a variety of consumer devices.

In this chapter, we will consider:

- The challenges and options for putting video codecs into practice, as in implementing encoders and decoders in software or in digital electronic hardware.
- How to evaluate and compare the performance of video codecs.
- How to get started with video coding by experimenting with public-domain video codec software.

Coding Video: A Practical Guide to HEVC and Beyond, First Edition. Iain E. Richardson.
© 2024 John Wiley & Sons Ltd. Published 2024 by John Wiley & Sons Ltd.
Companion website: www.wiley.com/go/richardson/codingvideo1

Please visit the book companion site, https://www.vcodex.com/coding-video-book, for download links and other resources related to this Chapter.

11.2 Implementing Video Codecs

A video encoder or decoder can be implemented in software, in hardware or in a mix of the two, by using dedicated hardware for certain computationally intensive functions and programmable software for other functions. Here are some examples:

- When you record a video on a mobile handset, your device captures and encodes the video in real time, saving the compressed version as a file. Encoding typically takes place in dedicated hardware, which is part of an integrated circuit or chip on the mobile device. When encoding in real time on a device with limited computational resources, computationally efficient encoding is very important.
- When you watch a video on a device, such as a TV, laptop or mobile handset, your device decodes and plays back the video in real time. Decoding once again takes place in dedicated hardware.
- When a streaming provider such as Netflix prepares a TV programme for streaming, the video source is typically encoded into multiple versions at different bitrates. Encoding may be carried out using multiple software codecs running on cloud computing instances [1]. Unless the TV programme is live, real-time encoding is not necessary, and it may be preferable to spend more time and/or computing resources to optimise the encoded video streams.

Consider the requirements for a video codec that conforms to a standard such as H.264 and H.265.

An encoder must produce a bitstream that conforms to the standard. This means that the bitstream output by the encoder must (1) follow the syntax structure specified in the standard and (2) be decodable by the decoding process specified in the standard, including any constraints on decoder processing and decoder memory.

A decoder must be capable of decoding a bitstream that conforms to the standard and must produce a decoded output – a series of decoded video frames – that is identical to the output of the decoding process specified in the standard.

When implementing a video codec in software or hardware, a primary consideration is to meet the conformance requirements of the chosen standard. Next, there are many design choices to be made in order to achieve an appropriate balance between criteria such as computational efficiency, compression performance, video quality and bit rate variation.

11.3 Software Implementation

Video encoding and decoding may be implemented entirely in a high-level software language such as C or C++. A key challenge for a software implementation is to carry out video coding that is fast enough for the application.

Video coding is inherently computationally intensive, due in part to the large amount of data that must be processed per frame of video. For example, each frame of a 4:2:0 format

1080p video clip contains 1920×1080 = 2,073,600 pixels, represented as 2,073,600 luma samples and 1,036,800 chroma samples. To operate in real time at 30 frames per second, a video encoder or decoder has to process over 93 million samples per second. A decoder has to carry out all of the steps specified in the standard, including entropy decoding, rescaling and inverse transform, prediction, reconstruction and filtering. An encoder has to carry out the corresponding encoding steps and also has to choose suitable modes and prediction parameters, including block sizes, prediction modes and motion parameters, to efficiently compress video.

Successive video coding standards have tended to be increasingly computationally intensive for both encoder and decoder. This is a deliberate trend, as each new standard aims to support increased compression efficiency by taking advantage of increases in processor and memory capacities. H.264/AVC was developed with the capabilities of typical consumer processors in the mid-2000s in mind. Of course, devices such as computers, mobile devices and TVs are significantly more powerful in the 2020s (around the time of writing) than they were in the 2000s, with increased computational capacity, more parallel processing capabilities and larger memories.

When is encoding or decoding 'fast enough'? The answer depends on the application. For example, offline encoding, such as encoding a video file in preparation for streaming, may not need to happen in real time, whereas face-to-face video calling applications such as Zoom or Skype require simultaneous encoding and decoding in real time.

Example: The HM Reference Software Codec

In common with other video coding standards, a software implementation of H.265/HEVC evolved in the process of developing the standard itself. The reference software implementation, known as the HEVC Test Model or HM, was itself published by the ITU-T as Recommendation H.265.2 [2]. Written in C, the HM software implementation includes an encoder and a decoder. The encoder can produce bitstreams that conform to Recommendation H.265; the decoder conforms to the decoding process described in H.265 and can decode H.265-conforming bitstreams. HM software can be downloaded as source code from the ITU website [3]. Similar reference implementations are available for video coding standards and formats published from the late 1990s onward.

The HM reference codec is intended as an illustration of how to perform encoding and decoding that conforms to the standard. It is required for H.265 conformance testing, as specified in ITU-T Recommendation H.265.1. It is a useful tool for experimentation and for developing another software or hardware implementation of H.265. When a software or hardware encoder or decoder is developed, the correct operation of the new implementation can be checked against the HM reference software (Figure 11.1).

Figure 11.2 shows the operation of the HM codec. Both encoder and decoder are file-based, which means that they take a file as an input and produce a file as an output. The HM encoder encodes an uncompressed video file to produce a compressed file that conforms to the H.265 standard. The HM decoder decodes a conforming H.265 file and produces a decoded output video file. As the figure illustrates, an encoder under test, such as a new encoder implementation, should produce a coded file that can be

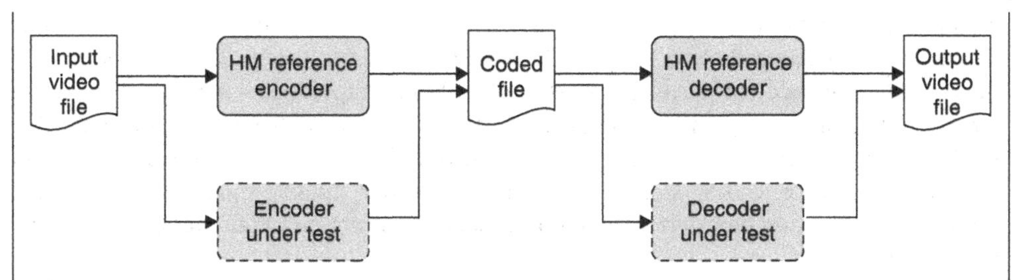

Figure 11.1 HM reference codec and test implementation

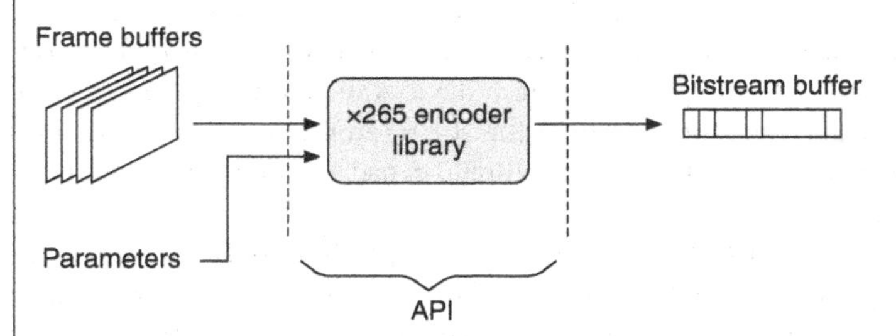

Figure 11.2 x265 encoder library

successfully decoded by the HM decoder. It does not need to produce an identical coded file to the HM encoder. A decoder under test should be capable of decoding a bitstream file produced by the HM encoder and should produce an identical output video file to the HM decoder.

The HM encoder and decoder are not specifically designed for fast operation. The HM encoder runs significantly slower than real time for higher-resolution video sources, when encoding using typical computer processors from the early 2020s.

Example: x265 HEVC Encoder

x265 is an open-source software implementation of an H.265-conforming encoder. Developed by MulticoreWare Inc., the source code for x265 is available to download according to the GNU GPL 2 license [4]. x265 is primarily intended for use as a software library for video encoding, although it can be used standalone in a similar way to the HM reference encoder. Figure 11.2 shows how x265 may be used as a software library. The interface to the library, an Application Programming Interface or API, is shown with dotted lines. In a typical scenario, a calling application provides the library with a series of frame buffers, each of which is a pointer to memory containing the samples of a video frame, and a bitstream buffer, which is a pointer to memory into which the library can write the encoded bitstream. The calling application sets up the encoder parameters including frame sizes and encoding decisions and calls the x265 library to encode the video frames. The encoded video data are written into the bitstream buffer.

x265 is relatively fast, taking advantage of processor resources such as Intel MMX instructions and parallelisation opportunities such as multiple processing cores that can speed up computationally expensive video encoding operations. It has a number of encoding presets, each of which is a predefined set of encoding parameters that allow the calling application to choose the trade-off between slower, computationally expensive encoding with higher compression and faster encoding with less compression. The x265 library is often used as part of the ffmpeg framework, described in Section 11.4, Experimentation.

11.3.1 Comparing HM and x265 Encoding

We can compare the performance of the HM and x265 HEVC encoders when each encodes the same video sequence, *Pedestrian Area*. Both encoders were used in file-based mode, with an uncompressed video file input and a bitstream file output, to encode 10 frames of this 1080p clip on an iMac computer with four 3.4 GHz Intel i5 processor cores, 32 GB of memory and a solid-state drive (SSD) hard disc[1]. The quantisation parameter (QP) settings for each encoder were adjusted to produce similar file sizes. x265 was tested with its Medium and Very Slow presets.

Video multimethod assessment fusion (VMAF) is a full-reference visual quality metric developed by Netflix [5]. A higher VMAF score indicates better visual quality. Out of the three tests listed in Table 11.1, the HM produces the best compression performance, as indicated by a file size of 70 kbytes and a VMAF score of 84.4. However, this comes at the expense of a very slow encoding time, around 150 s or 2.5 minutes per video frame. x265's very slow encoding preset takes around 3 seconds per frame and produces the same encoded file size as the HM test, with a slightly lower VMAF score of 81.9. Using the Medium or default preset, x265 encodes 10 frames in under 2 seconds, with a slightly larger file and a slightly lower VMAF score.

This example illustrates the trade-off between computational complexity and compression performance when encoding video. Here, the best compression performance comes at the expense of high computational complexity, in this case taking 2.5 minutes to encode each frame. The x265 encoder operating with the Very Slow preset offers a much more

Table 11.1 Comparing HM and x265 HEVC encoders

Encoder	x265 preset	Encoding time	File size	VMAF score
HM v16.2	–	1480 seconds	70 kbytes	84.4
x265	Medium	1.6 seconds	75 kbytes	79.9
x265	Very slow	34 seconds	70 kbytes	81.9

1 When encoding or decoding in file-based mode, a fast hard disc such as an SSD disc can significantly improve processing speeds compared with a slower hard disc such as an hard disc drive (HDD). This is because of the large volume of reads and writes needed to read and/or store an uncompressed video file.

realistic encoding time at the expense of a slight drop in compression performance compared to the HM encoder. The Medium preset is faster but still below real-time performance on the test computer. The fastest preset, Very Fast, encodes around 30 frames per second, which is approximately real-time performance for 30 fps video. Thus on this platform, x265's slower presets might be suitable for slower than real-time encoding when a smaller file size and better visual quality are more important than encoding speed, whereas the Very Fast preset might be suitable for real-time encoding such as for live streaming or videoconferencing.

11.4 Hardware Implementation

In many consumer electronic devices, video encoding and decoding are implemented in dedicated hardware. Here are a few examples:

- Apple's iPhone 13 and 14 models include the A15 chip, which has dedicated hardware support for H.265 and H.264 encoding and decoding.
- Some Samsung mobile devices include an Exynos processor. The Exynos 2200 can encode and decode a number of video coding formats in hardware, including H.264, H.265 and AV1.
- Intel's 7th and later generations of Core processors, which have been available since 2017, include integrated graphics processors with hardware encoding and decoding support for video formats including H.264 and H.265.
- Some NVIDIA graphics processing units (GPUs) can encode and/or decode H.264 and H.265 in hardware. Figure 11.3 shows a representation of an NVIDIA GPU chipset. Multiple CUDA graphics processing cores carry out graphics operations such as rendering

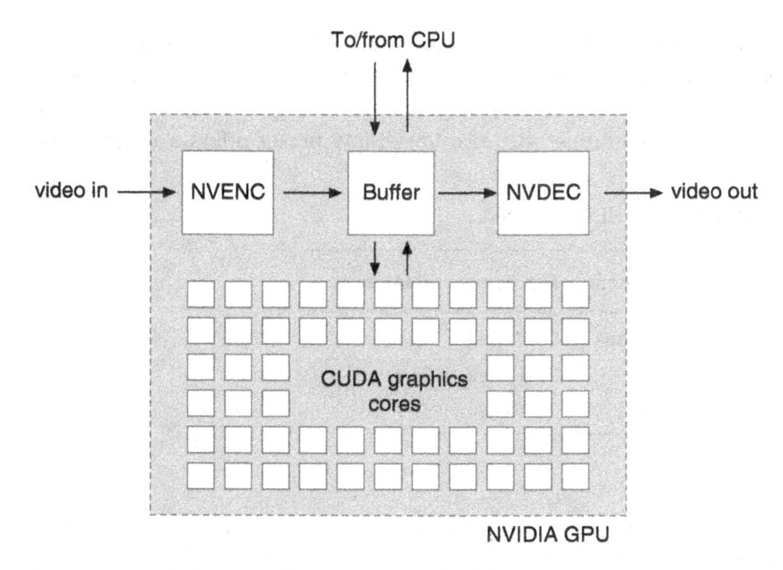

Figure 11.3 Schematic illustration of NVIDIA GPU architecture

in an efficient and parallelised way. Separate hardware modules labelled NVENC and NVDEC carry out video encoding and video decoding, respectively. This enables the NVIDIA GPU to process both graphics and video with minimal interaction and, therefore, minimal extra processing from the system's CPU [6].

Designing and implementing a dedicated hardware codec on a chip is a time-consuming and expensive process. As the examples above illustrate, semiconductor manufacturers include hardware encoder and decoder engines in successive generations of their consumer chipsets. When a new video coding standard is finalised, it can take several years before hardware support for the standard is integrated into the current generation of consumer products. The H.265 standard was first published in 2013. It was not until 4 years later, in 2017, that newer Intel Core processors included hardware encoding and decoding support for H.265.

Once they are widely available in consumer devices, hardware codecs can have certain advantages over software video codecs. Perhaps the most important advantage is the ability to encode and/or decode in real time with significantly less power consumption than the equivalent software implementation. When designing a consumer video service such as a videoconferencing or video streaming system, it may be preferable to use hardware video codecs in the consumer devices where possible, in order to make efficient use of computational resources and power consumption.

A streaming service provider may offer higher-resolution versions of video content using a newer, more efficient coding standard such as H.265 and may also provide lower-resolution fallback versions using an older standard such as H.264. This way, higher-quality video can be delivered efficiently if the consumer device has hardware support for H.265. Older consumer devices can always access lower-resolution versions using an H.264 hardware codec. Figure 11.4 shows an example of this. A streaming provider encodes a source video into multiple resolutions, including higher-resolution versions for clients with H.265 hardware decoding support and lower-resolution versions for clients with only H.264 hardware decoding support. Client 1 is an older device that only has H.264 support built in. It can choose a stream of up to 720p resolution. Client 2 has a newer chipset with both H.265 and H.264 hardware decoding capability and can choose to decode any of the lower or higher-resolution streams. Note that it is common for newer consumer device chipsets to include hardware support for multiple video coding standards.

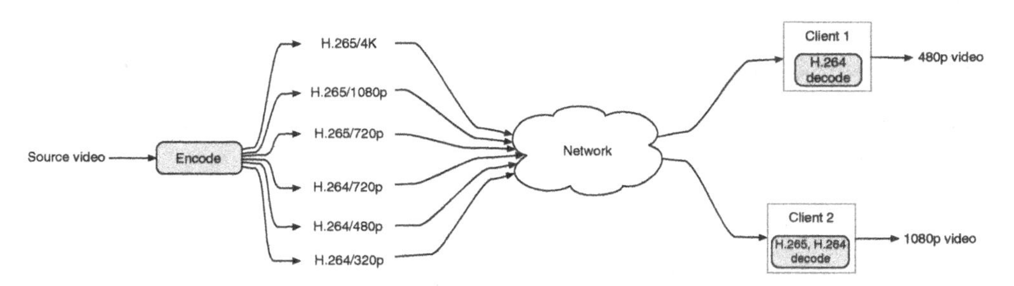

Figure 11.4 Streaming using multiple resolutions and two formats

Example: Software versus Hardware Encoding, Comparing x265 and VideoToolbox

VideoToolbox is a software framework provided by Apple that provides direct access to hardware video encoders and decoders available on MacOS and iOS devices. VideoToolbox encoding or decoding can be selected as an option using the popular ffmpeg video processing framework. It is therefore relatively straightforward to compare the HEVC encoding performance of × 265 operating in software and VideoToolbox operating in dedicated hardware on an Apple device.

Ffmpeg was run in a file-based mode in order to encode 100 frames of the 1080p video clip *Pedestrian Area* in HEVC format using (1) × 265 – Medium preset – and (2) VideoToolbox, with the encoding parameters of each codec chosen to give a similar compressed file size, that is, a similar encoded bitrate. The results are shown in Table 11.2.

In this test, VideoToolbox runs around 3 × faster than × 265. The real difference in processing speed may be more significant than this since file-based operation requires a large amount of uncompressed video data to be read from the input file, which typically takes up a significant amount of the encoding time. As well as being faster in real terms, a hardware codec takes the load away from the main CPU, leaving the system CPU free to perform other tasks. However, in terms of video quality, as estimated by VMAF, VideoToolbox is considerably worse than × 265. In fact, VideoToolbox needs to encode at almost double the compressed bitrate to achieve a similar visual quality to × 265.

In this example, if compression performance is more important and processing speed is less important, × 265 may be a better choice, whereas for time-critical applications such as real-time video encoding, a hardware codec, such as the codec(s) available through VideoToolbox, may be preferable.

Table 11.2 Comparing × 265 and VideoToolbox HEVC encoders

Encoder	Encoding time	File size	VMAF score
X265	5.8 seconds	549 kbytes	77.5
VideoToolbox	1.9 seconds	569 kbytes	52.9

11.5 Video Codec Performance

11.5.1 Measuring Codec Performance

The performance of a video codec can be evaluated in at least three important ways, which are as follows:

- **Compressed bitrate**: how large is the compressed file, and how many bits per second of video?
- **Decoded quality**: how good is the quality compared with the original, uncompressed version?

- **Computational complexity**: how long does it take to compress or decompress a video sequence? Can the compression happen in real time – can a 30 fps video clip be compressed at a rate of 30 fps?

Example:

Compare two compressed and decompressed versions of the *Jockey* test video clip, 1920 × 1080 pixel resolution, 100 frames. The uncompressed video file, stored in y4m format, is 311 Mbytes.

g) Compression using × 264, choosing QP = 22, so encoding at a relatively high quality. The file is compressed by around 133 ×. Ffmpeg encodes at around 34 frames per second.
h) Compression using × 265, QP = 22, again at a relatively high quality. The file is compressed by around 170 ×. Ffmpeg encodes at around 15.5 frames per second.

A still frame from the original (left), the decoded H.264 version (centre) and the decoded H.265 version (right) are shown in Figure 11.5. It is difficult to distinguish the decoded versions from the original video. According to the VMAF metric, the decoded visual quality of the two versions is approximately the same. In this example, the × 265 codec compresses the file around 1.3 × more than the × 264 codec, but it takes just over twice as long to do so[2].

Video compression performance can depend on a number of factors, including:

- The video coding standard or format. Each standard places certain limitations on the coding process, such as the maximum CTU size of 64 × 64 samples in H.265 or the fixed macroblock size of 16 × 16 samples in H.264. Within these limitations, compression performance can vary significantly. Each standard places a certain upper limit on overall performance and has certain implications for computational complexity.
- The choice of encoding parameters.

Jockey - section of original 1080p frame
311040681 bytes, 100 frames

Jockey - H.264, QP = 22, decoded
2323497 bytes, VMAF 99.770, 100 frames

Jockey - H.265, QP = 22, decoded
1832080 bytes, VMAF 99.766, 100 frames

Figure 11.5 Example: comparing H.264 and H.265 performance

2 With careful choice of encoding parameters, we might expect × 265 to compress around 1.7× – 2× more than × 264, probably at the expense of further increased computational complexity.

- The implementation platform, such as hardware or software, and the choice of implementation architecture.
- The video sequence to be compressed.
- The end user's subjective opinion of video quality.

When considering all these factors, choosing how to compress a video clip usually involves making some sort of compromise between the three performance factors of compression, quality and computation. For example:

- If computation is roughly constant, we can increase compression at the expense of reduced quality or increase quality at the expense of reduced compression. For example, increasing the QP will result in increased compression and reduced quality.
- If quality is roughly constant, we can increase compression at the expense of increased computation or reduce computation at the expense of reduced compression. Figure 11.5 shows two clips with the same decoded quality. The H.265 version has better compression at the expense of increased encoding time.
- If compression is roughly constant, we can increase quality at the expense of increased computation or reduce computation at the expense of reduced quality. If we set the same bitrate target for an H.264 encoder and an H.265 encoder, the H.264 encoder will probably encode lower-quality video but will use less computational time.

11.5.2 Measuring Rate-Distortion Performance

The rate versus distortion, or alternatively, the compression versus quality performance of a complete coded video sequence, can be represented as a single point on a rate-distortion curve such as the four rate-distortion curves shown in Figure 11.6. In this example, each curve corresponds to a particular video sequence encoded using H.264 using either CABAC or CAVLC entropy coding.

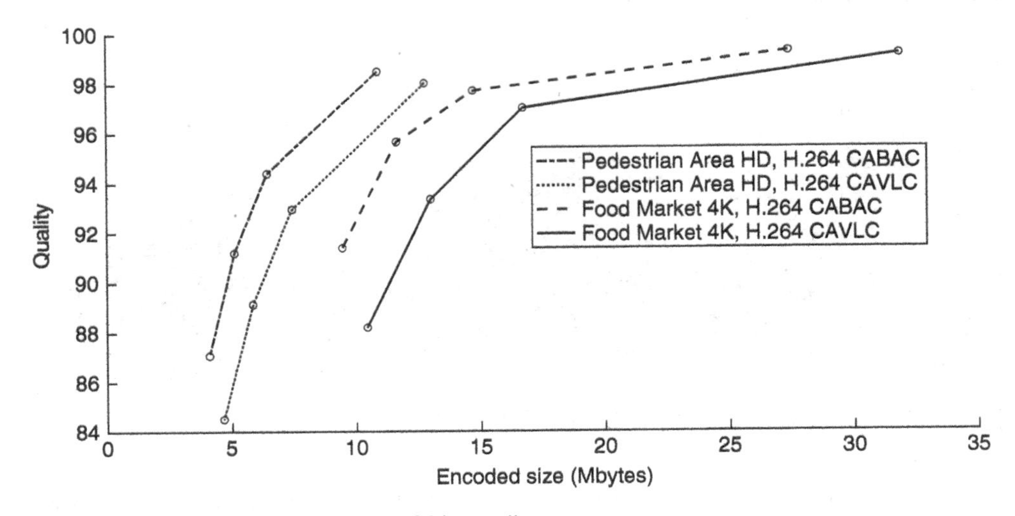

Figure 11.6 Rate-distortion curves, H.264 encoding

We can consider one of these curves, the 4K *FoodMarket* sequence encoded using CAVLC entropy coding and plotted using a solid line. Each point on the curve records the compressed file size or the average bitrate on the horizontal axis, and the decoded quality on the vertical axis, for one encoding run. Quality is estimated using the VMAF metric. By changing parameters, for example, by increasing or decreasing QP, we obtain a rate-distortion curve. The top-right point on this curve was encoded at a lower QP. As the QP is increased, the file size reduces and the quality drops, producing this characteristic curve shape.

Now consider *FoodMarket* encoded using CABAC. The rate-distortion curve has a similar shape but is higher up and to the left. If we consider any bitrate or file size, such as a file size of 15 Mbytes, the CABAC version always has a higher quality. If we consider any particular quality, for example, 95.0, the CABAC version always has a lower bitrate and hence a smaller file size. Comparing the two rate-distortion curves for FoodMarket makes it immediately clear that CABAC encoding provides better performance than CAVLC encoding in this case.

Consider the two encoded versions of the HD *Pedestrian Area* sequence. Each curve has the same overall shape, but now the curves have shifted to the left. This makes sense because the HD clip has a frame size that is a quarter of the 4K clip, so we would expect the encoded file sizes to be smaller.

From Figure 11.6, we can see that H.264 using CABAC provides better compression than H.264 using CAVLC for these two clips. What is not so clear is the overall performance difference. The Bjøntegaard Delta Metrics, first proposed by Bjøntegaard [7], estimate the average difference in bitrate and/or quality between two sets of rate-distortion measurements. Given two sets of rate-distortion points, such as Curve 1 and Curve 2 in Figure 11.7, the following steps are carried out:

1) Fit a curve to each set of points.
2) Calculate the area between the two curves, shown as the shaded area in Figure 11.7.
3) Calculate the average percentage difference in bitrate between the two curves: this is BD-Rate.
4) Calculate the average difference in the quality score between the two curves: this is BD-Quality.

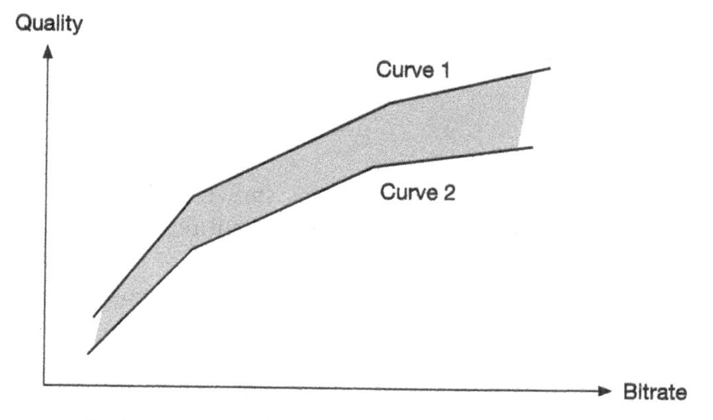

Figure 11.7 Bjøntegaard Delta-Rate or Delta-Quality estimation

The BD-Rate and BD-Quality metrics have been used since the early 2000s to estimate the performance gains of new video coding methods, such as a proposed method for a coding standard that is under development. BD-Rate or BD-Quality gives a single score that helps to evaluate a new method. For example, a proposal to improve a video codec might deliver a BD-Rate improvement of $x\%$ at a cost of an increase of $y\%$ in computational complexity.

11.5.3 Optimising Video Codec Performance

Video codec performance is typically a trade-off between compression that is compressed bitrate, decoded visual quality and computation. Optimising this trade-off is a complex challenge.

11.5.3.1 Computational Optimisation

Operations that are carried out frequently can have a significant impact on performance. In a typical codec, certain operations such as CABAC entropy coding, block coefficient processing and prediction processing are carried out for every coded block and are therefore likely to have a big effect on computational efficiency. Other operations such as parameter set processing are carried out less frequently and so may have less of an impact on computation.

When coding a basic unit such as a macroblock or CTU, an encoder has many choices available. These include the block sizes for prediction, transform and coding, the choice of prediction modes and parameters, the choice(s) of QP and the choice of filtering parameters. An encoder typically has to make these choices within the available computational constraints. Due to the range of available options, the computational complexity of video encoding can vary considerably, from real-time encoding on a mobile handset with limited computation to offline encoding many times slower than real time on a more powerful platform.

A video decoder processes an encoded bitstream in which all the coding choices have already been made, such that the computational complexity of video decoding is constrained by these coding choices. A video coding standard specifies certain combinations of Profile and Level that place upper limits on the computational complexity and memory requirements of a decoder. A decoder handling a video bitstream that was encoded using H.265 Main Profile at Level 2 knows the upper limit on computation that will be required to decode the bitstream in real time.

11.5.3.2 Rate-Distortion Optimisation

We will consider one point on a rate-distortion (R-D) curve as shown in Figure 11.8. What happens if we change the encoder decisions? Each choice of partition sizes, prediction modes, QP and so on may have an effect on the R-D performance of the current encoding process. One such decision might have a very small impact, but the cumulation of decisions across the entire video clip can push the R-D point in any direction. A good example is to test the presets for a codec such as x265. Moving from the Medium to the Very Slow preset, without changing QP, might be expected to shift the R-D performance of one point on the curve up and/or to the left, resulting in higher quality and/or a lower

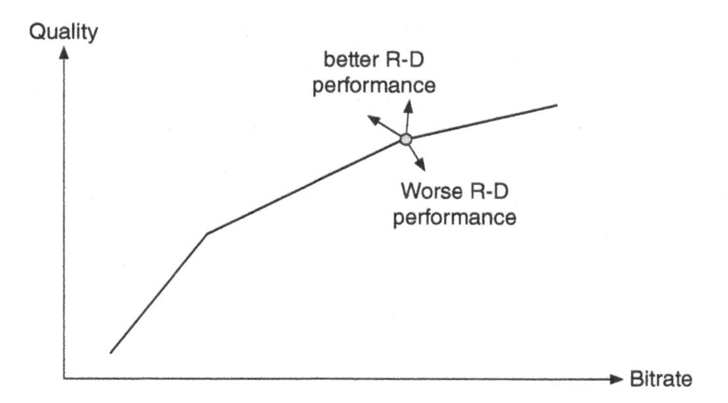

Figure 11.8 Rate-distortion curve, effect of decisions on one point

bitrate. This is because Very Slow should produce a smaller file size, which reduces bitrate and causes the R-D point to move to the left, and/or better quality, which causes the R-D point to move upwards. Moving from Medium to Very Fast might shift the R-D performance down and to the right, therefore coding at lower quality and/or higher bitrate. Adjusting the encoder behaviour for one QP value will tend to shift the point as shown in Figure 11.8. Adjusting the encoder behaviour and testing a range of QP values, will tend to shift the entire curve.

What does optimal or near-optimal rate-distortion performance look like? As the overall R-D performance improves, each point on the R-D curve shifts up and/or to the left. Optimal performance is going to be represented by an R-D curve that is as far up and left as is achievable for the current codec. This is sometimes known as the convex hull of R-D performance (see Figure 11.9). The perfect choice of encoding parameters, given the constraints of the particular video coding standard, results in the convex hull R-D curve shown in Figure 11.9. In practice, this is unlikely to be achievable and instead, practical coding decisions will result in sub-optimal R-D points for each set of coding parameters. Good performance is represented by an R-D curve that is as close as possible to the optimal convex hull, so as far up and to the left as possible.

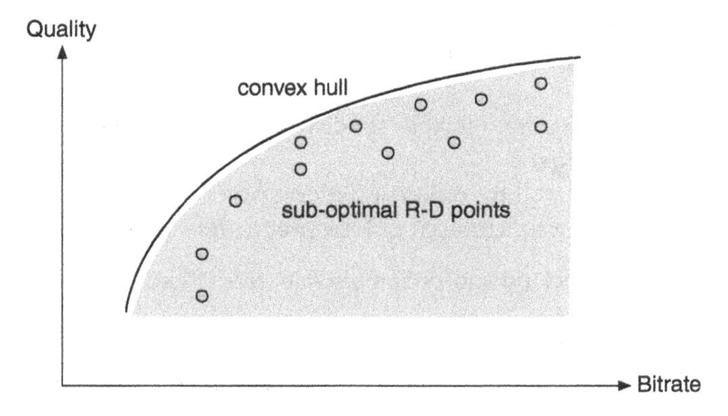

Figure 11.9 Optimal rate-distortion curve

An encoder can use this knowledge to improve its choices during the encoding process. Each decision, such as a choice of CTU partitioning into CUs, has an associated R-D effect. If the encoder can measure or estimate the rate and distortion associated with each decision, it can choose the decision that minimises bit rate and minimises distortion, or choose some combination of the two. Such a combination can be calculated as a rate-distortion cost, often expressed as a cost function, $J = D + \lambda R$, where D is distortion, R is rate and λ is a weighting parameter [8]. A lower value of J indicates a smaller combination of R and D, so an overall better R-D performance pushes the overall R-D performance up and to the left, towards the ideal convex hull. Hence, selecting a video coding choice from a set of options, such as partition selection and prediction choice, may be formulated as follows:

For each coding option

```
    Calculate or estimate R and D
    Calculate J=D+λR
```

Select the option with the smallest value of J

Example: Coding Decisions for a CTU

We can consider one CTU in an HEVC encoding process. The choice of optimal or near-optimal coding decisions can get very complicated.

There are many choices for encoding a CTU using the HEVC standard. Take, for example, a 64 × 64 CTU. Encoding choices for the CTU include the following:

- The CTU may be partitioned into CUs, for example, one 64 × 64 CU, four 32 × 32 CUs or further partitioning of each CU down to the minimum CU size, which may be set to 8 × 8 or larger.
- Each CU can be encoded using intra-prediction, inter-prediction with one reference picture and one motion vector per PU, inter-prediction with two references and two motion vectors per PU, or Skip mode, with no further information sent for the CU.
- Each CU is partitioned into PUs: one inter-predicted PU or intra-predicted U, two symmetrical horizontal or vertical inter-predicted PU partitions, two asymmetric horizontal or vertical inter-predicted PU partitions or four equal inter or intra-predicted PU partitions.
- Each CU is partitioned into TUs: one, four or a further subdivision down to the minimum TU size which may be 4 × 4 or larger.
- For each PU, the encoder can choose an intra prediction mode, or a choice of inter prediction motion vector(s) and reference pictures(s), or inter prediction merge mode.

Further encoder choices include deblocking and/or Sample Adaptive Offset filtering, transform and/or quantisation bypass.

The encoding choices for one CTU affect subsequent CTUs that may be predicted from this one, and/or that may inherit parameters from this one, for example, through the use of merge mode.

The optimal choice of all options for all CTUs in a sequence is probably unattainable on present-day computing platforms. In practice, even a slower than real-time encoder will make certain simplifications to manage computational complexity by:

- Choosing modes and parameters for each CTU and picture, wholly or partially independently of other CTUs and pictures.
- Using simplified metrics to estimate decoded quality.
- Estimating rate-distortion cost algorithmically, without actually encoding every possible decision.
- Pruning the tree of available decisions. Let us assume that a single 64 × 64 CU provides a lower rate-distortion cost than four 32 × 32 CUs. In this case, it is unlikely that further subdivisions of CUs, such as dividing one of the 32 × 32 CUs into four 16 × 16 CUs, will improve the rate-distortion cost, so the encoder may choose to skip testing any further subdivisions of CUs.
- Considering complexity in the optimisation process and minimising some combination of rate, distortion and complexity during encoding [9].
- Using two-pass encoding, in which a first pass collects statistics such as bit counts and estimated distortion without producing an output bitstream. A second pass uses the statistical information from the first pass to make efficient parameter choices during the actual coding of the bitstream.

Because of the huge number of possible encoding choices and the impact of combinations of choices on bitrate, quality and encoding speed, there is significant opportunity for optimisation and innovation in encoder mode decisions. This is one reason for the variation in performance between different encoder implementations of the same standard.

11.5.3.3 Software Optimisation

Both encoder and decoder can benefit from efficient software implementation. Much of the processing in an encoding or decoding pass involves repeated operations such as block transforms, additions and subtractions, quantisation and entropy coding. Contemporary processors can significantly improve the execution speed of such repeated operations if the software is designed to facilitate efficient repeated processing.

Since the 1990s, many processor architectures have incorporated instructions that are designed to improve the efficiency of operations that are carried out during multimedia encoding and decoding.

Many present-day processors support parallel processing via multiple parallel processing cores. HEVC and later standards include features that were specifically intended to support encoding and decoding on multi-core processors, such as Tiles and Wavefront Parallel Processing.

11.6 Getting Started with Experiments

In this book, I have introduced many of the techniques and processes that a video codec uses to efficiently compress and decompress video. A good way to understand more about these techniques is to experiment with the processes of coding and decoding video. In this section, I will provide some examples of how to get started with experimentation. Please visit the book companion site, https://www.vcodex.com/coding-video-book, for download links and other resources related to this section.

11.6.1 Installing Software

Many different software implementations of video coding standards are available; some are available in the public domain, whereas others require a commercial license. A good starting point is the ffmpeg cross-platform framework for processing, coding and streaming video and audio material. It includes a number of video codecs, including the ×264 and ×265 software implementations of H.264 and H.265. Ffmpeg is widely used, relatively fast, well-documented and frequently updated. It can be used in a file-based mode, for converting video files to or from compressed files, or used as a software library as part of a larger application. The popular processing and playback applications Handbrake and VLC provide a more user-friendly graphical interface to the ffmpeg library. In this section, we will be using ffmpeg in file-based mode.

The official ffmpeg site, https://ffmpeg.org/, provides ffmpeg as source code for compilation. Third-party sites provide pre-compiled binaries for installation on platforms such as MacOS, Windows and Linux, https://ffmpeg.org/download.html. MacOS and Linux users can use a package manager such as Brew to install the latest version of ffmpeg.

Once you have installed ffmpeg using a binary, a package manager or by compiling the source code, open a command prompt or terminal and type `'ffmpeg -h'` to get basic help options. The syntax for calling ffmpeg from the command line is generally:

```
ffmpeg [options] [[infile options] -i infile]... {[outfile
options] outfile}...
```

where infile is an input file and outfile is the file that ffmpeg produces. Numerous command line examples are provided below. On a Windows system, replace ffmpeg with ffmpeg.exe.

11.6.2 Source Video Material

It is possible to experiment with any video material available to you. However, it is possible that readily available video files on your computer or mobile device may have already been compressed using lossy compression. It can be preferable to start with uncompressed source material. Uncompressed video clips at a range of resolutions are available at https://media.xiph.org/video/derf/ and https://ultravideo.fi/#testsequences. These are typically provided in .y4m format or .yuv, in which the video samples are stored in a file as Y, Cr and Cb samples. The .y4m format has a header indicating the resolution and sampling format of the video, e.g. 1920×1080, 4:2:0 sampling. These files are large: a 500-frame, 1080p clip takes up 1.5Gbytes of storage. A few example screenshots[3] are shown in Figure 11.5. These

---

3 Quarter common intermediate format (QCIF), CIF and 4CIF are conventional terms for 176×144, 352×288 and 704×576 resolution, respectively. 4CIF is approximately the resolution of Standard Definition television.

Figure 11.10 Still frames from selected test video sequences, not to scale

screenshots are not illustrated to scale. For example, the 4K *Old Town Cross* clip is around 20× the horizontal resolution of the QCIF *Carphone* clip (Figure 11.10).

11.6.3 Adjusting Source Video Clips

Ffmpeg can be used to change various aspects of a test clip. You might choose to download a 4K test clip such as *Jockey* or *Old Town Cross* and then create versions of the clip at lower resolutions such as 1080p or 720p. Here are a few examples of how to use ffmpeg to manipulate a source clip in .yuv or .y4m format.

Convert a .yuv file to a .y4m file:

```
ffmpeg -s [resolution, e.g. 1920×1080] -r [framerate] -pix_
fmt yuv420p -i [source.yuv] [destination.y4m]
```

Adjust the resolution of any .y4m file to 1080 pixels vertical resolution:

```
ffmpeg -i [source.y4m] -vf scale=-1:1080 [destination.y4m]
```

Convert an input file to a file containing only N frames:

```
ffmpeg -i [source.y4m] -frames N [destination.y4m]
```

11.6.4 Encode, Decode and Play Video Clips

Experimenting with video coding requires the capability to encode/compress, decode/decompress and view or play back video clips.

To encode a y4m file using ffmpeg:

```
ffmpeg -i [source.y4m] -c:v [encoder library] [encoding
parameters] [output.mp4]
```

where [encoder library] is a choice of ffmpeg library for encoding using a certain format, such as libx264 for x264/H.264 encoding or libx265 for x265/H.265 encoding. The encoding parameters available depend on the chosen library. To encode using x265 with a video encoder QP of 38:

```
ffmpeg -i [source.y4m] -c:v libx265 -x265-params qp=38 [output_
QP38.mp4]
```

Here, the output is specified as a .mp4 file. This means that the H.265 bitstream is encapsulated in an MP4 container file. No audio has been specified in the command line, so the MP4 file (output_qp38.mp4) will contain no audio. It can be convenient to use a container file format such as MP4 so that other applications such as VLC player can readily open and play the file.

To decode a compressed file into an output, uncompressed y4m file:

```
ffmpeg -i [output.mp4] [decoded.y4m]
```

Ffmpeg will attempt to determine the codec format (e.g. H.265) and will apply the appropriate video decoder to produce an uncompressed file.

To play back a coded or uncompressed file, with minimal user controls:

```
ffplay [filename]
```

Many applications are available to play compressed video clips, and a smaller number of applications can play back uncompressed clips in .y4m or .yuv format. VLC player, which uses the ffmpeg framework, is a useful tool for playing compressed video files, and applications such as YUVviewer and Vooya can play back uncompressed clips.

11.6.5 Measure Rate-Distortion Performance

Ffmpeg calculates average bitrate while it is encoding the sequence and can be directed to calculate peak signal-to-noise ratio (PSNR) or other quality estimates during encoding. The rate-distortion performance of a codec can be measured over a range of bitrates. Note that it is not necessary to actually decode the clip(s), since ffmpeg can calculate the average PSNR of the decoded sequence during encoding. The process is best illustrated with an example.

Example: Comparing the Rate-Distortion Performance of x264 and x265

1) Obtain a source video sequence in .y4m format.
2) Encode 100 frames of the sequence using the x265 codec, libx265, recording the average or global PSNR:

```
ffmpeg -i [source.y4m] -frames 100 -c:v libx265
-x265-params qp=24 -psnr -y test265_qp24.mp4
```

3) Note the global PSNR and average bitrate from final output line, which will look something like this:

```
encoded 100 frames in 7.58s (13.20 fps), 11786.34 kb/s,
Avg QP:25.15⁴, Global PSNR: 43.517
```

4) Repeat with QP = 30, 36, 42
5) Repeat with the following command line, using the x264 codec, libx264:

```
ffmpeg -i source.y4m -frames 100 -c:v libx264 -x264-params
qp=24 -psnr -y test264_qp24.mp4
```

6) Note PSNR and average bitrate from final output line, which will look something like this:

```
PSNR Mean Y:42.606 U:44.794 V:44.767 Avg:43.217
Global:43.201 kb/s:15637.61
```

7) Repeat with QP = 30, 36, 42.

The results for four different QP values should look something like Table 11.3, of course with different bitrates and PSNRs depending on your source clip. These results were obtained for 100 frames of the *Jockey* video clip, at 1920 × 1080 resolution.

Plotting the bitrate and PSNR points on a graph gives a pair of classic rate-distortion curves (see Figure 11.11). Note that the bitrate and PSNR corresponding to the lowest QP are at the top-right of the graph because the lowest QP gives the highest bitrate and also the highest quality. As we increase QP, the points on the graph move down and to the left, which shows reducing bitrate and reducing quality as the QP increases. From the figure, you can clearly see the difference in performance between the two codecs. Consider the dotted horizontal line corresponding to a PSNR of 40dB. At this PSNR, × 265 produces a bitrate of around 3250 bits per second, whereas x265 produces a bitrate of around 5600 bits per second. Therefore, at an estimated quality corresponding to PSNR = 40 dB, × 265 produces a compressed output that is around 40% smaller than the output produced by × 264, i.e. × 265 compresses the video around 1.6 × more than × 264.

Table 11.3 Bitrate and PSNR, 'Jockey' clip, × 264 and × 265

QP	Bitrate (× 264)	PSNR (× 264)	Bitrate (× 265)	PSNR (× 265)
24	15638	43.2	11786	43.5
30	7214	41.2	4871	41.6
36	3997	38.8	2312	39.0
42	2328	36.0	1109	36.2

4 Note the average QP is 25.15, even though the target QP was 24. This is because the x265 encoder chooses by default to adjust the QP depending on the slice type, I, P or B.

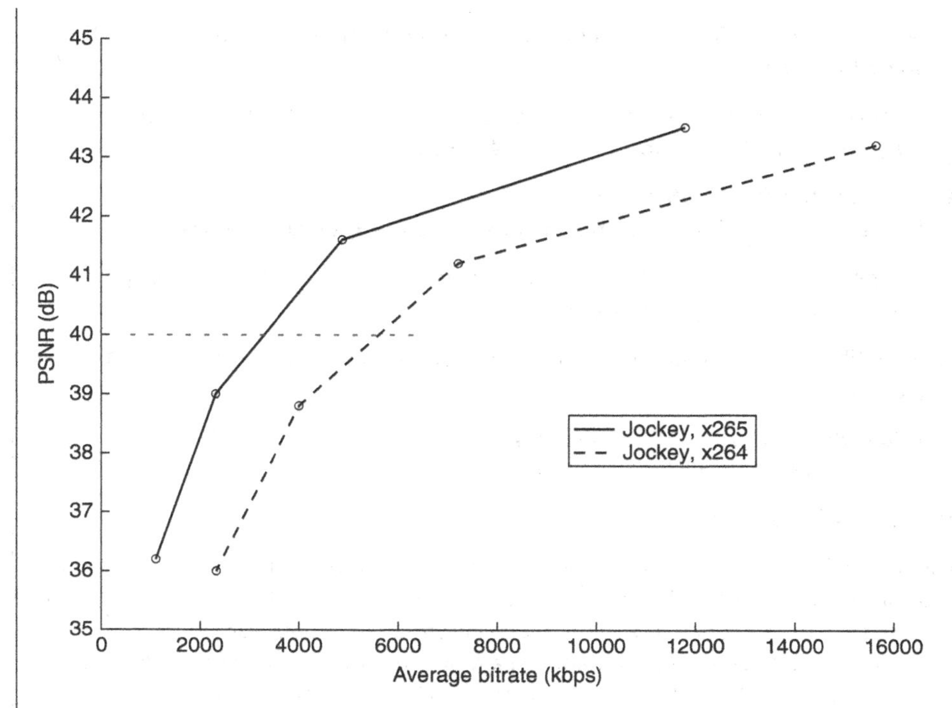

Figure 11.11 Bitrate versus PSNR, *Jockey*, x264 and x265

PSNR can be calculated easily but is a relatively crude estimate of visual quality. The following command line calculates the more sophisticated VMAF metric and should be executed after first encoding the source file, [source.y4m] into an encoded file, testX.mp4:

```
ffmpeg -i testX.mp4 -i [source.y4m] -lavfi libvmaf -f null -
```

This command calculates VMAF based on comparing the encoded file with the original, uncompressed file. VMAF is computationally complex, and this command may well take longer to execute than the original encoding process. The result is an average VMAF score, e.g. 85.5, in the range of 0 to 100, where 100 is the best possible quality, indicating no perceivable difference from the original.

11.6.6 Analysing Coded Bitstreams

Syntax analyser software applications can be very helpful when trying to interpret the coding decisions made by a video encoder and/or to understand how a video encoder chooses encoding modes, prediction modes, etc.

I have included screenshots from analyser software throughout the book. Here are screenshots from three commercially available analyser applications, reproduced with permission:

Elecard Streameye, H.264, H.265 and H.266 analyser [10] (Figure 11.12):

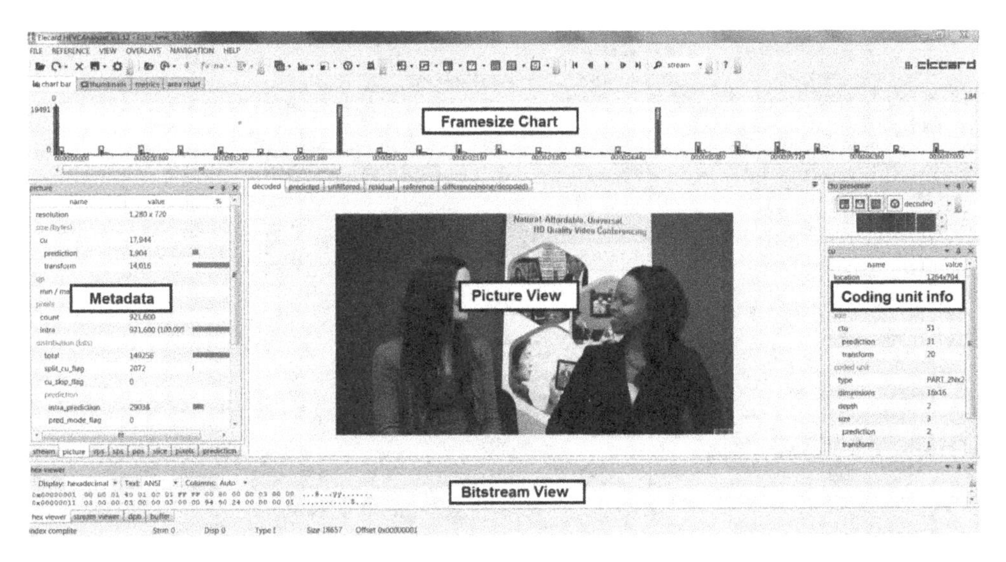

Figure 11.12 Elecard Streameye, screenshot of main window

Parabola Explorer, H.265 analyser [11] (Figure 11.13):

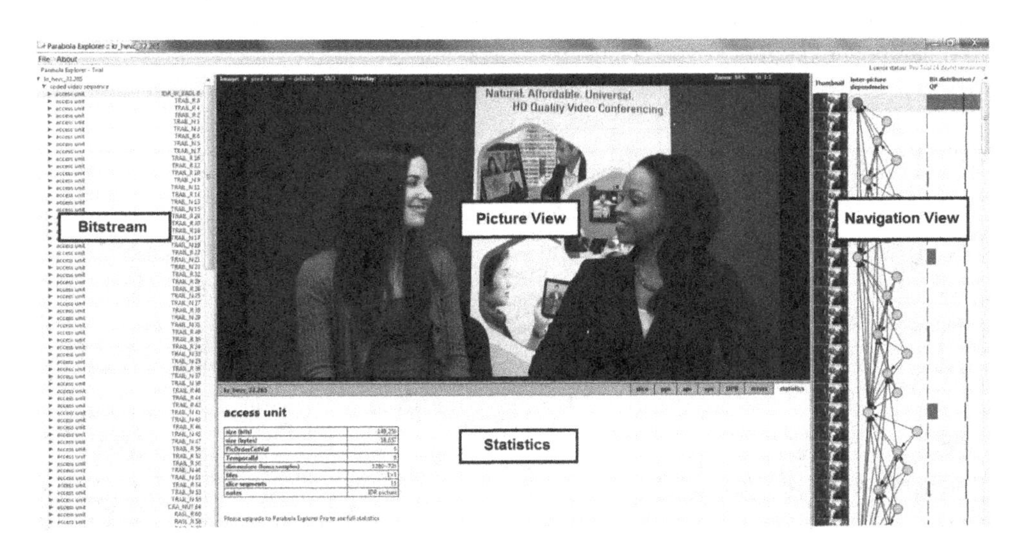

Figure 11.13 Parabola Explorer, screenshot of main window

Zond265, H.264 and H.265 analyser [12] (Figure 11.14):

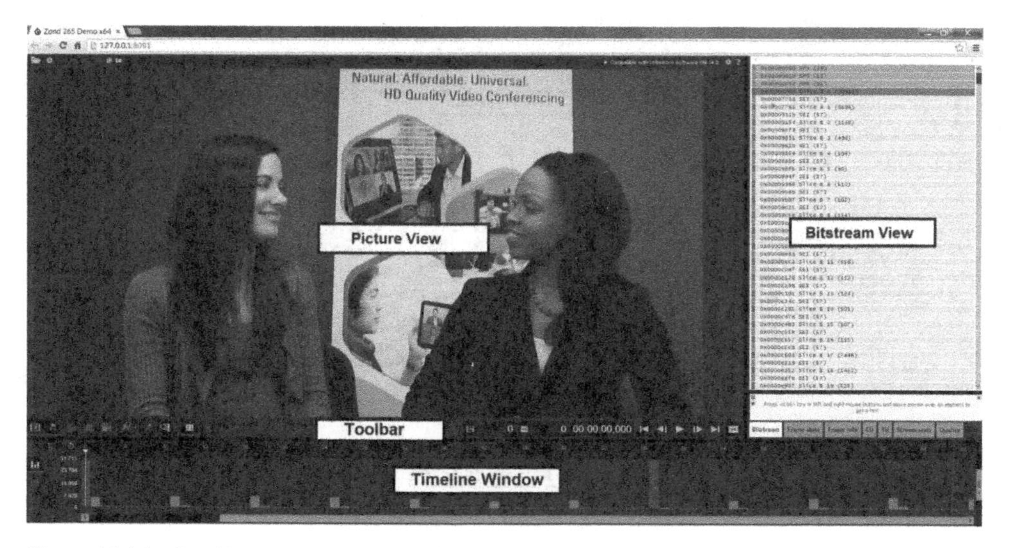

Figure 11.14 Zond265, screenshot of main window

Each of the analyser applications has the capability to overlay illustrations of syntax structures such as block partitions and prediction types. The following examples are screenshots from Parabola Explorer (Figures 11.15 and 11.16).

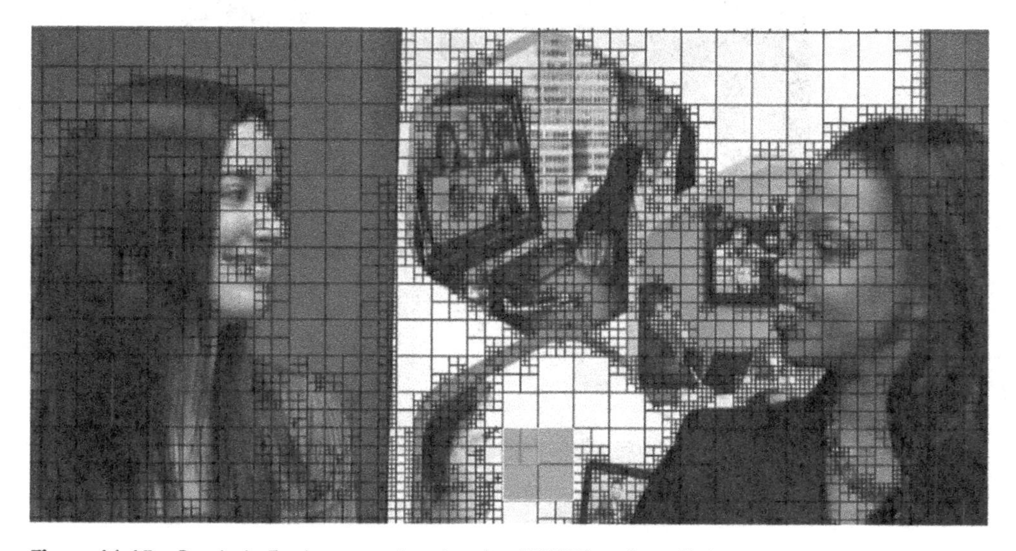

Figure 11.15 Parabola Explorer, overlay showing HEVC Transform Units

Figure 11.16 Parabola Explorer, overlay showing I, P, B and skipped Prediction Units

11.7 Conclusion

In this chapter, we have seen how video codecs can be implemented in software, hardware or a mixture of the two. Measuring and comparing the performance of video codecs is not always a straightforward task, but the techniques introduced here can help to determine which codec is best for a particular task. We have also looked at ways of experimenting with public-domain video codecs. Setting up and carrying out your own experiments to find out how all the different processes interact and contribute to coding video is a great way to learn more about the tools and techniques that are introduced in this book.

References

1 Aaron, A., Li, Z., Manohara, M. et al. (2015). Challenges in cloud based ingest and encoding for high quality streaming media. *2015 IEEE International Conference on Image Processing*, 1732–1736. doi: https://doi.org/10.1109/ICIP.2015.7351097.

2 Reference software for ITU-T H.265 high efficiency video coding (2016). ITU-T Recommendation H.265.2. Approved December 2016.

3 ITU-T H.265.2 reference software source code. www.itu.int/rec/T-REC-H.265.2-201612-I/en (published 2016, accessed 04 January 2024).

4 X265 open source video encoder. https://bitbucket.org/multicoreware/x265_git (accessed 04 January 2024).

5 Li, Z., Aaron, A., Katsavounidis, I., et al. (2016). Toward a practical perceptual video quality metric. *Netflix Technology Blog*. netflixtechblog.com/toward-a-practical-perceptual-video-quality-metric-653f208b9652 (accessed 06 January 2024).

6 NVIDIA Video Codec SDK overview. developer.nvidia.com/nvidia-video-codec-sdk (accessed 06 January 2024).

7 Bjøntegaard, G. (2001). Calculation of average PSNR differences between R-D curves. ITU-T VCEG document VCEG-M33. www.itu.int/wftp3/av-arch/video-site/0104_Aus/VCEG-M33.doc (accessed 06 January 2024).

8 Sullivan, G.J., Wiegand, T. (1998). Rate-distortion optimization for video compression. *IEEE Signal Processing Magazine* v15 (6): 74–90. doi: https://doi.org/10.1109/79.733497.

9 Kannangara, C.S., Richardson, I.E., Bystrom, M. et al. (2009). Complexity control of H.264/AVC based on mode-conditional cost probability distributions. *IEEE Transactions on Multimedia* 11 (3): 433–422. doi: https://doi.org/10.1109/TMM.2009.2012937.

10 Elecard streameye studio. www.elecard.com/products/video-analysis/streameye-studio (accessed 06 January 2024).

11 Parabola research analyser. parabolaresearch.com/explorer-hevc-analyzer.html (accessed 06 January 2024).

12 Zond265 analyser. www.solveigmm.com/en/products/zond/ (accessed 06 January 2024).

12

Conclusions

12.1 What This Book Has and Has Not Covered

We will briefly review what has been covered in this book. I started with an overview of the basic video codec system or model that has been shared by all the popular video coding standards since the early 1990s. I reviewed the historical developments that have occurred since the 1950s that have influenced today's video codecs and standards.

I examined the use of structures to partition coded video, including organising a coded video sequence, coding Groups of Pictures, splitting up each frame into slices, tiles and basic coding units and partitioning coding units for prediction, transform and entropy coding. I looked at intra prediction, in which each block is predicted from pixels in the same frame, and inter prediction, in which blocks are predicted from pixels in previously coded frames such as past and future frames. I reviewed transforms and quantisation, which are the processes of transforming blocks of pixel data into a spatial frequency domain and then reducing the precision of the spatial frequency data. I covered entropy coding, which converts the elements of a coded video sequence into a compressed binary bitstream, with a particular focus on arithmetic coding. I considered how in-loop filters can improve the performance of inter prediction. Finally, I reviewed the ways in which coded video is stored and communicated across today's networks and looked at how video codecs can be implemented in software or hardware.

Of course, there are many topics that I have chosen not to cover in this book. These include:

- Audio coding, a fundamental component of today's media applications. Audio compression shares some concepts with video coding such as using prediction and entropy coding and exploiting the properties of the data – digital audio samples – and the human listener. It is a complex and ever-evolving subject in its own right.
- Important aspects of modern-day usage of coded video, including digital rights management, content management and delivery, and many of the details of storage and transport.
- The variety of new and emerging paradigms of creating and consuming video, including stereoscopic, multiview, immersive video, screen content coding and computer-generated video.
- The wide range of innovative techniques for video coding that have not yet made it into video coding standards.

Coding Video: A Practical Guide to HEVC and Beyond, First Edition. Iain E. Richardson.
© 2024 John Wiley & Sons Ltd. Published 2024 by John Wiley & Sons Ltd.
Companion website: www.wiley.com/go/richardson/codingvideo1

12.2 Where Is Video Coding Going Next?

I am confident that video coding will continue to evolve and develop in ways that we cannot predict yet. Perhaps one day, there will be so much communication bandwidth available that video compression will no longer be necessary, as my visitor predicted in the 1990s (see the beginning of Chapter 1). Perhaps a more realistic endpoint will be if or when communication bandwidth becomes sufficient to make it unnecessary to continue pushing for more and more compression. However, until then, the three-way trade-off between compression, quality and computation makes it attractive to keep developing ways of improving the compression of video at the expense of increasing demands on processing power.

Newer standards such as Versatile Video Coding (VVC/H.266) and Essential Video Coding (EVC/ISO/IEC 23094-1) are at an early stage of deployment. As with any newly published video coding standard, it takes time for implementers to develop efficient software and hardware codecs. It also takes time for these to be deployed in consumer devices, such as the chipsets in mobile devices, and to be adopted by service providers. It remains to be seen whether these recently published standards will make a big impact on the services and industries that use video coding.

As computing power in consumer devices continues to advance, there is an ongoing opportunity to improve the performance of video compression by applying more computing power. For example, applying more sophisticated algorithms at the encoder side can increase compression, for example, by evaluating more prediction and coding modes for each block, at a cost of more computing power. Will we see further video coding standards being developed in the future? I would expect so, once the computing power available in client devices such as mobile handsets, TVs and computers increases to the point where it is worthwhile to do so.

Most video coding applications continue to code and display conventional single-screen, rectangular video. On the one hand, modalities such as stereoscopic video and immersive video have not made a significant impact on the market. On the other hand, the way in which we use video has changed considerably with the increase in adaptive bitrate streaming for delivering video, new ways of sharing and consuming user-generated video via apps such as YouTube and TikTok, and the ubiquity of recording and sharing video. Video coding needs to continue to evolve to support the changing ways in which we use video.

The recently published H.266 standard is vastly more sophisticated than its early ancestors, H.261 and MPEG-1 video, as we saw in Chapter 3, which discussed the history of video coding. However, H.266 continues to use the same basic model of block-based prediction, transform, quantisation and entropy coding that was found in those early standards, albeit with many innovations in each processing step. Other video coding models have been proposed over the years but, as of yet, have not found their way into video coding standards. Will we see a different coding model make its way into the mainstream, or will future video coding standards continue to refine the same basic model?

12.3 Where Should You Go Next?

Now that you have read this book, you might want to know more about how a particular process is implemented in a particular codec, you might want to measure the performance of a codec, or you might have an idea that you want to investigate further. What next? I discussed this question in Chapter 1. There are many excellent papers and resources on specific aspects of the video codec standards and these books about the H.265 standard [1, 2]. The standards themselves specify all aspects of the decoder or the bitstream in detail but can be somewhat challenging to interpret. Hopefully, the explanations and guidance in this book will help you to find what you might need to know in the video coding standards. I would encourage you to try things out for yourself, experimenting with the video codecs and concepts you are interested in. Chapter 11 suggests how to get started with the reference codecs or with frameworks such as ffmpeg. Experimentation can give you experience of the kind of performance to expect from different video codec implementations and is a good starting point for research and trying out new ideas. The book companion site, https://www.vcodex.com/coding-video-book, includes download links and other resources that will help you find out more and to begin your own experiments.

Finally, thank you for taking the time to read this book. I hope I have answered some of the questions you might have had about video coding and that the book has left you wanting to find out more. I wish you success with your own investigations in this challenging and fast-moving field.

The Netherlands, 2024

References

1 Wien, M. (2014). *High Efficiency Video Coding: Coding Tools and Specification*. Berlin: Springer. doi: https://doi.org/10.1007/978-3-662-44276-0.

2 Sze, V., Budagavi, M., and Sullivan, G. ed. (2014). *High Efficiency Video Coding: Algorithms and Architectures*. Berlin: Springer. https://doi.org/10.1007/978-3-319-06895-4.

Glossary

AU	Access Unit, a coded video picture
ALF	Adaptive Loop Filter
AMVP	Adaptive Motion Vector Prediction
AOMedia	Alliance for Open Media
API	Application Programming Interface
APS	Adaptation Parameter Set
AVC	Advanced Video Coding, also known as H.264
AV1	A video codec specification
B-Picture	Coded picture that can use bidirectional prediction
BAC	Binary Arithmetic Coding/Coder
BD	Bjøntegaard Delta, a performance comparison metric
BLA	Broken Link Access
BMFF	Base Media File Format
CABAC	Context-based Adaptive Binary Arithmetic Coding
CAVLC	Context Adaptive Variable Length Coding
CBR	Constant Bit Rate
CCITT	International Telegraph and Telephone Consultative Committee
CDN	Content Delivery Network
CIF	Common Intermediate Format
CPB	Coded Picture Buffer
CRA	Clean Random Access
CRC	Cyclic Redundancy Code
CTB	Coding Tree Block
CTU	Coding Tree Unit
CB	Coding Block
CU	Coding Unit
DASH	Dynamic Adaptive Streaming over HTTP
DCT	Discrete Cosine Transform
DFT	Discrete Fourier Transform
DHT	Discrete Hadamard / Walsh-Hadamard Transform
DPB	Decoded Picture Buffer

Coding Video: A Practical Guide to HEVC and Beyond, First Edition. Iain E. Richardson.
© 2024 John Wiley & Sons Ltd. Published 2024 by John Wiley & Sons Ltd.
Companion website: www.wiley.com/go/richardson/codingvideo1

DPCM	Differential Pulse Code Modulation
DSCQS	Double Stimulus Continuous Quality Scale
DTV	Digital television broadcasting
DVD	Digital Versatile Disk
DVR	Digital Video Recorder
EBU	European Broadcasting Union
EVC	Essential Video Coding, a video coding standard
FFMPEG	A software video coding framework
Gbits	Gigabits
GOP	Group of Pictures
GPU	Graphics Processing Unit
HD	High Definition
HDD	Hard Disk Drive
HLS	HTTP Live Streaming
HM	HEVC Model, a software video codec
HEVC	High Efficiency Video Coding, also known as H.265
HRD	Hypothetical Reference Decoder
HVS	Human Visual System
I-Picture	Coded picture that uses only intra-prediction
IDR	Instantaneous Decoder Refresh
IP	Intellectual Property
IPR	Intellectual Property Rights
IRAP	Intra Random Access Picture
ISDN	Integrated Services Digital Network
ISO/IEC	International Organisation for Standardisation / International Electrotechnical Commission
ITU-T	International Telecommunication Union Telecommunication Standardisation Sector
ITU-T BT.500	A standard for video quality measurement
JCT-VC	Joint Collaborative Team on Video Coding
JND	Just Noticeable Difference
JVET	Joint Video Experts Team
JVT	Joint Video Team
KLT	Karhunen-Loeve Transform
LCEVC	Low Complexity Enhancement Video Coding, a video coding standard
LFNST	Low-Frequency Non-Separable Transform
LMCS	Luma Mapping with Chroma Scaling
LPS	Least Probable Symbol
MKV	Matroska Video, a container format
MOS	Mean Opinion Score
MOSp	Predicted Mean Opinion Score
MPEG	Moving Picture Experts Group
MPM	Most Probable Mode
MPS	Most Probably Symbol
MSB	Most Significant Bit

MSE	Mean Squared Error
MV	Motion Vector
MVD	Motion Vector Difference
NAL Units or NALUs	Network Adaptation Layer Units
NR	No Reference
P-Picture	Coded picture that can use inter prediction from one reference frame
PB	Prediction Block
PCM	Pulse Code Modulation
PES	Packetized Elementary Stream
PPS	Picture Parameter Set
POC	Picture Order Count
PSNR	Peak Signal to Noise Ratio
PU	Prediction Unit
QCIF	Quarter Common Intermediate Format
QoS	Quality of Service
QP	Quantization Parameter
RADL	Random Access Decodable Leading Picture
RAND	Reasonable and Non-Discriminatory
RAP	Random Access Picture
RASL	Random Access Skipped Leading Picture
R-D	Rate-Distortion
RBSP	Raw Byte Sequence Payload
RPS	Reference Picture Set
RR	Reduced Reference
RTP	Real Time Protocol
RTSP	Real Time Streaming Protocol
SAO	Sample Adaptive Offset
SD	Standard Definition
SEPs	Standard Essential Patents
SMPTE	Society for Motion Picture and Television Engineers
SPS	Sequence Parameter Set
SSD	Solid State Drive
SSIM	Structural SIMilarity index
TB	Transform Block
TCP	Transmission Control Protocol
TS	Transport Stream
TU	Transform Unit
UDP	User Datagram Protocol
UHD	Ultra High Definition
URL	Uniform Resource Locator
VBR	Variable Bit Rate
VCEG	Video Coding Experts Group
VCL	Video Coding Layer
VLC	Variable Length Codeword/Variable Length Coding

VLD	Variable Length Decode
VMAF	Video Multimethod Assessment Fusion
VPS	Video Parameter Set
VP9	A video coding specification
VQM-VFD	Video Quality Model for Variable Frame Delay
VVC	Versatile Video Coding, also known as H.266
WMV9	Windows Media Video Version 9, a video coding specification
WPP	Wavefront Parallel Processing

Index

Coding Video: A Practical Guide to HEVC and Beyond, First Edition. Iain E. Richardson.
© 2024 John Wiley & Sons Ltd. Published 2024 by John Wiley & Sons Ltd.
Companion website: www.wiley.com/go/richardson/codingvideo1

n

o

p

t